P9-DCU-850

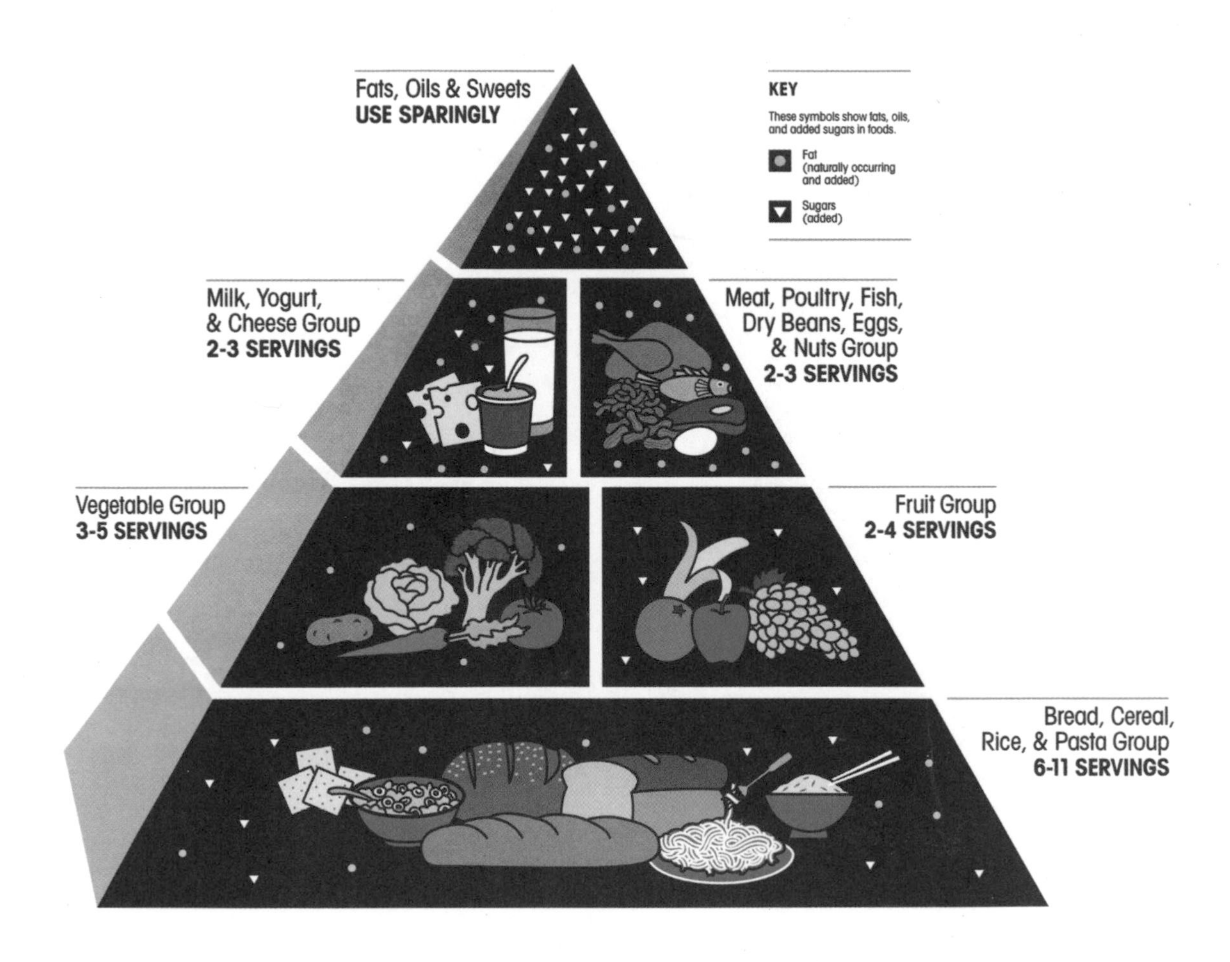
Fats, Oils & Sweets
USE SPARINGLY
KEY
These symbols show fats, oils, and added sugars in foods.
Fat (naturally occurring and added)
Sugars (added)
Milk, Yogurt, & Cheese Group
2-3 SERVINGS
Meat, Poultry, Fish, Dry Beans, Eggs, & Nuts Group
2-3 SERVINGS
Vegetable Group
3-5 SERVINGS
Fruit Group
2-4 SERVINGS
Bread, Cereal, Rice, & Pasta Group
6-11 SERVINGS

Nutrition for Foodservice and Culinary Professionals

Karen Eich Drummond
Ed.D., R.D., L.D.N., F.A.D.A., F.M.P.

Lisa M. Brefere
C.E.C., A.A.C.

Nutrition for Foodservice and Culinary Professionals

Fifth Edition

WILEY

John Wiley & Sons, Inc.

This book is printed on acid-free paper. 

Published by John Wiley & Sons, Inc., Hoboken, New Jersey
Published simultaneously in Canada

Library of Congress Cataloging-in-Publication Data:
Drummond, Karen Eich.
Nutrition for foodservice and culinary professionals / Karen Eich Drummond, Lisa M. Brefere.—5th ed.
p. cm.
Includes bibliographical references and index.
ISBN 0-471-41977-X (Cloth)
1. Nutrition. 2. Food service. I. Brefere, Lisa M. II. Title.

TX353.D78 2004
613.2—dc21 2002012152

Printed in the United States of America

10 9 8 7 6 5 4 3 2

In memory of my father,
Frank J. Eich

Contents

Chapter 2

Using Dietary Recommendations, Food Guides, and Food Labels to Plan Menus 35

Chapter 3

Carbohydrates 90

Chapter 6
Vitamins 215

Chapter 7
Water and Minerals 250

Part Two
Developing and Marketing Healthy Recipes and Menus 283

Chapter 8
Developing Healthy Menus and Recipes 285

Preface

This book is written for students in hotel, restaurant, and institution management programs and culinary programs. Practicing management and culinary professionals will find it useful as well. As with previous editions, this is meant to be a practical how-to book tailored to the needs of students and professionals. It is written for those who need to use nutritional principles to evaluate and modify menus and recipes, as well as to respond knowledgeably to customers' questions and needs. As in the fourth edition, co-author Lisa Brefere, C.E.C., A.A.C., lends her firsthand experiences applying nutrition to selecting, cooking, and menuing healthy foods in restaurants and foodservices. After all, we eat foods, not nutrients!

What's New

Following are the elements that make this book more user-friendly and up-to-date.

- **Updated discussions,** including dietary recommendations, added sugars, alternative sweeteners, fiber, fat, and essential fatty acids
- New section on the use of **starches as thickeners**
- New national standards for **organic foods**
- **More coverage** of vitamins and minerals, including more tables displaying the amount of a vitamin or mineral in selected foods
- **Updated text and tables** using the 2001 Dietary Reference Intakes for vitamins A and K, chromium, copper, iodine, iron, manganese, molybdenum, and zinc and the 2002 Dietary Reference Intakes for energy, carbohydrates, fiber, fat, and protein
- Updated discussion on **nutrition and heart disease,** including the TLC diet using the National Cholesterol Education Program 2001 clinical

guidelines, as well as discussion of the 2002 American Cancer Society nutrition guidelines

- Updated and expanded section on **biotechnology**
- Completely updated chapter on **weight management,** including updated statistics on the prevalence of overweight and obesity in the United States and the latest information on weight loss drugs and surgery
- New discussion on **Female Athlete Triad,** a newly recognized type of eating disorder
- Expanded **Chef's Tips** throughout the book
- Many new **tables** and **figures,** such as "Food Practices of World Religions" and "Reduced-Calorie Menu for Asian-American Cuisine"
- Expanded **Activities and Applications,** including using the **Interactive Healthy Eating Index** (at the website for the U.S. Department of Agriculture's Center for Nutrition Policy and Promotion) to do nutritional analysis of recipes and one-day diets

In addition, the popular **Nutrition Web Explorer** has been completely updated and now includes new, carefully selected websites.

Organization

The book is organized into three major parts, as before.

I. Fundamentals of Nutrition and Foods (Chapters 1–7)
II. Developing and Marketing Healthy Recipes and Menus (Chapters 8–10)
III. Nutrition's Relationship to Health and Life Span (Chapters 11–13)

Part One has two introductory chapters followed by chapters on specific nutrients: carbohydrates, lipids, protein, vitamins, and water and minerals. The first two chapters introduce basic nutrition concepts and explain how to use dietary recommendations, the Food Guide Pyramid, and food labels to plan menus.

Part Two, especially Chapter 8, discusses how to use ingredients, flavoring principles, healthy cooking methods and techniques, and presentation to produce tasty, healthy food with eye appeal. Chapters 9 and 10 go further to explain how to market healthy foods and beverages in restaurants, foodservices, and beverage operations.

Part Three looks at nutrition and health issues, including how nutrition is related to heart disease, cancer, diabetes, and obesity. Vegetarian diets and diets for weight loss are discussed in detail. The final chapter looks at nutrition over the life span, from pregnancy to the infant, child, adolescent, and older adult.

Pedagogical Aids

In addition to tables, charts, illustrations, recipes, and a glossary, the book includes many other pedagogical aids.

1. Each chapter begins with an **outline** to help students get the big picture first.
2. The outline is followed by an introduction and a bulleted list of **learning objectives.**
3. **Sidebars** appear in all chapters to give definitions of key terms.
4. Instead of a summary at the end of each chapter, a **Mini-Summary** is given after each chapter heading. This is intended to help students focus on what is important in each section of the chapter.
5. The heading **Chef's Tips** is used in many chapters. Chef's Tips include experienced advice on which foods go together, how to flavor foods, how to use foods' natural colors to create an attractive dish, which foods work well in which dishes, availability of new and different products, and other culinary techniques to make healthy food taste wonderful.
6. **Food Facts** and **Hot Topics** continue to appear in most chapters. Food Facts provide in-depth information on relevant food-related topics, such as different types of oils and margarines. Hot Topics discuss often-controversial subjects related to nutrition, such as biotechnology and functional foods.
7. At the end of each chapter, the **Check-Out Quiz** allows students to check their comprehension of the important concepts presented. Answers to this quiz are found in Appendix E.
8. After the quiz, you will find **Activities and Applications.** This section encourages students to analyze, evaluate, create, problem-solve, and apply chapter concepts. One set of activities, **Nutrition Web Explorer,** involves getting specific food- and nutrition-related information from carefully selected websites.

Total Diet Assessment CD-ROM software is available to package with this text (package ISBN 0-471-46238-1). This software makes it easy for students to perform a nutrient analysis of recipes, menus, and diets.

An *Instructor's Manual* (ISBN 0-471-31342-4) that includes class outlines, Student Activity Sheets, visual aids, and test questions and answers is available. Please contact your John Wiley & Sons representative for a copy. Also, please use the website www.wiley.com/college to download the *Instructor's Manual,* Test Bank, and PowerPoint slides.

The National Restaurant Association (NRA) Educational Foundation, in consultation with the authors, has developed a *Student Workbook* (ISBN

0-471-31270-3) for its ProMgmt. Certificate program. The workbook contains exercises and a study outline for each chapter, and a practice test of 80 multiple-choice questions. This practice test will assist students in preparing for the certificate examination.

In addition, an *Instructor's Guide* (ISBN 0-471-31321-1) is available to qualified adopters to complement and highlight the information in the textbook and *Student Workbook*.

Acknowledgments

We are grateful for the help of all the educators who have contributed to this and previous editions through their constructive comments. They include:

Marian Benz, Milwaukee Area Technical College
Keith E. Gardiner, Guilford Technical Community College
Debra Macchia, College of DuPage
Kevin Monti, Western Culinary Institute
Richard Roberts, Wake Technical Community College
Joan Vogt, Kendall College
Jane Ziegler, Cedar Crest College

Part One

Fundamentals of Nutrition and Foods

Chapter 1

Introduction to Nutrition

Americans are fascinated with food: choosing foods, reading newspaper articles on food, perusing cookbooks, preparing and cooking foods, checking out new restaurants, and, of course, eating foods. Why are we so interested in food? Of course, eating is fun, enjoyable, and satisfying, especially when we are eating with other people whose company we like. Beyond the physical and emotional satisfaction of eating, we may be concerned about how our food choices affect our health. Our choice of diet strongly influences whether we will get certain diseases (such as heart disease and cancer). Indeed, high costs are associated with poor eating patterns: heart disease alone costs the United States between $50 billion and $100 billion per year for medical treatment and lost wages. No doubt eating right contributes to our health and quality of life as we grow older.

This introductory chapter explores why we choose the foods we eat and then explores important nutrition concepts that build a foundation for the remaining chapters. It will help you to:

- Identify factors that influence food selection
- Define *nutrition, kilocalorie, nutrient,* and *nutrient density*
- Identify the classes of nutrients and their characteristics
- Describe four characteristics of a nutritious diet
- Define Dietary Reference Intakes and explain their function
- Compare the EAR, RDA, AI, and UL
- Describe the processes of digestion, absorption, and metabolism
- Explain how the digestive system works

Factors Influencing Food Selection

Why do people choose the foods they do? This is a very complex question, and there are many factors influencing what you eat, as you can see from this list:

- Flavor
- Other aspects of food (such as cost, convenience, nutrition)
- Demographics
- Culture and religion
- Health
- Social and emotional influences
- Food industry and the media
- Environmental concerns

Now we will look at many of these factors in depth.

Flavor

Flavor—An attribute of a food that includes its appearance, smell, taste, feel in the mouth, texture, temperature, and even the sounds made when it is chewed.

Taste—Sensations perceived by the taste buds on the tongue.

Taste buds—Clusters of cells found on the tongue, cheeks, throat, and roof of the mouth. Each taste bud houses 60 to 100 receptor cells. The body regenerates taste buds about every three days. These cells bind food molecules dissolved in saliva and alert the brain to interpret them.

The most important consideration when choosing something to eat is the flavor of the food. **Flavor** is an attribute of a food that includes its appearance, smell, taste, feel in the mouth, texture, temperature, and even the sounds made when it is chewed. Flavor is a combination of all five senses: taste, smell, touch, sight, and sound. From birth, we have the ability to smell and taste. Most of what we call taste is really smell, a fact we realize when a cold hits our nasal passages. Even though the taste buds are working fine, the smell cells are not, and this dulls much of food's flavor.

Taste comes from 10,000 **taste buds**—clusters of cells resembling the sections of an orange. Taste buds, found on the tongue, cheeks, throat, and roof of the mouth, house 60 to 100 receptor cells each. The body regenerates taste buds about every three days. They are most numerous in children under six, which may explain why youngsters are such picky eaters. These cells bind food molecules dissolved in saliva, and alert the brain to interpret them.

Although the tongue is often depicted as having regions that specialize in particular taste sensations—for example, the tip is said to detect sweetness—researchers know that taste buds for each sensation (sweet, salty, sour, and bitter) are actually scattered around the tongue. In fact, a single taste bud can have receptors for all four types of taste.

If you could taste only sweet, salty, sour, and bitter, how could you taste the flavor of cinnamon, chicken, or any other food? This is where smell comes in. Your ability to identify the flavors of specific foods requires smell.

The ability to detect the strong scent of a fish market, the antiseptic odor of a hospital, the aroma of a ripe melon, and thousands of other smells is possible thanks to a yellowish patch of tissue the size of a quarter high up in your nose. This patch is actually a layer of 12 million specialized cells, each sporting 10 to 20 hairlike growths called cilia that bind with the smell and send a message to the brain. Our sense of smell may not be as refined as that of dogs, who have billions of olfactory cells, but we can distinguish among about 10,000 scents.

You can smell foods in two ways. If you smell coffee brewing while you are getting dressed, you smell it directly through your nose. But if you are drinking coffee, the smell of the coffee goes to the back of your mouth and then up into your nose. To some extent, what you smell (or taste) is genetically determined.

All foods have texture, a natural texture granted by Mother Nature. It may be coarse or fine, rough or smooth, tender or tough. Whichever the texture, it influences whether you like the food. The natural texture of a food may not be the most desirable texture for a finished dish, so a cook may create another texture. For example, a fresh apple may be too crunchy to serve at dinner, so it is baked or sautéed for a softer texture. Or a cream soup may be too thin, so a thickening agent is used to increase the viscosity of the soup, or, simply stated, make it harder to pour.

Food appearance or presentation strongly influences which foods you choose to eat. Eye appeal is the purpose of food presentation, whether the food is hot or cold. It is especially important for cold foods because they lack the come-on of an appetizing aroma. Just the sight of something delicious to eat can start your digestive juices flowing.

Other Aspects of Food

Food cost is a major consideration. For example, breakfast cereals were inexpensive for many years. Then prices jumped, and it seemed that most boxes of cereal cost over $3.00. Some consumers switched from cereal to bacon and eggs because the bacon and eggs became less expensive. Cost is a factor in many of the purchasing decisions at the supermarket, whether one is buying dry beans at $0.39 per pound or fresh salmon at $8.99 per pound.

Convenience is more of a concern now than at any time in the past. Just think about the variety of foods you can purchase today that are already cooked or can simply be microwaved. Even if you desire ready-to-eat fruits and vegetables, supermarkets offer cut-up fruits, vegetables, and salads that need no further preparation. Of course, convenience foods are more expensive than their raw counterparts, and not every budget can afford them.

Everyone's food choices are affected by availability and familiarity. Whether it is a wide choice of foods at an upscale supermarket or a choice of only two restaurants within walking distance of where you work, you can eat only what is available. The availability of foods is very much influenced by how food is produced and distributed. For example, the increasing number of soft drink vending machines, particularly in schools and workplaces, has contributed to increasing soft drink consumption year-round. Fresh fruits and vegetables are perfect examples of foods that are most available (and at their lowest prices) when in season. Of course, you are more likely to eat fruits and vegetables, or any food for that matter, with which you are familiar and have eaten before.

The nutritional content of a food can be an important factor in deciding what to eat. You have probably watched people reading nutritional labels on a food package, or perhaps you have read nutritional labels yourself. Current estimates show that about 66 percent of Americans use nutrition information labels. Older people tend to read labels more often than younger people.

Demographics

Demographic factors that influence food choices include age, gender, educational level, income, and cultural background (discussed next). Women and older adults tend to consider nutrition more often than men or young adults when choosing what to eat. Older adults are probably more nutrition-minded because they have more health problems and are more likely to have to change their diet for health reasons. People with higher incomes and educational levels tend to think about nutrition more often when choosing what to eat.

Culture and Religion

Culture can be defined as the behaviors and beliefs of a certain social, ethnic, or age group. Culture strongly influences the eating habits of its members. Each culture has norms about which foods are edible, which foods have high or low status, how often foods are consumed, what foods are eaten together, when foods are eaten, and what foods are served at special events and celebrations (such as weddings). For example, some French people eat horsemeat, but Americans do not consider horsemeat acceptable to eat. Likewise, many common American practices seem strange or illogical to persons from other cultures. For example, what could be more unusual than boiling water to make tea and adding ice to make it cold again, sugar to sweeten it, and then lemon to make it tart? When immigrants come to live in the United States, their eating habits do gradually change, but they are among the last habits to adapt to the new culture.

Culture—The behaviors and beliefs of a certain social, ethnic, or age group.

For many people, religion affects their day-to-day food choices. For example, many Jewish people abide by the Jewish dietary laws, called the Kashrut. They do not eat pork, nor do they eat meat and dairy products together. Muslims also have their own dietary laws. Like the Jews, they will not eat pork. Their religion also prohibits drinking alcoholic beverages. For other people, religion influences what they eat mostly during religious holidays and celebrations. Religious holidays such as Passover are observed with appropriate foods. Table 1-1 explains the food practices for different religions.

Health

Have you ever dieted to lose weight? Most Americans are either trying to lose weight or keep from gaining it. You probably know that obesity and overweight can increase your risk of cancer, coronary heart disease, diabetes, and other health problems. What you eat influences your health. Even if you are healthy, you may choose foods based on a desire to prevent health problems and/or improve your appearance.

TABLE 1-1 Food Practices of World Religions	
Religion	**Dietary Practices**
Judaism	Kashrut: Jewish dietary law of keeping kosher. 1. *Meat and Poultry.* Permitted: Meat of animals with a split hoof who chew their cud (cattle, sheep, goats, deer); birds that are not birds of prey (chicken, turkey, goose, pheasant, duck). Not permitted: Pig and pork products, rabbit, birds of prey. All animals require a ritual slaughtering. All meat and poultry foods must be free of blood, which is done by soaking and salting the food or by broiling it. Meat must also be free of blood vessels and the sciatic nerve. 2. *Fish.* Permitted: Fish with fins and scales. Not permitted: Shellfish (scallops, oysters, clams), crustaceans (crab, shrimp, lobster), fishlike mammals (dolphin, whale), frog, shark, eel, catfish. Do not cook fish with meat or poultry. 3. Meals are dairy or meat, not both. It is also necessary to have two sets of cooking equipment, dishes, and silverware for dairy and meat. 4. All fruits, vegetables, grains, and eggs can be served with dairy or meat meals. 5. A processed food is kosher only if the package has a rabbinical authority's name or insignia.
Roman Catholicism	1. Abstain from eating meat on Fridays during Lent (the 40 days before Easter). 2. Fast (one meal is allowed) and abstain from meat on Ash Wednesday (beginning of Lent) and Good Friday (the Friday before Easter).
Eastern Orthodox Christianity	1. Numerous feast days and fast days. On fast days, no fish, meat, or other animal products (including dairy products) are allowed. Shellfish are allowed.
Protestantism	1. Food on religious holidays is largely determined by family's cultural background and preferences. 2. Fasting is uncommon.
Mormonism	1. Prohibit tea, coffee, and alcohol. Some Mormons abstain from anything containing caffeine. 2. Eat only small amounts of meat and base diet on grains. 3. Some Mormons fast once a month.
Seventh-Day Adventist Church	1. Many members are lacto-ovo-vegetarians (eat dairy products and eggs, but no meat or poultry). 2. Avoid pork and shellfish. 3. Prohibit coffee, tea, and alcohol. 4. Drink water before and after meals, not during. 5. Avoid highly seasoned foods and eating between meals.
Islam	1. All foods are permitted (halal) except for swine (pigs), four-legged animals that catch prey with their mouth, birds of prey that grab their prey with their claws, animals (except fish and seafood) that have not been slaughtered according to ritual, and alcoholic beverages. Use of coffee and tea is discouraged. 2. Muslims celebrate many feast and fast days. On fast days, they do not eat or drink from sunup to sundown.
Hinduism	1. Encourages eating in moderation. 2. Meat is allowed, but the cow is sacred and is not eaten. Also avoided are pork and certain fish. Many Hindus are vegetarian. 3. Many Hindus avoid garlic, onions, mushrooms, and red foods such as tomatoes. 4. Water is taken with meals. 5. Some Hindus abstain from alcohol. 6. Hindus have a number of feast and fast days.
Buddhism	1. Dietary laws vary depending on the country and the sect. Many Buddhists do not believe in taking life, so they are lacto-ovo-vegetarians (eat dairy products and eggs, but no meat or poultry). 2. Buddhists celebrate feast and fast days.

A knowledge of nutrition and a positive attitude toward nutrition may translate into nutritious eating practices. Just knowing that eating lots of fruits and vegetables may prevent heart disease does not mean that someone will automatically start eating more of these foods. For some people, knowledge is enough to stimulate new eating behaviors, but for most people, knowledge is not enough and change is difficult. There are many circumstances and beliefs that prevent change, such as a lack of time or money to eat right. But some people manage to change their eating habits, especially if they feel that the advantages (such as losing weight or preventing cancer) outweigh the disadvantages.

Social and Emotional Influences

People have historically eaten meals together, making meals important social occasions. Our food choices are influenced by the social situations we find ourselves in, whether in the comfort of our home or eating out in a restaurant. For example, social influences are involved when several members of a group of college friends are vegetarian. Peer pressure no doubt influences many food choices for children and young adults. Even as adults, we tend to eat the same foods that our friends and neighbors eat. This is due to cultural influences as well.

Food is often used to convey social status. For example, in a trendy, upscale New York City restaurant, you will find prime cuts of beef and high-priced wine.

Emotions are closely tied to some of our food selections. You may have been given something sweet to eat, such as cake or candy, whenever you were unhappy or upset. As an adult, you may gravitate to those kinds of foods, called comfort foods, when under stress.

Food Industry and the Media

The food industry very much influences what you choose to eat. After all, the food companies decide what foods to produce and where to sell them. They also use advertising, product labeling and displays, information provided by their consumer services departments, and websites to sell their products.

On a daily basis, the media (television, newspapers, magazines, radio, etc.) portray food in many ways: paid advertisements, articles on food in magazines and newspapers, or foods eaten on television shows. Much research has been done on the impact of television food commercials on children. Quite often the commercials succeed in getting children to eat foods such as cookies, candies, and fast foods. Television commercials are likely contributing to higher calorie and fat intakes.

The media also report frequently on new studies related to food, nutrition, and health topics. It is hard to avoid hearing sound bites such as "more fruits and vegetables lower blood pressure." Media reports may influence which foods people eat.

Environmental Concerns

Some people have environmental concerns, such as the use of chemical pesticides, so they often, or always, choose organically grown foods (which are grown without such chemicals—see Food Facts on page 31 for more information). Many vegetarians won't eat meat or chicken for ecological reasons, because livestock and poultry require so much land, energy, water, and plant food, which they consider wasteful.

Now that you have a better understanding of why we eat the foods we do, we can look at some basic nutrition concepts and terms.

■ **MINI-SUMMARY**

Table 1-2 summarizes factors that influence what we eat.

TABLE 1-2 Factors Influencing What You Eat
Flavor
Taste
Smell
Appearance
Texture
Other Aspects of Food
Cost
Convenience
Availability
Familiarity
Nutrition
Demographics
Age
Gender
Educational level
Income
Culture and Religion
Traditional foods and food habits
Special events and celebrations
Religious foods and food practices

TABLE 1-2 (continued)

Health
Health status and desire to improve health
Desire to improve appearance
Nutrition knowledge and attitudes
Social and Emotional Influences
Social status
Peer pressure
Emotional status
Food associations
Food Industry and the Media
Food industry
Food advertising
Food portrayal in media
Reporting of nutrition/health studies
Environmental Concerns
Use of synthetic fertilizers and pesticides
Wastefulness of fattening up livestock/poultry

Basic Nutrition Concepts

Nutrition

Nutrition—A science that studies nutrients and other substances in foods and in the body and how these nutrients relate to health and disease. Nutrition also explores why you choose particular foods and the type of diet you eat.

Nutrients—The nourishing substances in food that provide energy and promote the growth and maintenance of your body.

Diet—The food and beverages you normally eat and drink.

Nutrition is a science, which means that it is a branch of knowledge dealing with a body of facts. Compared with some other sciences, such as chemistry, nutrition is a young science. Many nutritional facts revolve around nutrients, such as carbohydrates. **Nutrients** are the nourishing substances in food that provide energy and promote the growth and maintenance of your body. In addition, nutrients aid in regulating body processes, such as heart rate and digestion, and in supporting the body's optimum health.

Nutrition researchers look at how nutrients relate to health and disease. Almost daily we are bombarded with news reports that something in the food we eat, such as fat, is not good for us—that it may indeed cause or complicate conditions such as heart disease or cancer. Researchers look closely at the relationships between nutrients and disease, as well as the processes by which you choose what to eat and the balance of foods and nutrients in your diet.

In summary, nutrition is a science that studies nutrients and other substances in foods and in the body, and how these nutrients relate to health and disease. Nutrition also explores why you choose the foods you do and the type of **diet** you eat. *Diet* is a word that has several meanings. Anyone who has tried to lose weight has no doubt been on a diet. In this sense, *diet* means weight-reducing diet and is often thought of in a negative way. But a more general definition of *diet* is the foods and beverages you normally eat and drink.

Kilocalories

Kilocalorie—A measure of the energy in food, specifically the energy-yielding nutrients.

Food energy, as well as the energy needs of the body, is measured in units of energy called **kilocalories.** The number of kilocalories in a particular food can be determined by burning a weighed portion of the food and measuring the amount of heat (or kilocalories) it produces. A kilocalorie, also called a Calorie (notice the upper case "C"), raises the temperature of 1 kilogram of water 1 degree Celsius. Just as 1 kilogram contains 1000 grams, 1 kilocalorie contains 1000 calories (notice the lower case "c"). When you read in a magazine that a cheeseburger has 350 calories, understand that it is really 350 *kilo*calories. The American public has been told for years that an apple has 80 calories, a glass of regular milk has 120 calories, and so on, when the correct term is NOT calories, but kilocalories. The media shortened the term kilocalories to calories, which is incorrect. This book uses the term kilocalorie and its abbreviations, kcalorie and kcal, throughout each chapter.

Basal metabolism—The minimum energy needed by the body for vital functions when at rest and awake.

Thermic effect of food—The energy needed to digest and absorb food.

The number of kcalories you need is based on three factors: your energy needs when your body is at rest and awake (referred to as **basal metabolism**), your level of physical activity, and the energy you need to digest and absorb food (referred to as the **thermic effect of food**). Basal metabolic needs include energy needed for vital bodily functions when the body is at rest but awake. For example, your heart is pumping blood to all parts of your body, your cells are making proteins, and so on. Your basal metabolic rate (BMR) depends on the following factors:

1. **Gender.** Men have a higher BMR than women because men have a higher proportion of muscle tissue (muscle requires more energy for metabolism than fat does).
2. **Age.** As people age, they generally gain fat tissue and lose muscle tissue. BMR declines about 2 percent per decade after age 30.
3. **Growth.** Children, pregnant women, and lactating women have higher BMRs.
4. **Height.** Tall people have more body surface (than shorter people) and lose body heat faster. Their BMR is therefore higher.
5. **Temperature.** BMR increases in both hot and cold environments, in order to keep the temperature inside the body constant.
6. **Fever and stress.** Both of these increase BMR. Fever raises BMR by 7 percent for each 1 degree Fahrenheit above normal. The body reacts to

stress with the secretion of hormones that speed up metabolism so the body can respond quickly and efficiently.

7. **Exercise.** Exercise increases BMR for several hours afterward.
8. **Smoking and caffeine.** Both cause increased energy expenditure.
9. **Sleep.** Your BMR is at its lowest when you are sleeping.

Basic metabolic rate also decreases when you diet or eat fewer kcalories than normal. Basal metabolic rate accounts for the largest percentage of energy expended—about two-thirds for individuals who are not very active.

Your level of physical activity strongly influences how many kcalories you need. Table 1-3 shows the kcalories burned per hour for a variety of activities. The number of kcalories burned depends on the type of activity, how long and how hard it is performed, and the individual's size. The larger your body, the more energy you use in physical activity. Aerobic activities such as walking, jogging, cycling, and swimming are excellent ways to burn calories if they are brisk enough to raise heart and breathing rates. Physical activity accounts for 25 to 40 percent of total energy needs.

TABLE 1-3 Kcalories Spent per Hour in Physical Activity

Activity	KCalories Burned*
Bicycling, 6 mph	240 kcalories
Bicycling, 12 mph	410 kcalories
Cross-country skiing	700 kcalories
Jogging, 5-1/2 mph	740 kcalories
Jogging, 7 mph	920 kcalories
Jumping rope	750 kcalories
Running in place	650 kcalories
Running, 10 mph	1280 kcalories
Swimming, 25 yards/minute	275 kcalories
Swimming, 50 yards/minute	500 kcalories
Tennis, singles	400 kcalories
Walking, 2 mph	240 kcalories
Walking, 3 mph	320 kcalories
Walking, 4-1/2 mph	440 kcalories

* The kcalories burned in a particular activity vary in proportion to one's body weight. For example, a 100-pound person burns 1/3 fewer kcalories, so you would multiply the number of kcalories by 0.7. For a 200-pound person, multiply by 1.3.

Source: National Heart, Lung, and Blood Institute and the American Heart Association. 1993. *Exercise and Your Heart: A Guide to Physical Activity.*

The thermic effect of food is the smallest contributor to your energy needs: from 5 to 10 percent of the total. In other words, for every 100 kcalories you eat, 5 to 10 are used for digestion, absorption, and metabolism of nutrients, our next topic.

Nutrients

As stated, nutrients provide energy or kcalories, promote the growth and maintenance of the body, and/or regulate body processes. There are about 50 nutrients that can be arranged into six classes, as follows:

1. Carbohydrates
2. Fats (the proper name is lipids)
3. Protein
4. Vitamins
5. Minerals
6. Water

Each nutrient class performs different functions in the body, as shown in Table 1-4.

Foods rarely contain just one nutrient. Most foods provide a mix of nutrients. For example, bread is often thought of as providing primarily carbohydrates, but it is also an important source of certain vitamins and minerals. Food contains more than just nutrients. Depending on the food, it may contain colorings, flavorings, caffeine, phytochemicals (minute substances in plants that may protect health), and other substances.

Energy-yielding nutrients—Nutrients that can be burned as fuel to provide energy for the body, including carbohydrates, fats, and proteins.

Carbohydrates, lipids, and protein are called **energy-yielding nutrients** because they can be burned as fuel to provide energy for the body. They provide kcalories as follows:

Carbohydrates:	4 kcalories per gram
Lipids:	9 kcalories per gram
Protein:	4 kcalories per gram

TABLE 1-4 Functions of Nutrients

Nutrients	Provide Energy	Promote Growth and Maintenance	Regulate Body Processes
Carbohydrates	X		
Lipids	X	X	X
Protein	X	X	X
Vitamins		X	X
Minerals		X	X
Water		X	X

(A gram is a unit of weight in the metric system; there are about 28 grams in 1 ounce.) Vitamins, minerals, and water do not provide energy or calories.

Micronutrients—Nutrients needed by the body in small amounts, including vitamins and minerals.

Macronutrients—Nutrients needed by the body in large amounts, including carbohydrates, lipids, and proteins.

Organic—In chemistry, any compound that contains carbon.

Inorganic—In chemistry, any compound that does not contain carbon.

Carbohydrates—A large class of nutrients, including sugars, starch, and fibers, that function as the body's primary source of energy.

Lipids—A group of fatty substances, including triglycerides and cholesterol, that are soluble in fat, not water, and that provide a rich source of energy and structure to cells.

Proteins—Major structural parts of the body's cells that are made of nitrogen-containing amino acids assembled in chains, particularly rich in animal foods.

The body needs vitamins and minerals in small amounts, so these nutrients are called **micronutrients** (*micro* means small). In contrast, the body needs large amounts of carbohydrates, lipids, and protein so they are called **macronutrients** (*macro* means large).

Another way to group the classes of nutrients is to look at them from a chemical point of view. In chemistry, any compound that contains carbon is called **organic.** If a compound does not contain carbon, it is called **inorganic.** Carbohydrates, lipids, proteins, and vitamins are all organic. Minerals and water are inorganic.

Carbohydrates are a large class of nutrients, including sugars, starches, and fibers, that function as the body's primary source of energy. *Sugar* is most familiar in its refined forms, such as table sugar or high-fructose corn syrup, and is used in soft drinks, cookies, cakes, pies, candies, jams, jellies, and other sweetened foods. Sugar is also present naturally in fruits and milk (even though milk does not taste sweet). *Starch* is found in breads, breakfast cereals, pastas, potatoes, and beans. Both sugar and starch are important sources of energy for the body. *Fiber* can't be broken down or digested in the body, so it is excreted. It therefore does not provide energy for the body. Fiber does a number of good things in the body such as improve the health of the digestive tract. Good sources of fiber include legumes (dried beans and peas), fruits, vegetables, whole-grain foods such as whole-wheat bread or cereal, nuts, and seeds.

Lipids are a group of fatty substances, including triglyerides and cholesterol, that are soluble in fat, not water, and that provide a rich source of energy and structure to cells. The most familiar lipids are fats and oils, which we find in butter, margarine, vegetable oils, mayonnaise, and salad dressings. Lipids are also found in the fatty streaks in meat, the fat under the skin of poultry, the fat in milk and cheese (except skim milk and products made with it), baked goods such as cakes, fried foods, nuts, and many processed foods, such as canned soups and frozen dinners. Most breads, cereals, pasta, fruits, and vegetables have little or no fat. Triglycerides are the major form of lipids. They provide energy for the body as well as a way to store energy as fat.

Most of the kcalories we eat are from carbohydrates or fats. Only about 15 percent of total kcalories are from **protein.** This doesn't mean that protein is less important. On the contrary, protein is the main structural component of all the body's cells. It is made of units called amino acids, which are unique in that they contain nitrogen. Besides its role as an important part of cells, protein also regulates body processes and can be burned to provide energy (although the body prefers to burn carbohydrates and lipids). Protein is present in significant amounts in foods from animal sources, such as beef, pork, chicken, fish, eggs, milk, and cheese. Protein appears in plant foods, such as grains, beans, and vegetables in smaller quantities. Fruits contain only very small amounts of protein.

Vitamins—Noncaloric, organic nutrients found in a wide variety of foods that are essential in small quantities to regulate body processes, maintain the body, and allow growth and reproduction.

Minerals—Noncaloric, inorganic chemical substances found in a wide variety of foods needed to regulate body processes, maintain the body, and allow growth and reproduction.

There are 13 different **vitamins** in food. Vitamins are noncaloric, organic nutrients found in a wide variety of foods. They are essential in small quantities to regulate body processes, to maintain the body, and to allow growth and reproduction. Instead of being burned to provide energy for the body, vitamins work as helpers. They assist in the processes of the body that keep you healthy. For example, vitamin A is needed by the eyes for vision in dim light. Vitamins are found in fruits, vegetables, grains, meat, dairy products, and other foods. Unlike other nutrients, many vitamins are susceptible to being destroyed by heat, light, or other agents.

Minerals are also required by the body in small amounts and do not provide energy. Like vitamins, they work as helpers in the body and are found in a variety of foods. Some minerals, such as calcium and phosphorus, become part of the body's structure by building bones and teeth. Unlike vitamins, minerals are indestructible and inorganic.

Although deficiencies of energy or nutrients can be sustained for months or even years, a person can survive only a few days without water. Experts rank water second only to oxygen as essential to life. Water plays a vital role in all bodily processes and makes up just over half of your body's weight. It supplies the medium in which various chemical changes of the body occur and aids digestion and absorption, circulation, and lubrication of body joints. For example, as a major component of blood, water helps deliver nutrients to body cells and removes waste to the kidneys for excretion.

It's been said many times, "You are what you eat." This is certainly true; the nutrients you eat can be found in your body. As mentioned, water is the most plentiful nutrient in the body, accounting for about 60 percent of your weight. Protein accounts for about 15 percent of your weight, fat for 20 to 25 percent, and carbohydrates only 0.5 percent. The remainder of your weight includes minerals, such as calcium in bones, and traces of vitamins.

Essential nutrients—Nutrients that either cannot be made in the body or cannot be made in the quantities needed by the body; therefore, we must obtain them through food.

Most, but not all, nutrients are considered **essential nutrients.** Essential nutrients either cannot be made in the body or cannot be made in the quantities needed by the body; therefore, we must obtain them through food. Carbohydrates (in the form of glucose), vitamins, minerals, water, and some lipids and some parts of protein are considered essential.

Nutrient Density

All foods were not created equal in terms of the kcalories and nutrients they provide. Some foods, such as milk, contribute much calcium to your diet, especially when you compare them to other beverages, such as soft drinks. Your typical can of cola (12 fluid ounces) contributes large amounts of sugar (40 grams, or about 10 teaspoons), no vitamins, and virtually no minerals. When you compare calories, you will find that skim milk (at 86 kcalories per cup) packs fewer calories than cola (at 97 kcalories per cup). Therefore, we can say that milk is more "nutrient-dense" than cola, meaning that milk contains more nutrients per kcalorie than colas.

Nutrient density—A measure of the nutrients provided in a food per kcalorie of the food.

The **nutrient density** of a food depends upon the amount of nutrients it contains and the comparison of that to its caloric content. In other words, nutrient density is a measure of the nutrients provided per kcalorie of the food. As Figure 1-1 shows, broccoli offers many nutrients for its few calories. Broccoli is considered to have a high nutrient density because it is high in nutrients relative to its caloric value. Vegetables and fruits are examples of nutrient-dense foods. In comparison, a cupcake contains many more kcalories and few nutrients. By now, you no doubt recognize that some foods, such as candy bars, have a low nutrient density, meaning that they are low in nutrients and high in kcalories. These foods are called **empty-kcalorie foods** because the kcalories they provide are "empty" (that is, they deliver few nutrients). The next section will tell you more about what a nutritious diet is.

Empty-kcalorie foods—Foods that provide few nutrients for the number of kcalories they contain.

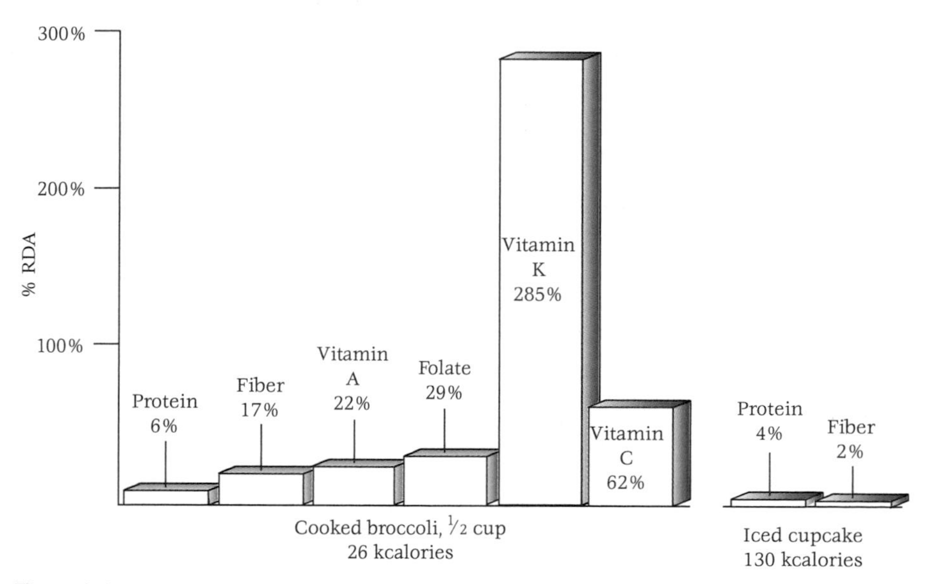

Figure 1-1
Nutrition density comparison*

*The nutrients examined were protein, fiber, and vitamins that contributed at least 10 percent of the RDA.

■ **MINI-SUMMARY**

Nutrition is a science that studies nutrients and other substances in foods and in the body, and how these nutrients relate to health and disease. The number of kcalories (a measure of the energy in food) you need is based on three factors—your energy needs when your body is at rest and awake (basal metabolism), your level of physical activity, and the energy you need to digest and absorb food (thermic effect of foods). Nutrition also explores the reasons you choose the foods you do and the type of diet you eat. Nutrients are the nourishing substances in food, providing energy and promoting the growth and maintenance of the body. In addition, nutrients regulate the many body processes and support the body's optimum health and growth. The six classes of nutrients are carbohydrates, fats (properly called lipids), protein, vitamins, minerals, and water. Carbohydrates, fats, and proteins, are macronutrients, while vitamins and minerals are micronutrients. Their characteristics are summarized in Figure 1-2. Nutrient density is a measure of the nutrients provided per kcalorie of a food.

Characteristics of a Nutritious Diet

Adequate diet—A diet that provides enough kcalories, essential nutrients, and fiber to keep a person healthy.

Moderate diet—A diet that avoids excessive amounts of kcalories or any particular food or nutrient.

Nutrient-dense foods—Foods that contain many nutrients for the kcalories they provide.

Balanced diet—A diet in which foods are chosen to provide kcalories, essential nutrients, and fiber in the right proportions.

A nutritious diet has four characteristics. It is:

1. Adequate
2. Balanced
3. Moderate
4. Varied

Your diet must provide enough nutrients, but not too many. This is where adequate and moderate diets fit in. An **adequate diet** provides enough kcalories, essential nutrients, and fiber to keep you healthy, whereas a **moderate diet** avoids excessive amounts of kcalories or any particular food or nutrient. In the case of kcalories, for example, consuming too many leads to obesity. The concept of moderation allows you to occasionally indulge in high-kcalorie, high-fat foods such as french fries or premium ice cream.

Although it may sound simple to eat enough, but not too much, of the necessary nutrients, surveys show that most adult Americans find this hard to do. One of the best ways to overcome this problem is to select nutrient-dense foods. As stated earlier, **nutrient-dense foods** contain many nutrients for the kcalories they provide.

Next, you need a **balanced diet.** Eating a balanced diet means choosing foods that provide kcalories, essential nutrients, and fiber in the right pro-

Carbohydrates—A large class of nutrients including sugars, starches, and fibers that are the body's primary source of energy.

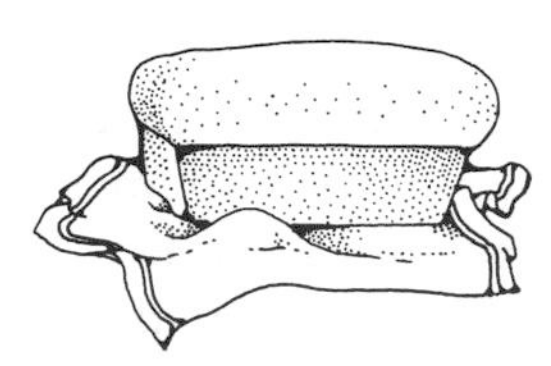

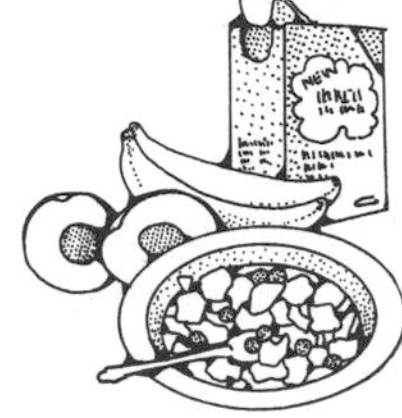

Lipids (fats)—A group of fatty substances including triglycerides and cholesterol that are soluble in fat, not water, and that provide a rich source of energy and structure to cells.

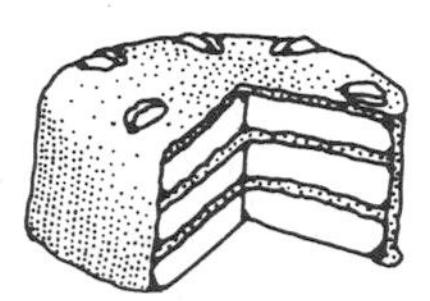

Proteins—Major structural part of body's cells composed of nitrogen-containing amino acids assembled in chains, particularly rich in animal foods.

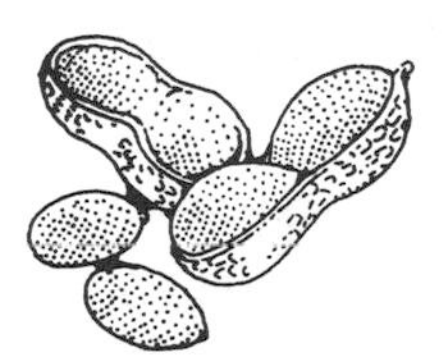

Vitamins—13 noncaloric nutrients found in a wide variety of foods (especially fruits and vegetables) that are essential in small quantities to regulate body processes, to maintain the body, and to allow growth and reproduction.

Minerals—Noncaloric, inorganic chemical substances found in a wide variety of foods needed to regulate body processes, to maintain the body, and to allow growth and reproduction.

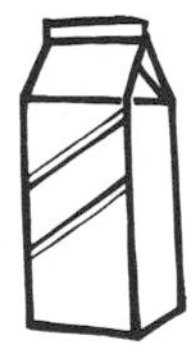

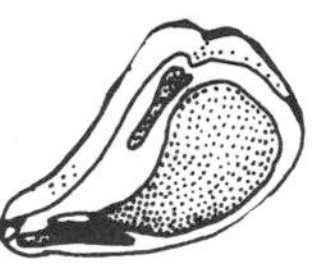

Water—Inorganic nutrient that plays a vital role in all bodily processes and makes up just over half of the body's weight.

Figure 1-2

Six classes of nutrients

portions—neither too much nor too little. For example, if you drink a lot of soft drinks, you will be getting too much sugar and possibly not enough calcium, a mineral found in milk. This is a particular concern for children whose bones are growing and who are more likely than ever before to be obese. The typical American diet is unbalanced. We eat more fried foods and fatty meats than we need, and we drink too much soda. At the same time we eat too few fruits, vegetables, and whole grains. A balanced diet is also likely to be adequate and moderate.

Varied diet—A diet in which you eat a wide selection of foods to get necessary nutrients.

Last, you need a **varied diet**—in other words, you need to eat a wide selection of foods to get the necessary nutrients. If you imagine everything you eat for one week piled in a grocery cart, how much variety is in your cart from week to week? Do you eat the same bread, the same brand of cereal, the same types of fresh fruit, and so on, every week? Do you constantly eat favorite foods? Do you try new foods? A varied diet is important because it makes it more likely that you will get the essential nutrients in the right amounts. Our next topic, the Dietary Reference Intakes, gets specific about the right amounts we need of most nutrients.

■ MINI-SUMMARY

A nutritious diet is adequate, moderate, balanced, varied, and packed with nutrient-dense foods.

Nutrient Recommendations: Dietary Reference Intakes

Dietary Reference Intake (DRI)—Nutrient standards that include four lists of values for dietary nutrient intakes of healthy Americans and Canadians.

The **Dietary Reference Intakes (DRI)** expand and replace what you may have known as the Recommended Dietary Allowances in the United States and the Recommended Nutrient Intakes in Canada. The **Recommended Dietary Allowance (RDA)** is the amount of a nutrient that meets the known nutrient needs of practically all healthy persons. The DRIs are developed by the Standing Committee on the Scientific Evaluation of Dietary Reference Intakes of the Food and Nutrition Board (a unit of the Institute of Medicine, part of the National Academy of Sciences), with involvement by Canadian scientists.

DRIs are estimates of nutrient intakes to be used for planning and evaluating diets. The DRIs are greatly expanded from the original RDAs, and include the original RDAs as well as three new values.

1. **Estimated Average Requirement (EAR).** The dietary intake value that is estimated to meet the requirement of half the healthy individuals in

Recommended Dietary Allowance (RDA)—The dietary intake value that is sufficient to meet the nutrient requirements of 97 to 98 percent of all healthy individuals in a group.

Estimated Average Requirement (EAR)—The dietary intake value that is estimated to meet the requirement of half the healthy individuals in a group.

Adequate Intake (AI)—The dietary intake that is used when an RDA cannot be based on an Estimated Average Requirement.

Tolerable Upper Intake Level (UL)—The maximum intake level above which risk of toxicity would increase.

Estimated Energy Requirement (EER)—The dietary energy intake measured in kcalories that is needed to maintain energy balance in a healthy adult.

a group. At this level of intake, the remaining 50 percent would not have its needs met. The EAR is used to assess the nutritional adequacy of intakes of groups or populations and in nutrition research. An EAR is set only when there is conclusive scientific research.

2. **Recommended Dietary Allowance (RDA).** The dietary intake value that is sufficient to meet the nutrient requirements of 97 to 98 percent of all healthy individuals in a group. The RDA is based on the EAR. If there is not enough scientific evidence to justify setting an EAR, no RDA value is given. The RDA is a goal for individuals.
3. **Adequate Intake (AI).** The dietary intake that is used when an RDA cannot be based on an estimated average requirement (EAR). AI is based on an approximation of nutrient intake for a group (or groups) of healthy people. For example, there is no EAR or RDA for calcium, only an AI. An AI is given when there is insufficient scientific research to support an RDA. Both the RDA and the AI may be used as goals for individual intake or to assess individual intake.
4. **Tolerable Upper Intake Level (UL).** The maximum intake level above which risk of toxicity increases. Intakes below the UL are unlikely to pose a risk of adverse health effects in healthy people. For most nutrients, this figure refers to total intakes from food, fortified food, and nutrient supplements. UL cannot be established for some nutrients, due to inadequate research.

The DRIs vary depending upon age and gender, and there are DRIs for pregnant and lactating women. The DRIs are meant to help healthy people maintain health and prevent disease. They are not designed for seriously ill people, whose nutrient needs may be much different.

The 2002 Dietary Reference Intake report established an **Estimated Energy Requirement (EER)** for healthy individuals. EER is the dietary energy intake measured in kcalories that is needed to maintain energy balance in a healthy adult so he/she does not gain or lose weight. Your actual EER depends on your age, gender, weight, height, and level of physical activity. There is no RDA or Tolerable Upper Intake Level for kcalories because these concepts do not apply to energy and would lead to weight gain.

Table 1-5 shows the EER for men and women 30 years of age. To find your EER, first find your height (in inches) in the first left-hand column (inches is in the parentheses). Next, select your PAL or physical activity level from the second column. Are you sedentary, low active, active, or very active? You will notice a new term, BMI, in the next two columns. Body Mass Index (BMI) is a method of measuring degree of obesity that uses your height and weight. A BMI between 18.5 and 24.99 means that you have a healthy weight and are neither overweight nor underweight. Pick the weight you are closer to (weight in pounds in parentheses) and then note the BMI for this weight in the third or fourth column heading. For the last step, you need to

TABLE 1-5 Estimated Energy Requirements (EER) for Men and Women 30 Years of Age[a]

Ht (m [in])	PAL[b]	Weight for BMI of 18.5 (kg [lb])	Weight for BMI of 24.99 (kg [lb])	EER, Men (kcal/day)		EER, Women (kcal/day)	
				BMI of 18.5	BMI of 24.99	BMI of 18.5	BMI of 24.99
1.45 (57)	Sedentary	38.9 (86)	52.5 (116)	1,777	1,994	1,564	1,691
	Low active			1,931	2,172	1,734	1,877
	Active			2,127	2,399	1,946	2,108
	Very active			2,450	2,771	2,201	2,386
1.50 (59)	Sedentary	41.6 (92)	56.2 (124)	1,848	2,080	1,625	1,762
	Low active			2,009	2,267	1,803	1,956
	Active			2,215	2,506	2,025	2,198
	Very active			2,554	2,898	2,291	2,489
1.55 (61)	Sedentary	44.4 (98)	60.0 (132)	1,919	2,167	1,688	1,834
	Low active			2,089	2,365	1,873	2,037
	Active			2,305	2,615	2,104	2,290
	Very active			2,660	3,027	2,382	2,593
1.60 (63)	Sedentary	47.4 (104)	64.0 (141)	1,993	2,257	1,752	1,907
	Low active			2,171	2,464	1,944	2,118
	Active			2,397	2,727	2,185	2,383
	Very active			2,769	3,160	2,474	2,699
1.65 (65)	Sedentary	50.4 (111)	68.0 (150)	2,068	2,349	1,816	1,982
	Low active			2,254	2,566	2,016	2,202
	Active			2,490	2,842	2,267	2,477
	Very active			2,880	3,296	2,567	2,807
1.70 (67)	Sedentary	53.5 (118)	72.2 (159)	2,144	2,442	1,881	2,057
	Low active			2,338	2,670	2,090	2,286
	Active			2,586	2,959	2,350	2,573
	Very active			2,992	3,434	2,662	2,917
1.75 (69)	Sedentary	56.7 (125)	76.5 (169)	2,222	2,538	1,948	2,134
	Low active			2,425	2,776	2,164	2,372
	Active			2,683	3,078	2,434	2,670
	Very active			3,108	3,576	2,758	3,028
1.80 (71)	Sedentary	59.9 (132)	81.0 (178)	2,301	2,635	2,015	2,211
	Low active			2,513	2,884	2,239	2,459
	Active			2,782	3,200	2,519	2,769
	Very active			3,225	3,720	2,855	3,141
1.85 (73)	Sedentary	63.3 (139)	85.5 (188)	2,382	2,735	2,083	2,290
	Low active			2,602	2,994	2,315	2,548
	Active			2,883	3,325	2,605	2,869
	Very active			3,344	3,867	2,954	3,255
1.90 (75)	Sedentary	66.8 (147)	90.2 (199)	2,464	2,836	2,151	2,371
	Low active			2,693	3,107	2,392	2,637
	Active			2,986	3,452	2,693	2,971
	Very active			3,466	4,018	3,053	3,371
1.95 (77)	Sedentary	70.3 (155)	95.0 (209)	2,547	2,940	2,221	2,452
	Low active			2,786	3,222	2,470	2,729
	Active			3,090	3,581	2,781	3,074
	Very active			3,590	4,171	3,154	3,489

[a]For each year below 30, add 7 kcal/day for women and 10 kcal/day for men. For each year above 30, subtract 7 kcal/day for women and 10 kcal/day for men.

[b]PAL = Physical activity level.

Source: Adapted with permission from the *Dietary References Intakes for Energy, Carbohydrates, Fiber, Fat, Protein, and Amino Acids (Macronutrients).* © 2002 by the National Academy of Sciences. Courtesy of the National Academy Press, Washington, D.C.

select which of the four right-hand columns is appropriate for you based on gender and whether your weight is closer to the BMI of 18.5 or 24.99. Once you know which column to use, go down to the kcalorie level for the activity level you selected.

Acceptable Macronutrient Distribution Range (AMDR)—A range of intakes for a particular nutrient that is associated with reduced risk of chronic disease while providing adequate intake.

The 2002 Dietary Reference Intake report also established **Acceptable Macronutrient Distribution Ranges (AMDR)** for carbohydrate, fat, and protein. AMDR is defined as the range of intakes for a particular nutrient that is associated with reduced risk of chronic disease while providing adequate intake, and is expressed as a percentage of total kcalorie intake. For example, adults (and children over 1 year old) should obtain 45 to 65 percent of their total kcalories from carbohydrates. The AMDR for adults is 20 to 35 percent of total kcalories from fat, and 10 to 35 percent of total kcalories from protein. The wide range allows for more flexibility in dietary planning for healthy people.

The DRIs are used to assess dietary intakes as well as plan diets. The RDA and AI are useful when planning diets for individuals. The EAR can be used to plan diets for groups to ensure that most people get enough nutrients and also to assess the number of people with inadequate intakes within a group.

The Dietary Reference Intakes Committee has periodically issued new reference values for groups of nutrients since 1997. The first five reports are used in this book and appear in Appendix B. A report is expected in 2003 for sodium, potassium, chloride, other electrolytes, and water.

■ MINI-SUMMARY

The DRI includes four dietary intake values: EAR (value estimated to meet requirements of half the healthy individuals in a group), RDA (value estimated to meet requirements of 97 to 98 percent of healthy individuals in a group), AI (the dietary intake used when there is not enough scientific basis for an EAR or RDA), and UL (maximum intake). The DRIs also include Estimated Energy Requirements and Acceptable Macronutrient Distribution Ranges. The DRIs are useful when assessing and planning diets of individuals and groups.

What Happens When You Eat

Digestion, Absorption, and Metabolism

To become part of the body, food must first be digested and absorbed. **Digestion** is the process by which food is broken down into its components in the mouth, stomach, and small intestine with the help of digestive

Digestion—The process by which food is broken down into its components in the mouth, stomach, and small intestine with the help of digestive enzymes.
Enzymes—Catalysts in the body.
Absorption—The passage of digested nutrients through the walls of the intestines or stomach into the body's cells. Nutrients are then transported through the body via the blood of lymph systems.
Metabolism—All the chemical processes by which nutrients are used to support life.
Anabolism—The metabolic process by which body tissues and substances are built.
Catabolism—The metabolic processes by which large, complex molecules are converted to simpler ones.
Gastrointestinal tract—A hollow tube running down the middle of the body in which digestion of food and absorption of nutrients takes place.
Oral cavity—The mouth.

enzymes. Protein is digested, or broken down, into its building blocks, called amino acids; complex carbohydrates are reduced to simple sugars, such as glucose; and fat molecules are broken down into fatty acids.

Before the body can use any nutrients present in food, they must pass through the walls of the stomach or intestines into the body's tissues, a process called **absorption.** Nutrients are absorbed into either the blood or the lymph, two fluids that circulate throughout the body delivering needed products to the cells and picking up wastes. Blood is composed mostly of water, red blood cells (which carry and deliver oxygen to the cells), white blood cells (which are important in resistance to disease, called immunity), nutrients, and other components. Lymph is similar to blood but has no red blood cells. It goes into areas where there are no blood vessels to feed the cells.

Within each cell, **metabolism** takes place. Metabolism refers to all the chemical processes by which nutrients are used to support life. Metabolism has two parts: the building up of substances (called **anabolism**) and the breaking down of substances (called **catabolism**). Within each cell, nutrients such as glucose are split into smaller units in a catabolic reaction that releases energy. The energy is either converted to heat to maintain body temperature or used to perform work within the cell. During anabolism, substances such as proteins are built from their amino-acid building blocks.

Gastrointestinal Tract

Once we have smelled and tasted our food, our meal goes on a journey through the **gastrointestinal tract** (also called the digestive tract), a hollow tube running down the middle of your body (Figure 1-3). The top of the tube is your mouth, which is connected in turn to your pharynx, esophagus, stomach, small intestine, large intestine, rectum, and anus, where solid wastes leave the body.

The digestive system starts with the mouth, also called the **oral cavity.** Your tongue and teeth help with chewing. The tongue, which extends across the floor of the mouth, moves food around the mouth during chewing. Your 32 permanent teeth grind and break down food. Chewing is important because it breaks up the food into smaller pieces so it can be swallowed. **Saliva,** a fluid secreted into the mouth from the salivary glands, contains important digestive enzymes and lubricates the food so that it may readily pass down the esophagus. Enzymes are substances that speed up chemical reactions. Digestive enzymes help break down food into forms of nutrients that can be used by the body. Enzymes in the saliva start the digestion of carbohydrate. The tongue rolls the chewed food into a **bolus** (or ball) to be swallowed.

The **pharynx** is a passageway about 5 inches long that connects the oral and nasal cavities to the esophagus and the air tubes to the lungs. When

Figure 1-3
Human digestive tract

Mouth:	Tastes food. Chews food. Makes saliva.
Pharynx:	Directs food from mouth to esophagus.
Esophagus:	Passes food to stomach.
Stomach:	Makes enzyme that breaks down protein. Makes hydrochloric acid. Churns and mixes food. Acts like holding tank.
Small intestine:	Makes enzymes. Digests most of food. Absorbs nutrients across villi into blood and lymph.
Large intestine:	Passes waste to be excreted. Reabsorbs water and some minerals. Absorbs vitamins made by bacteria.
Rectum:	Stores feces.
Anus:	Keeps rectum closed. Opens for elimination.

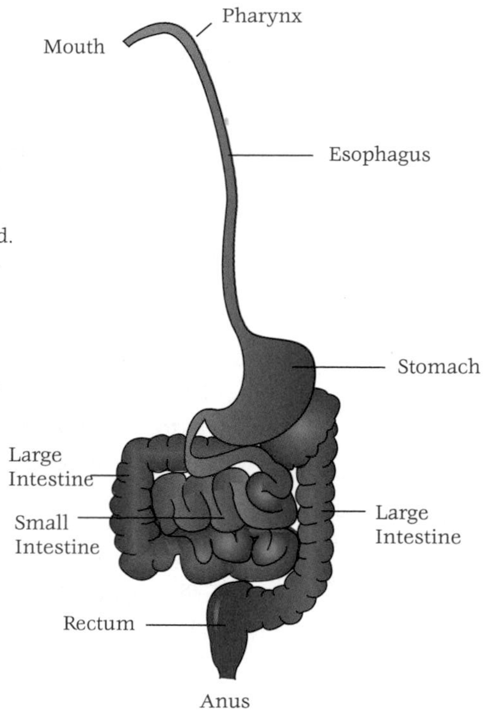

Saliva—A fluid secreted into the mouth from the salivary glands, which contains important digestive enzymes and lubricates the food so that it may readily pass down the esophagus.

Bolus—A ball of chewed food that travels from the mouth through the esophagus to the stomach.

Pharynx—A passageway that connects the oral and nasal cavities to the esophagus and air tubes to the lungs.

Epiglottis—The flap that covers the air tubes to the lungs so that food does not enter the lungs during swallowing.

Esophagus—The muscular tube that connects the pharynx to the stomach.

Peristalsis—Involuntary muscular contraction that forces food through the entire digestive system.

swallowing occurs, a flap of tissue, the **epiglottis,** covers the air tubes so that food does not get into the lungs. Food now enters the **esophagus,** a muscular tube that leads to the stomach. Food is propelled down the esophagus by **peristalsis,** rhythmic contractions of muscles in the wall of the esophagus. You might think of this involuntary contraction that forces food through the entire digestive system as squeezing a marble (the bolus) through a rubber tube. Peristalsis also helps break up food into smaller and smaller particles.

Food passes from the esophagus through the **lower esophageal (**or **cardiac) sphincter,** a muscle that relaxes and contracts (in other words, opens and closes) to move food from the esophagus into the stomach. The **stomach,** a J-shaped muscular sac that holds about 4 cups (or 1 liter) of food, is lined with a mucous membrane. Within the folds of the mucous membrane are digestive glands that make **hydrochloric acid** and an enzyme to break down proteins. Hydrochloric acid aids in protein digestion, destroys harmful bacteria, and increases the ability of calcium and iron to be absorbed. Because hydrochloric acid can damage the stomach, the stomach protects itself with a thick lining of mucus. Also, acid is produced only when we are eating or thinking about eating.

Lower esophageal (cardiac) sphincter—A muscle that relaxes and contracts to move food from the esophagus into the stomach.

Stomach—J-shaped muscular sac that holds about 4 cups of food and prepares food chemically and mechanically so it can be further digested and absorbed.

Hydrochloric acid—A strong acid made by the stomach that aids in protein digestion, destroys harmful bacteria, and increases the ability of calcium and iron to be absorbed.

Chyme—A semiliquid mixture in the stomach that contains partially digested food and stomach secretions.

Pyloric sphincter—A muscle that permits passage of chyme from the stomach to the small intestine.

Small intestine—The digestive tract organ that extends from the stomach to the opening of the large intestine.

Duodenum—The first segment of the small intestine, about 1 foot long.

From the top part of the stomach, food is slowly moved to the lower part, where the stomach churns it with the hydrochloric acid and digestive enzymes. The stomach has the strongest muscles and thickest walls of all the organs found in the gastrointestinal tract. The food is now called **chyme** and has a semiliquid consistency. Chyme is next passed into the first part of the small intestine in small amounts (the small intestine can't process too much food at one time) through the **pyloric sphincter,** which operates like the lower esophageal sphincter. Liquids leave the stomach faster than solids, and carbohydrate or protein foods leave faster than fatty foods. The stomach absorbs few nutrients, but it does absorb alcohol. It takes one and one-half to four hours after you have eaten for the stomach to empty.

The **small intestine,** about 10 to 12 feet long, has three parts: the **duodenum,** the **jejunum,** and the **ileum.** The small intestine was so named because its diameter is smaller (about 1 inch) than that of the large intestine (about 2-1/2 inches), not because it is shorter. Actually, the small intestine is longer.

The duodenum, about 1 foot long, receives the digested food from the stomach as well as enzymes from other organs in the body, such as the pancreas. The liver provides **bile,** a substance that is necessary for fat digestion. Bile is stored in the gall bladder and released into the duodenum when fat is present. The pancreas provides bicarbonate, a substance that neutralizes stomach acid. The small intestine itself produces digestive enzymes.

On the folds of the duodenal wall (and throughout the entire small intestine) are tiny, fingerlike projections called **villi.** Under a microscope you will see hairlike structures on the villi. These are called **microvilli.** The villi and microvilli increase the surface area of the small intestine and therefore allow for more absorption of nutrients into the body. The cells that cover the surface of the villi work so hard that they are replaced every six days.

The muscular walls of the small intestine mix the chyme with the digestive juices and bring the nutrients into contact with the villi. Most nutrients pass through the villi of the duodenum and jejunum into either the blood or lymph vessels, where they are transported to the liver and to the body cells. The duodenum connects with the second section of the small intestine, the jejunum, which connects to the ileum. Most digestion is completed in the first half of the small intestine; whatever is left goes into the large intestine. Food is in the small intestine for about 7 to 8 hours, and about 18 to 24 hours in the large intestine.

The **large intestine** (also called the colon) is about 5 feet long and extends from the end of the ileum to a cavity called the rectum. One of the functions of the large intestine is to receive the waste products of digestion and pass them on to the rectum. Waste products are the materials that were not absorbed into the body. The large intestine does absorb water, some

Jejunum—The second portion of the small intestine between the duodenum and the ileum.
Ileum—The final segment of the small intestine.
Bile—A substance made by the liver that is stored in the gallbladder and released when fat enters the small intestine because it helps digest fat.

minerals (such as sodium and potassium), and a few vitamins made by bacteria residing there. The **rectum** stores the waste products until released as solid feces through the **anus,** which opens to allow elimination.

■ MINI-SUMMARY

Before the body can use the nutrients in food, the food must be digested and the nutrients absorbed through the walls of the small or large intestine into either the blood or the lymph system. Within each cell, metabolism (all the chemical processes by which nutrients are used to support life) takes place. Metabolism has two parts: anabolism (building up) and catabolism (breaking down). Figure 1-3 summarizes food digestion and absorption.

Check-Out Quiz

1. Match the nutrients with their functions/qualities. The functions/qualities may be used more than once.

Nutrients	Functions
Carbohydrate	Provides energy
Lipid	Promotes growth and maintenance
Protein	Supplies the medium in which chemical changes of the body occur
Vitamins	Works as main structure of cells
Minerals	Regulates body processes
Water	

Villi—Tiny, fingerlike projections in the wall of the small intestines that are involved in absorption.
Microvilli—Hairlike projections on the villi that increase the surface area for absorbing nutrients.
Large intestine—The part of the gastrointestinal tract between the small intestine and the rectum.

2. Match the Dietary Reference Intake values with their definition.

DRI Value	Definition
RDA	Value for kcalories
AI	Maximum safe intake level
UL	Value that meets requirements of 50 percent of individuals in a group
EAR	Value that meets requirements of 97 to 98 percent of individuals
EER	Value used when there is not enough scientific data to support an RDA

Rectum—The last section of the large intestine in which feces, the waste products of digestion, is stored until elimination.

Anus—The opening of the digestive tract through which feces travels out of the body.

3. Match the terms on the left with their definitions on the right.

Term	Definition
Absorption	Process of building substances
Enzyme	Involuntary muscular contraction
Anabolism	Substance that speeds up chemical reactions
Peristalsis	Process of breaking down substances
Catabolism	Process of nutrients entering the tissues from the gastrointestinal tract

4. Which digestive organ passes waste to be excreted and reabsorbs water and minerals?
   a. stomach b. small intestine
   c. large intestine d. liver
5. Which nutrient supplies the highest number of calories per gram?
   a. carbohydrate b. fat c. protein d. vitamin pills
6. Flavor is a combination of all five senses.
   a. True b. False
7. Women have a higher basal metabolic rate than men.
   a. True b. False
8. Hydrochloric acid aids in protein digestion, destroys harmful bacteria, and increases the ability of calcium and iron to be absorbed.
   a. True b. False
9. The nutrient density of a food depends upon the amount of nutrients it contains and the comparison of that to its caloric content.
   a. True b. False
10. The DRIs are designed for both healthy and sick people.
    a. True b. False

Activities and Applications

1. How Many Calories Do You Need Each Day?

Use the following two steps to calculate the number of calories you need.

A. To determine your basal metabolic needs, multiply your weight in pounds by 10.9 if you are male, 9.8 if you are female. (These numbers are based on a BMR factor of 1.0 kcalorie per kilogram of body weight per hour for men and 0.9 for women.) Example: 150-pound woman × 9.8 = 1470 kcalories

B. To determine how much you use each day for physical activity, first determine your level of activity.

Very light activity—You spend most of your day seated or standing.

Light activity—You spend part of your day up and about, such as in teaching or cleaning house.
Moderate activity—You engage in exercise for an hour or so at least every other day, or your job requires some physical work.
Heavy activity—You engage in manual labor, such as construction.

Once you have picked your activity level, you need to multiply your answer in A by one of the following numbers.

Very light (men and women): Multiply by 1.3
Light (men): Multiply by 1.6
Light (women): Multiply by 1.5
Moderate (men): Multiply by 1.7
Moderate (women): Multiply by 1.6
Heavy (men): Multiply by 2.1
Heavy (women): Multiply by 1.9
Example: A woman with light activity.
1470 calories × 1.5 = 2205 calories needed daily

Compare the number of kcalories you need with your Estimated Energy Requirement using Table 1-5. The results should be similar.

2. Factors Influencing What You Eat
Answer the following questions to try to understand the factors influencing what you eat.

A. How many meals and snacks do you eat each day and when are they eaten?
B. What are your favorite foods?
C. What foods do you avoid eating and why?
D. Rate the importance of each of these factors when selecting foods (1 = very important, 3 = somewhat important, 5 = not important)
Cost
Convenience
Availability
Familiarity
Nutrition
E. Are you usually willing to try a new food?
F. What holidays do you and your family celebrate? What foods are served?
G. Do your food habits differ from those of your family? Your friends? Your co-workers? If yes, describe how your food habits are different and why you think this is so.
H. What foods, if any, do you eat to stay healthy or improve appearance?
I. How much do you know about nutrition? How important is good nutrition to you?

J. Do you eat differently when you are with others than when alone?
K. Which foods do you eat when you are under stress?
L. Which foods do you eat when you are sick?
M. Do you think that food advertising affects what you eat? Describe.
N. Do you prefer organic fruits and vegetables? Why or why not?
O. Are you a vegetarian, and if so, why did you choose this eating style?

3. Taste and Smell

Pick one of your favorite foods, eat it normally, and then take a bite of it while holding your nose. How does it taste when you can't smell very well? What influence does smell have on taste?

4. Nutrient-Dense Foods

Pick one food that you ate yesterday that could be considered nutrient-dense. Also pick one food that would *not* be considered nutrient-dense. Compare the nutrition labels. Explain why you chose these foods.

Nutrition Web Explorer

U.S. Government Healthfinder www.healthfinder.gov

This government site can help you find information on virtually any health topic. On their home page, enter "cancer and diet" in the Search box. Then click on the article entitled "Diet: Food Choice Recommendations for Reducing Risk of Cancer." Which food groups should you eat more of to reduce your risk of cancer? Which food groups should you eat less of?

McDonald's www.mcdonalds.com

Visit the McDonald's website to add up the kcalories in this meal combination: Big Mac, Super Size Fries, and Super Size Coke (double the small drink). On their home page, click on "USA" under "Select a Country." On the USA web page, click on "Food," then click on "McDonald's USA Nutrition Facts." Compare the number of kcalories in this meal to your EER.

National Academy of Sciences www.nas.edu

As noted in this chapter, the Food and Nutrition Board, which develops the DRI, is part of the National Academy of Sciences. Visit its web page and click on "Health and Medicine." Next, click on "Food and Nutrition," which lists its most recent nutrition publications. Determine if any new DRIs have been published; if so, click on the new publication and read the Summary.

Food Facts *Food Basics*

Whole foods are foods as we get them from nature. Examples include milk, eggs, meats, poultry, fish, fruits, vegetables, dried beans and peas, and grains. Many whole foods are also considered **fresh foods.** Fresh foods are raw foods that have not been processed (such as canned or frozen), or heated. Fresh foods also do not contain any preservatives. Examples of fresh foods include fresh fruits and vegetables, fresh meats, poultry, and fish.

Organic foods are foods that have been grown without most conventional pesticides, fertilizers, herbicides, antibiotics, or hormones, and without genetic engineering or irradiation. Organic farmers use, as examples, animal and plant manures to increase soil fertility and crop rotation to decrease pest problems. The goal of organic farming is to preserve the natural fertility and productivity of the land. This chapter's Hot Topic goes into more depth about organic foods.

Processed foods have been prepared using a certain procedure: cooking (such as frozen pancakes), freezing (frozen dinners), canning (canned vegetables), dehydrating (dried fruits), milling (white flour), culturing with bacteria (yogurt), or adding vitamins and minerals (enriched foods). In some cases, processing removes and/or adds nutrients.

When processing adds nutrients, the resulting food is either an **enriched** or a **fortified food.** For example, when whole wheat is milled to produce white flour, nutrients are lost. By law, white flour must be enriched with several vitamins and iron to make up for some of these lost nutrients. A food is considered enriched when nutrients are added to it to replace the same nutrients that are lost in processing.

Milk is often fortified with vitamin D because there are few good food sources of this vitamin. A food is considered fortified when nutrients are added that were not present originally, or nutrients are added that increase the amount already present. For example, orange juice does not contain calcium, so when calcium is added to orange juice, the product is called calcium-fortified orange juice. Probably the most notable fortified food is iodized salt. Iodized salt was introduced in 1924 to combat iodine deficiencies.

Whereas the food supply once contained mostly whole farm-grown foods, today's supermarket shelves are stocked primarily with processed foods. Some of these are minimally processed, such as canned peaches. Many processed foods contain parts of whole foods, and often have added ingredients such as sugars, or sugar or fat substitutes. For instance, cookies are made with eggs and white flour from grains. Then sugar, shortening, and nutrients are added. Highly processed foods, such as many breakfast cereals, cookies, crackers, sauces, soups, baking mixes, frozen entrées, pasta, snack foods, and condiments, are staples nowadays.

As you can probably guess, your best bet nutritionally is to consume more whole foods and fewer highly processed foods.

Hot Topic Organic Foods

Organic farming is one of the fastest growing segments of U.S. agriculture. The number of organic farmers is increasing by about 12 percent per year. The Organic Foods Production Act (2000) offers a national definition for the term *organic,* as well as the methods, practices, and substances that can be used in producing and handling organic crops and livestock. Common organic foods include fruits, vegetables, and cereals. Meat, poultry, and eggs can also be organic. This Hot Topic uses a question-and-answer format to help you learn about what organic foods are and what they aren't.

1. What is organic food?

Organic food is produced by farmers who emphasize the use of renewable resources and the conservation of soil and water to enhance environmental quality for future generations. Organic meat, poultry, eggs, and dairy products come from animals that are given no antibiotics or growth hormones. Organic food is produced without using most conventional pesticides, petroleum-based fertilizers or sewage sludge-based fertilizers, bioengineering (also called biotechnology), or ionizing radiation (also called irradiation). Before a product can be labeled "organic," a government-approved certifier inspects the farm where the food is grown to make sure the farmer is following all the rules necessary to meet U.S. Department of Agriculture (USDA) organic standards. Companies that handle or process organic food before it gets to your local restaurants and supermarkets must be certified as well.

2. How often are farms producing organic foods inspected?

Annual inspections are conducted of each farm producing organic foods. Government-approved certifiers must be notified by a producer immediately of any changes affecting an operation's compliance with the regulations, such as application of a prohibited pesticide to a field. Also, the USDA or the certifying agent are allowed to conduct unannounced inspections at any time to adequately enforce the regulations. The Organic Foods Production Act also requires that residue tests be performed to help in enforcement of the regulations. Certifying agents and the USDA will conduct residue tests of organically produced products when there is reason to believe that they have been contaminated with prohibited substances.

3. What makes organic fruits and vegetables different from nonorganic fruits and vegetables?

The organic crop production standards state that:

A. The land will have no prohibited substances applied to it for at least three years before harvesting of an organic crop.
B. The use of genetic engineering, ionizing radiation, and sewage sludge is prohibited.
C. Soil fertility and crop nutrients will be managed through tillage and cultivation practices, crop rotations, and cover crops, supplemented with animal and crop waste materials and allowed synthetic materials.
D. Preference will be given to the use of organic seeds and other planting stock, but a farmer may use nonorganic seeding and planting stock under specified conditions.
E. Crop pests, weeds, and diseases will be controlled primarily through management practices including physical, mechanical, and biological controls.

4. What makes meat, milk, and eggs organic?
Animals for slaughter must be raised under organic management from the last third of gestation, or no later than the second day of life for poultry. Livestock must be fed 100 percent organic feed. Organically raised animals may not be given hormones to promote growth, or antibiotics for any reasons (unless an animal is sick or injured, in which case the animal can't be sold as organic). Preventive management practices, including the use of vaccines, are used to keep animals healthy. Also, livestock may be given allowed vitamin and mineral supplements.

All organically raised animals must have access to the outdoors, including access to pasture. They may be temporarily confined only for reasons such as health and safety.

5. How are organic foods labeled?
There are three categories of labeling.

A. Foods labeled "100 percent organic" must contain only organically produced ingredients (excluding water and salt).
B. Foods labeled "organic" must consist of at least 95 percent organically produced ingredients (excluding water and salt). Any remaining ingredients must consist of nonagricultural products approved on the national list or agricultural products that are not commercially available in organic form.
C. Processed foods labeled "made with organic ingredients" must contain at least 70 percent organic ingredients, and they can list up to three of the organic ingredients or food groups on the principal display panel. For example, soup made with at least 70 percent organic ingredients and only organic vegetables may be labeled either "soup made with organic peas, potatoes, and carrots," or "soup made with organic vegetables."

Figure 1-4
The logo for the USDA National Organic Program. Courtesy of the U.S. Department of Agriculture

Processed foods that contain less than 70 percent organic ingredients cannot use the term "organic" anywhere on the principal display panel. However, they may identify the specific ingredients that are organically produced on the ingredients statement.

Foods that are 100 or 95 percent organic may display the USDA organic symbol, shown in Figure 1-4. The symbol is not required.

Chapter 2

Using Dietary Recommendations, Food Guides, and Food Labels to Plan Menus

Dietary recommendations— Guidelines that discuss specific foods and food groups to eat for optimal health.

Dietary recommendations have been published for the healthy American public for almost 100 years. Early recommendations centered on encouraging intake of certain foods to prevent deficiencies, fight disease, and enhance growth. Although deficiency diseases have been virtually eliminated, they have been replaced by diseases of dietary excess and imbalance—problems that now rank among the leading causes of illness and death in the United States. Diseases such as heart disease and cancer touch the lives of most Americans and generate substantial health-care costs. More recent dietary guidelines have therefore centered on modifying the diet, in most cases cutting back on certain foods, to prevent lasting degenerative conditions such as heart disease.

This chapter looks at dietary recommendations, food guides, and food labels. It will help you to:

- List the Dietary Guidelines for Americans
- Recommend ways to implement each Dietary Guideline
- List the nutritional goals of the Food Guide Pyramid
- Describe how the Food Guide Pyramid illustrates variety, proportionality, and moderation
- Name the five food groups in the Food Guide Pyramid and give examples of their respective subgroups
- Plan a menu using the Food Guide Pyramid
- List the information required on a food label
- Describe the information found on a Nutrition Facts panel
- Explain the uses of nutrient and health claims and outline the requirements the claims must meet
- Discuss the relationship between portion size and the Food Guide Pyramid

Dietary Recommendations and Food Guides

Whereas dietary recommendations discuss specific foods to eat for optimum health, **food guides** tell us the amounts of foods we need to eat for a nutritionally adequate diet. Their primary role, whether in the United States or around the world, is to communicate an optimum diet for overall health of the population. Food guides are based on current dietary recommendations, the nutrient content of foods, and the eating habits of the targeted population. The use of the Food Guide Pyramid, a food guide, has been very successful in the United States. The pyramid shape easily con-

Food guides—Guidelines that tell us the kinds and amounts of foods that constitute a nutritionally adequate diet; they are based on current dietary recommendations, the nutrient content of foods, and the eating habits of the targeted population.

veys the concept of variety and the relative amounts to eat of the various food groups. However, because of cultural differences in communication and other cultural norms, the pyramid is not necessarily the graphic of choice for food guides worldwide. Other countries, such as Canada, have a four-banded rainbow, with each color representing one of its four food groups.

The Food Guide Pyramid is based on the Dietary Guidelines for Americans and nutrient recommendations. To better understand the Food Guide Pyramid, let's first look at the Dietary Guidelines for Americans.

Dietary Guidelines for Americans

Dietary Guidelines for Americans—A set of dietary recommendations for Americans that is periodically revised.

The most recent set of U.S. dietary recommendations, the **Dietary Guidelines for Americans** (fifth edition), was published in 2000. Updated every five years by a joint advisory committee of the U.S. Department of Agriculture and the Department of Health and Human Services, the Dietary Guidelines for Americans are used by the federal government to plan food programs. They provide sound advice to help people make food choices for a healthy, active life, and reflect a consensus of the most current science and medical knowledge available. These guidelines are for healthy Americans age 2 years and over—not for younger children and infants, whose dietary needs differ. Dietary Guidelines for Americans reflect the recommendations of nutrition authorities, who agree that enough is known about diet's effect on health to encourage certain dietary practices. The guidelines should be applied to diets consumed over several days rather than to single meals or foods.

Dietary recommendations are quite different from the Dietary Reference Intakes (DRIs). Whereas the DRIs deal with specific nutrients, dietary recommendations discuss specific foods and food groups that will help individuals meet the DRIs. The DRIs also tend to be written in a technical style. Dietary recommendations are generally written in easy-to-understand terms. Following are the Dietary Guidelines for Americans.

Aim for Fitness

- Aim for a healthy weight.
- Be physically active each day.

Build a Healthy Base

- Let the Pyramid guide your food choices.
- Choose a variety of grains daily, especially whole grains.
- Choose a variety of fruits and vegetables daily.
- Keep food safe to eat.

Choose Sensibly

- Choose a diet that is low in saturated fat and cholesterol and moderate in total fat.
- Choose beverages and foods to moderate your intake of sugars.
- Choose and prepare foods with less salt.
- If you drink alcoholic beverages, do so in moderation.

The "ABC" approach for the Dietary Guidelines focuses on "Aim, Build, Choose—for Good Health." By following these guidelines, you can promote good health and reduce your risk of chronic diseases such as heart disease, certain cancers, diabetes, stroke, and osteoporosis. These diseases are leading causes of death and disability among Americans. Your food choices, your lifestyle, your environment, and your genes all affect your well-being.

Following is a discussion of the 10 Dietary Guidelines for Americans.

1. **Aim for a healthy weight.** Being overweight or obese increases your risk for high blood pressure, high blood cholesterol, heart disease, stroke, diabetes, certain types of cancer, arthritis, and breathing problems. A healthy weight is key to a long, healthy life. Chapter 11 discusses how to use a concept called Body Mass Index to evaluate your weight.

Advice for Today

- Build a healthy base by eating vegetables, fruits, and grains (especially whole grains) with little added fat or sugar.
- Select sensible portion sizes.
- Get regular physical activity to balance calories from the foods you eat.
- Keep in mind that even though heredity and the environment are important influences, your behaviors help determine your body weight.

2. **Be physically active each day.** Regular physical activity has many benefits: helps build and maintain healthy bones, muscles, and joints; builds endurance and muscular strength; helps manage weight; lowers risk factors for cardiovascular disease, colon cancer, and type 2 diabetes; helps control blood pressure; and promotes psychological well-being and self-esteem. People of all ages need to be active.

Advice for Today

- Aim to accumulate at least 30 minutes (adults) or 60 minutes (children) of moderate physical activity most days of the week, preferably daily.

- Choose physical activities that fit in with your daily routine, recreational or structured exercise programs, or both.
- Stay active throughout your life.
- Help children get at least 60 minutes of physical activity daily.
- Consult your health-care provider before starting a new vigorous physical-activity plan if you have a chronic health problem, or if you are over 40 (men) or 50 (women).

3. **Let the Pyramid guide your food choices.** Different foods contain different nutrients and other healthful substances. No single food can supply all the nutrients in the amounts you need. For example, oranges provide vitamin C and folate but no vitamin B_{12}; cheese provides calcium and vitamin B_{12} but no vitamin C. To make sure you get all the nutrients and other substances you need for health, build a healthy base by using the Food Guide Pyramid (Figure 2-1) as a starting point. Choose the recommended number of daily servings from each of the five major food groups, as discussed in the next section.
4. **Choose a variety of grains daily, especially whole grains.** Foods made from grains (such as wheat, rice, and oats) help form the foundation of a nutritious diet. They provide vitamins, minerals, carbohydrates, and other substances that are important for good health. Grain products are generally low in fat. Whole grains differ from refined grains in the amount of fiber and nutrients they provide. Whole grains, such as whole wheat, oatmeal, or brown rice, contain more fiber, vitamins, and minerals than refined grains.

Advice for Today

- Eat six or more servings of grain products daily. Include two to three servings of whole-grain foods daily for their good taste and their health benefits.
- Eat foods made from a variety of whole grains—such as whole wheat, oats, and whole-grain corn—every day.
- Prepare or choose grain products with little added saturated fat and a moderate or low amount of added sugars.

5. **Choose a variety of fruits and vegetables daily.** Eating plenty of fruits and vegetables may help protect you against many chronic diseases. It also promotes healthy bowel function. Fruits and vegetables provide essential vitamins and minerals, fiber, and other substances that are important for good health. Most people, including children, eat fewer servings of fruits and vegetables than are recommended. To promote your health, eat a variety of fruits and vegetables—at least two servings of fruits and three servings of vegetables—each day. Different fruits and vegetables are rich in different nutrients. Most fruits and vegetables are

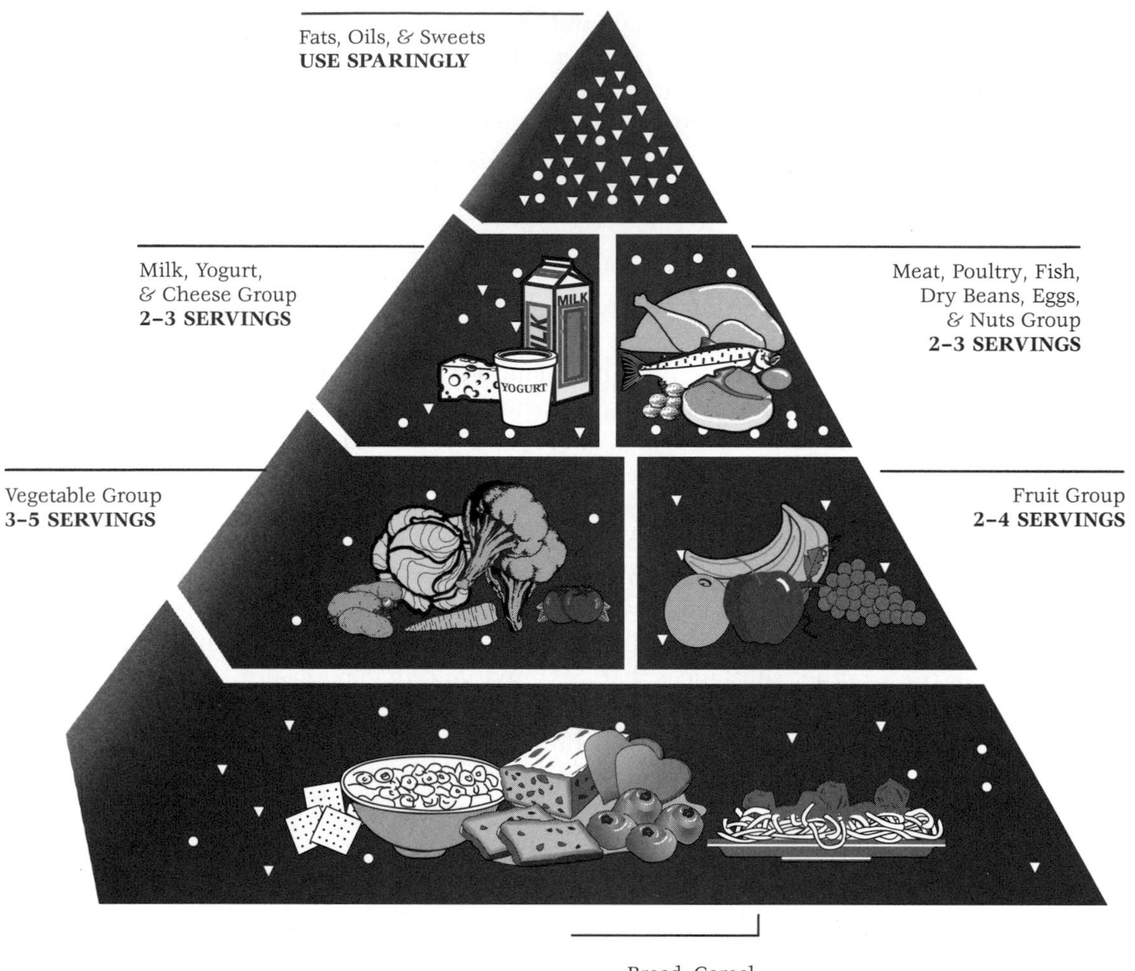

Figure 2-1

The food guide pyramid—a guide to daily food choices.

Source: U.S. Department of Agriculture and U.S. Department of Health and Human Services.

naturally low in fat and kcalories and are filling. Some are high in fiber, and many are quick to prepare and easy to eat.

Advice for Today

- Choose whole or cut-up fruits and vegetables rather than juices most often. Juices contain little or no fiber.
- Try many colors and kinds of fruits and vegetables.
- Choose dark green leafy vegetables, orange fruits and vegetables, and cooked dry beans and peas often.

6. **Keep food safe to eat.** Foods that are safe from harmful bacteria, viruses, parasites, and chemical contaminants are vital for healthful eating.

Advice for Today

- **Clean.** Wash hands and surfaces often.
- **Separate.** Separate raw, cooked, and ready-to-eat foods while shopping, preparing, or storing.
- **Cook.** Cook foods to a safe temperature.
- **Chill.** Refrigerate perishable foods promptly.
- **Check and follow the label.**
- **Serve safely.** Keep hot foods hot and cold foods cold.
- **When in doubt, throw it out.**

7. **Choose a diet that is low in saturated fat and cholesterol and moderate in total fat.** Fats supply energy and essential fatty acids, and they help the body absorb the fat-soluble vitamins A, D, E, and K. Some kinds of fat, especially saturated fats, increase the risk for coronary heart disease by raising blood cholesterol. In contrast, unsaturated fats (found mostly in vegetable oils) do not increase blood cholesterol.

Advice for Today

- Limit use of solid fats, such as butter, hard margarines, lard, and partially hydrogenated shortenings. Use vegetable oils as a substitute.
- Choose fat-free or low-fat dairy products, cooked dry beans and peas, fish, and lean meats and poultry.
- Eat plenty of grain products, vegetables, and fruits daily.

8. **Choose beverages and foods to moderate your intake of sugars.** Sugars are carbohydrates and a source of energy. Foods containing sugars (and also starches) promote tooth decay. Bacteria in the mouth use sugars and starches to produce the acid that causes tooth decay. The more often you eat foods with sugar or starch, and the longer these foods remain in your mouth before you brush your teeth, the greater

your risk for tooth decay. Foods high in sugar can also contribute to overweight when you eat more calories than you use up.

Advice for Today

- Take care not to let soft drinks or other sweets (such as cakes, cookies, pies, and candy) crowd out other foods you need to maintain health, such as low-fat milk or other good sources of calcium.
- Between meals, eat few foods or beverages containing sugars or starches. If you do eat them, brush your teeth afterward.

9. **Choose and prepare foods with less salt.** Many people can reduce their chances of developing high blood pressure by consuming less salt. In the body, sodium—which you get mainly from salt—plays an essential role in regulating fluids and blood pressure. Many studies in diverse populations have shown that a high sodium intake is associated with higher blood pressure. Salt is found mainly in processed and prepared foods.

Advice for Today

- Choose fruits and vegetables often. They contain very little salt unless it is added in processing. They are also rich in potassium, which may help decrease blood pressure.
- Use herbs, spices, and fruits to flavor foods, and cut the amount of salty seasonings by half.
- Read the Nutrition Facts label to compare and help identify foods lower in sodium—especially prepared foods.

10. **If you drink alcoholic beverages, do so in moderation.** Alcoholic beverages supply calories but few nutrients. They are harmful when consumed in excess. Taking more than one drink per day for women or two drinks per day for men can raise the risk for motor vehicle crashes, other injuries, high blood pressure, stroke, violence, suicide, and certain types of cancer. Alcohol consumption during pregnancy increases the risk of birth defects.

Advice for Today

- If you choose to drink alcoholic beverages, limit intake to one drink per day for women or two for men. Take with meals to slow alcohol absorption.
- Avoid drinking before or when driving, or whenever it puts you at risk.

Food Guide Pyramid

Food Guide Pyramid—A food guide developed by the U.S. Department of Agriculture to help healthy Americans follow the Dietary Guidelines for Americans.

The **Food Guide Pyramid** (Figure 2-1) was developed to help healthy Americans follow the Dietary Guidelines for Americans. This guide was developed by the U.S. Department of Agriculture (USDA) to:

1. Promote overall health for Americans 7 years of age and older (there is a separate Food Guide Pyramid for Young Children for ages 2 to 6)
2. Incorporate the most up-to-date research
3. Focus on the total diet
4. Be useful, flexible, and practical

Many years of research and testing went into the development of the Food Guide Pyramid.

The Food Guide Pyramid was a major change from the previous guide, known as the "Basic Four," which divided foods into four basic groups. The Basic Four was a foundation diet. This means that it was intended only to meet part of the caloric needs and part of the Dietary Reference Intakes for nutrients. The assumption was that people would eat more food than was recommended by the Basic Four. In this way, they would get enough calories and nutrients. At that time, not much was known about the importance of fiber or the relationship between high intakes of certain food components and disease. Little was said about the selection of fat and added sugars or about appropriate caloric intake in the Basic Four food guide.

Instead of being based on just a foundation diet, the Pyramid is based on the total diet. This means that the Pyramid shows how to get enough nutrients and, at the same time, how to avoid excesses of certain food components. Those components high in fat and/or added sugars and low in nutrient density—such as butter, margarine, oils, sugars, jam, and soft drinks—are separated from the Pyramid's five major food groups and placed in the Pyramid's small tip. In addition, symbols are used within the major food groups to show that certain foods are also high in fat and/or added sugars. The Pyramid is now the official food guide for the United States.

The nutritional goals of the Pyramid are as follows:

1. Energy: provide 1300 to 3000 kcalories.
2. Protein, vitamins, and minerals: provide 100 percent of the DRI for people over 2 years of age.
3. Fiber: increase intake.
4. Total fat: limit to 30 percent or less of total calories.
5. Saturated fat: limit to less than 10 percent of total calories.
6. Cholesterol: limit to 300 mg or less.
7. Sodium: limit to 2400 mg or less.

8. Added sugars: not to exceed caloric needs. When enough food has been eaten from the major food groups, the rest of the calories can come from fat and added sugars.

As you can see, the first three goals are related to nutritional adequacy, whereas the remaining goals are related to moderation.

The Food Guide Pyramid emphasizes moderation and adequacy to reduce your risk of conditions or diseases such as the following:

- High blood pressure
- Stroke
- Heart disease
- Certain cancers
- Osteoporosis
- Diabetes

Chapter 11 discusses nutrition and health issues.

Variety, Proportionality, and Moderation. The key concepts of the Pyramid include variety, proportionality, and moderation. The Pyramid was designed around the variety of food eaten by many Americans. To meet the Pyramid's nutritional goals, it is important to eat from all five food groups and to choose a variety of foods within each group. Members of certain ethnic groups and people such as vegetarians, who do not eat foods from all of the food groups, may need special advice on how to meet their nutrient needs.

Proportionality—A concept of eating relatively more foods from the larger food groups at the base of the Food Guide Pyramid and fewer foods from the smaller food groups nearer the top of the Pyramid.

Proportionality means eating relatively more foods from the larger food groups at the base of the Pyramid and fewer foods from the smaller food groups nearer the top of the Pyramid. The Pyramid illustrates proportionality by the shape of the Pyramid itself, the relative size of the food group sections, and the recommended number of servings for each group. The Pyramid emphasizes grains, fruit, and vegetables and deemphasizes animal products such as meat and milk. Proportionality does not mean that some food groups are more important than others. Each of the food groups provides some, but not all, of the necessary nutrients. To meet nutrient requirements while eating a typical American diet, no one group can replace another group.

Moderation means not eating any foods to excess. The Pyramid illustrates moderation by recommending a certain number of servings for each food group, encouraging variety (which helps one avoid eating any foods to excess), and advising sparing use of the foods at the tip of the Pyramid (fats, oils, and sweets).

Food Groups. The five nutrient-dense food groups in the Pyramid are the following:

1. Bread, Cereal, Rice, and Pasta Group
2. Vegetable Group
3. Fruit Group
4. Milk, Yogurt, and Cheese Group
5. Meat, Poultry, Fish, Dry Beans, Eggs, and Nuts Group

When placing foods in these groups, the USDA grouped foods primarily by the nutrients they provide. Table 2-1 lists the major nutrients in each food group. Typical use of a food in meals and how it was grouped in past guides were also considered.

Subgroups within the major food groups (Table 2-2) emphasize foods that are particularly good sources of dietary fiber or of certain vitamins and minerals that are low in the diets of many Americans. Eating more foods from certain subgroups is recommended. For example, the bread, cereal, rice, and pasta group has two subgroups: enriched and whole grains. USDA recommends several servings a day of whole grains, which provide more fiber, vitamins, and minerals than enriched grains.

The milk, yogurt, and cheese group is divided into two subgroups: low-fat milk products and other milk products with more fat or sugar. The USDA recommends choosing primarily low-fat or nonfat items. It also recommends choosing meat alternatives, such as beans, several times a week, because they are rich in dietary fiber and minerals.

The vegetable group has five subgroups: dark green leafy, deep yellow, starchy, dry beans and peas, and other. Dark green leafy vegetables and dry beans and peas are particularly important. The fruit group has two subgroups: citrus, melons, and berries, and other. Citrus, melons, and berries are emphasized for their high vitamin C content. The USDA recommends a dark green vegetable for vitamin A and a vitamin-C-rich fruit every day. Table 2-2 lists some food examples in each food group and subgroup.

The tip of the Pyramid includes fats, oils, and sweets. This group does not count as a major food group because these foods provide energy but little else nutritionally. Servings of these foods are optional. Examples of these foods include butter, margarine, salad dressing, sugar, jelly, honey, regular soft drinks, and candy bars. Foods such as doughnuts, cakes, and cookies are counted in the bread group. Potato chips are counted in the vegetable group. The extra fat and sugar found in these foods is counted as additional fat and/or added sugar in the diet.

Some food items can be difficult to classify. For example, grouping of corn products depends on the form in which corn is used. Sweet corn is counted as a starchy vegetable; popcorn and cornmeal products such as corn tortillas are counted as grain products; hominy is grouped with starchy vegetables, and hominy grits with grain products. Snack and dessert items such as cakes, cookies, ice cream, french-fried potatoes, potato chips, and so forth count with the food group of their major ingredient, for example, the bread,

TABLE 2-1 Major Nutrients in Food Groups

Food Group	Major Nutrients
Bread, Cereal, Rice, and Pasta	Complex carbohydrates B vitamins—thiamin, riboflavin, niacin, and folate Minerals—iron Fiber
Vegetables	Carbohydrate Vitamins—A, C, and folate Minerals—iron and magnesium Fiber
Fruit	Carbohydrate Vitamins—A and C Minerals—potassium Fiber
Milk, Yogurt, and Cheese	Protein Fat Vitamins—riboflavin, A, and D (if fortified) Minerals—the best source of calcium
Meat, Poultry, Fish, Dry Beans, Eggs, and Nuts	Protein Fat B vitamins—niacin, thiamin, and B_{12} Minerals—iron and zinc

Source: The Food Guide Pyramid. Cooperative Extension, University of California, Davis. 1998.

dairy, or vegetable group. However, use of these higher-fat items must be limited to keep total fat intake to the recommended level.

Dry beans and peas (called legumes) can count either as a meat alternate or as a starchy vegetable (they should not be counted twice in the same menu). These foods are good sources of protein and other nutrients provided by the meat group, such as iron and zinc, and have long been recommended as inexpensive alternatives to meat. Dry beans and peas are also high in carbohydrates and are good sources of vitamins, minerals, and

TABLE 2-2 Variety from the Food Groups

BREAD, CEREAL, RICE, PASTA					
Whole-Grain		Enriched		Grain Products with More Fat and Sugar	
Brown rice	Pumpernickel bread	Bagels	Italian bread	Biscuits	Danish
Buckwheat groats	Whole-grain cereals	Cornmeal	Macaroni	Cake (unfrosted)	Doughnuts
Bulgur	Rye bread and crackers	Crackers	Noodles	Cookies	Muffins
Corn tortillas	Whole-wheat bread rolls, crackers	English muffins	Pancakes and waffles	Cornbread	Pie crust
Graham crackers	Whole-wheat pasta	Farina	Pretzels	Croissants	Tortilla chips
Granola	Whole-wheat cereals	Flour tortillas	Ready-to-eat cereals		
Oatmeal		French bread	White rice		
Popcorn		Grits	Spaghetti		
		Hamburger and hot dog rolls	White bread and rolls		
FRUITS					
Citrus, Melons, Berries			Other Fruits		
Blueberries	Honeydew melon	Strawberries	Apples	Guavas	Pineapples
Cantaloupe	Kiwi fruit	Tangerines	Apricots	Grapes	Plantains
Citrus juices	Lemons	Watermelons	Asian pears	Mangoes	Plums
Cranberries	Oranges	Ugli fruit	Bananas	Nectarines	Prickly pears
Grapefruit	Raspberries		Cherries	Papayas	Prunes
			Dates	Passion fruit	Raisins
			Figs	Peaches	Rhubarb
			Fruit juices	Pears	Star fruit
VEGETABLES					
Dark-Green Leafy			Deep Yellow	Starchy	
Beet greens	Dandelion greens	Romaine lettuce	Carrots	Breadfruit	Lima beans
Broccoli	Endive	Spinach	Pumpkin	Corn	Potatoes
Chard	Escarole	Turnip greens	Sweet potatoes	Green peas	Rutabagas
Chicory	Kale	Watercress	Winter squash	Hominy	Taro
Collard greens	Mustard greens				

TABLE 2-2 *(continued)*

VEGETABLES					
Dry Beans and Peas (Legumes)		**Other Vegetables**			
Black beans	Lima beans (mature)	Artichokes	Cauliflower	Green peppers	Snow peas
Black-eyed peas	Mung beans	Asparagus	Celery	Lettuce	Summer squash
Chickpeas (garbanzos)	Navy beans	Bean and alfalfa sprouts	Chinese cabbage	Mushrooms	Tomatoes
Kidney beans	Pinto beans	Beets	Cucumbers	Okra	Turnips
Lentils	Split peas	Brussels sprouts	Eggplants	Onions (mature and green)	Vegetable juices
		Cabbage	Green beans	Radishes	Zucchini
MEAT, POULTRY, FISH, AND ALTERNATES					
Meat, Poultry, and Fish				**Alternates**	
Beef	Ham	Pork	Veal	Eggs	Peanut butter
Chicken	Lamb	Shellfish	Luncheon meats, sausage	Dry beans and peas (legumes)	Tofu
Fish	Organ meats	Turkey		Nuts and seeds	
MILK, YOGURT, AND CHEESE					
Low-fat Milk Products		**Other Milk Products with More Fat or Sugar**			
Buttermilk	Low-fat or nonfat plain yogurt	Cheddar cheese	Frozen yogurt	Ice milk	Swiss cheese
Low-fat cottage cheese	Skim milk	Chocolate milk	Fruit yogurt	Process cheeses and spreads	Whole milk
1% and 2% milk		Flavored yogurt	Ice cream	Puddings made with whole milk	
FATS, SWEETS, AND ALCOHOLIC BEVERAGES					
Fats		**Sweets**			**Alcoholic Beverages**
Bacon, salt pork	Mayonnaise	Candy	Jam	Popsicles and ices	Beer
Butter	Mayonnaise-type salad dressing	Corn syrup	Jelly	Sherbets	Liquor
Cream (dairy, nondairy)	Salad dressing	Frosting (icing)	Maple syrup	Soft drinks and colas	Wine
Cream cheese	Shortening	Fruit drinks	Marmalade		
			Molasses		

TABLE 2-2 (continued)

FATS, SWEETS, AND ALCOHOLIC BEVERAGES					
Fats		Sweets			Alcoholic Beverages
Lard	Sour cream	Gelatin desserts	Table syrup	Sugar (white and brown)	
Margarine	Vegetable oil	Honey			

Source: "Using the Food Guide Pyramid: A Resource for Nutrition Educators" by A. Shaw, L. Fulton, C. Davis, and M. Hogbin. U.S. Department of Agriculture: Food, Nutrition, and Consumer Services; Center For Nutrition Policy and Promotion.

dietary fiber. To increase use of these nutrient-dense foods, the Food Guide Pyramid suggests including dry beans and peas as a vegetable selection several times a week, instead of considering them only as meat alternates.

Symbols for fat and/or added sugar are included in all food groups on the Pyramid to show that these substances are found in all the food groups. Small circles represent added or naturally occurring fat. Upside-down triangles represent added sugar, but not naturally occurring sugar as found in fruit. The relative concentrations of fat and added sugars in the food groups are shown by the number of symbols in each group. The greatest number of symbols are in the tip.

Number of Servings. The Food Guide Pyramid suggests foods and number of servings for the total diet. If more kcalories are needed than are provided by the lower numbers of servings in the ranges, additional servings from the major food groups are suggested, along with modest increases in amounts of total fat and added sugars. Increasing amounts of grain products, vegetables, and fruit helps keep higher-kcalorie diets moderate in fat and also provides additional vitamins, minerals, and dietary fiber—nutrients that are low in many American diets.

Table 2-3 shows sample food patterns for a day at three calorie levels, covering the ranges of servings suggested by the Pyramid. It also indicates some age/sex groups for whom those kcalorie levels may be appropriate. The sample food patterns are not prescriptions, but rather illustrations of healthy proportions in the diet. Specific numbers of servings may vary somewhat from day to day.

Table 2-4 shows one day's menu adapted for three kcalorie levels—1600, 2200, and 2800 kcalories. Those with higher kcalorie needs take larger portions of some meal items and can supplement their meals with more snacks.

TABLE 2-3 Sample Food Patterns for a Day at Three Calorie Levels

1600 kcalories is about right for many sedentary women and some older adults.

2200 kcalories is about right for most children, teenage girls, active women, and many sedentary men. Women who are pregnant or breast-feeding may need somewhat more.

2800 kcalories is about right for teenage boys, many active men, and some very active women.

	About 1600	About 2200	About 2800
Bread Group Servings	6	9	11
Fruit Group Servings	2	3	4
Vegetable Group Servings	3	4	5
Meat Group	5 ounces	6 ounces	7 ounces
Milk Group Servings	2–3*	2–3*	2–3*
Total fat (grams)[a]	53	73	93
Total added sugars (teaspoons)[a]	6	12	18

* Women who are pregnant or breast-feeding, teenagers, and young adults to age 24 need 3 servings.

[a] Values for total fat and added sugars include fat and added sugars that are in food choices from the five major food groups as well as fat and added sugars from foods in the Fats, Oils, and Sweets group.

Source: "Using the Food Guide Pyramid: A Resource for Nutrition Educators" by A. Shaw, L. Fulton, C. Davis, and M. Hogbin. U.S. Department of Agriculture: Food, Nutrition, and Consumer Services; Center For Nutrition Policy and Promotion.

Portion or Serving Sizes. Serving sizes specified by the Pyramid are listed in Table 2-5, and an expanded list of serving sizes appears in Appendix C. For ease of use, the number of different serving sizes for foods in each food group was kept to a minimum. Food guide servings are based on food "as eaten." That is, meats are cooked and trimmed of fat and bone. Vegetables are rinsed, trimmed, and cooked, or eaten raw as appropriate. Rice, pasta, and cereal grains such as oatmeal are cooked.

The serving size for all fruit juices is 3/4 cup, rather than varying from 1/3 to 3/4 cup based on the carbohydrate content of the specific juice. For most food groups, the amount to count as a serving is comparable to the amount typically reported in food-consumption surveys—for example, 1/2 cup of cooked vegetable, or 1 cup of leafy raw salad greens. For foods in the bread group, portions typically reported (e.g., 1 cup of rice or pasta, 1 whole hamburger bun) more nearly equate to two servings from the Pyramid. For this group, the familiar serving size used in previous guides (e.g., 1 slice of bread or 1/2 cup of rice or pasta) was retained for the Pyramid.

For meat, poultry, and fish, the portion sizes reported in surveys vary widely depending on the type of meat and the eating occasion. For example, dinner portions are typically 3 ounces or more, while amounts used in a sandwich are 1 to 2 ounces. The Pyramid suggests that the two to three servings from the meat group should total 5 to 7 ounces per day. Common

TABLE 2-4 One Day's Menu and Food Group Servings at Three Kcalorie Levels

Item	Kcalorie Level 1600	2200	2800
BREAKFAST			
Cantaloupe	1/4 medium	1/4 medium	1/4 medium
Whole-wheat pancakes	2	2	3
Blueberry sauce	1/4 cup	1/4 cup	6 tablespoons
Margarine		1 teaspoon	2 teaspoons
Turkey patty		1-1/2 ounces	1-1/2 ounces
Milk	skim, 1 cup	skim, 1 cup	2%, 1 cup
LUNCH			
Chili-stuffed baked potato	3/4 cup chili, 1 potato	3/4 cup chili, 1 potato	3/4 cup chili, 1 potato
Low-fat, low-sodium cheddar cheese		3 tablespoons	3 tablespoons
Spinach-orange salad	1 cup	1 cup	1 cup
Wheat crackers	6	6	6
Grapes			12
Fig bars			2
Milk		skim, 1 cup	2%, 1 cup
DINNER			
Apricot-glazed chicken	1 breast half	1 breast half	1 breast half
Rice-pasta pilaf	3/4 cup	3/4 cup	3/4 cup
Steamed zucchini			1/2 cup
Tossed salad	1 cup	1 cup	1 cup
Reduced-calorie Italian dressing	1 tablespoon	1 tablespoon	
Regular Italian dressing			1 tablespoon
Hard roll(s)	1 small	2 small	2 small
Margarine		2 teaspoons	2 teaspoons
Vanilla ice milk	1/2 cup	1/2 cup	1/2 cup
SNACKS			
Fig bar	1		
Skim milk	3/4 cup		
Apple		1/2 medium	1/2 medium
Soft pretzel		1 large	1 large
Lemonade			1 cup
2% milk			1 cup

Source: "Using the Food Guide Pyramid: A Resource for Nutrition Educators" by A. Shaw, L. Fulton, C. Davis, and M. Hogbin. U.S. Department of Agriculture: Food, Nutrition, and Consumer Services; Center For Nutrition Policy and Promotion.

TABLE 2-5 The Pyramid Guide to Daily Food Choices

Food Group	Suggested Daily Servings	What Counts as a Serving
Bread, Cereal, Rice, Pasta Whole-grain Enriched	6 to 11 servings from entire group (Include several servings of whole-grain products daily.)	1 slice of bread 1/2 hamburger bun or English muffin a small roll, biscuit, or muffin 5 to 6 small or 3 to 4 large crackers 1/2 cup cooked cereal, rice, or pasta 1 ounce ready-to-eat cereal
Fruits Citrus, melon, berries Other fruits	2 to 4 servings from entire group	a whole fruit such as a medium apple, banana, or orange a grapefruit half a melon wedge 3/4 cup juice 1/2 cup berries 1/2 cup chopped, cooked, or canned fruit 1/4 cup dried fruit
Vegetables Dark green leafy Deep yellow Dry beans and peas (legumes) Starchy Other vegetables	3 to 5 servings (Include all types regularly; use dark-green leafy vegetables and dry beans and peas several times a week.)	1/2 cup cooked vegetables 1/2 cup chopped raw vegetables 1 cup leafy raw vegetables, such as lettuce or spinach 3/4 cup vegetable juice
Meats, Poultry, Fish, Dry Beans and Peas, Eggs, and Nuts	2 to 3 servings from entire group	Amounts should total 5 to 7 ounces of cooked lean meat, poultry without skin, or fish per day. Count 1 egg, 1/2 cup cooked beans, or 2 tablespoons peanut butter as 1 ounce of meat.
Milk, Yogurt, Cheese	2 servings (3 servings for women who are pregnant or breast-feeding, teenagers, and young adults to age 24.)	1 cup milk 8 ounces yogurt 1-1/2 ounces natural cheese 2 ounces process cheese

TABLE 2-5 *(continued)*

Food Group	Suggested Daily Servings	What Counts as a Serving
Fats, Sweets, and Alcoholic Beverages	Use fats and sweets sparingly. If you drink alcoholic beverages, do so in moderation.	

Note: The guide to daily food choices described here was developed for Americans who regularly eat foods from all the major food groups listed. Some people such as vegetarians and others may not eat one or more of these types of foods. These people may wish to contact a dietitian or nutritionist for help in planning food choices.

Source: "Using the Food Guide Pyramid: A Resource for Nutrition Educators" by A. Shaw, L. Fulton, C. Davis, and M. Hogbin. U.S. Department of Agriculture: Food, Nutrition, and Consumer Services; Center For Nutrition Policy and Promotion.

portions of meat alternatives, such as one egg, or 2 tablespoons of peanut butter, or 1/2 cup of cooked dry beans or peas, are equivalent in protein and most vitamins and minerals to 1 ounce of lean meat.

For foods in the Fats, Oils, and Sweets category, no serving size or numbers of servings are listed. The amounts of these foods that can be included depend on the fat and added sugars provided as part of the specific food items selected from the major food groups. For example, a medium croissant counts as two servings from the bread group but provides 12 grams of fat, compared with 2 grams of fat provided by two slices of plain bread. Thus, if a croissant is selected, the amount of spreads and dressings used should be reduced to compensate for the extra fat provided by the croissant (equivalent to about 2 teaspoons of butter or margarine) to keep total fat in the menu below 30 percent of total kcalories.

Table 2-6 shows how to count food group servings in one day's menu. Note the following points:

- A large portion of a food item counts as more than one serving. For example, the whole toasted raisin English muffin at breakfast counts as two servings from the bread group. A smaller portion counts as part of a serving—the 1/2 cup of skim milk at breakfast counts as half a serving from the milk group.
- Desserts and snacks contribute to food group servings. In this menu, plain cookies (gingersnaps), fruit (pineapple chunks for dessert at dinner), crackers, cheese, vegetable juice, and a half sandwich contribute substantially to food group servings and nutrient intake for the day.
- The relatively high-fat entrée at lunch (Taco Salad) and the cheese for a snack are balanced by a low-fat breakfast, a low-fat entree for dinner (Pork and Vegetable Stir-fry), and a selection of fruit and lower-fat cookies for desserts.
- Reduced-fat and reduced-sodium products can also help keep the fat and sodium levels in check. This menu uses low-fat, low-sodium cheese and

TABLE 2-6 Counting Food Group Servings in One Day's Menu at 2200 Calories

Recipe	Bread	Vegetable	Fruit	Milk	Meat oz.	Fat grams	Kcalories
BREAKFAST							
Medium grapefruit, 1/2			1			trace	41
Medium banana			1			1	108
Ready-to-eat cereal flakes, 1 ounce	1					trace	111
Toasted raisin English muffin	2					1	138
Soft margarine, 2 teaspoons						8	68
Skim milk, 1/2 cup				1/2		trace	43
LUNCH							
Taco salad, 1 serving						19	455
unsalted tortilla chips	3/4						
tomato puree and greens		1-1/2					
low-fat, low-sodium cheddar cheese				1/2			
beef and beans					2-1/2		
Medium gingersnaps, 2	1					2	101
DINNER							
Pork and vegetable stir-fry, 1 serving						9	370
rice	1-1/2						
vegetables		1					
pork					3		
Cooked broccoli, 1/2 cup		1				trace	26
Small white rolls, 2	2					3	167
Soft margarine, 2 teaspoons						8	68
Minted pineapple chunks, juice-pack, 1/2 cup			1			trace	75
SNACKS							
Wheat crackers, 6	1					4	86
Cheddar cheese, 1-1/2 ounces				1		14	171
Turkey sandwich, 1/2						4	137
rye bread	1						
turkey					1		
lettuce leaf							
mayonnaise-type salad dressing, fat-free, 1/2 tablespoon							
No-salt-added tomato juice, 3/4 cup		1				trace	31
TOTAL	10-1/4	4-1/2	3	2	6-1/2	73	2196

Source: "Using the Food Guide Pyramid: A Resource for Nutrition Educators" by A. Shaw, L. Fulton, C. Davis, and M. Hogbin. U.S. Department of Agriculture: Food, Nutrition, and Consumer Services; Center For Nutrition Policy and Promotion.

unsalted tortilla chips in the Taco Salad, low-kcalorie mayonnaise-type dressing in the turkey sandwich, and no-salt-added tomato juice.

In order to keep kcalories to the target level (2200), sources of added sugars in this menu are limited to the cookies at lunch.

Many foods that Americans eat are mixtures of foods from several food groups—pizza, beef stew, and macaroni and cheese, for example. These mixed dishes or combination foods contain items from more than one group, and often provide a wide variety of nutrients. For example, macaroni and cheese counts as servings from both the bread group (macaroni) and the milk group (cheese).

Most recipes contain foods from more than one food group. For example, Pork and Vegetable Stir-fry (Figure 2-2) uses 1 pound of boneless pork loin, which is expected to yield 12 ounces of meat, or four 3-ounce servings. The recipe also uses 3-1/2 cups of fresh vegetables, which, after cooking, will yield about four 1/2-cup servings. Finally, the recipe includes 3 cups of cooked rice, or four 3/4-cup servings. Each serving will therefore provide 3 ounces from the meat group, 1/2 cup from the vegetable group, and 3/4 cup from the bread group.

The Pyramid recommends a number of servings of a certain size daily. However, the number and size of servings a person eats from each food group usually varies from day to day. Therefore, the Pyramid actually applies to the amount of food a person eats over several days, not just one day.

Planning Menus Using the Food Guide Pyramid. Planning menus gives you the opportunity to include a variety of foods from each food group, especially foods from subgroups that provide nutrients often low in American diets. It also provides the chance to balance fat and sodium to maintain healthful levels over time.

Table 2-7 shows a five-day menu designed to meet the Food Guide Pyramid guidelines and provide 2200 calories.

- Bread, Cereal, Rice, and Pasta—While there are many products to choose from, most people eat less than the minimum of six servings per day, and choices tend to be enriched, rather than whole grains. The Whole-Wheat Cornmeal Muffins and the Whole-Wheat Pancakes illustrate some whole-grain products. Rice-Pasta Pilaf illustrates the use of a grain mixture as an attractive side dish, and provides part of a vegetable serving as well. Recipes for Chocolate Mint Pie, Peach Crisp, and Lemon Pound Cake show that desserts can contribute to grain servings, too.
- Vegetables—Although most people report having some vegetables each day, they are often potatoes, especially french fries. The Food Guide Pyramid encourages consumption of a variety of different vegetables, with

Pork and Vegetable Stir-fry

Category: Entrée   **Yield: 4 servings, 1 cup meat mixture, 1/4 cup sauce and 3/4 cup rice each**

INGREDIENTS

Boneless pork loin, lean	1 pound	Lemon juice	1/4 cup
Tarragon leaves	1/2 teaspoon	Carrots, sliced	1 cup
Pepper	1/4 teaspoon	Fresh mushrooms, sliced	1 cup
Garlic powder	1/4 teaspoon	Celery, sliced	1 cup
Salt	1/4 teaspoon	Onions, chopped	1/2 cup
Cornstarch	2 teaspoons	Rice, cooked	3 cups
Water	1 cup		

STEPS

1. Partially freeze meat. Trim fat and slice meat across the grain into 1/4-inch-thick slices.
2. Combine seasonings. Sprinkle mixture over meat.
3. Combine cornstarch, water, and lemon juice. Set aside.
4. Heat nonstick frying pan. Add meat and stir-fry until brown, about 5 minutes. Drain meat, remove to another container, and cover to keep warm.
5. In same pan, stir-fry carrots 5 minutes or until tender-crisp. Add remaining vegetables and stir-fry 2 minutes. Add meat, and cornstarch mixture. Bring to a boil. Cook, stirring constantly, until thickened.
6. Serve over rice.

NUTRITIONAL ANALYSIS:

Kcalories:	Protein (gm):	Fat (gm):	Carbo (gm):	Sodium (mg):
370	29	9	42	240

EACH SERVING PROVIDES:

3 ounces from meat group
1 serving from vegetable group
1-1/2 servings from bread group

Figure 2-2

Recipe for Pork and Vegetable Stir-fry.

Source: "Using the Food Guide Pyramid: A Resource for Nutrition Educators" by A. Shaw, L. Fulton, C. Davis, and M. Hogbin, U.S. Department of Agriculture: Food Nutrition, and Consumer Services; Center for Nutrition Policy and Promotion.

special emphasis on dark green leafy vegetables and cooked dry beans and peas, and urges preparation in lower-fat ways. The Corn and Zucchini Combo, Spinach-Orange Salad, and Confetti Coleslaw illustrate the use of vegetables in attractive, lower-fat ways. Other recipes—Chili-Baked Potato, Pork and Vegetable Stir-fry, Creole Fish Fillet—suggest ways to increase the use of vegetables as part of main dishes. In some recipes, vegetables add flavor or serve as extenders to make larger portions—the Breakfast Pita Sandwich or Tuna Sprouts Sandwich. Fresh vegetables add crunch to the Turkey Pasta Salad. Versatile legumes can count as vegetables or as meat alternatives, as in Split Pea Soup or Lentil Stroganoff.

- Fruit—Fruit is particularly underconsumed by Americans. In recent USDA food-consumption surveys, only a little over half the adults reported having fruit or fruit juice on any given day. Even fewer low-income people reported eating any fruit. The recipes included here illustrate the use of fruit in a variety of ways. The Blueberry Sauce makes a tasty, nutritious substitute for syrup; fruit can flavor and enhance meat in a main dish, as in the Apricot-Glazed Chicken. It can be a colorful part of a main-dish salad, as in the Turkey Pasta Salad, or in the Spinach-Orange Salad. It also makes a great low-fat dessert, as in the Strawberry Yogurt Parfait or Peach Crisp. The menus also include a variety of whole fruits, fruit juices, and canned fruit as part of meals and snacks.
- Milk, Yogurt, and Cheese—Milk products are often underconsumed by adults, especially fluid milk. The menus show the use of a variety of milk products in addition to fluid milk that contribute to servings from this group: cheese, ice milk, yogurt, and frozen yogurt. Recipes for Strawberry Yogurt Parfait and Chocolate Mint Pie illustrate milk's use in attractive low-fat desserts.
- Meat, Poultry, Fish, and Alternatives—The main-dish and sandwich recipes illustrate the use of a variety of meats and alternates. The recipes use lean meats, low-fat preparation techniques, and herbs and spices for flavoring to reduce sodium. Servings of meat, poultry, or fish average 3 ounces in main-dish recipes; addition of vegetables and grains make larger portions. The lentils in Lentil Stroganoff provide meat equivalents for a meatless main dish.

Use the following questions to ensure that your menu follows the Food Guide Pyramid.

1. Does a day's menu on the average provide at least the lower number of servings from each of the major food groups?
2. Does the menu have two or more servings of whole-grain breads and cereals each day?
3. Does the menu include several servings of each of the vegetable subgroups: dark green leafy vegetables (such as spinach, broccoli, romaine lettuce), deep yellow vegetables (carrots, sweet potatoes), dry beans and peas (kidney beans, lentils), starchy vegetables (potatoes, corn), and other vegetables each week?
4. Do most vegetables and fruits have their skins and seeds (baked potatoes with skin, berries, or apples or pears with peels)?

TABLE 2-7 Five Days' Menus at 2200 Kcalories

Day 1	Day 2	Day 3	Day 4	Day 5
		BREAKFAST		
Orange juice . . . . . . 3/4 c	Grapefruit juice . . . . 3/4 c	Grapefruit 1/2	Fresh sliced strawberries . . . . 1/2 c	Cantaloupe . . . 1/4 melon
Oatmeal . . . . 1/2 c	Breakfast pita . . . . 1 sandwich	Banana. . . . . . . 1 medium	Whole-grain cereal flakes . . . . 1 oz	Turkey patty. . . . . . 1-1/2 oz
White toast . . . . . 2 slices	2% fat milk. . . . . 1 c	Ready-to-eat cereal flakes . . . . 1 oz	Toasted plain bagel . . . . 1 medium	Whole-wheat pancakes 2
Margarine. . . . . 2 tsp		Toasted English muffin with raisins. 1	Cream cheese. . . . 1 tbsp	Blueberry sauce . . . 1/4 c
Jelly. . . . . 1 tsp		Margarine. . . . . 2 tsp	2% fat milk. . . . . 1 c	Margarine. . . . . 1 tsp
2% fat milk. . . . . 1/2 c		Skim milk . . . . 1/2 c		Skim milk . . . . 1 c
		LUNCH		
Split pea soup. . . . . . . 1 c	Turkey pasta salad . . . . 1-1/2 c	Taco salad	Broiled chicken fillet sandwich 1	Chili-stuffed baked potato 1
Quick tuna and sprouts sandwich 1	Tomato wedges on lettuce leaf . . . . . 1 serving	greens . . . . 1 c	Mayonnaise . . . . 1 pkt	Low-fat, low-sodium cheddar cheese . . 3 tbsp
Mixed green salad . . . 1 c	Hard rolls 2	chili . . . . 3/4 c	Confetti coleslaw. . . 1/2 c	Spinach-orange salad. . . . . 1 c
Reduced-calorie Italian dressing . . . . 1 tbsp	Margarine. . . . . 2 tsp	Gingersnaps 2	Fresh orange 1	Wheat crackers. 6
Chocolate mint pie . . . . 1 serving	Oatmeal cookies 4		2% fat milk. . . . . 1 c	Skim milk . . . . 1 c
	2% fat milk. . . . . 1 c			

DINNER				
Savory sirloin . . . . 3 oz Corn and zucchini combo . . . . 3/4 c Tomato and lettuce salad . . . . 1 serv French dressing. . . 1 tbsp Whole-wheat rolls . . . . . 2 Margarine. . . . . 1 tsp Yogurt-strawberry parfait . . . . 1 c	Creole fish fillets . . . . 4 oz Small new potatoes with skin. . . . . 2 Cooked green peas . . . . 1/2 c with margarine. . . 1 tsp Whole-wheat cornmeal muffins . . . . 2 Margarine. . . . . 2 tsp Peach crisp . . . . 1/2 c	Pork and vegetable stir-fry mixture . . . . 1 c rice . . . . 3/4 c Cooked broccoli . . . 1/2 c White rolls . . . . 2 Margarine. . . . . 2 tsp Minted pineapple chunks . . . . 1/2 c	Lentil stroganoff mixture . . . . 1-1/2 c noodles. . . . . 3/4 c Cooked whole green beans . . . . 1/2 c with margarine. . . 1 tsp Tomato and cucumber salad. . . . . 1 serv Reduced-calorie vinaigrette dressing. . . . . 1 tbsp Pumpernickel roll. . . . . . 1 Margarine. . . . . 1 tsp Honeydew . . . . 1/8 melon	Apricot-glazed chicken. . . . . 3 oz Rice-pasta pilaf . . . . 3/4 c Tossed salad . . . . 1 c Reduced-calorie Italian dressing . . . . 1 tbsp Hard rolls . . . . 2 Margarine . . . . 2 tsp Vanilla ice milk. . . . . 1/2 c
SNACKS				
Graham crackers . . . . 6 squares 2% milk . . . . 1 c Peanut butter. . . . . 2 tbsp Fresh peach. . . . . 1 Carrot sticks . . . . 7–8 medium	Bagel . . . . 1 medium Margarine. . . . . 2 tsp Fresh pear 1	Wheat crackers. . . . . . . 6 Cheddar cheese . . . . 1-1/2 oz Turkey sandwich. . . . . 1/2 No-salt-added tomato juice . . . . 3/4 c	No-salt-added vegetable juice . . . . 3/4 c Roast beef sandwich . . 1 2% milk . . . . 1 c	Soft pretzel . . . . . . 1 large Fresh apple 1/2

Source: "Using the Food Guide Pyramid: A Resource for Nutrition Educators" by A. Shaw, L. Fulton, C. Davis, and M. Hogbin. U.S. Department of Agriculture: Food, Nutrition, and Consumer Services; Center For Nutrition Policy and Promotion.

Pirámide

Del día con el sabor popular mexicano

A Food Guide Pyramid with a Mexican Flavor

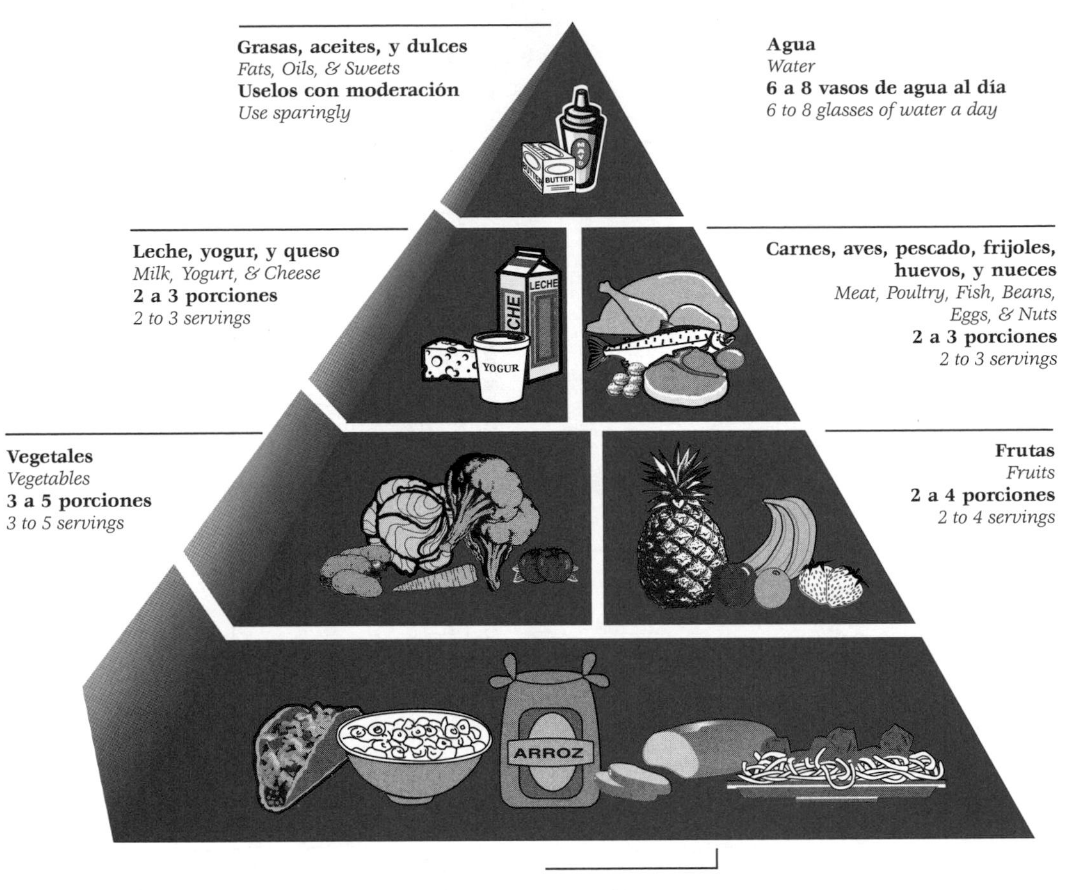

Al preparar nutritivos alimentos mexicanos con sus niños o nietos, usted les ayuda a:
—mejorar su alimentación, y
—lograr que esos alimentos sigan siendo parte de su cultura.
Preparing healthy and nutritious Mexican foods with your children and grandchildren will improve their diet and help ensure that these foods remain part of the culture.

Figure 2-3
A food guide pyramid with a Mexican flavor.

Source: U.S. Department of Agriculture and U.S. Department of Health and Human Services.

5. Are foods (especially desserts) high in fat, sugar, and/or sodium balanced with choices lower in these nutrients?

In developing the food guide, the typical use of foods by Americans was an important factor in establishing food groups and in developing nutrient profiles for each food group. The Pyramid has been adapted to reflect the customs of numerous ethnic and cultural groups within the United States, including Mexicans (Figure 2-3). It has also been adapted to meet the needs of young children and older adults (see Chapter 13), as well as vegetarians (see Chapter 11). Figures 2-4 to 2-6 show the Asian, Mediterranean, and Latin American Diet Pyramids. These pyramids were developed by the Oldways Preservation and Exchange Trust to illustrate traditional food

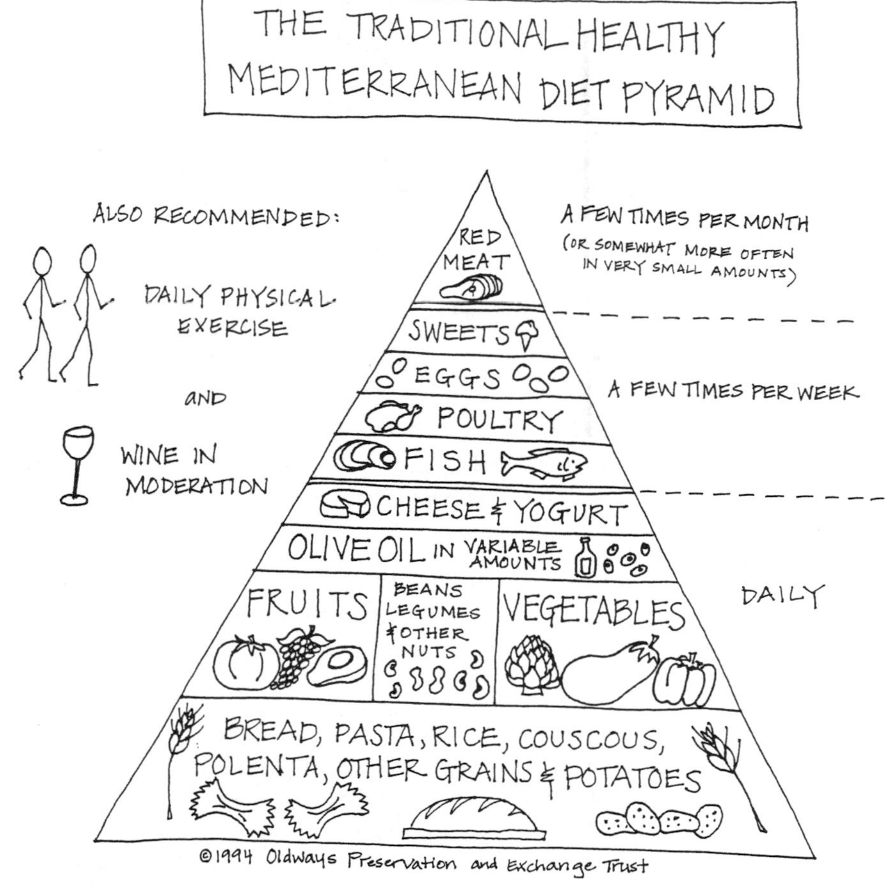

Figure 2-4

The Traditional Healthy Mediterranean Diet Pyramid. © 1995 Oldways Preservation & Exchange Trust. Courtesy of Oldways Preservation & Exchange Trust, Cambridge, MA

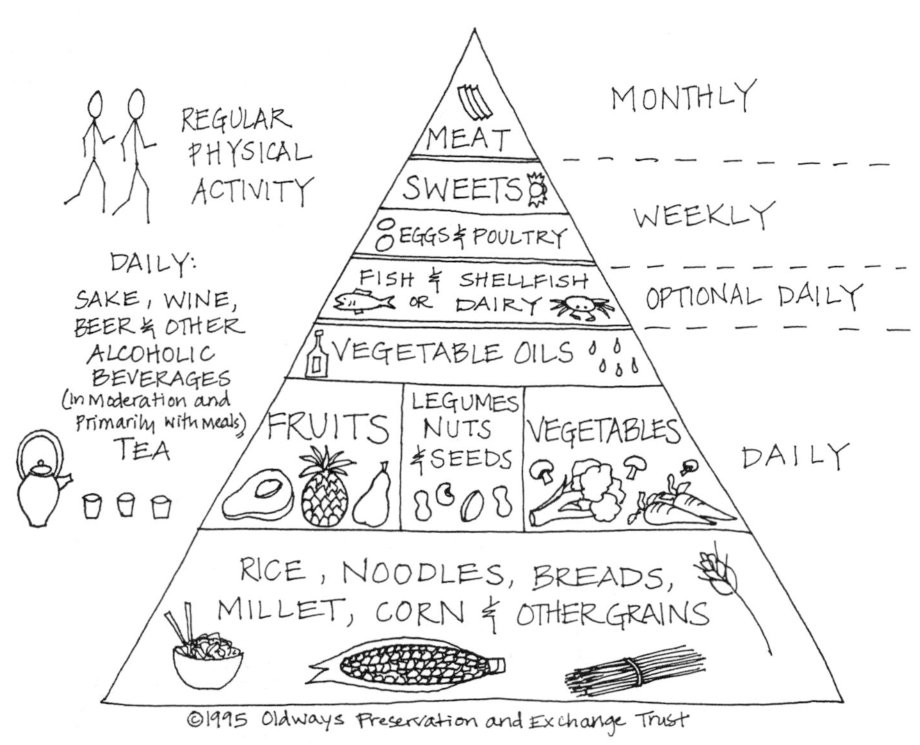

Figure 2-5

The Traditional Healthy Asian Diet Pyramid. © 1994 Oldways Preservation & Exchange Trust. Courtesy of Oldways Preservation & Exchange Trust, Cambridge, MA

patterns that are healthy. These pyramids illustrate proportions rather than specific amounts of foods to eat. Like the Food Guide Pyramid, they emphasize grains, fruits, and vegetables.

■ MINI-SUMMARY

Dietary recommendations have been published for the healthy American public for almost 100 years. The Food Guide Pyramid, with its five food groups (Figure 2-1), is based on the Dietary Guidelines for Americans and nutrient recommendations. The key concepts of the Pyramid include variety, proportionality, and moderation. By using the number of servings and portion sizes in the Pyramid as well as additional Pyramid guidelines, you can plan healthy menus.

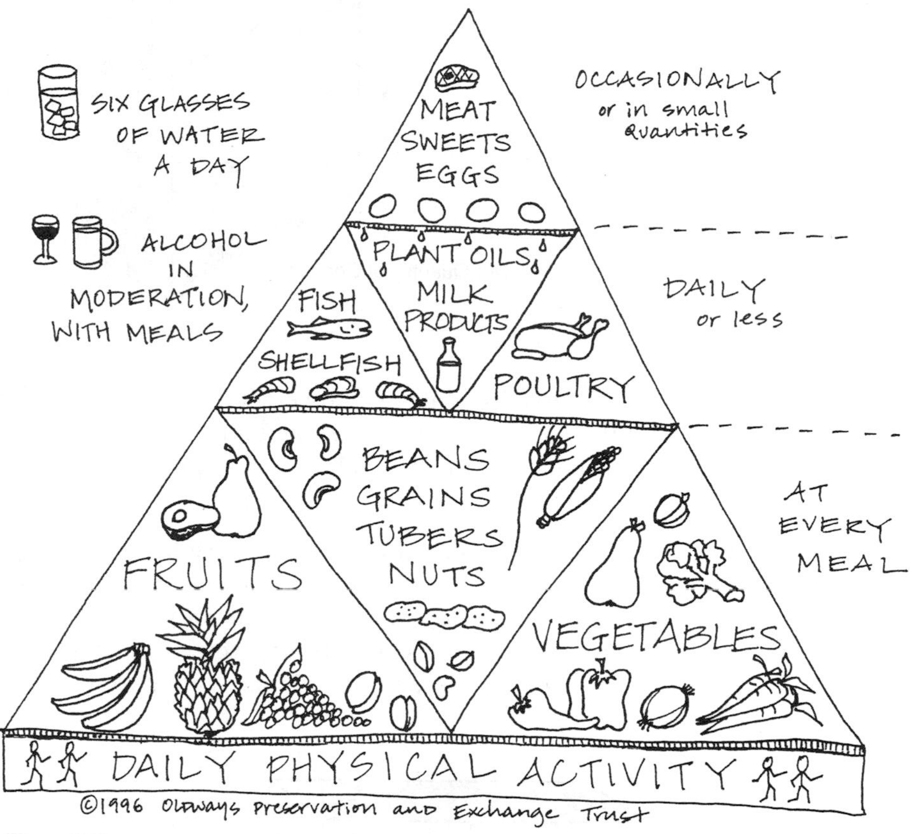

Figure 2-6

The Traditional Healthy Latin American Diet Pyramid. © 1996–1998 Oldways Preservation & Exchange Trust. Courtesy of Oldways Preservation & Exchange Trust, Cambridge, MA

Food Labels

Since 1938, the federal government has required basic information on food labels (Figure 2-7). The Food and Drug Administration (FDA) regulates labels on all packaged foods except for meat, poultry, and egg products—foods regulated by the U.S. Department of Agriculture (USDA). The amount of information on food labels varies, but all food labels must contain at least:

- The name of the food
- A list of ingredients

Figure 2-7
Location of Nutrition Facts

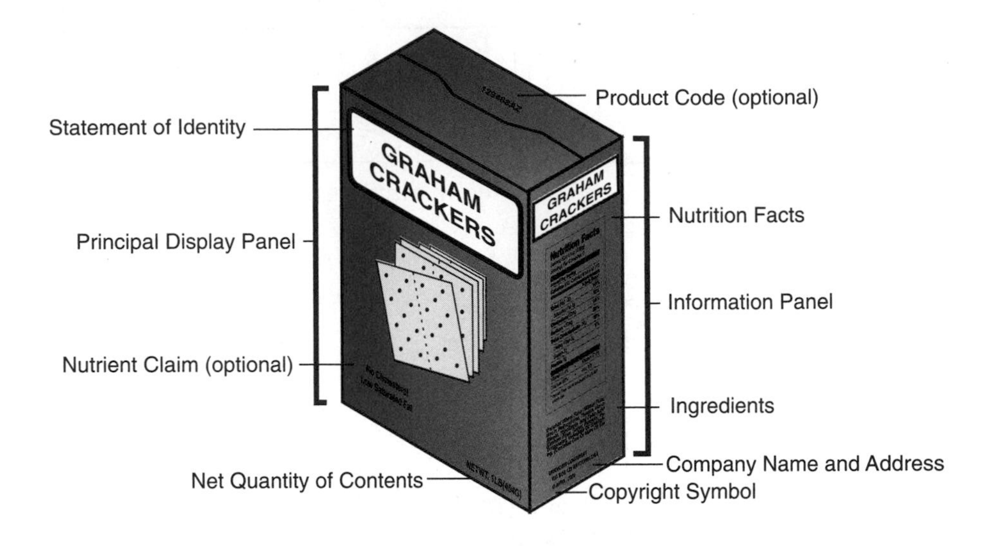

- The net contents or net weight—the quantity of the food itself without the packaging (in English and metric units)
- The name and place of business of the manufacturer, packer, or distributor
- Nutrition information is also required for most foods, our next topic.

For most foods, all ingredients must be listed on the label and identified by their common names to help consumers identify ingredients that they are allergic to or want to avoid for other reasons. The ingredient that is present in the largest amount, by weight, must be listed first. Other ingredients follow in descending order according to weight (Figure 2-8).

Figure 2-8
Food label

Cereal Flako's
with Raisins
Ingredients: Whole Wheat Kernels, Raisins, Malt Flavoring, Salt
Made By
Flakes, Inc.
Hometown, Michigan 48002
Net Wt. 15 oz. (425 grams)

Nutrition Facts

Figure 2-9 is a sample Nutrition Facts panel from a package of macaroni and cheese. Serving size and number of servings in the package are the first stop when you read the Nutrition Facts. Just how big is a serving? Serving sizes are provided in familiar units, such as cups or pieces, followed by the metric amount (the number of grams). A serving of applesauce would read "1/2 cup (114 g)." The household measure is easier to understand, but the metric measure gives a more precise idea of the amount. For example, "114 g" means 114 grams, a measure of weight. There are 28 grams in 1 ounce. The label helps you get familiar with metrics, too.

Serving sizes are designed to reflect the amounts people actually eat. Compare the serving size, including how many servings there are in the food package, to how much you actually eat. The size of the serving on the food package influences all the nutrient amounts listed on the top part of the label. In the sample label, one serving of macaroni and cheese equals 1 cup. If you ate the whole package, you would eat 2 cups, and that doubles the kcalories and other nutrient numbers.

If you check the serving size of different brands of macaroni and cheese, you'll see that the sizes are similar. That means you don't need to be a math

Figure 2-9
Nutrition Label

Nutrition Facts
Serving Size 1 cup (228g)
Servings Per Container 2

Amount Per Serving
Calories 250 Calories from Fat 110

	% **Daily Value***
Total Fat 12g	**18**%
Saturated Fat 3g	**15**%
Cholesterol 30mg	**10**%
Sodium 470mg	**20**%
Total Carbohydrate 31g	**10**%
Dietary Fiber 0g	**0**%
Sugars 5g	
Protein 5g	

Vitamin A 4%	•	Vitamin C 2%
Calcium 20%	•	Iron 4%

* Percent Daily Values are based on a 2,000 calorie diet. Your daily values may be higher or lower depending on your calorie needs:

	Calories:	2,000	2,500
Total Fat	Less than	65g	80g
Sat Fat	Less than	20g	25g
Cholesterol	Less than	300mg	300mg
Sodium	Less than	2,400mg	2,400mg
Total Carbohydrate		300g	375g
Dietary Fiber		25g	30g

whiz to compare two foods. Likewise, it's easy to see the kcalorie and nutrient differences between similar servings of canned fruit packed in syrup versus the same fruit in natural juices.

The next stop on the Nutrition Facts panel is the Calories per Serving category, which lists the total kcalories in one serving, as well as the kcalories from fat. In the example, there are 250 kcalories in one serving of this macaroni and cheese. How many kcalories from fat are there in one serving? Answer: 110 kcalories, which means almost half of the food's kcalories come from fat. If you ate the whole package, you would consume 500 kcalories and 220 would come from fat.

Nutrients are listed next. Information about some nutrients is required: total fat, saturated fat, cholesterol, sodium, total carbohydrates, dietary fiber, sugars, protein, vitamin A, vitamin C, calcium, and iron. Information about additional nutrients may be given voluntarily, but it is required in two cases: if a claim is made about the nutrients on the label, or if the nutrients are added to the food. For example, fortified breakfast cereals must give Nutrition Facts for any added vitamins and minerals.

The nutrients listed first (total fat, saturated fat, cholesterol, sodium) are the ones Americans generally eat in adequate amounts or even too much. Americans often don't get enough of some of the other nutrients listed: fiber, vitamins A and C, calcium, and iron. You can use the food label to help limit those nutrients you want to cut back on, and also increase those nutrients you want to consume in greater amounts.

Daily Value—A set of nutrient-intake values developed by the Food and Drug Administration that are used as a reference for expressing nutrient content on nutrition labels.

Nutrient amounts are listed in two ways: in metric amounts (in grams) and as a percentage of the **Daily Value.** Developed by the Food and Drug Administration, Daily Values are a set of nutrient-intake values that are used as a reference for expressing nutrient content on nutrition labels (Table 2-8). The DRIs can't be used on nutrition labels because they are set for specific age and gender categories, and the nutrition label applies to everyone.

The percentage of the Daily Value (%DV) is based on a 2000-kcalorie diet. Therefore, the Daily Value may be a little high, a little low, or right on target for you. The percentage of the Daily Value shows you how much of the Daily Value is in one serving of food. For example, in Figure 2-6, the %DV for total fat is 18 percent and for dietary fiber is 0 percent. When one serving of macaroni and cheese contains 18 percent of the DV for Total Fat, that means you have 82 percent of your fat allowances left for all the other foods you eat that day.

When you are looking at the %DV on a food label, use this guide. Foods that contain 5 percent or less of the Daily Value for a nutrient are generally considered low in that nutrient. Foods that contain 20 percent or more of the Daily Value for a nutrient are generally considered high in that nutrient. One cup of macaroni and cheese contains 18 percent of the Daily

TABLE 2-8 Daily Values*

Nutrient	Daily Value
Carbohydrate	300 grams
Fiber	25 grams
Cholesterol	300 milligrams
Fat	65 grams
Saturated fat	20 grams
Protein	50 grams
Vitamin A	5000 International Units
Vitamin D	400 International Units
Vitamin E	18 milligrams
Vitamin K	80 micrograms
Vitamin C	60 milligrams
Thiamin	1.5 milligrams
Riboflavin	1.7 milligrams
Niacin	20 milligrams
Vitamin B_6	2 milligrams
Folate	400 micrograms
Vitamin B_{12}	6 micrograms
Biotin	0.3 milligrams
Pantothenic acid	10 milligrams
Calcium	1000 milligrams
Chloride	3400 milligrams
Chromium	120 micrograms
Copper	2 milligrams
Iodine	150 micrograms
Iron	18 milligrams
Magnesium	400 milligrams
Manganese	2 milligrams
Molybdenum	75 micrograms
Phosphorus	1000 milligrams
Potassium	3500 milligrams
Selenium	70 micrograms
Sodium	2400 milligrams
Zinc	15 milligrams

* Designed for adults and children over 4 years of age and based on 2000 kcalorie diet.

Source: Food and Drug Administration.

Value for fat, which is just below 20 percent. Therefore, the macaroni and cheese is pretty high in fat, particularly if you eat 1-1/2 to 2 cups.

You can use the %DV to help you make dietary trade-offs with other foods throughout the day. You don't have to give up a favorite food to eat a healthy diet. When a food you like is high in fat, balance it with foods that are low in fat at other times of the day.

The values listed for total carbohydrate include all carbohydrates, including dietary fiber and sugars listed below it. The sugar values include naturally present sugars, such as lactose in milk and fructose in fruits, as well as those added to the food, such as table sugar and corn syrup. The label can claim no sugar added but still have naturally occurring sugars. An example is fruit juice.

Neither Sugars nor Protein lists a %DV on the Nutrition Facts panel. No Daily Value has been established for sugar because no specific recommendations have been made for the total amount of sugars to eat in a day. For protein, a %DV is required to be listed if a claim is made for protein, such as "high in protein." Otherwise, unless the food is meant for use by infants and children under 4 years old, none is needed. Current scientific evidence indicates that inadequate protein intake is not a public health concern for adults and children over 4 years old.

The values listed for total fat refer to all the fat in the food. Only total fat and saturated fat information is required on the label, because high intakes of both are linked to high blood cholesterol, which is linked to increased risk of coronary heart disease. Listing the amount of polyunsaturated and monounsaturated fats in the food is voluntary. In 1999, the Food and Drug Administration announced plans to require food manufacturers to include the amount of trans fatty acids on nutrition labels. Trans fatty acids are liquid fats that have been turned into solid fats. They are often found in margarine and baked goods, and they raise the risk of heart disease. Trans fatty acid information will be on food labels in 2003.

The Daily Value for calcium is 1000 milligrams (mg). Experts advise adolescents, especially girls, to consume 1300 mg and postmenopausal women to consume 1200 mg of calcium daily. The daily target for teenagers should therefore be 130%DV, and the daily target for postmenopausal women should be 120%DV.

The table at the bottom of the food label shows the Daily Values for certain nutrients at both the 2000- and 2500-calorie levels. For example, if you eat a 2000-calorie diet, you should eat less than 65 grams of fat in all the foods you eat in a day. If you consume 2500 calories per day, the amount of cholesterol and sodium you eat in a day are not different from others eating 2000 calories per day.

Nutrient content claims—Claims on food labels about the nutrient composition of a food, regulated by the Food and Drug Administration.

Nutrient Claims

Nutrient content claims, such as "good source of calcium" or "fat-free," can appear on food packages only if they follow legal definitions (Table 2-9). For example, a food that is a good source of calcium must provide 10 to 19 percent of the Daily Value for calcium in one serving. Phrases such as

TABLE 2-9 Nutrient Content Claims—A Dictionary

Nutrient (Content Claim)	Definition (Per Serving)
Calories	
Calorie free	less than 5 kcalories
Low calorie	40 kcalories or less
Reduced or fewer calories	at least 25% fewer kcalories*
Sugar	
Sugar free	less than 0.5 gram sugars
Reduced sugar or less sugar	at least 25% less sugars*
No added sugar	no sugars added during processing or packing, including ingredients that contain sugars, such as juice or dry fruit
Fat	
Fat free	less than 0.5 gram fat
Low fat	3 grams or less of fat
Reduced or less fat	at least 25% less fat*
Trans fat free+	less than 0.5 gram of trans fat and less than 0.5 gram saturated fat
Saturated fat free	less than 0.5 gram saturated fat and less than 0.5 gram of trans fat
Low saturated fat	1 gram or less saturated fat (and less than 0.5 gram trans fat)+
Reduced saturated fat	at least 25% less saturated fat* (and at least 25% less saturated fat and trans fat combined)
Cholesterol	
Cholesterol free	less than 2 milligrams cholesterol and 2 grams or less of saturated fat (and trans fat combined)+
Sodium	
Sodium free	less than 5 milligrams sodium
Very low sodium	35 milligrams or less sodium
Low sodium	140 milligrams or less sodium
Reduced or less sodium	at least 25% less sodium*
Light in sodium	50% less*
Fiber	
High fiber	5 grams or more
Good source of fiber	2.5 to 4.9 grams
More or added fiber	at least 2.5 grams more
Other Claims	
High, rich in, excellent source of	20% or more of Daily Value*
Good source	10% to 19% of Daily Value*
More	10% or more of Daily Value*
Fresh	raw, unprocessed, or minimally processed, with no added preservatives

TABLE 2-9 (continued)

Nutrient (Content Claim)	Definition (Per Serving)
Healthy	Low in fat and saturated fat, contains no more than 20% of the Daily Value for sodium and cholesterol, contains at least 10% of the Daily Value for one of the following: vitamin A or C, calcium, iron, protein, fiber (fresh, canned, or frozen fruits and vegetables and enriched breads and cereals are exempt from 10% rule)
Light	One of the following: one-third fewer kcalories or 50% less fat*; low-calorie, low-fat, containing 50% less sodium than normally present; or light in color and texture (such as light brown sugar)
Lean (meat and poultry only)	Less than 10 grams fat, 4.5 grams saturated fat (and trans fat combined),† and 95 milligrams cholesterol
Extra lean (meat and poultry only)	Less than 5 grams fat, 2 grams saturated fat (and trans fat combined),† and 95 milligrams cholesterol

* Compared with a standard serving size of the traditional food.
† Proposed in 1999.
Source: Food and Drug Administration.

"sugar-free" describe the amount of a nutrient in a food but don't tell exactly how much. These nutrient content claims differ from Nutrition Facts, which do list specific nutrient amounts.

If a food label contains a descriptor for a certain nutrient but the food contains other nutrients at levels known to be less healthy, the label would have to bring that to consumers' attention. For example, if a food making a low-sodium claim is also high in fat, the label must state "see back panel for information about fat and other nutrients."

Health claims—Claims on food labels that state certain foods or food substances—as part of an overall healthy diet—may reduce the risk of certain diseases.

Health Claims

The Nutrition Labeling and Education Act of 1990 provided, for the first time, the authority to allow food labels to carry claims about the relationship between the food and specific diseases or health conditions. This was a major shift in labeling philosophy. These **health claims** state that certain foods or food substances—as part of an overall healthy diet—may reduce the risk of certain diseases. Examples include calcium and osteoporosis, and dietary saturated fat and cholesterol and the risk of coronary heart disease. Although food manufacturers may use health claims approved by the FDA to market their products, the intended purpose of health claims is to benefit consumers by providing information on healthful eating patterns that may help reduce the risk of heart disease, cancer, osteoporosis, high blood pressure, dental cavities, or certain birth defects.

Health claims may show links between the following nutrients and conditions (see Table 2-11 for sample claims and more information):

1. Calcium and osteoporosis
2. Sodium and hypertension (high blood pressure)
3. Dietary fat and cancer
4. Dietary saturated fat and cholesterol, and risk of coronary heart disease
5. Fiber-containing grain products, fruits, and vegetables, and cancer
6. Fruits, vegetables, and grain products that contain fiber, particularly soluble fiber, and risk of coronary heart disease
7. Fruits and vegetables, and cancer
8. Folate and neural-tube birth defects
9. Dietary-sugar alcohol and dental caries (cavities)
10. Dietary soluble fiber, such as that found in whole oats and psyllium seed husk, and coronary heart disease
11. Soy protein and risk of coronary heart disease
12. Plant sterol/stanol esters and risk of coronary heart disease
13. Whole-grain foods and risk of heart disease and certain cancers
14. Potassium and risk of high blood pressure and stroke

In addition to these pre-approved claims, the FDA now accepts petitions directly from food manufacturers seeking to make health claims on their labels. In order to receive FDA approval, manufacturers must show that their health claims meet a scientific "weight of the evidence" standard, including support by a credible body of scientific evidence.

■ MINI-SUMMARY

All food labels must contain the name of the product; the net contents or net weight; the name and place of business of the manufacturer, packer, or distributor; a list of ingredients in order of predominance by weight; and nutrition information (Figure 2-6). Daily Values are nutrient standards used on food labels to allow nutrient comparisons among foods. Any nutrient or health claims on food labels must comply with Food and Drug Administration regulations and definitions as outlined in this chapter.

TABLE 2-10 Health Claims

Health Claim	Foods Displaying Claim	Sample Claim	Requirements
1. **Calcium and osteoporosis.** Low calcium intake is one risk factor for osteoporosis, a condition of lowered bone mass, or density. Lifelong adequate calcium intake helps maintain bone health by increasing as much as genetically possible the amount of bone formed in the teens and early adult life, and by helping to slow the rate of bone loss that occurs later in life.	Low-fat and fat-free milks, yogurts, calcium-fortified citrus drinks.	"Regular exercise and a healthy diet with enough calcium helps teen and young adult white and Asian women maintain good bone health and may reduce their high risk of osteoporosis later in life."	Food or supplement must be "high" in calcium and must not contain more phosphorus than calcium.
2. **Sodium and hypertension (high blood pressure).** Hypertension is a risk factor for coronary heart disease and stroke deaths. The most common source of sodium is table salt. Diets low in sodium may help lower blood pressure and related risk in many people.	Fruits, vegetables, unsalted tuna, low-fat milk and yogurt, sherbet, ice milk, cereal.	"Diets low in sodium may reduce the risk of high blood pressure, a disease associated with many factors."	Requirements: Foods must meet criteria for "low sodium."
3. **Dietary fat and cancer.** Diets high in fat increase the risk of some types of cancer, such as cancers of the breast and colon. While scientists don't know how total fat intake affects cancer development, low-fat diets reduce the risk.	Fruits, vegetables, low-fat milk products, cereals, pasta.	"Development of cancer depends on many factors. A diet low in total fat may reduce the risk of some cancers."	Foods must meet criteria for "low fat." Fish and game meats must meet criteria for "extra lean."
4. **Dietary saturated fat and cholesterol, and risk of coronary heart disease.** Diets high in saturated fat and cholesterol increase total and LDL (bad) blood cholesterol levels, and thus the risk of coronary heart disease. Diets low in saturated fat and cholesterol decrease the risk.	Fruits, vegetables, skim and low-fat milks, cereals, whole-grain products and pastas.	"While many factors affect heart disease, diets low in saturated fat and cholesterol may reduce the risk of this disease."	Foods must meet criteria for "low saturated fat," "low cholesterol," and "low fat." Fish and game meats must meet criteria for "extra lean."
5. **Fiber-containing grain products, fruits, and vegetables, and cancer.** Diets low in fat and rich in fiber-	Whole-grain breads and cereals, fruits, and vegetables.	"Low-fat diets rich in fiber-containing grain products, fruits, and vegetables may reduce	Foods must meet criteria for "low fat" and, without fortification, be a

TABLE 2-10 *(continued)*

Health Claim	Foods Displaying Claim	Sample Claim	Requirements
containing grain products, fruits, and vegetables may reduce the risk of some types of cancer. The exact role of total dietary fiber, fiber components, and other nutrients and substances in these foods is not fully understood.		the risk of some types of cancer, a disease associated with many factors."	"good source" of dietary fiber.
6. **Fruits, vegetables, and grain products that contain fiber, particularly soluble fiber, and risk of coronary heart disease.** Diets low in saturated fat and cholesterol and rich in fruits, vegetables, and grain products that contain fiber, particularly soluble fiber, may reduce the risk of coronary heart disease.	Fruits, vegetables, and whole-grain breads and cereals.	"Diets low in saturated fat and cholesterol and rich in fruits, vegetables, and grain products that contain some types of dietary fiber, particularly soluble fiber, may reduce the risk of heart disease, a disease associated with many factors."	Foods must meet criteria for "low saturated fat," "low fat," and "low cholesterol." They must contain, without fortification, at least 0.6 gram of soluble fiber per reference amount, and the soluble fiber content must be listed.
7. **Fruits and vegetables, and cancer.** Diets low in fat and rich in fruits and vegetables may reduce the risk of some cancers. Fruits and vegetables are low-fat foods, and may contain fiber or vitamin A and vitamin C.	Fruits and vegetables.	"Low-fat diets rich in fruits and vegetables (foods that are low in fat and may contain dietary fiber, vitamin A, or vitamin C) may reduce the risk of some types of cancer, a disease associated with many factors. Broccoli is high in vitamins A and C, and it is a good source of dietary fiber."	Foods must meet criteria for "low fat" and, without fortification, be a "good source" of fiber, vitamin A, or vitamin C.
8. **Folate and neural-tube birth defects.** Defects of the neural tube (a structure that develops into the brain and spinal cord) occur within the first six weeks after conception, often before the pregnancy is known. The U.S. Public Health Service recommends that all women of childbearing age in the United States consume	Enriched cereal grain products, some legumes (dried beans and peas), fresh leafy green vegetables, oranges, grapefruit, many berries, some dietary supplements, and fortified breakfast cereals.	"Healthful diets with adequate folate may reduce a woman's risk of having a child with a brain or spinal-cord birth defect."	Foods must meet or exceed criteria for "good source" of folate—that is, at least 40 micrograms of folate per serving (at least 10 percent of the Daily Value). A serving of food cannot contain more than

TABLE 2-10 *(continued)*

Health Claim	Foods Displaying Claim	Sample Claim	Requirements
0.4 mg of folate daily to reduce their risk of having a baby affected with spina bifida or other neural-tube defects.			100 percent of the Daily Value for vitamin A and vitamin D, because of their potential risk to fetuses.
9. **Dietary-sugar alcohol and dental caries (cavities).** Between-meal eating of foods high in sugar and starches may promote tooth decay. Sugarless candies made with certain sugar alcohols do not.	Sugarless candy and gum.	"Frequent between-meal consumption of foods high in sugar and starches promotes tooth decay. The sugar alcohols in this food do not promote tooth decay."	Foods must meet the criteria for "sugar free."
10. **Dietary soluble fiber, such as that found in whole oats and psyllium seed husk, and coronary heart disease.** When included in a diet low in saturated fat and cholesterol, soluble fiber may affect blood lipid levels, such as cholesterol, and thus lower the risk of heart disease. However, because soluble dietary fibers constitute a family of very heterogeneous substances that vary greatly in their effect on the risk of heart disease, the FDA has determined that sources of soluble fiber for this health claim need to be considered case by case. To date, the FDA has reviewed and authorized two sources of soluble fiber eligible for this claim: whole oats and psyllium seed husk.	Oatmeal cookies, muffins, breads, and other foods made with rolled oats, oat bran, or whole oat flour, hot and cold breakfast cereals containing whole oats or psyllium seed husk; dietary supplements containing psyllium seed husk.	"Three grams of soluble fiber from oatmeal daily in a diet low in saturated fat and cholesterol may reduce the risk of heart disease. This cereal has two grams per serving.	Foods must meet criteria for "low saturated fat," "low cholesterol," and "low fat." Foods that contain whole oats must contain at least 0.75 gram of soluble fiber per serving. Foods that contain psyllium seed husk must contain at least 1.7 grams of soluble fiber per serving. The claim must specify the daily dietary intake of the soluble fiber source necessary to reduce the risk of heart disease and the contribution that one serving of the product makes toward that intake level. Foods bearing a psyllium seed husk health claim must also bear a label statement concerning the need

TABLE 2-10 (continued)

Health Claim	Foods Displaying Claim	Sample Claim	Requirements
			to consume them with adequate amounts of fluid.
11. **Soy protein and risk of coronary heart disease.** Soy protein (about 25 grams a day) included in a diet low in saturated fat and cholesterol may reduce the risk of coronary heart disease by lowering blood cholesterol levels.	Soy beverages, tofu, tempeh, soy-based meat alternatives, soy flour.	"Diets low in saturated fat and cholesterol that include 25 grams of soy protein a day may reduce the risk of heart disease. One serving of this food provides ____ grams of soy protein."	Only foods that contain at least 6.25 grams of soy protein per serving can use this claim. Food must be low in cholesterol, low in saturated fat, and low-fat (except that foods made from whole soybeans that contain no fat in addition to that in the whole soybeans are exempt).
12. **Plant sterol/stanol esters and risk of coronary heart disease.** Plant sterol esters and plant stanol esters may reduce the risk of coronary heart disease by lowering blood cholesterol levels. Plant sterol and stanol esters work by blocking the absorption of cholesterol from the diet. Plant sterols are present in small quantities in many fruits, vegetables, nuts, seeds, cereals, legumes, and other plant sources. Plant stanols occur naturally in even smaller quantities from some of the same sources. Manufacturers of spreads and some other products have intentionally added plant sterol or stanol esters to these foods and marketed them to people at risk of coronary heart disease.	Spreads, salad dressings, snack bars, and dietary supplements.	"Diets low in saturated fat and cholesterol that include two servings of foods that provide a daily total of at least 3.4 grams of plant stanol esters in two meals may reduce the risk of heart disease. A serving of this food supplies ____ grams of plant stanol esters." "Foods that contain at least 0.65 gram per serving of vegetable oil sterol esters, eaten twice a day with meals for a daily total intake of at least 1.3 grams, as part of a diet low in saturated fat and cholesterol, may reduce the risk of heart disease. A serving of this product supplies ___ grams of vegetable oil sterol esters."	A food must contain at least 0.65 gram of plant sterol esters per serving of spreads and salad dressings or at least 1.7 grams plant stanol esters per serving of spreads, salad dressings, snack bars, and dietary supplements. A food must also be low in saturated fat and cholesterol. Spreads and salad dressings that exceed 13 grams of fat per 50 grams must bear the statement "see nutrition information for fat content."

TABLE 2-10 *(continued)*

Health Claim	Foods Displaying Claim	Sample Claim	Requirements
13. **Whole-grain foods and risk of heart disease and certain cancers.***	Whole-wheat bread, whole-wheat pasta, whole-grain cereals.	"Diets rich in whole-grain foods and other plant foods and low in total fat, saturated fat, and cholesterol may reduce the risk of heart disease and certain cancers."	To qualify, a food must contain 51 percent or more of its weight as whole-grain ingredients, be low-fat, and contain a certain amount of fiber based on weight.
14. **Potassium and risk of high blood pressure and stroke.***	Orange juice.	Diets containing foods that are a good source of potassium and that are low in sodium may reduce the risk of high blood pressure and stroke.	The food must be a good source of potassium and low in sodium, total fat, saturated fat, and cholesterol.

*These claims were developed by food manufacturers and based on authoritative statements by federal scientific bodies. The wording has been approved by the Food and Drug Administration.

Source: Food and Drug Administration.

Portion Size Comparisons

Portion size is an important concept for anyone involved in preparing, serving, and consuming foods. Serving sizes vary from kitchen to kitchen, but American serving sizes have been steadily increasing. In comparison with the Food Guide Pyramid portion sizes, as well as those served in many European countries, our portion sizes are huge. It wasn't that long ago when a "large" soft drink was typically 16 fluid ounces. Now, that's often the "small" size.

What you may consider to be one serving of the bread group may actually be three or four servings. For example, the Food Guide Pyramid considers 1 ounce of bread, about one slice, to be one serving. A typical New York–style bagel is about 4 ounces, or about four servings using the Food Guide Pyramid.

You may have noticed that the portion sizes in the Food Guide Pyramid do not always match the serving sizes found on food labels. This is because the purpose of the Food Guide Pyramid is not the same as nutrition labeling. The Food Pyramid was designed to be very simple to use. Therefore, the USDA specified only a few serving sizes for each food group so that they could be remembered easily. Food labels have a different purpose: to allow the consumer to compare the nutrients in equal amounts of foods. To compare the nutrient amounts in equal amounts of pasta, the portion size on the label is 2 ounces of uncooked pasta (about 56 grams), which will cook up to about

1 cup of spaghetti or as much as 2 cups of a large shaped pasta, such as ziti. Using the Pyramid, the portion size for cooked pasta is only half a cup.

In many cases, the portion sizes are similar on labels and in the food guide, especially when expressed as household measures. For foods falling into only one major group, such as fruit juices, the household measures provided on the label (1 cup or 8 fluid ounces) can help you relate the label serving size to the Pyramid serving size. For mixed dishes, food-guide serving sizes may be used to visually estimate the food item's contribution to each food group as the food is eaten—for example, the amounts of bread, vegetable, and cheese contributed by a portion of pizza.

Use these following tips to help you determine what portion sizes really look like.

- A handful = about 1 to 2 ounces of snack food
- A thumb = about 1 ounce of cheese
- A Ping-Pong ball = about 2 tablespoons
- A fist = about 1 cup
- Palm of your hand or a deck of cards = 3 ounces of meat

Keep in mind that these are only approximations.

■ MINI-SUMMARY

The portion sizes in the Food Guide Pyramid do not always match the serving sizes found on food labels. This is because the purpose of the Food Guide Pyramid is not the same as the purpose of nutrition labeling. Portion sizes in the United States have been increasing.

Check-Out Quiz

1. What are the Food Guide Pyramid's daily nutritional goals? Draw a line from the name of the item to the Pyramid's daily goal for that item.

Item	Pyramid's Daily Goal
Energy	30 percent or less of kcalories
Added sugar	100 percent of RDA/DRIs
Fiber	2400 mg or less
Total fat	Increase intake
Saturated fat	Don't exceed caloric needs
Cholesterol	Less than 10 percent of kcalories
Protein/vitamins/minerals	1300–3000
Sodium	300 mg or less

2. What are the Food Guide Pyramid's serving sizes? Draw a line from the food to the Pyramid's serving size. Each serving size may be used more than once.

Food	Serving Size
Apple	2–3 ounces
Fruit juice	1/2 cup
Bread	3/4 cup
Cold cereal	1 cup
Cooked vegetables	2
Raw vegetables	4 tablespoons
Milk or yogurt	1 medium
Chicken	
Cooked kidney beans	
Eggs	
Peanut butter	
Raw leafy vegetables	
Cooked rice or pasta	

3. Which food group(s) is a good source of protein?
   a. bread, cereal, rice, and pasta
   b. vegetable
   c. milk, yogurt, and cheese
   d. meat, poultry, fish, dry beans, eggs, and nuts
4. Which food group(s) provides one or more B vitamins (thiamin, riboflavin, niacin, folate, B_{12})?
   a. bread, cereal, rice, and pasta
   b. vegetable
   c. milk, yogurt, and cheese
   d. meat, poultry, fish, dry beans, eggs, and nuts
5. Upside-down triangles on the Food Guide Pyramid represent:
   a. added fat
   b. added sugar
   c. Fats, Oils, and Sweets group
   d. added sodium
6. Food Guide Pyramid serving sizes are always the same as food-label serving sizes.
   a. True b. False
7. Daily Values are based on an 1800-calorie diet.
   a. True b. False
8. Claims such as "good source of calcium" or "fat-free" on food labels are examples of health claims.
   a. True b. False

9. Health claims on labels are regulated by the federal government.
   a. True b. False
10. Portion sizes in the United States have been decreasing.
   a. True b. False

Activities and Applications

1. Checking Out Nutrient and Health Claims

Look at food labels from two of the following sections of the supermarket. Write down nutrient claims (such as "low-fat") given on at least two different foods from each section. Don't forget: fresh fruits and vegetables, meat, poultry, and seafood don't have labels—look for nutrition information nearby. Also look at the label to see which nutrition facts support this claim.

Produce
Frozen foods
Fresh meats, poultry, and fish
Dairy
Cereals
Cookies

During your search, also find one food item with a health claim and write it down. Use Table 2-10 to determine the category of the health claim. Then, look on the label to make sure the food meets all the requirements of that category as described in Table 2-10.

2. Label Reading at Breakfast

Look closely at the Nutrition Facts for each food you normally eat for breakfast, such as cereal, milk, and juice. Add up the %DVs for fat, saturated fat, cholesterol, sodium, total carbohydrates, vitamin A, vitamin C, calcium, and iron. How nutritious is your breakfast?

3. Label Comparison

Below is the nutrition label information from regular mayonnaise and low-fat mayonnaise dressing. Examine the labels, then answer these questions

A. Which label is the regular mayonnaise? How do you know that?
B. Does either mayonnaise contain significant amounts of vitamins and minerals?
C. What is the percent of total calories coming from fat in both products? (for example, in product A, divide 10 by 25 and multiply by 100 to get 40 percent.)
D. Which product contains sugar?

E. Which product contains more saturated fat and cholesterol? Why do you think that is so?

PRODUCT A	PRODUCT B
NUTRITION FACTS	**NUTRITION FACTS**
Serving size 1 tablespoon	Serving size 1 tablespoon
AMOUNT PER SERVING	**AMOUNT PER SERVING**
Calories 25	Calories 100
Calories from fat 10	Calories from fat 99
Total Fat 1 g	Total Fat 11 g
Saturated 0 g	Saturated 2 g
Polyunsaturated 0.5 g	Polyunsaturated 6 g
Monounsaturated 0 g	Monounsaturated 3 g
Cholesterol 0 mg	Cholesterol 5 mg
Sodium 140 mg	Sodium 180 mg
Total Carbohydrate 4 g	Total Carbohydrate 0 g
Sugars 3 g	Sugars 0 g
Protein 0 g	Protein 0 g
Not a significant source of dietary fiber, vitamin A, vitamin C, calcium, or iron.	Not a significant source of dietary fiber, vitamin A, vitamin C, calcium, or iron.

4. Menu/Diet Evaluation

Obtain a cycle menu (a menu rotated at specific time intervals, such as two or four weeks) from a college dining hall, school foodservice, or other foodservice. Evaluate the menu using the Food Guide Pyramid and the questions starting on page 57. Use the form "Determining Number of Food Guide Pyramid Servings in a One-Day Menu or Diet" in Table 2-11 for at least two days of the menu to establish how many servings of each pyramid group are served each day. You can also use Table 2-11 to write down everything you eat for one day, then determine how many servings you had from each pyramid group, and compare those number to the Pyramid recommendations.

5. Interactive Healthy Eating Index

The Interactive Healthy Eating Index is an online dietary assessment tool that measures how well your diet complies with the Dietary Guidelines for Americans and the Food Guide Pyramid. The program was developed and is run by the federal government. To log into the site, all you need to provide is a user name, age, and gender. Your age and gender are used to customize your results. http://147.208.9.133/Default.asp

6. Fat in Snacks

Obtain a single serving of one of your favorite snacks. Write down the percentage of kcalories from fat the snack contains and write it on an index

TABLE 2-11 Determining Number of Food Guide Pyramid Servings in a One-Day Menu or Diet

Amount (cup, fluid ounces, etc.)	Food/Beverage	Grain	Veg	Fruit	Dairy	Meat	Fats	Sweets
	Totals:							

card. In class, you are to line up from lowest percentage of kcalories from fat to highest percentage. Once the class is in the correct order, each of you will identify your food and the percentage of kcalories from fat.

7. Mystery Food

Bring to class the Nutrition Facts panel from a food product. Exchange your Nutrition Facts panel with a partner and examine the panel from your partner carefully. One at a time, you are each to guess what food category it falls into. Once you have the correct food category (your partner can tell you if you are right or not), guess what food it is (again with feedback from your partner).

Nutrition Web Explorer

Food Guide Pyramid www.nal.usda.gov/fnic

On the home page for the Food and Nutrition Information Center of the USDA, click on "Food Guide Pyramid." Then click on "The Interactive Food Pyramid." Click on each food group and write down the eating tips given for each group.

American Dietetic Association www.eatright.org

Visit the ADA website and get their "Daily Tip." Also use the dietitian locator to find a list of dietitians in your area.

Quackwatch www.quackwatch.com

Visit this website and click on "25 Ways to Spot It" under "Quackery." What are 10 ways to spot quackery?

International Dietary Guidelines www.nal.usda.gov/fnic/dga/

Visit this government website about the Dietary Guidelines for Americans. On the home page, click on "Dietary Guidelines from Around the World" and read about the diets of people in another country.

Food Label Quiz www.cfsan.fda.gov/~dms/flquiz1.html

Take the "Test Your Food Label Knowledge Quiz" at this government website.

Food Facts *Computerized Nutrient Analysis*

Computer software is the standard for analyzing the amount of nutrients in a recipe. Little wonder, when a task that used to take a half hour or more is done in a matter of minutes. Before computers, it was necessary to look up the nutritional content of each recipe ingredient and record it on a piece of paper. If the amount of the ingredient was not the same as was listed in the reference book, you would have to do some multiplication or division on all the nutrient values to come up with the right numbers. Then, after looking up all the ingredients, you would add up all your columns to get totals. In the final step, you would divide the totals by the yield of the recipe to get the amount of nutrients per serving. Sounds complicated! It sure is, and very time-consuming, too.

The computer has done a lot to speed up this process and increase its accuracy. Computerized nutrient-analysis programs contain nutrient information from many different resources. When the name of an ingredient is typed in, the computer lists similar ingredients so that you can choose exactly which one is appropriate. Then you click on or type in the amount of the ingredient you want to be used in the analysis, such as 1 cup. After inputting all the ingredients, you can ask the computer to divide the results by the yield, such as 12 portions. Then the computer will tell you exactly how much of each nutrient (and what percentage of the RDA or AI) is contained in one portion. Most nutrient-analysis programs can also give you a percentage breakdown of kcalories from protein, fat, carbohydrate, and alcohol. Of course, these figures can be printed out on a printer and/or stored in the computer's memory. A sample recipe analysis is given in Figure 2-10.

When considering a nutrient-analysis program, consider the following:

- What do you need the program to do? What different functions can the program perform, and how many of these functions do you need?
- What kind of computer system do you have available to run this software? Be sure you have enough hard-disk space, random access memory (RAM), and an appropriate-speed microprocessor.
- How large is the nutrient database? What is the source of the data? The database may contain from several thousand to more than 30,000 foods. Most databases use the U.S. Department of Agriculture's *Handbook #8* and manufacturer's information.
- Can foods be added to the database? It's also a good idea to check how many foods can be added.
- How many nutrients are provided for each food?
- How is output presented (graphs, tables, pie charts, etc.), and how easily can it be printed?
- How easy is it to use this program?
- What's the price?
- What service and support is available once you purchase the program? Is online help available? Is there an annual update fee? How much does it cost?

Breakfast–3 pancakes, 1 oz. sausage, 2 T syrup, 1 cup orange juice

Recipe Nutrient Analysis   Yield: 1.00 (1.00 SERVING)   Category: No Category

Recipe Food ID: 29   Source: Custom   Goal: DAILY VALUES/RDI - ADULT/CHILD

Nutrient	Value	Goal	% Goal
Weight (gm)	430.350		
Kilocalories (kcal)	566.180	2000.000	28%
Protein (gm)	12.503	50.000	25%
Carbohydrate (gm)	88.334	300.000	29%
Fat, total (gm)	19.654	65.000	30%
Alcohol (gm)	0.000		
Cholesterol (mg)	84.617	300.000	28%
Saturated Fat (gm)	5.372	20.000	27%
Monounsaturated Fat (gm)	4.888		
Polyunsaturated Fat (gm)	6.702		
MFA 18:1, Oleic (gm)	2.786		
PFA 18:2, Linoleic (gm)	4.533		
PFA 18:3, Linolenic (gm)	0.608		
PFA 20:5, EPA (gm)	0.000		
PFA 22:6, DHA (gm)	0.006		
Sodium (mg)	686.569	2400.000	29%
Potassium (mg)	647.280	3500.000	18%
Vitamin A (RE)	116.946	1000.000	12%
Vitamin A (IU)	738.724	5000.000	15%
Beta-carotene (μg)	0.000		
Vitamin C (mg)	124.689	60.000	208%
Calcium (mg)	283.126	1000.000	28%
Iron (mg)	2.792	18.000	16%
Vitamin D (ug)	0.000	10.000	0%
Vitamin D (IU)	0.000	400.000	0%
Vitamin E (ATE)	0.223	20.000	1%
Vitamin E (IU)		30.000	
Alpha-tocopherol (mg)	0.099		
Thiamin (mg)	0.456	1.500	30%
Riboflavin (mg)	0.398	1.700	23%
Niacin (mg)	2.786	20.000	14%
Pyridoxine/Vit B_6 (mg)	0.151	2.000	8%
Folate (μg)	118.464	400.000	30%
Cobalamin/Vit B_{12} (μg)	0.251	6.000	4%
Biotin (μg)	0.800	300.000	0%
Pantothenic Acid (mg)	0.942	10.000	9%
Vitamin K (μg)	0.248	80.000	0%
Phosphorus (mg)	227.020	1000.000	23%
Iodine (μg)		150.000	
Magnesium (mg)	46.320	400.000	12%

Nutrient	Value	Goal	% Goal
Zinc (mg)	0.778	15.000	5%
Copper (mg)	0.250	2.000	13%
Manganese (mg)	0.298	2.000	15%
Selenium (mg)	0.017	0.070	24%
Fluoride (μg)			
Chromium (mg)		0.120	
Molybdenum (μg)		75.000	
Dietary Fiber, total (gm)	1.624	25.000	6%
Soluble Fiber (gm)	0.000		
Insoluble Fiber (gm)	0.000		
Crude Fiber (gm)	0.250		
Sugar, total (gm)	52.544		
Glucose (gm)	14.826		
Galactose (gm)	0.000		
Fructose (gm)	9.440		
Sucrose (gm)	15.002		
Lactose (gm)	0.000		
Maltose (gm)	4.522		
Tryptophan (mg)	96.160		
Threonine (mg)	290.020		
Isoleucine (mg)	358.420		
Leucine (mg)	617.060		
Lysine (mg)	388.260		
Methionine (mg)	175.020		
Cystine (mg)	144.640		
Phenylalanine (mg)	385.980		
Tyrosine (mg)	283.520		
Valine (mg)	409.180		
Arginine (mg)	435.760		
Histidine (mg)	180.720		
Alanine (mg)	316.500		
Aspartic Acid (mg)	664.800		
Glutamic Acid (mg)	1871.640		
Glycine (mg)	248.040		
Proline (mg)	765.760		
Serine (mg)	446.060		
Moisture (gm)	288.930		
Ash (gm)	4.036		
Caffeine (mg)	0.000		

1 SERVING

% of Kcals

Protein	9%
Carbohydrate	62%
Fat, total	31%
Alcohol	0%

Exchanges

Bread/Starch
Fruit
Other Carbohydrate
Milk - Skim
Milk - Low-Fat
Milk - Whole
Vegetable
Meat - Very Lean
Meat - Lean
Meat - Medium-Fat
Meat - High-Fat
Fat

3/19/2000   *First DataBank Nutritionist Five™*

Figure 2-10

Sample computerized nutrient analysis output © 2000 First Data Bank, Inc., a wholly owned subsidiary of the Hearst Corporation. Reprinted with permission.

Many companies offer demonstration software at no cost. This is a real benefit, because you can try out the program before buying it. Following are a number of companies who offer software for nutrient analysis.

The CBORD Group: www.cbord.com
Computrition: www.computrition.com
DINE Systems: www.dinesystems.com
ESHA Research: www.esha.com
First Databank: www.firstdatabank.com
Nutribase: www.nutribase.com
Vision Software Technologies: www.vstech.com

The United States Department of Agriculture offers its National Nutrient Database for Standard Reference (Release 15) at its website (www.nal.usda.gov/fnic/foodcomp/). Release 15 is also available at this website to download onto a handheld personal digital assistant (PDA).

Hot Topic Quack! Quack!

The U.S. Surgeon General's Report on Nutrition and Health defines food quackery as "the promotion for profit of special foods, products, processes, or appliances with false or misleading health or therapeutic claims." Have you ever seen advertisements for supplements that are guaranteed to help you lose weight, or herbal remedies to prevent serious disease? If a product's claim seems just too good to be true, it probably is too good to be true. The problem with quackery is not just loss of money—you can be harmed as well. When you listen to a quack, you usually stop your regular medical treatment and don't receive, or even seek, further care from a legitimate medical professional.

Nutrition is brimming with quackery, in part because nutrition is such a young science. Questions on many fundamental nutrition issues, such as the relationship between sodium and hypertension, are far from being resolved, yet the media publicizes the results of research studies long before those results can be said to really prove a scientific theory. Unfortunately, because much research is only in its early stages, the public has been bombarded with conflicting ideas about issues that relate directly to two very important parts of their lives: their health and their eating habits. This conflict leaves the public confused about the truth and vulnerable to dubious health products (most often nutrition products) and practices—on which people spend $10 billion to $30 billion annually.

Much misinformation proliferates also because, in some states, anyone can call him- or herself a dietitian or nutritionist. In addition, one may even buy mail-order B.S., M.S., or Ph.D. degrees in nutrition from "schools" in the United States. In all states, nutrition books that are entirely bogus can be published and sold in bookstores, dressed up to look like legitimate health books.

A quack is someone who makes excessive promises and guarantees for a nutrition product or practice that is said to enhance your physical and mental health by, for example, preventing or curing a disease, extending your life, or improving some facet of performance. Health schemes and misinformation proliferate because they thrive on wishful thinking. Many people want easy answers to their medical concerns, such as a quick and easy way to lose weight. Often, claims appear to be grounded in science. Here's how to recognize quacks.

1. Their products make claims such as:
   - Quick, painless, and/or effortless

- Contains special, secret, foreign, ancient, or natural ingredients
- Effective cure-all for a wide variety of conditions
- Exclusive product not available through any other source

2. They use dubious diagnostic tests, such as hair analysis, to detect supposed nutritional deficiencies and illnesses. Then they offer you a variety of nutritional supplements, such as bee pollen or coenzymes, as remedies against deficiencies and disease.
3. They rely on personal stories of success (testimonials) rather than on scientific data for proof of effectiveness.
4. They use food essentially as medicine.
5. They often lack any valid medical or health-care credentials.
6. They come across more as salespeople than as medical professionals.
7. They offer simple answers to complex problems.
8. They claim that the medical community or government agencies refuse to acknowledge the effectiveness of their products or treatments.
9. They make dramatic statements that are refuted by reputable scientific organizations.
10. Their theories and promises are not written in medical journals using a peer-review process, but appear in books written only for the lay public.

Keep in mind that there are few, if any, sudden scientific breakthroughs. Science is evolutionary, even downright slow, not revolutionary.

But where can you find accurate nutritional information? In the United States, over 50,000 registered dietitians (R.D.s) comprise the largest and most visible group of professionals in the nutrition field. Registered dietitians are recognized by the medical profession as the legitimate providers of nutrition care. They have specialized education in human anatomy and physiology, chemistry, medical nutrition therapy, foods and food science, the behavioral sciences, and foodservice management. Registered dietitians must complete at least a bachelor's degree from an accredited college or university, a program of college-level dietetics courses, a supervised practice experience, and a qualifying examination. Continuing education is required to maintain R.D. status. Registered dietitians work in private practice, hospitals, nursing homes, wellness centers, business and industry, and many other settings. Most are members of the American Dietetic Association, and most are licensed or certified by the state in which they live. Over 40 states have licensure or certification laws that regulate dietitians/nutritionists.

In addition to using the expertise of an R.D., you can ask some simple questions that will help you judge the validity of nutrition information seen in the media or heard from friends.

1. What are the credentials of the source? Does the person have academic degrees in a scientific or nutrition-related field?
2. Does the source rely on emotions rather than scientific evidence, or use sensationalism to get a message across?
3. Are the promises of results for a certain dietary program reasonable or exaggerated? Is the program based on hard scientific information?
4. Is the nutrition information presented in a reliable magazine or newspaper, or is the information published in an advertisement or a publication of questionable reputation?
5. Is the information someone's opinion or the result of years of valid scientific studies with possible practical nutrition implications?

Much nutrition information that we see or read is based on scientific research. It is helpful to first understand how research studies are designed, as well as pitfalls in each design, so that you can evaluate a study's results. The following three types of studies are commonly used in research.

Laboratory studies use animals, such as mice or guinea pigs, or tissue samples in test tubes to find out more about a process that occurs in people to determine if a substance might be beneficial or hazardous in humans, or to test the effect of a treatment. A major advantage of using laboratory animals is that researchers can control many factors that they can't control in human studies. For instance, researchers can make sure that comparison groups are genetically identical and the conditions to which they are exposed are the same as well. However, mice and other animals are not the same as humans, so results from these studies can't automatically be generalized to humans. For example, laboratory studies have indicated that the artificial sweetener saccharin caused cancer in mice, but this has never been proven for humans.

Another type of research, called *epidemiological research,* looks at how disease rates vary among different populations and also factors associated with disease. Epidemiological studies rely on observational data from human populations, so they can only suggest a relationship between two factors; they cannot establish that a particular factor causes a disease. This type of observational study may compare factors found among people with a disease, such as cancer, to factors among a comparable group without that disease, or may try to identify factors associated with diseases that develop over time within a population group. Researchers may find, for example, fewer cases of osteoporosis in women who take estrogen after menopause.

A third type of research goes beyond using animals or observational data and uses humans as subjects. *Clinical trials* refer to studies that assign similar participants randomly to two groups. One group receives the experimental treatment; the other does not. Neither the researchers nor the participants know who is in which group. For example, a clinical trial to test the effects of estrogen after menopause would randomly assign each participant to one of two groups. Both groups would take a pill, but for one group this pill would be a dummy pill, called a placebo. Clinical studies are used to assess the effects of nutrition-education programs and medical nutrition therapy. Unlike epidemiological studies, clinical studies can observe cause-and-effect relationships.

When reading or listening to a news account of a particular study, it is helpful to have a few key questions in the back of your mind not only to help evaluate the merits of the study, but also to determine whether it is applicable to you. Look to news reports to address the following:

1. How does this work fit with the body of existing research on the subject? Even the most well-written article does not have enough space to discuss all relevant research on an issue. Yet it is extremely important for the article to address whether a study is confirming previous research and therefore adding more weight to scientific beliefs, or whether the study's results and conclusions take a wild departure from current thinking on the subject.
2. Could the study be interpreted to say something else? Scientists often reach different conclusions when commenting on the same or similar data. Look for varying conclusions from experts, because certain issues they address may be important when putting the findings into context.
3. Are there any flaws in how the study was undertaken that should be considered when making conclusions? The more experts are quoted, or provide background, in a news story, the more likely potential flaws will be described.
4. Are the study's results generalizable to other groups? Not all research incorporates all types of people: men, women, older adults, or people of various ethnicities. Also, a study may have been conducted on animals and not humans. If study results are applicable only to a narrow group of people, it should be reported as such.

Websites that will help you separate fact from fiction:

www.quackwatch.com (Quack Watch)
www.ncahf.org (National Council Against Health Fraud)

With permission, this *Hot Topic* used sections of "If It Sounds Too Good to Be True . . . It Probably Needs a Second Look" from *Food Insight*, published by the International Food Information Council Foundation, March/April 1999.

Chapter 3

Carbohydrates

Carbohydrate—A large class of nutrients, including sugars, starch, and fibers, that function as the body's primary source of energy.
Photosynthesis—A process during which plants convert energy from sunlight into energy stored in carbohydrate.
Simple carbohydrates—Sugars, including monosaccharides and disaccharides.
Complex carbohydrates (polysaccharides)—Long chains of many sugars, including starches and fibers.

Carbohydrate literally means hydrate (water) of carbon, a name derived from the investigations of early chemists, who found that heating sugars for a long period of time in an open test tube produced droplets of water on the sides of the tube and a black substance, carbon. Later chemical analysis of sugars and other carbohydrates indicated that they all contain at least carbon, hydrogen, and oxygen.

Carbohydrates are the major components of most plants, making up from 60 to 90 percent of their dry weight. In contrast, animals and humans contain a comparatively small amount of carbohydrates. Plants are able to make their own carbohydrates from the carbon dioxide in air and water taken from the soil in a process known as **photosynthesis.** Photosynthesis converts energy from sunlight into energy stored in carbohydrates, which the plant uses to grow and be healthy. Animals are incapable of photosynthesis and, therefore, depend on plants as a source of carbohydrates. Green plants, such as wheat or broccoli, supply the carbohydrates in our diets.

Carbohydrates are separated into two categories: simple and complex. **Simple carbohydrates,** also called sugars, include both natural and refined sugars. Carbohydrates are much more than just sugars, though, and include the **complex carbohydrates** starch and fiber. Another name for complex carbohydrates is **polysaccharides** (*poly-* means many), a good name for starch and most fibers because both are long chains of many sugars.

After completing this chapter, you should be able to:

- Identify the functions of carbohydrates
- Distinguish between monosaccharides and disaccharides and give examples of each
- Identify foods high in natural sugars and added sugars and fiber
- List the potential health risks of consuming too much sugar
- Identify food sources of starch and list the uses of starch in cooking
- Distinguish between the two types of dietary fiber and list examples of food containing each
- Identify foods as being made from whole grains or refined grains
- Describe the health benefits of a high-fiber diet and identify foods high in fiber
- Describe how carbohydrates are digested, absorbed, and metabolized by the body
- State the dietary recommendations for carbohydrates
- Discuss the storage, cooking, and menuing of grains, legumes, and pasta
- Identify characteristics of grain starches and root starches

Functions of Carbohydrates

Carbohydrates are the primary source of the body's energy, supplying about 4 kcalories per gram. Protein and fat can be burned for energy by our cells, but the body uses carbohydrates first, in part because carbohydrates are the

Glycogen—The storage form of glucose in the body; it is stored in the liver and muscles.

Ketone bodies—A group of organic compounds that cause the blood to become too acidic as a result of fat being burned for energy without any carbohydrates present.

Ketosis—Excessive level of ketone bodies in the blood and urine.

most efficient energy source. In fact, the central nervous system (which includes the brain and nerve cells) and red blood cells rely almost exclusively on glucose, a simple carbohydrate, for energy. Some glucose can be stored in the liver and muscles in a form called **glycogen.**

If there are not enough carbohydrates for energy, the body can burn either fat or protein, but this is not desirable. When fat is burned for energy without any carbohydrates present, the process is incomplete and results in the production of **ketone bodies,** which start to accumulate in the blood. An excessive level of ketone bodies can cause the blood to become too acidic (called **ketosis**), a condition that interferes with the transport of oxygen in the blood. Ketosis can cause dehydration and may even lead to a fatal coma. Carbohydrates are important to help the body use fat efficiently. About 50 to 100 grams of carbohydrates are needed daily to prevent ketosis.

Carbohydrates also spare protein from being burned for energy so that protein can be better used to build and repair the body. You need at least 100 to 150 grams of carbohydrates daily to spare protein from being burned for fuel and to provide glucose to the central nervous system. This amount represents what you minimally need, not what is desirable (about two times more). We obtain from 50 to 60 percent of our energy intake from carbohydrates. Therefore, if you eat 2000 calories per day, you take in 1000 to 1200 calories of carbohydrates, which represents 250 to 300 grams.

Carbohydrates are part of various materials found in the body, such as connective tissues, some hormones and enzymes, and genetic material.

Fiber promotes the normal functioning of the intestinal tract and is associated with a reduced risk of developing cardiovascular disease and obesity. Certain forms of dietary fiber can help lower blood cholesterol and also control blood glucose in people with diabetes (a disease characterized by high blood glucose levels).

■ MINI-SUMMARY

Carbohydrates are the primary source of the body's energy. The central nervous system and red blood cells rely almost exclusively on glucose for energy. Carbohydrates are also important to help the body use fat efficiently. When fat is burned for energy without any carbohydrates present, the process is incomplete and could result in ketosis. Carbohydrate also spares protein from being burned for energy. You need at least 100 to 150 grams of carbohydrate daily to prevent ketosis, spare protein, and provide glucose to the central nervous system and red blood cells. Carbohydrates are part of various materials found in the body. Fiber promotes the normal functioning of the intestinal tract, may reduce the risk of developing cardiovascular disease and obesity, helps lower blood cholesterol levels, and helps control blood glucose in people with diabetes.

Sugars

Monosaccharides—Single sugars such as glucose or fructose.

Disaccharides—Double sugars such as sucrose.

Glucose—The most significant monosaccharide, the body's primary source of energy.

Blood glucose level (blood sugar level)—The amount of glucose found in the blood; glucose is vital to the proper functioning of the body.

Fructose—A monosaccharide found in fruits and honey.

Galactose—A monosaccharide found linked to glucose to form lactose or milk sugar.

Sucrose—A disaccharide commonly called table sugar, granulated sugar, or simply sugar.

Maltose—A disaccharide made of two glucose units bonded together.

Lactose—A disaccharide found in milk and milk products that is made of glucose and galactose.

Simple carbohydrates include **monosaccharides,** or single sugars, and **disaccharides,** or double sugars (*mono-* means one and *di-* means two). The term *sugar* refers to both monosaccharides and disaccharides collectively. The chemical names of the six sugars to be discussed all end in *-ose,* which means sugar.

Monosaccharides include the simple sugars glucose, fructose, and galactose, which are the building blocks of other carbohydrates, such as disaccharides and starch.

In photosynthesis, plants make glucose, which provides energy for growth and other plant activities. **Glucose,** also called dextrose, is the most important monosaccharide because it is the human body's number-one source of energy. Most of the carbohydrates you eat are converted to glucose in the body. The concentration of glucose in the blood, referred to as the **blood glucose level** or **blood sugar level,** is vital to the proper functioning of the human body. Glucose is found in fruits such as grapes, in honey, and in trace amounts in many plant foods.

Fructose, the sweetest natural sugar, is also found in honey as well as in fruits. Fructose is about 1.3 times as sweet as sucrose. Fructose and glucose are the most common monosaccharides in nature.

The last single sugar, **galactose,** does not occur alone in nature but is linked to glucose to make milk sugar, also called lactose, a disaccharide.

Most naturally occurring carbohydrates contain two or more monosaccharide units linked together. Disaccharides, the double sugars, include sucrose, maltose, and lactose. They each contain glucose (Figure 3-1). **Sucrose** is the chemical name for what is commonly called cane sugar, table sugar, granulated sugar, or simply sugar. It is refined from sugar cane or sugar beet juice and used mainly to sweeten foods. As Figure 3-1 indicates, sucrose is simply two common single sugars—glucose and fructose—linked together. Although the primary source of sucrose in the American diet is refined sugar, sucrose does occur naturally in small amounts in many fruits and vegetables. Table sugar is more than 99 percent pure sugar and provides virtually no nutrients for its 16 kcalories per teaspoon.

Maltose, which consists of two bonded glucose units, does not occur in nature to any appreciable extent. It is fairly abundant in germinating (sprouting) seeds and is produced in the manufacture of beer.

The last disaccharide, **lactose,** is commonly called milk sugar. It is found naturally only in milk (which is about 5 percent lactose) and in certain other dairy products. Unlike most carbohydrates, which are in plant products, lactose is one of the few carbohydrates associated exclusively with animal products. Milk is not thought of as sweet because lactose is one of the lowest-ranking sugars in terms of sweetness (Table 3-1).

Figure 3-1
Monosaccharides and disaccharides

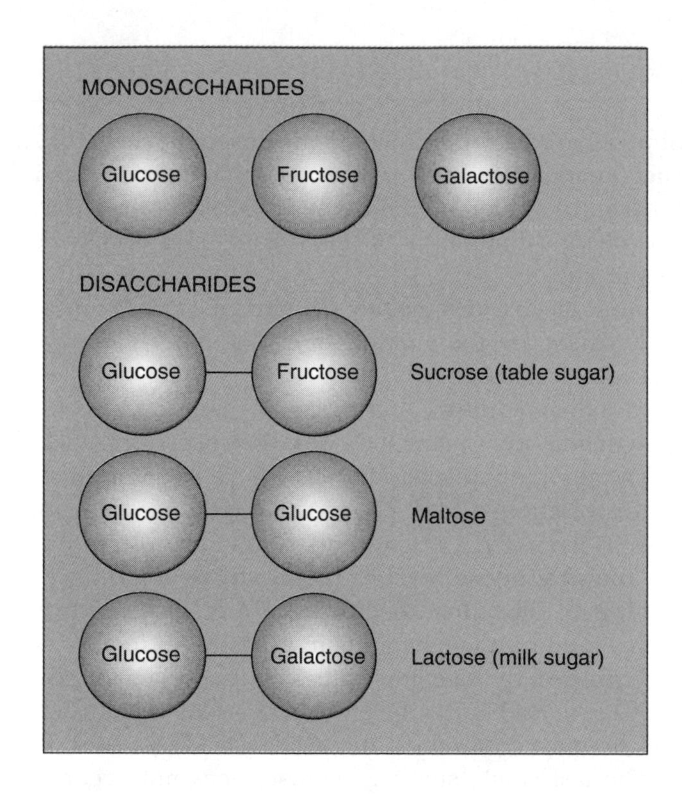

TABLE 3-1 Relative Sweetness of Sugars and Alternative Sweeteners

Name	Sweetness Compared to Sucrose
Sugars	
Lactose	0.2
Glucose	0.7
Sucrose	1.0
High-fructose corn syrup (42% fructose)	1.0
High-fructose corn syrup (55% fructose)	1.1
Fructose	1.3
Alternative Sweeteners	
Aspartame (Nutrasweet, Equal)	160–220
Acesulfame-K (Sunette)	200
Saccharin (Sweet 'N Low)	300
Sucralose (Splenda)	600
Neotame	7000–13,000

Sources: American Dietetic Association. "Position of The American Dietetic Association: Use of Nutritive and Nonnutritive Sweeteners." *Journal of The American Dietetic Association* 98 (1998): 580–586.

Coulston, Ann, and Rachel K. Johnson. "Sugar and Sugars: Myths and Realities." *Journal of The American Dietetic Association* 102(2002): 351–353.

Sugars in Food

Sugar occurs naturally in some foods, such as fruits and milk. Fruits are an excellent source of natural sugar, but be aware that some canned fruits contain much added sugar. Canned fruits are packed in one of three styles: in fruit juice, in light syrup, or in heavy syrup. Both light syrup and heavy syrup have added sugar. Heavy syrup contains the most added sugar (about 4 teaspoons of sugar to 1/2 cup of fruit). Dried fruits, such as raisins, are more concentrated sources of natural sugar than fresh fruits because dried fruits contain much less water.

Although a natural sugar, honey (made by bees) is primarily fructose and glucose, the same two components of table sugar. Therefore, by the time they are absorbed, honey and table sugar are the same thing. Although they are different in flavor and texture, the body can't tell the difference between natural and refined sugars. Honey and sugar contribute only energy and no other nutrients in significant amounts. Because honey is more concentrated, it has 50 percent more calories than an equal volume of sugar.

Lactose, or milk sugar, is present in large amounts in milk, ice cream, ice milk, sherbet, cottage cheese, cheese spreads and other soft cheeses, eggnog, and cream. Hard cheeses contain only traces of lactose.

Added sugars—Sugars added to a food for sweetening or other purposes; they do not include the naturally occurring sugars in foods such a fruit or milk.

High-fructose corn syrup—Corn syrup that has been treated with an enzyme that converts part of the glucose it contains to fructose.

Sugar alcohols—Sugarlike compounds that occur naturally in small amounts in fruits and vegetables; they are used to sweeten sugar-free candies, cookies, and chewing gum.

Added sugars (Table 3-2) include corn syrup and other sweeteners added to foods in processing, such as soft drinks, as well as sugars added to foods at the table. Added sugars do not include the naturally occurring sugars in foods such as fruit or milk.

Added sugars perform several functions in foods besides sweetening. They prevent spoilage in jams and jellies and perform several functions in baking, such as browning the crust and retaining moisture in baked goods so they stay fresh. Sugar also acts as a food for yeast in breads and other baked goods that use yeast for leavening.

High-fructose corn syrup is corn syrup that has been treated with an enzyme to convert part of the glucose it contains to fructose. The reason for changing the glucose to fructose lies in the fact that fructose is twice as sweet as glucose. High-fructose corn syrup is therefore sweeter, ounce for ounce, than corn syrup, so smaller amounts can be used (making it cheaper). It is used to sweeten almost all regular soft drinks and is frequently used in canned juices, fruit drinks, sweetened teas, cookies, jams and jellies, syrups, and sweet pickles.

The added sugar content of various foods is listed in Table 3-3. Keep in mind, when you look at Sugars on the Nutrition Facts panel, that this number includes naturally occurring sugars and added sugars.

If you chew sugarless gums, you may know that they often contain xylitol, sorbitol, or another sugar alcohol. **Sugar alcohols** are sugarlike compounds that occur naturally in small amounts in fruits and vegetables.

TABLE 3-2 Common Forms of Refined Sugars

Form of Sugar	Description
Granulated sugar (sucrose)	Most important and most used sugar product on the market. Made from beet sugar or cane sugar, which are identical in chemical composition.
Powdered or confectioners' sugar	Granulated sugar that has been pulverized. Available in several degrees of fineness, designated by the number of Xs following the name. 6X is the standard confectioners' sugar and is used in icing and toppings.
Brown sugar	Sugar crystals contained in a molasses syrup with natural flavor and color—91 to 96 percent sucrose. Sold in 4 grades—the higher the grade, the darker the brown sugar, the more flavor.
Turbinado sugar	Raw sugar that has been partially refined and washed.
Syrups	
Corn syrup	Made from cornstarch. Mostly glucose with some maltose. Only 75 percent as sweet as sucrose. Less expensive than sucrose. Used in baked goods and canned goods.
High-fructose corn syrup	Corn syrup treated with an enzyme that converts glucose to fructose, which results in a sweeter product. Used in soft drinks, baked goods, jelly, syrups, fruits, and desserts.
Maple syrup	A concentrated sucrose solution made from mature sugar-maple-tree sap that flows in the spring. Mostly replaced by pancake syrup—a mixture of sucrose and artificial maple flavorings.
Molasses	Thick syrup left over after making sugar from sugar cane. Brown in color with a high sugar concentration.
Honey	Sweet syrupy fluid made by bees from the nectar collected from flowers and stored in nests or hives as food. Made of fructose and glucose.

They contain fewer calories per gram than sucrose (from 1.5 to 3 kcalories per gram) and are used mainly to sweeten sugar-free candies, cookies, and chewing gums. Sugar alcohols are metabolized and absorbed slowly. They are not as sweet as sucrose, so a greater quantity is used in products such as chewing gum, breath mints, and hard candies. The main advantage of these sugarless products is that they do not promote tooth decay, as does sugar. A disadvantage is that large amounts of sugar alcohols can cause diarrhea, and foods using the sugar alcohols sorbitol or mannitol must be labeled "excess consumption may have a laxative effect."

Some of the sugar alcohols on the market are sorbitol, xylitol, mannitol, and isomalt. Sorbitol is 60 percent as sweet as sucrose and is used in such products as sugarless hard and soft candies and chewing gums. Xylitol is almost as sweet as table sugar and is popular in chewing gum and candies. Mannitol is 70 percent as sweet as sucrose and is used in chewing gum. Isomalt is about half as sweet as sucrose and can be used in candies and chewing gums, as well as in baked goods and frostings because it does not break down when heated. Lactitol and maltitol are two additional sugar alcohols.

TABLE 3-3 Added Sugars in Food

Food Group	Added Sugars (in teaspoons)
Bread, Cereal, Rice, and Pasta	
Bread, 1 slice	0
Muffin, 1 medium	1
Cookies, 2 medium	1
Danish pastry, 1 medium	1
Doughnut, 1 medium	2
Ready-to-eat cereal, sweetened, 1 ounce	1+ (depends on brand)
Pound cake, nonfat, 1 ounce	2
Angel food cake, 1/12 tube cake	5
Cake, frosted, 1/16 average	6
Pie, fruit, 2 crust, 1/6 8″ pie	6
Fruit	
Fruit, canned in juice, 1/2 cup	0
Fruit, canned in light syrup, 1/2 cup	2
Fruit, canned in heavy syrup, 1/2 cup	4
Milk, Yogurt, and Cheese	
Milk, plain, 1 cup	0
Chocolate milk, reduced-fat (2%)	3
Low-fat yogurt, plain, 8 ounces	0
Low-fat yogurt, flavored, 8 ounces	5
Low-fat yogurt, fruit, 8 ounces	7
Ice cream, ice milk, or frozen yogurt, 1/2 cup	3
Chocolate shake, 10 fluid ounces	9
Other	
Sugar, jam, or jelly, 1 teaspoon	1
Syrup or honey, 1 tablespoon	3
Chocolate bar, 1 ounce	3
Fruit sorbet, 1/2 cup	3
Gelatin dessert, 1/2 cup	4
Sherbet, 1/2 cup	5
Cola, 12 fluid ounces	9
Fruit drink, ade, 12 fluid ounces	12

Source: U.S. Department of Agriculture, "The Food Guide Pyramid," Home and Garden Bulletin Number 252. 1996.

Added Sugars and Health

As Americans have been trying to eat less fat, their consumption of added sugars has skyrocketed, fueled by soda consumption (accounting for 33 percent of all added sugars), table sugar, syrups, sweets, sweetened grains, and regular fruitades/drinks. The amount of added sugars in the U.S. food supply increased from 27 teaspoons per person in 1970 to 32 teaspoons in 1996, an increase of 23 percent. The Food Guide Pyramid recommends

12 teaspoons of sugar in a 2200-calorie diet. Is all this sugar good for you? Let's take a look at how it affects health.

Obesity. Although there is no research stating that added sugars alone cause obesity, added sugars are undoubtedly a major factor in rising obesity rates among adults and children. Over the past two decades, Americans have been eating more kcalories, and most of those kcalories are coming from carbohydrates, especially from soft drinks. Just add one 12-ounce can of soft drinks to your diet every day for a year, and you will gain 15 pounds!

Other high-sugar foods, such as cookies and candy, are almost always teamed up with fat and high in kcalories. These foods are also typically low in nutrients and are therefore referred to as **empty kcalories.** For example, the cupcake pictured in Figure 3-2 supplies 170 kcalories with virtually no nutrients. If you look at foods with natural sugars, such as fruits and milk, you will notice that they also contain many essential nutrients. The foods highest in added sugars are, unfortunately, not nearly so rich in other nutrients.

Empty kcalories—Foods that provide few nutrients for the number of calories they contain.

High consumption of beverages and foods rich in added sugars contribute to overweight and obesity when too many calories are consumed. Being overweight or obese increases your risk for high blood pressure, high blood cholesterol, heart disease, stroke, diabetes, certain types of cancer, arthritis, and breathing problems. High consumption of added sugars also makes it difficult to get in all the required micronutrients (vitamins and minerals) that are necessary, which is a real concern for children. Children especially need to eat a varied and adequate diet to ensure proper growth.

Figure 3-2
Example of empty calories

Homemade Cupcake with Icing

Information per serving	
Serving size =	1 Cupcake
Kcalories	170
Protein	2 grams
Carbohydrates	30 grams
Fat	5 grams
Sodium	110 milligrams
% Daily Value	
Vitamin A	0%
Vitamin C	0%
Calcium	0%
Iron	4%

Diabetes—A disorder of carbohydrate metabolism characterized by high blood sugar levels and inadequate or ineffective insulin.

Insulin—A hormone that increases the movement of glucose from the bloodstream into the body's cells.

Diabetes. Obesity is more closely linked to diabetes than to any other health problem. **Diabetes** is a disorder in which the body does not metabolize carbohydrates properly. It results from having inadequate or ineffective insulin. **Insulin** is a hormone that increases the movement of glucose from the bloodstream into the body's cells, where it is used to produce energy. People with untreated diabetes have high blood sugar levels. Treatment for diabetes is individualized to the patient and includes a balanced diet that supports a healthy weight and physical activity, as well as insulin if needed. When overweight people with diabetes lose weight, the disease is usually more controllable. Diabetes is discussed in more depth in Chapter 11.

Heart Disease. A moderate intake of sugars does not increase heart disease risk. However, in some individuals, a high sugar intake changes the blood fat levels in such a way that the risk for heart disease increases.

Hypoglycemia—A symptom in which blood sugar levels are low.

Postpranial hypoglycemia—Low blood sugar that occurs generally one to four hours after meals and includes symptoms such as shakiness, sweating, and dizziness.

Fasting hypoglycemia—Low blood sugar that occurs after not eating for eight or more hours.

Attention Deficit Hyperactivity Disorder (ADHD)—A developmental disorder of children characterized by impulsiveness, distractibility, and hyperactivity.

Lactase—An enzyme needed to split lactose into its components in the intestines.

Hypoglycemia. Hypoglycemia is the term used to describe an abnormally low blood glucose level. It occurs most often in people who have diabetes and take insulin. If they take too much insulin, eat too little food, and/or exercise a lot, their blood sugar level may drop to the point where they are hypoglycemic. In people without diabetes, the most common form is **postprandial hypoglycemia.** It occurs generally one to four hours after meals and has symptoms such as quickened heartbeat, shakiness, weakness, anxiety, sweating, and dizziness, mimicking anxiety or stress symptoms. It may be caused when a rapid rise in blood glucose after a meal causes a temporary overproduction of insulin, which pulls too much sugar out of the bloodstream.

A second type of hypoglycemia, **fasting hypoglycemia,** is rare. It has numerous causes, including pancreatic cancer, and can be serious. Its symptoms occur after not eating for eight or more hours, so it usually occurs during the night or before breakfast.

A diet for people with hypoglycemia includes regular, well-balanced meals with moderate amounts of refined sugars and sweets. Protein, fat, and fiber can moderate swings in blood glucose levels.

Hyperactivity in Children. During the 1980s, reports came out linking sugar intake with hyperactivity in children. Unfortunately, the study results were unclear as to whether sugar caused the hyperactivity or the hyperactivity caused the children to eat sugar. Extensive research since then has failed to show that high sugar intake causes hyperactivity or **attention deficit hyperactivity disorder (ADHD).**

Lactose Intolerance. Lactose (milk sugar) is a problem for certain people who lack or, more commonly, don't have enough of the enzyme **lactase.** Lactase is needed to split lactose into its components in the small intestine.

If lactose is not split, it travels to the colon, where bacteria ferment it and produce short-chain fatty acids and gas. These by-products do not normally cause any problems or discomfort in small amounts. However, if a lot of lactose travels to the colon, symptoms such as gas, abdominal distention, and diarrhea often occur within about 30 minutes to 2 hours after ingesting milk products. The symptoms normally clear up within two to five hours. This problem, called **lactose intolerance,** seems to be an inherited problem especially prevalent among Asian Americans, Native Americans, African Americans, and Latinos, as well as some other population groups.

Lactose intolerance— An intolerance to milk and most milk products due to a deficiency of the enzyme lactase. Symptoms often include flatulence and diarrhea.

Treatment for lactose intolerance requires a diet that is limited in lactose, which is present in large amounts in milk, ice cream, ice milk, sherbet, cottage cheese, eggnog, and cream. Most individuals can drink small amounts of milk without any symptoms, especially if it is taken with food. Lactose-reduced milk and some other lactose-reduced dairy products are available in supermarkets, as is the enzyme lactase (which is also sold in pharmacies). Lactase can be added to milk to reduce the lactose content. Eight fluid ounces of lactose-reduced milk contain only 3 grams of lactose, compared with 12 grams in regular milk. Reducing the lactose content of milk by 50 percent is often adequate to prevent symptoms of lactose intolerance. Although lactose-reduced milk and other lactose-digestive aids are available, they may not be necessary when lactose intake is limited to 1 cup of milk (or the equivalent) or less a day.

Yogurt is usually well tolerated because it is cultured with live bacteria that digest lactose. This is not always the case with frozen yogurt, because most brands do not contain nearly the number of bacteria found in fresh yogurt (there are no federal standards for frozen yogurt at this time). Also, some yogurts have milk solids added to them that can cause problems. Many hard cheeses contain very little lactose and usually do not cause symptoms because most of the lactose is removed during processing or digested by the bacteria used in making cheese.

People who have difficulty digesting lactose report tremendous variation in which lactose-containing foods they can eat and even the time of day they can eat them. For example, one individual may not tolerate milk at all, whereas another can tolerate milk as part of a big meal. The ability to tolerate lactose is not an all-or-nothing phenomenon. As people with lactase deficiency usually decrease their intake of dairy products and thus their calcium intake, they should try different dairy products to see what they can tolerate.

Dental Caries. The only negative health effect of sugar that most health experts agree on is that sugar (and starches too) do contribute to the development of **dental caries,** or cavities. The more often sugars and starches—even small amounts—are eaten and the longer they are in the mouth before the teeth are brushed, the greater the risk for tooth decay. Dental

Dental caries— Tooth decay.

caries are a major cause of tooth loss. This is so because every time you eat something sweet, the bacteria living on your teeth ferment the carbohydrates, which produces acid. This acid eats away at the teeth for 20 to 30 minutes, and cavities eventually develop. The deposit of bacteria, protein, and polysaccharides that forms on the teeth in the absence of toothbrushing during a period of 12 to 24 hours is called **plaque.** Without good tooth-brushing habits, plaque may cover all surfaces of the teeth.

Plaque—Deposits of bacteria, protein, and polysaccharides found on teeth that contribute to tooth decay.

Other factors influence how much impact foods will have on the development of dental caries. The sequence of eating foods in a meal, the foods' form (liquid or solid and sticky), and combinations of foods also influence dental caries. At meals, if an unsweetened food, such as cheese, is eaten after a sugared food, the plaque will be less acidic, so less acid eats away at the teeth. Cheese also stimulates more saliva, which helps to wash away acids. This is why eating sugary or starchy foods as frequent between-meal snacks is more harmful to teeth than having them at meals. Sticky carbohydrate foods, such as raisins or caramels, cause more problems than liquid carbohydrate foods because they stick to the teeth and provide a constant source of fermentable carbohydrates for the bacteria until washed away. Liquids containing sugars have been considered less harmful to teeth than solid sweets because they clear the mouth quickly.

Food such as dried fruits, breads, cereals, cookies, crackers, and potato chips increase the chances of dental caries when eaten frequently. Foods that do not seem to cause cavities include cheese, peanuts, sugar-free gum, some vegetables, meats, and fish. To prevent dental caries, brush your teeth often, floss your teeth once a day, try to limit sweets to mealtime, limit between-meal snacks, and see your dentist regularly.

■ MINI-SUMMARY

Simple carbohydrates are sugars and include both natural and refined sugars. Monosaccharides are the building blocks of other carbohydrates and include glucose (the main source of the body's energy), fructose (found in fruits), and galactose (found in milk sugar). Disaccharides include sucrose (table sugar), maltose, and lactose (milk sugar). Refined sugars, such as high-fructose corn syrup, are used to sweeten soft drinks, breakfast cereals, candy, baked goods such as cakes and pies, syrups, and jams and jellies. Added sugar consumption has risen dramatically. Added sugars are a major factor in rising obesity rates. Obesity is closely linked to the development of diabetes. In some individuals, a high sugar intake changes blood fat levels in such a way that the risk for heart disease increases. Sugar contributes to dental decay but not to hyperactivity in children. Lactose (milk sugar) is a problem for people with lactose intolerance. They experience abdominal cramps, bloating, and diarrhea about 30 minutes to 2 hours after ingesting milk products. Treatment for lactose intolerance requires a diet that is limited in lactose, which is present in large amounts in milk, ice cream, ice milk, sherbet, cottage cheese, eggnog, and cream. Most individuals can drink small amounts of milk without any symptoms, especially if it is taken with food.

Starch

Starch—A complex carbohydrate made up of a long chain of glucoses linked together; found in grains, legumes, vegetables, and some fruits; the straight form is called amylose and the branched form is called amylopectin.

Plants such as peas store glucose in the form of **starch.** Starch is made of many chains of hundreds to thousands of glucoses linked together. The chains may be straight or have tree-like branches (Figure 3-3). The straight form of starch is called *amylose,* and the branched form is called *amylopectin.*

Just as plants store glucose in the form of starch, your body stores glucose as glycogen. Like starch, glycogen is a polysaccharide. It is a chain of glucose units, but the chains are longer and have more branches than starch (Figure 3-3). Glycogen is stored in two places in the body: the liver and the muscles. An active 150-pound man has about 400 kcalories stored in his liver glycogen and about 1400 kcalories stored in his muscle glycogen. When the blood sugar level starts to dip and more energy is needed, the liver converts glycogen into glucose, which is then delivered by the bloodstream. Muscle glycogen does not supply glucose to the bloodstream but is used strictly to supply energy for exercise.

Starch in Food

Starch is found only in plant foods. Cereal grains, the fruits or seeds of cultivated grasses, are rich sources of starch and include wheat, corn, rice, rye, barley, and oats. Cereal grains are used to make breads, baked goods, break-

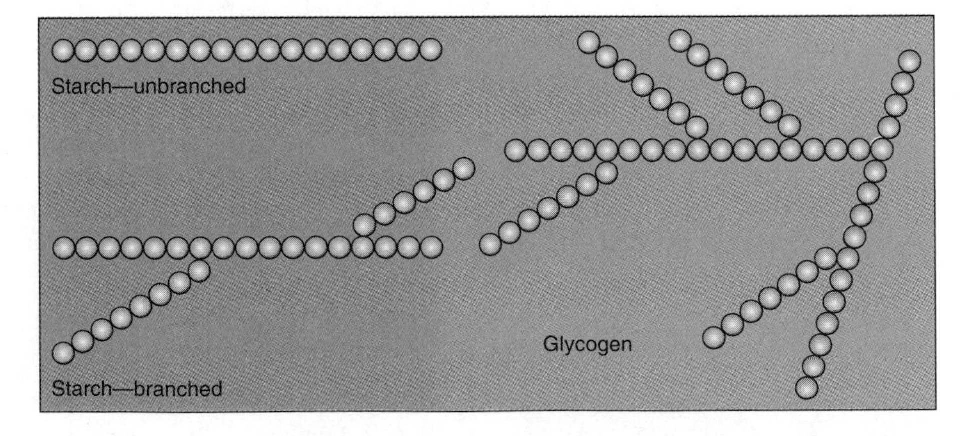

Figure 3-3
The structures of starch and glycogen

fast cereals, and pastas. Starches are also found in root vegetables, such as potatoes, and in dried beans and peas, such as navy beans.

Starchy foods in general are not flavorful if eaten raw, so most are cooked to make them taste better and be more digestible. Starchy foods are used extensively as thickeners in cooking, because starch undergoes a process called **gelatinization** when heated in liquid. When starches gelatinize, granules absorb water and swell, making the liquid thicken. Around the boiling point, the granules have absorbed a lot of water and they burst, letting starch out into the liquid. When this occurs, the liquid quickly becomes still thicker. Gelatinization is a process unique to starches, so you find them frequently used as thickeners in soups, sauces, gravies, puddings, and other foods.

Gelatinization—A process in which starches, when heated in liquid, absorb water and swell in size.

Starch and Health

Starch creates the same problem as sugar in the mouth and therefore contributes to tooth decay. Starch from whole-grain sources is preferable to starch found in refined grains such as white flour. This is discussed in detail in the following section.

■ **MINI-SUMMARY**

Starch is a storage form of glucose found in plants. The body stores glucose as glycogen in the liver and muscles. Glycogen is important to maintain normal blood sugar levels. Starches are found in cereal grains, breads, baked goods, cereals, pastas, root vegetables, and dried beans and peas. Starches are commonly used as thickeners because they gelatinize. Like sugar, starch contributes to dental caries.

Fiber

Dietary fiber can be defined as the edible/nondigestible component of carbohydrates and lignin naturally found in plant food. Like starch, most fibers are chains of bonded glucose units, but what's different is that the units are linked with a chemical bond that our digestive enzymes can't break down. In other words, most fiber passes through the stomach and intestines unchanged and is excreted in the feces.

Dietary fiber—The edible, nondigestible part of carbohydrates and lignin naturally found in plant food.

Fiber is found only in plant foods; it does not appear in animal foods. One way to classify fibers is by how well they dissolve in water. **Soluble fiber** swells in water, like a sponge, into a gel-like substance. **Insoluble fiber** also swells in water, but not nearly to the extent of soluble fiber.

Soluble fiber—A classification of fiber that includes gums, mucilages, pectin, and some hemicelluloses; they are generally found around and inside plant cells.

Insoluble fiber—A classification of fiber that includes cellulose, lignin, and the remaining hemicelluloses; they generally form the structural parts of plants.

The soluble fibers include gums, mucilages, pectin, and some hemicelluloses. They are generally found around and inside plant cells. The insoluble fibers include cellulose, lignin, and the remaining hemicelluloses. They generally form the structural parts of plants. The amount of fiber in a plant varies among plants and may vary within a species or variety, depending on growing conditions and the plant's maturity at harvest. Lignin is an insoluble fiber but is not technically a carbohydrate.

Fiber in Food

Fiber is abundant in plants, so legumes (dried beans, peas, and lentils), fruits, vegetables, whole grains, nuts, and seeds provide fiber (Table 3-4). Fiber is not found in meat, poultry, fish, dairy products, and eggs.

Foods containing soluble fiber are as follows.

- Many fruits and vegetables, such as apples, grapes, citrus fruits, and carrots (fruit juices are not good sources of fiber)
- Some cereal grains, such as oats and barley
- Beans and peas, such as kidney beans, pinto beans, chickpeas, split peas, and lentils

Soluble fiber is found both in and around the cells of plants, where it acts to keep the plant stuck together.

Insoluble fiber includes the structural parts of plants, such as skins and the outer layer of the wheat kernel. You have seen insoluble fiber in the skin of whole-kernel corn and in celery strings. It is found in the following foods:

- Wheat bran
- Whole grains, such as whole wheat and brown rice, and products made with whole grains, such as whole-wheat bread
- Many vegetables and fruits
- Beans and peas
- Seeds

Most foods contain both soluble and insoluble fiber.

Endosperm—In cereal grains, a large center area high in starch.

Germ—In cereal grains, the area of the kernel rich in vitamins and minerals that sprouts when allowed to germinate.

Whole Grains

The full name for grains is cereal grains. Cereal grains are the seeds of cultivated grasses such as wheat, corn, rice, rye, barley, and oats, among others. All cereal grains have a large center area high in starch known as the **endosperm.** The endosperm also contains some protein. At one end of the endosperm is the **germ,** the area of the kernel that sprouts when allowed to germinate. The germ is rich in vitamins and minerals, and contains

TABLE 3-4 Fiber Content of Selected Foods

Food and Portion Size	Grams of Dietary Fiber
All-bran cereal, 1/2 cup	9.7
Prunes, stewed 1/2 cup	8.2
Red kidney beans, canned, 1/2 cup	8.2
Baked beans with pork, 1/2 cup	6.6
Cowpeas, common (blackeyes, southern), boiled, 1/2 cup	5.6
Lentil soup, 1 cup	5.6
Raspberries, frozen, sweetened, 1/2 cup	5.5
Chickpeas, canned 1/2 cup	5.3
Post Shredded Wheat Cereal, 2 biscuits	5.3
Fast food french fries, 1 medium	4.7
Baked potato, with skin, 1	4.6
Green peas, frozen, boiled, 1/2 cup	4.4
Kellogg's Raisin Bran cereal, 1/2 cup	4.1
Bulgur, cooked, 1/2 cup	4.1
Mixed vegetables, frozen, boiled, 1/2 cup	4.0
Apples, raw with skin, 1	3.7
Oranges, all varieties, 1	3.1
Macaroni and cheese, canned, 1 cup	3.0
Instant oatmeal, plain, 1 packet	3.0
Sweet potato, canned, 1/2 cup	2.9
Sunflower seeds, dry roasted, 1/4 cup	2.9
Pecans, 20 halves (1 ounce)	2.8
Broccoli, frozen, chopped, boiled, 1/2 cup	2.8
Peanuts, all types, 1 ounce	2.6
Carrots, boiled, 1/2 cup	2.6
Popcorn, air-popped, 2 cups	2.4
Spaghetti, cooked, 1 cup	2.4
Bread, whole-wheat, 1 slice	1.9
Brown rice, long-grain, cooked, 1/2 cup	1.8
Raw carrots, 1/2 cup	1.7
Grapes, red or green, raw, 1 cup	1.6
Muffins, blueberry, 1	1.5
Raisins, seedless, 1/4 cup	1.5
Applesauce, 1/2 cup	1.5
Fruit cocktail, canned, 1/2 cup	1.2
Summer squash, 1/2 cup	1.1
Oatmeal bread, 1 slice	1.1
Peanut butter, smooth, 1 tablespoon	0.9
Bread, white, 1 slice	0.6
Apple juice, 1 cup	0.3
Vanilla pudding, 1/2 cup	0.1
Milk, 1 cup	0.0
Meat, poultry, fish	0.0

Source: U.S. Department of Agriculture, Agricultural Research Service, 2001, U.S.D.A. Nutrient Database for Standard Reference, Release 14. Nutrient Data Laboratory Home Page, http://www.nal.usda.gov/fnic/ food.

Bran—In cereal grains, the part that covers the grain and contains much fiber and other nutrients.

some oil. The **bran,** containing much fiber and other nutrients, covers both the endosperm and the germ. The seed contains everything needed to reproduce the plant: the germ is the embryo, the endosperm contains the nutrients for growth, and the bran protects the entire seed (Figure 3-4).

Most grains undergo some type of processing or milling after harvesting to allow them to cook more quickly and easily, to make them less chewy, and to lengthen their shelf life. Grains such as oats and rice have an outer husk or hull that is tough and inedible, so it is removed. Other processing steps might include polishing the grain to remove the bran and germ (as in making white flour), cracking the grain (as in cracked wheat), or steaming the grain (as in bulgur) to shorten the cooking time. The process of rolling or grinding a grain, such as oats, also shortens the cooking time.

Whole grain—A grain that contains the endosperm, germ, and bran.

Refined or milled grain—A grain in which the bran and germ are separated (or mostly separated) from the endosperm.

Whenever the fiber-rich bran and the vitamin-rich germ are left on the endosperm of a grain, the grain is called a **whole grain.** Examples of whole grains include whole wheat, whole rye, oatmeal, whole oats, whole cornmeal, whole hulled barley, popcorn, and brown rice. Read the Food Facts in this chapter for more information on a variety of grains.

If the bran and germ are separated (or mostly separated) from the endosperm, the grain is called a **refined** or **milled grain.** Whereas whole-wheat flour is made from the whole grain, white flour (also called wheat flour) is made only from the endosperm of the wheat kernel. Whole-wheat flour does not stay fresh as long as white flours. This is due to the presence of the germ, which contains oil. When the oil turns rancid, or deteriorates, the flour will turn out a poor-quality product.

In baking, whole-grain flours produce breads that are denser and chewier. For example, breads made with only whole-wheat flour are more compact and heavier than breads made with only white flour. This is because the strands of gluten in the whole-wheat bread are cut by the sharp edges of the bran flakes. Some bakers prefer to use some white flour to strengthen the bread.

Many consumers prefer white bread because it lacks the dark color and dense texture of whole-wheat bread, but there are also quite a few health-conscious consumers who appreciate a good-quality whole-grain bread.

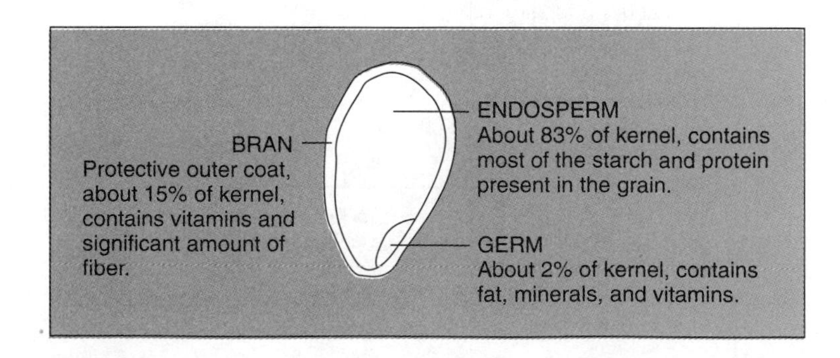
BRAN
Protective outer coat, about 15% of kernel, contains vitamins and significant amount of fiber.
ENDOSPERM
About 83% of kernel, contains most of the starch and protein present in the grain.
GERM
About 2% of kernel, contains fat, minerals, and vitamins.

Figure 3-4
A grain of wheat

When you compare the nutrients in whole grains and refined grains, whole grains are always a far more nutritious choice. They surpass refined grains in their fiber, vitamin, and mineral content. When wheat is refined, 22 nutrients and most of the fiber are removed. With whole wheat you get more vitamin E, vitamin B_6, pantothenic acid, magnesium, zinc, potassium, copper, and, of course, fiber. By federal law, refined grains are enriched with five nutrients that are lost in processing: thiamin, riboflavin, niacin, folate (folic acid), and iron. Whole-grain foods also contain **phytochemicals,** substances in plants that may reduce the risk of cancer and heart disease when eaten often.

Phytochemicals—Minute substances in plants that may reduce risk of cancer and heart disease when eaten often.

It can be quite difficult to determine from the name of a product whether it is indeed whole-grain. A multigrain bagel, for instance, is probably made with refined grains. On the other hand, whole-wheat bread (but not wheat bread, stoned-wheat bread, or seven-grain bread) is a 100% whole-grain product. The only way to be sure the product is whole-grain is to check the ingredient list. The first ingredient should be a whole grain, such as whole wheat or oatmeal.

Food Labeling and Fiber

In 2002, the Food and Nutrition Board proposed that the current system of labeling for dietary fiber be replaced by two values: dietary fiber and functional fiber. Dietary fiber is, as already defined, the edible, nondigestible part of carbohydrates and lignin naturally found in plant food. **Functional fiber** includes fiber sources shown to have similar health effects as dietary fiber, but are extracted from natural sources or are synthetic. For example, pectin is extracted from citrus peel and used as an ingredient in another food. The total fiber in any food product will be the total of its dietary fiber and its functional fiber (if any is added).

Functional fiber—Fiber sources shown to have similar health effects as dietary fiber but are extracted from natural sources or are synthetic.

Fiber and Health

What can fiber do for you? Numerous epidemiological (population-based) studies have found that diets low in saturated fat and cholesterol and high in fiber are associated with a reduced risk of certain cancers, diabetes, digestive disorders, and heart disease. However, this doesn't mean that fiber reduces the risk. Also, since high-fiber foods also often contain antioxidant vitamins, phytochemicals, and other substances that may offer protection against these diseases, researchers can't say for certain that fiber alone is responsible for the reduced health risks. Findings on the health effects of fiber show that it may play a role in the following:

- **Digestive disorders.** Because insoluble fiber aids digestion and adds bulk to stool, it hastens passage of fecal material through the intestines, thus helping to prevent or alleviate constipation. Fiber also may help reduce

Diverticulosis—A disease of the large intestine in which the intestinal walls become weakened, bulge out into pockets, and at times become inflamed.
Hemorrhoids—Enlarged veins in the lower rectum.
Carcinogen—Cancer-causing substance.

the risk of **diverticulosis,** a condition in which small pouches form in the colon wall (usually from the pressure of straining during bowel movements). People who already have diverticulosis often find that increased fiber consumption can alleviate symptoms, which include constipation and/or diarrhea, abdominal pain, and flatulence. A diet high in insoluble fiber is also used to prevent or treat **hemorrhoids,** enlarged veins in the lower rectum that usually result from straining during bowel movements.

- **Colon cancer.** Epidemiological studies have generally noted an association between low total fat intake, high fiber intake, and reduced incidence of colon cancer. The exact mechanism for reducing the risk is not known, but one possibility is that insoluble fiber adds bulk to stool, which in turn dilutes **carcinogens** (cancer-causing substances) and speeds their transit through the lower intestines and out of the body. More research needs to be done before we know for sure that fiber decreases the risk of colon cancer.
- **Heart disease.** Clinical studies show that a diet low in saturated fat and cholesterol and high in fruits, vegetables, beans, and grain products that contain soluble fiber can lower blood cholesterol levels, and therefore lower the risk of heart disease. As it passes through the gastrointestinal tract, soluble fiber binds to dietary cholesterol, helping the body to eliminate it. This reduces blood cholesterol levels, a major risk factor for heart disease. In particular, the fiber of whole oats or psyllium seed husk can lower total and LDL (low-density lipoprotein) blood cholesterol in diets that include these foods at appropriate levels.
- **Diabetes.** As with cholesterol, soluble fiber traps carbohydrates to slow their digestion and absorption. In theory, this may help prevent wide swings in blood sugar level throughout the day.
- **Obesity.** Because most fiber is indigestible and passes through the body virtually intact, it provides few calories. And since the digestive tract can handle only so much bulk at a time, fiber-rich foods are more filling than other foods—so people may tend to eat less and feel full longer.

There are certainly many health benefits to eating more fiber, but too much fiber can cause intestinal discomfort, gas, and diarrhea, and possibly even plug up the intestinal tract. When adding fiber to the diet, do so gradually to give the intestinal tract time to adapt, and drink lots of fluid.

■ MINI-SUMMARY

Dietary fiber is the nondigestible part of carbohydrate and lignin in plant foods. Most fibers are chains of glucose units. Soluble fiber is found in foods such as fruits, vegetables, oats, barley, and beans. Insoluble fiber is found in foods such as wheat bran and whole grains. Whole grains contain more fiber and nutrients than refined grains. Total fiber in any food is the total of its dietary fiber and its functional fiber (if any), terms that will eventually be on food labels. Eating plenty of fiber-containing foods, such as whole grains, promotes proper bowel function and may also help you feel full with fewer calories and protect you against chronic diseases such as heart disease and diabetes. Slowly add fiber to your diet and drink lots of fluid.

Digestion, Absorption, and Metabolism of Carbohydrates

Cooking carbohydrate foods makes them easier to digest. As mentioned previously, starches gelatinize, making them easier to chew, swallow, and digest. Cooking usually breaks down fiber in fruits and vegetables, also making them easier to chew, swallow, and digest.

Before carbohydrates can be absorbed through the villi of the small intestines, they must be broken down into monosaccharides, or one-sugar units. Starch digestion begins in the mouth, where an enzyme, salivary amylase, starts to break down some starch into small polysaccharides and maltose. In the stomach, salivary amylase is inactivated by the stomach acid. Next, the intestine completes the breakdown of starch into maltose, which is then split by an enzyme (maltase) into two glucose units. Glucose can now be absorbed.

Through the work of three enzymes made in the intestinal wall (see Table 3-5), all sugars are broken down into the single sugars: glucose, fructose, and galactose. They are then absorbed and enter the bloodstream, which carries them to the liver. In the liver, fructose and galactose are converted to glucose or further metabolized to make glycogen or fat. Glucose will go where it is most needed: into the bloodstream or to be made into glycogen or fat. The hormone insulin makes it possible for glucose to enter body cells, where it is used for energy or stored as glycogen.

TABLE 3-5 Carbohydrate Digestion

Site	Name of Enzyme	Carbohydrate Acted Upon	Products Formed
Mouth	Salivary amylase	Starch	Small polysaccharides, maltose
Small intestine	Pancreatic amylase	Starch	Small polysaccharides, maltose
	Sucrase	Sucrose	Glucose, fructose
	Lactase	Lactose	Glucose, galactose
	Maltase	Maltose	Glucose

Fiber cannot be digested, or broken down into its components by enzymes, so it continues down to the large intestine to be excreted. Although human enzymes can't digest most fibers, some bacteria in the large intestine can digest soluble fiber. As they digest it, the bacteria produce gas and small fat particles that are absorbed. The fats contribute some calories.

■ MINI-SUMMARY

During digestion, various enzymes break down starch and sugars into monosaccharides, which are then absorbed. In the liver, fructose and galactose are converted into glucose or further metabolized to make glycogen or fat. Fiber can't be digested by human enzymes, but soluble fiber can be digested by intestinal bacteria, producing gas and fat fragments that are absorbed.

Dietary Recommendations for Carbohydrate

In 2002, Dietary Reference Intakes were released for carbohydrate. The RDA for carbohydrate (Table 3-6) is 130 grams each day for children (from 1 year old) and adults. It is based on the minimum amount of carbohydrates needed to supply the brain with enough glucose. We normally eat much more than 130 grams of carbohydrate to meet our total energy needs. The 2002 DRI report also issued recommendations (Adequate Intakes) for total fiber (Table 3-6). An AI for total fiber is set at 38 and 25 grams/day for men and women respectively, ages 19 to 50 years. The AI for total fiber is based on 14 grams/1000 kcalories.

The 2002 Dietary Reference Intake report also established Acceptable Macronutrient Distribution Ranges (AMDR) for carbohydrate (along with fat and protein). AMDR is expressed as a percentage of total calorie intake. Consuming below or above the AMDR could cause insufficient intake of essential nutrients and increased risk of chronic diseases such as heart disease or diabetes. Adults and children (over 1 year old) should obtain 45 to 65 percent of their calories from carbohydrates. The wide range allows for more flexibility when planning diets for healthy people.

The Dietary Guidelines for Americans advise using sugars in moderation. The Food Guide Pyramid suggests that you try to limit your added sugars to 6 teaspoons a day if you eat about 1600 kcalories, 12 teaspoons at 2200 kcalories, or 18 teaspoons at 2800 kcalories. That works out to 6 to 10 percent of kcalories from added sugars, far below the 16 percent that many Americans eat (and drink). The World Health Organization suggests limit-

TABLE 3-6 Dietary Reference Intake Values for Carbohydrates and Total Fiber

	RDA Carbohydrate		AI Total Fiber	
Age*	Male	Female	Male	Female
1–3 years	130 g	130 g	19 g	19 g
4–8 years	130 g	130 g	25 g	25 g
9–13 years	130 g	130 g	31 g	26 g
14–18 years	130 g	130 g	38 g	36 g
19–30 years	130 g	130 g	38 g	25 g
31–50 years	130 g	130 g	38 g	25 g
Over 50 years	130 g	130 g	30 g	21 g
Pregnancy				
14–18 years		175 g		28 g
19–50 years		175 g		28 g
Lactation		210 g		29 g

*Note that infants from 1 to 6 months have an AI of 60 g of carbohydrate/day, infants from 7 to 12 months have an AI of 95 g carbohydrate/day. There is no AI for fiber for infants from 0 to 12 months old due to insufficient scientific evidence.

Source: Adapted with permission from the *Dietary References Intakes for Energy, Carbohydrates, FIber, Fat, Protein, and Amino Acids (Macronutrients).* © 2002 by the National Academy of Sciences. Courtesy of the National Academy Press, Washington, D.C.

ing added sugars to 10 percent of total calories. The Institute of Medicine's Food and Nutrition Board (2002) suggests that added sugars not exceed 25 percent of total kcalories.

The Food Guide Pyramid recommends that adults eat at least three servings of vegetables and 2 servings of fruits daily, and at least 6 servings of grain products, such as breads, cereals, pasta, and rice, with an emphasis on whole grains. Unfortunately, most Americans eat one serving or less of whole-grain food each day. Two to three servings would be much healthier. The Daily Value for carbohydrate is 300 grams, or 60 percent of total calories.

■ MINI-SUMMARY

The RDA for carbohydrate is 130 grams/day. The AI for fiber is based on 14 grams/1000 calories. The AMDR for carbohydrate for children (over 1 year) and adults is 45 to 65 percent of total calories. Added sugars should not exceed 25 percent of total kcalories. The Dietary Guidelines and Food Guide Pyramid advise using sugars in moderation. The Food Guide Pyramid recommends three servings of vegetables, 2 servings of fruits, and at least 6 servings of grains, with an emphasis on whole grains.

Ingredient Focus: Grains, Legumes, Pasta, and Thickeners

Carbohydrates are found, to varying degrees, in all the groups of the Food Guide Pyramid. This section will discuss grains, legumes, and pasta. Fruits and vegetables are discussed in Chapter 6, and milk and milk products in Chapter 4.

Grains

Nutritionally, grains such as rice have much to offer (Table 3-7). They are low or moderate in calories, high in starch and fiber (if whole-grain), low in fat, moderate in protein, and full of vitamins and minerals. In addition, they are inexpensive and can be quite profitable. Both traditional grains, such as rice, and newly popular grains, such as quinoa, are being featured more often on the menu, and not simply as a side dish, but also in main dishes. The Food Facts section in this chapter talks about many grains, except rice, which is discussed next.

Grains should be stored in their original packaging or in an airtight container. Store them in a cool, dry area. All grains can be refrigerated, which is a good idea if the kitchen is particularly hot and humid. Refrigeration is very important for whole grains, which do contain some oil that can go rancid. Store whole grains in the refrigerator for up to six months. Grains should be rinsed well before cooking to remove dust and dirt. Quinoa in

TABLE 3-7 Nutrition Information for Cooked Long-Grain Rice (per 1/2-cup serving)

	Brown	Regular—Milled White (enriched)	Parboiled (enriched)	Precooked White (enriched)
Kilocalories	111	131	100	80
Carbohydrate (g)	23	28	22	17
Protein (g)	3	3	2	2
Total fat (g)	1	0.3	0.2	0.1
Saturated fatty acids (g)	0	0	0	0
Cholesterol (mg)	0	0	0	0
Fiber (g)	1.7	0.5	0.5	0.6
Sodium (mg)	3.0	2.0	2	2

Source: U.S. Department of Agriculture.

particular must be thoroughly rinsed, because its natural coating has a bitter taste if not removed.

Rice, perhaps the first grain ever cultivated by humans, is a semiaquatic member of the grass family. Its edible seed is the staple grain for over half the world's population. Brown rice is the whole grain and, of course, more nutritious than white rice. Among white rices, the most nutritious is parboiled or converted rice. Converted rice is a specially processed long-grain rice that has been partially cooked under steam pressure, dried, and then milled to remove the outer hull and bran. The parboiling process results in a grain that is more nutritious and particularly separate and fluffy.

Brown rice has a nutty flavor and chewy texture compared with white rice, which has been milled to remove the bran that distinguishes brown rice. Either rice is classified by its shape: short-grain, medium-grain, or long-grain. Long-grain white rice is four to five times as long as it is wide. Its cooked grains are separate and fluffy, and are used for side dishes, entrées, salads, pilaf, and so on. Medium-grain rice is a little shorter and plumper than long-grain rice. After cooking, the rice is moister and more tender, and has a greater tendency to cling together than long-grains. Both medium-grain and short-grain white rice are good choices for making creamy dishes, such as rice pudding, risotto, molds, or croquettes. The shorter the grain, the more tender and clinging it becomes as it cooks. The boiled rice used in Japanese cooking is short-grain.

There are also numerous specialty rices.

- **Basmati rice** is an aromatic variety of extra-long-grain rice with a nutty flavor that is a very important ingredient in India. It is available both brown and white, and has a firm consistency.
- From Thailand and other parts of Southeast Asia comes **jasmine rice,** a long-grain white rice that somewhat resembles basmati rice. Jasmine rice is excellent in cold salads because it stays fluffy after cooling.
- **Texmati rice** is a cross between basmati and long-grain white rice. It is grown in the United States. It has a nutty flavor and is available both white and brown.
- In Italy, the rice of choice is **arborio rice,** a short-grain rice that is a must for making the classic Italian dish risotto. Arborio rice is very starchy and can absorb a great deal of liquid without becoming soggy, so it is ideal for dishes such as risotto, paella, and jambalaya, which need slow, gentle cooking. There are a number of varieties available.
- **Glutinous rice** is a short-grain rice used in some Chinese and Japanese desserts. It is very starchy and sticks together when cooked. It comes black (unhulled) or white (polished). This rice is used in sushi and many other Asian dishes.
- **Japonica rice** is a Japanese rice that sticks together when cooked. It is a good choice for Asian dishes.

- **Purple rice,** also called black Thai rice, cooks up purple. It is cooked risotto style.

Wild rice is not a true rice, but the seed of a grass that grows wild in the marshes of the Great Lakes region. It is dark brown to almost black in color, and has a nutty flavor. Because true wild rice is so expensive, some kitchens use cultivated wild rice. The cultivated variety is coarser in texture and not quite as flavorful.

Some popular rice dishes include the following:

- **Jambalaya**—a traditional Louisiana rice dish, highly seasoned, and flavored with sausage, ham, seafood, pork, chicken, or other meat
- **Paella**—a traditional Spanish dish of saffron-flavored rice, shellfish, chicken, chorizo, vegetables, and seasonings
- **Pilaf**—a light and fluffy rice dish originating in the Middle East; the rice is often sautéed in oil and then cooked in a broth with onions, various spices, and sometimes raisins or meat
- **Risotto**—a rich and creamy Italian rice dish in which the rice is browned in fat or oil with onions and then cooked in broth, often flavored with Parmesan cheese
- **Arroz con pollo**—a Spanish dish mixing rice with chicken

These dishes are all good examples of how to use other ingredients, or a certain cooking method, to add taste and flavor to grains, which often have a subtle taste alone. Like most grains, rice is prepared by cooking in liquid until tender and the liquid is absorbed. Cooking times for rice, as well as other grains, appear in Table 3-8.

CHEF'S TIPS

- Grains work very well as main dishes when mixed with each other or with lentils. For example, couscous and wheat berries are attractive, as is barley with quinoa. To either dish you could add lentils, vegetables, and seasonings.
- Orzo works well with barley or quinoa.
- Rice and beans is a very popular and versatile dish using grains and legumes. For appearance, mix purple rice with white beans, or wild rice with cranberry beans.
- Don't forget to serve whole-grain cereals as simple as oatmeal, and dress them up with fruit and spices.

Legumes

Legumes include all sorts of dried beans, peas, and lentils. Dried beans are among the oldest of foods and are an important staple for millions of people in other parts of the world. Beans were once considered to be worth their weight in gold—the jeweler's carat owes its origin to a pealike bean on the east coast of Africa.

From a nutritional point of view, legumes are a hit. They are:

- High in complex carbohydrates
- High in fiber
- Low in fat (only a trace, except in a couple of cases)
- Cholesterol-free
- A good source of vitamins and minerals
- Low in sodium

Besides being so nutritious, they are very cost-effective.

Many varieties of beans may be found on the grocery shelf. Here are popular varieties and their uses.

- **Adzuki beans**—These beans are small and reddish-brown in color, with a mild nutty flavor and soft texture. They originally came from China and Japan. They are colorful with rice and in many other dishes as well.
- **Anasazi beans**—Originally grown by Native Americans in the Southwest, these white beans are kidney-shaped and spotted with maroon. They are used in Mexican and Southwestern dishes, such as refried beans.
- **Black beans** (also called turtle beans)—These beans have an oval shape and a black skin. They are used in thick soups, chili, and spicy dishes, and in Central American, South American, and Caribbean cuisine.
- **Black-eyed peas** (also called black-eyed beans or cowpeas)—These beans are small, oval-shaped, and creamy white with a black spot on one side. They are often used as a main-dish vegetable, such as in the classic Southern dish Hoppin' John. Black-eyed peas are excellent in salads, stews, and soups.
- **Cannellini**—These white kidney beans are popular in many Italian dishes. They work well in soups or salad and can be pureed, due to their creamy texture, with herbs, garlic, and other flavorings.
- **Cranberry beans**—These light pink beans with beige spots lose their pink color during cooking. They are popular for baked beans and Italian dishes.
- **Fava beans**—These large, flat brown beans have an earthy flavor and are used in Mediterranean dishes such as falafel.
- **Flageolet**—These flat, oval beans are either green or white and grow in a green pod. They are popular in soups or as a side dish.
- **Garbanzo beans** (also called chickpeas)—These beans are nut-flavored and commonly pickled in vinegar and oil for salads. They are shaped like acorns, beige to yellow in color, and crunchy in texture. They can be used in soups or pasta salads, pureed with tahini and lemon to make hummus, or in Indian-style curry.
- **Great Northern beans**—Larger than, but similar to, pea beans, these beans are used in soups, salads, casserole dishes, and especially in baked beans. They have a mild flavor.

TABLE 3-8 Cooking Information for Grains

Grain	Appearance	Flavor	Soaking Required	Cooking Time	Cups Liquid for Cooking	Cups Yield	Uses	Storage
Amaranth	Golden	Sweet, nutlike	No	25 minutes	2-1/2	3-1/2	Hot cereal, pilaf, in baking, can be popped as a snack	Airtight container, in cool place for many months, otherwise refrigerate for 5 months
Barley, pearl	White-tan	Mild, nutty	No (but will reduce cooking time)	35–40 minutes	3	3-1/2	Soups, casseroles, stews, cooked cereals, side dishes, pilafs	Airtight container— 6–9 months at room temperature
Barley, whole hulled	Brownish gray	Nutty, chewy	Yes	60–90 minutes	3	4	Same as above	Airtight container— 1 month at room temperature, 4–5 months in refrigerator
Buckwheat, whole white	Brown-white	Mild	No	20 minutes	2	2-1/2	Side dishes	Airtight container— 1–2 months, better stored in refrigerator
Buckwheat, roasted (kasha)	Brown	Distinct, nutty, chewy	No	10–15 minutes	2	2-1/2	Soups, side dishes, salads, pilaf, stuffing, hot cereal	Same as above

Corn, whole hominy	Yellow or white	Sweet, creamy texture	No	2-1/2–3 hours	2-1/2	3	Soups, stews, casseroles, hot cereal, puddings, baked goods	Airtight container—1 month at room temperature, 5 months in refrigerator
Corn, hominy grits	Whitish gray	Distinct	No	20–25 minutes	4	3	Hot breakfast cereal	Airtight container, many months at room temperature
Millet	Bright gold color, small	Like corn, crunchy	No	30–35 minutes	2	3	Soups, casseroles, meat loaves, porridge, croquettes, pilaf, salads, stuffing, side dishes	Airtight container, 6 months at room temperature
Oats, steel-cut	Off-white	Mild, pleasant	No	45–60 minutes	2	2	Hot cereal	Airtight container—1 month at room temperature, 6 months in refrigerator
Quinoa	Pale yellow	Nutty	No	12–15 minutes	2	2-1/2	In place of rice	Airtight container in cool place for 1 month, otherwise refrigerate for 5 months
Rice, regular-milled long grain	White	Mild	No	15–20 minutes	2	3	Side dishes, casseroles, stews, soups, stuffing, salads	Airtight container—many months at room temperature

TABLE 3-8 (continued)

Grain	Appearance	Flavor	Soaking Required	Cooking Time	Cups Liquid for Cooking	Cups Yield	Uses	Storage
Rice, regular-milled, medium or short grain	White	Mild	No	20–25 minutes	1-1/2	3	Same as above	Same as above
Rice, parboiled	White	Mild	No	20–25 minutes	2–2-1/2	3–4	Same as above	Same as above
Rice, brown	Tan-brown	Nutty	No	40–50 minutes	2-1/2	4	Same as above	Airtight container—1 month at room temperature, 6 months in refrigerator
Rice, wild	Dark brown	Nutty	No (but rinse it)	30–45 minutes	3	3-1/2–4	Side dishes, stuffing, casseroles	Airtight container–many months at room temperature
Rice, basmati	White	Nutty, spicy	Yes (rinse also)	25 minutes	1-1/2	3	Side dishes, casseroles	Airtight container–1 month at room temperature, 6 months in refrigerator

Jasmine rice	White	Aromatic	No	15–20 minutes	2	3	Side dishes, casseroles, stews, soups	Same as above
Texmati Rice	White	Nutty	No	15–20 minutes	2	3	Same as above	Same as above
Rye, whole berries	Brown, oval	Distinct rye flavor	No	1-1/2 hours	3	3	Hot cereal, side dishes	Airtight container — 1 month at room temperature, 5 months in refrigerator
Wheat, bulgur	Dark brown	Nutty	No	20–25 minutes	2-1/2	2	Salads, soups, breads, desserts, with rice, meat dishes, in place of rice pilaf, stuffing	Airtight container in cool place, or refrigerator for 5–6 months
Wheat, whole berries	Deep brown	Nutty, crunchy	Yes (1 cup to 3-1/2 cups cold water)	1 hour	3	2	Salads, meat loaves, croquettes, breads, side dishes	Airtight container in cool place up to 1 month, up to 5 months in refrigerator

- **Kidney beans**—These beans are large and have a red color and kidney shape. They are popular for chili con carne and red beans and rice, and add zest to salads, soups, and Mexican dishes.
- **Lima beans**—Although not widely known as dry beans, lima beans make an excellent main-dish vegetable and can be used in casseroles. They are broad and flat and come in three sizes: large, regular, and baby.
- **Mung beans**—These small, ground beans are green or yellow and popular in Asian cuisine.
- **Navy beans**—Also called pea beans, navy beans are white beans that are smaller than Great Northern beans. They are often used in soups and baked beans.
- **Pinto beans**—These beans are of the same species as the kidney and red beans. Beige-colored and speckled, they turn a uniform pink when cooked. They are used often in salads, refried beans, and chili.
- **Red and pink beans**—Pink beans have a more delicate flavor than red beans. Both are used in many Mexican dishes and chili.

Beans should be of similar color and size. When beans get old, they lose their color. Look for beans and other legumes that have no obvious defects, such as cracks or pinholes that may indicate insect damage, and make sure there is no field debris (such as stones or twigs) in the bag. Once a package of beans has been opened, transfer them to an airtight container and store in a cool, dry spot—but not in the refrigerator. They can be stored at room temperature for one year.

Before soaking beans to rehydrate them, wash them carefully and pick over to remove any foreign particles. After soaking beans, be sure to discard the soaking water. Substances in the soaking water contribute to indigestion and can cause flatulence. Beans, like other legumes, are simmered in liquid (see Table 3-9 for cooking times). Casseroles, stews, stuffings, sandwich spreads, salads, and soups can all be prepared with beans as the central ingredient.

Dry peas are an interesting and versatile food that adds variety to meals. Dry peas may be green or yellow and may be bought either split or whole. Whole dry peas are available, but many cooks prefer to start with the half-circles of split peas. Green dry peas have a more distinct flavor than yellow dry peas. Split peas are popular in soups and are also great in side dishes, such as salads and pilafs.

Dry peas are served in many ways—with grains or as side dishes, or they can be puréed and made into dips, patties, croquettes, stuffed peppers, and even soufflés. They go well with vegetables, pasta, fish, meat, poultry, and more. They can go in soups, salads, side dishes, main dishes, and casseroles.

The lentil is an Old World legume that is disc-shaped and about the size of a pea. Thousands of years old, lentils were perhaps the first of the convenience foods—they do not require soaking. Lentils come in colors such as green, red, black, or brown. Lentils may come whole or split (split lentils cook faster).

When cooking lentils, you can add seasonings or flavorful ingredients to the cooking water, since lentils absorb flavors well. Do not add acid ingredients, such as tomatoes or lemon juice, until later, since they will slow cooking. Salt should be added at the end of the cooking time for the same reason. Drain, if necessary, and they are ready. Cooked lentils may be stored in liquids such as broth, fruit juice, or salad dressing to boost flavor.

Lentils are an excellent partner with many foods. They make wonderful side dishes and go well in soups, stews, sauces, stuffings, and salads. Consider them as you would potatoes or rice. Lentil purée, not unlike peanut butter, can be used on bread or muffins, and in sandwiches, dips, spreads, Mexican dishes, and vegetable fillings.

CHEF'S TIPS

- When choosing legumes for a dish, think color and flavor. Make sure the colors you pick will look good when the dish is complete. Also think of other ingredients you will use for flavor. In a salad, for example, black-eyed peas (black and white) go well with flageolets and red adzuki beans. To add a little more color and develop the flavor, you might add chopped tomatoes, fresh cilantro (Chinese parsley), and haricots verts (green beans) or fresh corn.
- Bigger beans, such as gigante white beans, hold their shape well and lend a hearty flavor to stews, ragouts, and salads.
- Chickpeas can be puréed, as in hummus, and used as a dip, a spread, or a sandwich filling; layered with grilled vegetables; or as a filling for pasta, crêpes, or twice-baked potatoes.
- You can cook together several types of beans, such as cranberry, turtle, and white beans, in stock flavored with herbs, vinegar, and carrots.
- A number of dried beans are also available fresh: cannellini, cranberry beans, fava beans, black-eyed peas, flageolets, lima beans, mung beans, and soybeans. If you can afford them, they are excellent products. They are plumper in size and have a fresher flavor than dry beans that you rehydrate.
- Use whole lentils, such as black or French green lentils, in grain dishes or salads because whole lentils hold their shape better. Use split lentils, such as brown, red, or yellow lentils, in soups, where they help thicken the liquid and shape is not as important.

Pasta

When we hear the word *pasta,* we relate it quickly to Italian cuisine. Pasta has been eaten for over 5,000 years and is very closely associated with Italian cooking. Pasta (the name is from the Italian word for paste) is an edible dough made from flour and water that is rolled and cut into one of over 150 pasta shapes found in the United States (Figure 3-5).

TABLE 3-9 Cooking Information for Legumes

Bean, Pea, or Lentil	Size/ Shape/ Color	Flavor	Soaking Required	Cooking Time	Cups Liquid for Cooking	Yield[a]	Uses
Adzuki beans	Small, reddish brown	Nutty, sweet	Yes	1–1-1/2 hours	3	2	Rich, Asian cooking
Anasazi beans	Kidney shaped, white with maroon	Rich, meaty	Yes	2 hours	3	2	Chili and other Mexican dishes
Black beans (turtle beans)	Small, pea-shaped, black	Full, mellow	Yes	1-1/2 hours	4	2	Mediterranean cuisine, soups (black bean soup), chilis, salads, with rice
Black-eyed peas (cowpeas, black-eyed beans)	Small, oval, creamy white with black spot	Earthy, absorb other flavors	No	50–60 minutes	3	2	Casseroles, with rice, with pork, Southern dishes
Chickpeas (garbanzo beans, ceci)	Round, tan, large	Nutty	Yes	2-1/2 hours	4	4	Salads, soups, casseroles, hors d'oeuvres, hummus and other Middle East dishes
Fava beans, whole	Large, round, flat, off-white or tan	Full	Yes	3 hours	2-1/2	4	Soups, casseroles, salads
Great Northern beans	Large, oval, white	Mild	Yes	1-1/2 hours	3-1/2	2	Soups, casseroles, baked beans, and mixing with other varieties

Kidney beans	Large, kidney-shaped, red or white (red is much more common)	Rich, meaty, sweet	Yes	1–1-1/2 hours	3	2	Chili, casseroles, salads, soups, a favorite in Mexican and Italian cooking
Lentils	Small, flat, disk-shaped, green, red, or brown, split or whole	Mild, earthy	No	30–45 minutes	2	2-1/4	Soups, stews, salads, casseroles, stuffing, sandwiches, spreads, with rice
Lima beans	Flat, oval, cream or greenish, large or baby size	Large—full Baby—mild	Yes	1-1/2 hours (large) 1 hour (baby)	2	1-1/4	Soups, casseroles, side dishes
Navy beans (pea beans)	Small to medium, round to oval, white	Mild	Yes	1-1/2 hours	3	2	Baked beans, soups, salads, side dishes, casseroles
Peas, split	Small, flat on one side, green or yellow	Rich, earthy	No	30 minutes	3	2-1/4	Soups, casseroles
Peas, whole	Small-medium, round, yellow or green	Rich, earthy	Yes	40 minutes	3	2-1/4	Soups, casseroles, Scandinavian dishes
Pinto beans	Medium, kidney-shaped, pinkish brown	Rich, meaty	Yes	1-1/2 hours	3	2	A favorite for chili, refried beans, and in other Mexican cooking
Pink beans	Medium, oval, pinkish brown	Rich, meaty	Yes	1 hour	3	2	Popular in barbecue-style dishes
Soybeans	Medium, oval-round, creamy yellow	Distinctive	Yes	3-1/2 hours or more	3	2	Soups, stews, casseroles

[a] From 1 cup of uncooked beans, peas, or lentils.

Figure 3-5
Forms of pasta. Courtesy National Pasta Association

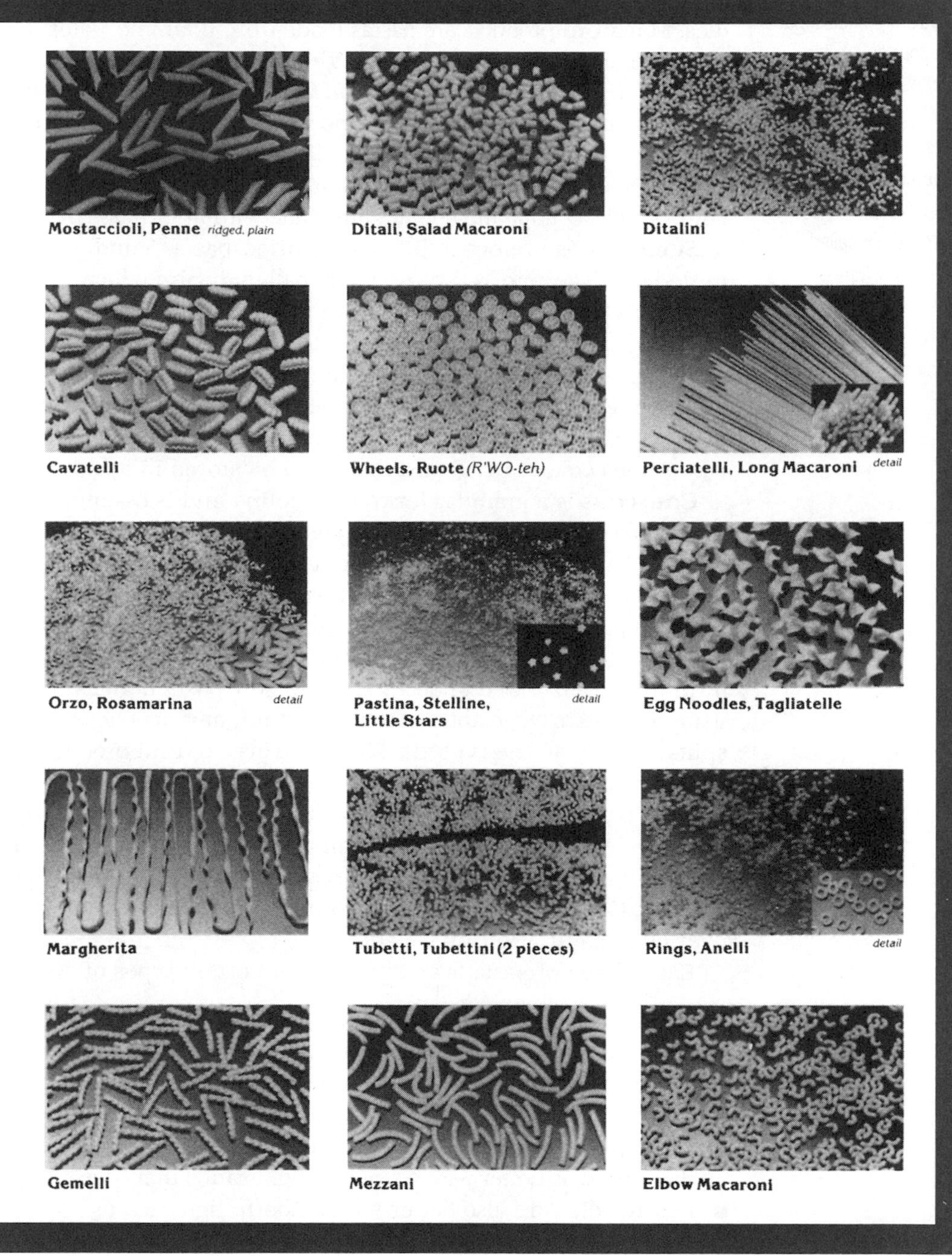
Mostaccioli, Penne ridged, plain
Ditali, Salad Macaroni
Ditalini
Cavatelli
Wheels, Ruote (R'WO-teh)
Perciatelli, Long Macaroni detail
Orzo, Rosamarina detail
Pastina, Stelline, Little Stars detail
Egg Noodles, Tagliatelle
Margherita
Tubetti, Tubettini (2 pieces)
Rings, Anelli detail
Gemelli
Mezzani
Elbow Macaroni

Macaroni—Pastas made from flour and water.

Noodles—Pastas made from flour, water, and egg solids.

Semolina—The roughly milled endosperm of a type of wheat called durum wheat.

Couscous—A granular form of semolina; like a tiny pasta.

Pasta may be dried or fresh. Dried pasta includes both macaroni and noodles. **Macaroni** products are pastas made from flour and water. These include spaghetti, elbow macaroni, lasagne, ziti, and other shapes. Many are available made with whole-wheat flour, or half whole-wheat flour and half white flour. **Noodles** are also made from flour and water, but, by law, must contain 5.5 percent egg solids. Noodles are usually flat, like a ribbon, and come in different widths. Noodles can also contain flour made from products such as legumes. For example, cellophane noodles are made with flour made from mung beans.

Semolina is preferred for making dried pasta. Semolina is the roughly milled endosperm of a type of wheat called durum wheat. Durum wheat is known as a very hard wheat, meaning that it has a high protein content. Semolina is used almost exclusively for making pasta. Less expensive pasta products are made from a softer flour. High-quality pasta should be brittle and yellow in color, and holds its shape well when cooked. Poor-quality pasta is often a whitish-gray color, and becomes soft and loses its shape when it is cooked. Dried pasta needs to be stored in a cool, dry place.

Couscous is a granular form of semolina and is essentially a tiny pasta. It is cooked by soaking and then steaming. Couscous is a staple of North African and some Middle Eastern cuisines. It is also available precooked, like instant rice. Couscous is often used in place of rice.

Fresh pasta is more perishable and expensive than dried pasta. It is available as dough or in shapes. For example, ravioli is a soft dough, stuffed with a filling. Spaetzle and gnocchi are also soft pasta doughs. Flavored pastas usually contain vegetables, such as red tomato, artichoke, beet, carrot, or spinach (only a little is used). They are very colorful products, and the vegetables can add a subtle flavor.

Most pasta is high in starch, low in fat, and moderate in protein content. Pasta is cooked in a generous amount of boiling water. Dried pasta requires a much longer cooking time than fresh pasta, which cooks in a few minutes at most. Cook pasta until it is al dente, or firm to the bite. Cooked pasta is over 60 percent water by weight.

Each shape of pasta is appropriate for certain types of dishes.

- Elbow macaroni is good in salads and soups because it retains its shape.
- Tube pastas (such as elbow macaroni) or pastas with a hollow space (such as shells) work well with meat or vegetable sauces, as they trap the sauce in their spaces.
- Fresh pasta, because it is softer in texture and absorbs sauce more readily than dried, is better with a smooth, light sauce that coats the pasta evenly.
- Flat noodles are also better with smooth, light sauces.
- Delicate pasta should be served with delicate sauces, and hearty pasta is best with hearty sauces.

Many shapes of pasta are suitable for filling, and all kinds of fillings may be used. Here is another challenge to the cook's ingenuity. Pasta sauces are discussed in Chapter 8.

Asian noodles are another type of pasta that is also excellent in main dishes and side dishes. **Bean thread vermicelli,** also called cellophane noodles, is a very starchy, thin noodle made from mung bean flour. It works well in soups, stews, and ragouts. **Rice vermicelli** is also very starchy and can be used much like bean thread vermicelli. When soaked in water, it can be used in stir-fries. **Soba noodles** are flat, thin, brownish gray noodles made from buckwheat flour. Whereas the vermicelli noodles are almost all starch, soba noodles are high in protein. They are served in hot broth, or cold with a dipping sauce.

CHEF'S TIPS

- When buying dried pasta, always buy a high-quality product.
- Pasta and beans work well together, such as in Pasta e Fagioli (see recipe on page 332).
- Pasta is easy to prepare, comes in many shapes and sizes, and can serve as the base for a wide variety of entrées and side dishes.
- Pasta dishes can include many vegetables. For example, spray a nonstick pan with olive oil and sauté roasted peppers, oven-dried tomatoes, lightly steamed broccoli florets, and roasted fresh garlic. Add a reduced broth with herbs, then add cooked rigatoni and toss the pasta with the vegetables. This can serve as the base for a 4-ounce serving of fish, chicken, veal, or pork paillard (a cutlet that has been pounded flat and then grilled or sautéed).

Thickeners

As mentioned earlier in the chapter, starches come in two forms: amylose and amylopectin. Starches behave differently when cooked depending on the amount of amylose or amylopectin they contain. Grain starches, such as wheat or cornstarch, are high in amylose, while the root starches (arrowroot, tapioca) are lower in amylose and higher in amylopectin.

Grain starches (wheat flour, cornstarch, oat flour) have several characteristics in common:

- They are clear during gelatinization but slightly cloudy once cooled. (When flour is used, the product is cloudy during cooking due to other components of the flour.)
- Such starches thicken just below the boiling point and maintain consistency during holding.
- Products made with them are very firm when cooled off.
- Foods made with grain starches can be cooled and reheated without thinning as long as the food product is not stirred when cool.
- These starches cannot be used in a food product that will be frozen and rethawed due to a complete loss of quality.

Root starches (arrowroot, tapioca) have a different set of characteristics:

- They stay clear when hot or cold.
- Such starches thicken quicker than grain starches but also thin out if overheated or overstirred.
- Foods made with them are not nearly as firm as those made with grain starches when cooled off.
- Root starches work well in food products that are frozen and thawed.

Modified starches and waxy cornstarch (99 percent amylopectin) are used extensively by the frozen-food industry.

These are important considerations when choosing a thickener. For example, when making a stir-fry, you could use cornstarch because it is clear when hot. When making a filling for a cream pie, you could also use a high-amylose starch because it will be firm to cut and the opaqueness will not affect the pie's appearance when cold. However, if you are making a peach pie, you don't want the opaqueness, so you might choose a root starch. When cooking foods that will be frozen and thawed, be sure to pick a root starch.

Other thickeners to consider are purées of starchy vegetables, legumes, and ground nuts.

Check-Out Quiz

1. Match the food below with the nutrient(s) it is rich in.

Food	*Nutrient*
White bread	Added sugars
Whole-wheat bread	Natural sugars
Apple juice	Fiber
Baked beans	Starch
Milk	
Bran flakes	
Sugar-frosted oats	
Cola drink	
Broccoli	

2. Honey is better for you than sugar.
   a. True
   b. False
3. Carbohydrates spare protein from being burned for energy so that protein can be used to build and repair the body.
   a. True
   b. False
4. Maltosc is made up of galactose and glucose.
   a. True
   b. False

5. Mannitol is an example of a sugar alcohol.
   a. True
   b. False
6. Obesity is more closely linked to diabetes than any other health problem.
   a. True
   b. False
7. All types of carbohydrates contribute to dental decay.
   a. True
   b. False
8. Seven-grain bread is an example of a whole-grain product.
   a. True
   b. False
9. Because insoluble fiber aids digestion and adds bulk to stool, it hastens passage of fecal material through the gut, thus helping to prevent or alleviate constipation.
   a. True
   b. False
10. Legumes are low in fiber and sodium.
   a. True
   b. False

Activities and Applications

1. Carbohydrate Basics

Check off under the appropriate column(s) when you think the food contains a significant amount of sugar, starch, and/or fiber.

Food	Sugar	Starch	Fiber
1. Hamburger	______	______	______
2. Chicken wing	______	______	______
3. Flounder	______	______	______
4. Boiled egg	______	______	______
5. American cheese	______	______	______
6. Sour cream	______	______	______
7. White bread	______	______	______
8. Whole-wheat bread	______	______	______
9. Chocolate cake	______	______	______
10. Macaroni	______	______	______
11. Brown rice	______	______	______
12. Split peas	______	______	______
13. Peanuts	______	______	______
14. Fresh orange	______	______	______
15. Broccoli	______	______	______

2. Self-Assessment

List how many servings of the following foods you normally eat daily.

Added Sugars	Number of Servings
Sugar in coffee or tea	______
Sweetened beverages	______
Sweetened breakfast cereals	______
Candy	______
Commercially made baked goods, including cakes, pies, cookies, and doughnuts	______
Jam, jelly, pancake syrup	______
Complex Carbohydrates	**Number of Servings**
Breads and rolls	______
Ready-to-eat and cooked cereals	______
Pasta, rice, other grains	______
Dried beans and peas	______
Potatoes	______
Fruits	______
Vegetables	______

How do you rate? Do you get at least six servings per day of breads, rolls, cereals, pasta, rice, and other grains? Is at least one serving a whole grain? Do you get a daily serving of dried beans and peas? Do you get at least five servings per day of fruits and vegetables combined? Compare the number of servings you have daily of foods high in refined sugars and those high in complex carbohydrates. You should not be choosing too many foods from the "refined sugar" part of the chart until you eat enough servings of grains, fruits, and vegetables. The idea is to push complex-carbohydrate intake and minimize refined sugars.

3. Whole Grain or Refined Grain?

Read the following ingredient labels for breads. Which one is white bread and which one is whole-grain bread?

#1 Made from: Unbromated unbleached enriched wheat flour, corn syrup, partially hydrogenated soybean oil, molasses, salt, yeast, raisin juice concentrate, potato flour, wheat gluten, honey, vinegar, mono- and diglycerides, cultured corn syrup, unbleached wheat flour, xanthan gum, and soy lecithin.

#2 Made from: Stone-ground whole-wheat flour, water, high-fructose corn syrup, wheat gluten, yeast, honey, salt, molasses, partially hydrogenated soybean oil, raisin syrup, soy lecithin, mono- and diglycerides.

4. How Many Teaspoons of Sugar?

One teaspoon of sugar weighs 4 grams. Determine how many teaspoons of sugar are in each of the following foods, as described on their nutrition labels. Which food contains more sugar? Which food contains more fiber?

NUTRITION FACTS	**NUTRITION FACTS**
Amount per serving	Amount per serving
Calories 230	Calories 140
Calories from Fat 140	Calories from Fat 20
Total Fat 16 g	Total Fat 2.5 g
Saturated 6 g	Saturated 0.5 g
Polyunsaturated 1 g	Polyunsaturated 1.0 g
Monounsaturated 7 g	Monounsaturated 0.5 g
Cholesterol 74 g	Cholesterol 0 g
Sodium 180 mg	Sodium 180 mg
Total Carbohydrate 28 g	Total Carbohydrate 21 g
Dietary Fiber 0 g	Dietary Fiber 5 g
Sugars 24 g	Sugars 4 g
Protein 21 g	Protein 8 g

In a group of two people, determine the number of teaspoons of sugar in a soft drink or other beverage by using the number of grams of sugar on the Nutrition Facts label. Next, measure out the amount of sugar and place it in a clear glass next to the beverage. Check out how much sugar is in the drinks your other classmates analyzed.

5. Alternative Sweetener Sleuth

Check your refrigerator and cupboards to see what kind of foods, and how many, contain nonnutritive sweeteners. Look for the words *Equal, aspartame, saccharin, acesulfame potassium, Sunette, Sweet One, sucralose, Splenda,* and *neotame.*

6. Baking with Alternative Sweeteners

Use the websites for Sweet 'N Low and Splenda to get some recipes using alternative sweeteners.

www.sweetnlow.com
www.splenda.com

The first website is for the company that manufactures Sweet 'N Low, the alternative sweetener containing saccharin. The second website is for the company that makes Splenda, the alternative sweetener containing sucralose.

On their home pages, click on "Recipes" and choose a recipe. How are these baking recipes different from recipes using sugar? If possible, make a recipe, such as brownies, using one of these alternative sweeteners and then compare it in taste, texture, and appearance to the same product made with sugar.

Nutrition Web Explorer

Joslin Diabetes Center www.joslin.org/education/library/index.shtml

Joslin Diabetes Center is an excellent site to learn almost anything about diabetes. At their library's home page, click on "Meal Planning Using Carbohydrate Counting." Read through it completely, then complete the exercise "Meal Planning Practice."

National Institutes of Health on Hypoglycemia www.nlm.nih.gov/medlineplus/hypoglycemia.html

On this government website, click on "Hypoglycemia- Interactive Tutorial" to learn more about this topic.

National Institutes of Health on Lactose Intolerance www.niddk.nih.gov/health/digest/pubs/lactose/lactose.htm

Use this informative site to learn more about the causes and treatment of lactose intolerance. Find out how many grams of lactose are in reduced-fat milk and compare that to yogurt.

Food Facts *Amber Waves of Grain*

The variety of grains is astonishing. Read on to learn more about these high-carbohydrate foods.

Barley. When the term *barley* is used in a recipe, it usually refers to pearl barley, a white variety that has had the inedible husk or hull, germ, and bran removed. Pearl barley can be boiled and used in soups, casseroles, stews, stuffings, cooked cereals, as a side dish, or as pilaf. Cooked barley can be sautéed with vegetables such as onions and mushrooms. Its taste is mild and nutty, and its texture is chewy.

Buckwheat. Although buckwheat has many grainlike characteristics, it is from an entirely different family and is actually a fruit. Two forms of buckwheat are available for purchase: whole white buckwheat and roasted hulled buckwheat, also called kasha. Kasha can be purchased whole or ground. Roasting gives the buckwheat kernels a distinct, nutty flavor. Kasha is best mixed with less flavorful grains. Kasha can be used in soups, stuffings, side dishes, and salads. Whole white buckwheat has a mild flavor and can be used to replace rice or pasta.

Millet. Although the millet raised in the United States is used mostly to feed animals and birds, millet has been eaten by people in other parts of the world since Old Testament times. Millet is an important dietary component for many Africans, Indians, and Chinese. No wonder; it contains a high-quality protein. Millet for cooking has had the inedible hull (which birds love) removed, as well as an outer bran layer that is also inedible. It can be boiled and used in soups, casseroles, meat loaves, porridge, croquettes, pilaf, salads, stuffings, or as a side dish with chopped onions and fresh herbs, such as basil.

Oats. Oats are a unique grain: When they are milled, only the inedible hull is removed, and the bran and germ are left with the kernel. Whichever form of oats you buy, you are getting whole-grain nutrition. Rolled oats, or "old-fashioned oats," are made by cutting up raw oats into a product that is then steamed, shaped into flakes, and dried. Rolled oats require only five minutes' cooking time. Quick oats start out as rolled oats but are sliced finer and slightly precooked, so they cook quicker—in about one minute. Instant oats are cut even smaller and result in a product that needs only to be mixed with boiling water.

Rye. Although rye is known most for its flour, you can buy whole rye berries that can be cooked in liquid into a hot cereal or side dish. Rye flakes, the equivalent of rolled oats, are also available and can be used as a hot cereal or side dish. They have a tangy taste.

Wheat. Wheat, the most important food grown in the world, is a rich source of nourishment. It is, of course, used to make flours, breads, cereals, and pastas.

Whole-wheat grains that have been steamed, dried, and ground into small pieces are called **bulgur.** Depending on how it is processed, some or all of the bran may be removed. Check the bulgur you want to purchase to see if the dark brown bran is still on. Bulgur has a nutty flavor that is excellent by itself or mixed with rice. Its uses are numerous—from salads to soups, from breads to desserts. It is also a nutritious extender and thickener for meat dishes and soups.

Wheat bran is available either processed (as in breakfast cereal) or unprocessed, and is used as an ingredient in cooking or baking. It makes a high-fiber addition to baked goods, such as breads and muffins, and can be substituted for bread crumbs in most recipes.

Wheat germ is separated from the wheat grain and can be purchased either toasted or raw. Wheat germ has a nutty, crunchy texture and can be added to cereal, pancakes, baked goods, casseroles, salads, and breading.

Wheat berries, the actual whole-wheat kernels, are also available cracked, as cracked wheat, in coarse, medium, or fine qualities. Cracked wheat is particularly popular in breads.

Amaranth. Amaranth is one of the newer grains to arrive on the market, yet it has been around for at least 5,000 years. An important part of the Aztec diet, it contains a high-quality protein and is rich in calcium. Amaranth seeds are tiny and yellow-brown in color. A spicy grain with a slightly peppery taste, amaranth seeds can be cooked with other grains or to make pilaf. Amaranth cooks up soupy instead of fluffy.

Quinoa. Quinoa is a tiny, pale yellow seed that is technically not a cereal grain but a dried fruit. Whereas amaranth was popular with the Aztecs, quinoa was a staple of the Incas in Peru. Like amaranth, quinoa is rich in complete protein (unlike other plant foods) and calcium. Also, quinoa, unlike grains in general, contains an appreciable amount of oil (7 grams per 8 ounces). The best-quality quinoa is altiplano quinoa from Bolivia or Peru. It can be used in any recipe to replace rice.

Hot Topic Alternative Sweeteners

The introduction of diet soft drinks in the 1950s sparked the widespread use of alternative sweeteners. Alternative sweeteners, also called artificial sweeteners, contain either no or very few kcalories. Several different ones will be discussed here.

Saccharin. Saccharin, discovered in 1879, has been consumed by Americans for more than 100 years. Its use in foods increased slowly until the two world wars, when its use increased dramatically due to sugar shortages. Saccharin is about 300 times sweeter than sucrose and is excreted unchanged directly into the urine. It is used in a number of foods and beverages, and when combined with aspartame, its sweetness is intensified. Saccharin by itself has a bitter aftertaste. It is sold in liquid, tablet, packet, and bulk form.

In 1977, the Food and Drug Administration (FDA), which regulates the use of food additives, proposed a ban on its use in foods and allowed its sale as a tabletop sweetener only as an over-the-counter drug. This proposal was based on studies that showed the development of urinary bladder cancer in rats fed the equivalent of 800 cans of diet soft drinks a day. The surge of public protest against this proposal (there were no other alternative sweeteners available at that time) led Congress to postpone the ban. In 2000 the National Toxicology Program decided that saccharin should no longer be considered a cancer-causing chemical. In 2001 the U.S. Congress repealed the warning labels that had been required on saccharin-containing products.

Aspartame. In 1965, aspartame, a low-calorie sweetener, was also discovered accidentally. After being tested in more than 100 scientific studies in animals and humans, it was approved by the FDA in 1981. Aspartame is marketed in the United States under the brand name NutraSweet and as Equal tabletop sweetener. It is 200 times sweeter than sucrose and has an acceptable flavor with no bitter aftertaste.

Aspartame is made by joining two protein components, aspartic acid and phenylalanine, and a small amount of methanol. Aspartic acid and phenylalanine are building blocks of protein. Methanol is found naturally in the body and in many foods, such as fruit and vegetable juices. In the intestinal tract, aspartame is broken down into its three components, which are metabolized in the same way as if they had come from food. Aspartame contains 4 kcalories per gram, but so little of it is needed that the calorie content is negligible.

Aspartame is used as a tabletop sweetener, to sweeten many prepared foods, and in simple recipes that do not require lengthy heating or baking. Aspartame's components separate when heated over time, resulting in a loss of sweetness. It is best used at the end of cooking. Aspartame can be found in diet soft drinks, powdered drink mixes, cocoa mixes, pudding and gelatin mixes, frozen desserts, and fruit spreads and toppings. If you drink canned diet soft drinks, chances are they are sweetened with aspartame. Fountain-made diet soft drinks are more commonly sweetened with a blend of aspartame and saccharin, because saccharin helps provide increased stability.

The stability of aspartame in liquid in storage, the safety of the products of its metabolism, and symptoms possibly related to its use have provoked concerns and much research. Available evidence suggests that aspartame consumption is safe over the long term and is not associated with serious health effects.

The FDA uses the concept of an Acceptable Daily Intake (ADI) for many food additives, including aspartame. The ADI represents an intake level that, if maintained each day throughout a person's lifetime, would be considered safe by a wide margin. The ADI for aspartame has been set at 50 milligrams per kilogram of body weight. To take in the ADI for a 150-pound adult, someone would have to drink twenty 12-ounce cans of diet soft drinks daily.

The only individuals for whom aspartame is a known health hazard are those who have the disease phenylketonuria (PKU), because they are unable to metabolize phenylalanine. For this reason, any product containing aspartame carries a warning label. Some other people may also be sensitive to aspartame and need to limit their intake.

Acesulfame-K. In 1988, the FDA approved a new noncaloric sweetener, acesulfame potassium, or acesulfame-K, for use in dry food products and as a powder or tablet to be used as a tabletop sweetener. It is marketed under the brand names Sunette and Sweet One tabletop sweetener. It is about as sweet as aspartame but is more stable and can be used in baking. Acesulfame-K is approved for use in soft drinks, chewing gums, candy, dry beverage mixes, gelatins, puddings, and baked goods. One major beverage maker mixes acesulfame-K with aspartame to sweeten one of its diet sodas. Its taste is reportedly clean and sweet, with no aftertaste in most products.

Acesulfame-K passes through the digestive tract unchanged. The sweetener is used in over 50 countries, including France, Britain, and Russia.

Sucralose. After reviewing more than 110 animal and human-safety studies conducted over 20 years, the FDA approved sucralose in the late 1990s for use as a tabletop sweetener and in soft drinks, baked goods, gum, syrup, gelatins, frozen dairy desserts, and various products made with fruit.

Known by its trade name, Splenda, sucralose is 600 times sweeter than sugar. It tastes like sugar because it is made from table sugar. But it cannot be digested, so it adds no calories to food. Because sucralose is so much sweeter than sugar, it is bulked up with maltodextrin, a starchy powder, so it will measure more like sugar. It has a good shelf life and doesn't degrade when exposed to heat. Numerous studies have shown that it does not affect blood glucose levels, making it an option for people with diabetes.

Neotame. In 2002, the Food and Drug Administration approved neotame for use as a general-purpose sweetener in a wide variety of food products. Neotame is a high-intensity sweetener that is manufactured by the NutraSweet Company, the same company that manufactures aspartame. Depending on its food application, neotame is about 7,000 to 13,000 times sweeter than sugar. It is a white crystalline powder that is heat stable and can be used as a tabletop sweetener as well as in cooking. Examples of uses for which neotame has been approved include baked good, nonalcoholic beverages (including soft drinks), chewing gum, confections and frostings, frozen desserts, gelatins and puddings, jams and jellies, processed fruits and fruit juices, toppings, and syrups. Neotame has a clean, sweet taste in foods.

In addition to these alternative sweeteners, several more may appear on supermarket shelves in the future.

Alitame. Alitame (the brand name is Aclame) is a sweetener made from amino acids (parts of proteins). It is 2,000 times sweeter than sucrose. The manufacturer has petitioned the Food and Drug Administration to use it in a variety of food and beverages. It is approved for use in Mexico, Australia, and China.

Cyclamate. Discovered accidentally in 1937, cyclamate was introduced into beverages and foods in the early 1950s. By the 1960s it dominated the noncaloric sweetener market. It is 30 times sweeter than sucrose and is not metabolized by most people. Cyclamate was banned in 1970 after studies showed that large doses of it, given with saccharin, were associated with increased risk of bladder cancer. Cyclamate is still banned in the United States but is approved and used in more than 50 other countries worldwide. Cyclamate is again under consideration by the Food and Drug Administration for use in specific products, such as tabletop sweeteners and nonalcoholic beverages. It is stable at hot and cold temperatures and has no aftertaste.

D-Tagatose. This product is made from lactose. It is about as sweet as sucrose but has about half the calories. D-tagatose is awaiting approval by the Food and Drug Administration.

Stevioside. Stevioside is an extract from the leaves of stevia, a South American plant. The product is 300 times sweeter than sucrose. It is used in some countries, such as Japan, but a review by the Food and Drug Administration concluded that there were some concerns, so it did not approve its use in the United States. Similar conclusions were reached in Canada and the European Union. You can buy this product as a dietary supplement because supplements are not required to submit to testing before being put on the market.

Thaumatin. Thaumatin (brand name Talin) is a mixture of sweet-tasting proteins from a West African fruit. It is about 2,000 to 3,000 times sweeter than sucrose. It has a licoricelike aftertaste and cannot be used in products that are heated. It is approved for use in the European Union and Japan.

It is important to have a variety of nonnutritive sweeteners in the marketplace, especially for people with diabetes. Though artificially sweetened products are not magic foods that will melt pounds away, they can be a helpful part of an overall weight-control program that includes exercise and a moderate diet.

Chapter 4

Lipids: Fats and Oils

Lipid—A group of fatty substances, including triglycerides and cholesterol, that are soluble in fat, not water, and that provide a rich source of energy and structure to cells.
Fat—A lipid that is solid at room temperature.
Oil—A lipid that is usually liquid at room temperature.
Triglyceride—The major form of lipid in food and in the body; it is made of three fatty acids attached to a glycerol backbone.

The word *fat* is truly all-purpose. We use it to refer to the excess pounds we carry, the blood component that is associated with heart disease, and the greasy foods in our diet that we feel we ought to cut out. To be more precise about the nature of fat, we need to look at fat in more depth.

To begin, **lipid** is the chemical name for a group of compounds that includes fats, oils, cholesterol, and lecithin. Fats and oils are the most abundant lipids in nature and are found in both plants and animals. A lipid is customarily called a **fat** if it is a solid at room temperature, and it is called an **oil** if it is a liquid at the same temperature. Lipids obtained from animal sources are usually solids, such as butter or beef fat, whereas oils are generally of plant origin. Therefore, we commonly speak of animal fats and vegetable oils, but we also use the word *fat* to refer to both fats and oils, which is what we will do in this chapter.

Like carbohydrates, lipids are made of carbon, hydrogen, and oxygen. Unlike most carbohydrates, lipids are not long chains of repeating units. Most of the lipids in foods (over 90 percent), and also in the human body, are in the form of **triglycerides.** Therefore, when we talk about fat in food or in the body, we are really talking about triglycerides. This chapter will help you to:

- Describe lipids and list their functions in foods and in the body
- Describe the relationship between triglycerides and fatty acids
- Define *saturated, monounsaturated,* and *polyunsaturated fats* and list foods in which each are found
- Describe trans fatty acids and give examples of foods in which they are found
- Identify the two essential fatty acids, list their functions in the body, and give examples of foods in which they are found
- Define *cholesterol* and *lecithin,* list their functions in the body, identify where they are found in the body, and give examples of foods in which they are found
- Describe how fats are digested, absorbed, and metabolized
- Discuss the relationship between lipids and health conditions such as heart disease and cancer
- State recommendations for dietary intake of fat, saturated fat, trans fat, monounsaturated fat, polyunsaturated fat, and cholesterol
- Discuss the purchasing, storage, preparation, cooking, and menuing of meat, poultry, fish, and shellfish
- Describe rancidity
- Distinguish between the percentage of fat by weight and the percentage of kcalories from fat
- Calculate the percentage of kcalories from fat for a food item

Functions of Lipids

Fats have many vital purposes in the body, where they account for 15 to 25 percent or more of your weight. Fat is an essential part of all cells. At least 50 percent of your fat stores are located under the skin, where fat provides insulation (because fat doesn't conduct heat well), a cushion around critical organs (like shock absorbers), and optimum body temperature in cold weather.

Most cells store only small amounts of fat, but specific cells, called fat cells or adipose cells, can store loads of fat and actually increase 20 times in weight. If your fat cells are completely filled with fat and you need to store more fat, your body can even produce new fat cells. Fat cells are a compact way to store lots of energy. Remember that 1 gram of fat yields 9 kcalories, compared to 4 kcalories for 1 gram of carbohydrate or protein. Fats provide much of the energy to do the work in your body, especially work involving your muscles. Fat spares protein from being burned for energy, so protein can do its own important jobs.

Fat is an important part of all cell membranes. Fat also transports the fat-soluble vitamins (A, D, E, and K) throughout the body. Certain fat-containing foods also provide the body with two fatty acids that are considered **essential fatty acids** because the body can't make them. Fatty acids are the major component of triglycerides and are discussed next. The essential fatty acids are needed for normal growth and development in infants and children. They are used to maintain the structural parts of cell membranes, and they have a role in the proper functioning of the immune system. From the essential fatty acids, the body makes hormonelike substances that control a number of body functions, such as blood pressure and blood clotting.

Essential fatty acids—Fatty acids that the body cannot produce, making them necessary in the diet: linoleic acid and linolenic acid.

Lipids also include cholesterol and lecithin. Their functions will be discussed later in this chapter.

In foods, fats enhance taste, flavor, aroma, crispness (especially in fried foods), juiciness (especially of meat), and tenderness (especially in baked goods). Fats such as cooking oils do a wonderful job of carrying many flavors, such as in an Indian curry. Fats also provide a smooth texture and a creamy feeling in the mouth. The love of fatty foods cuts across all ages (just watch a preschooler devour french fries or an elderly adult eat a piece of chocolate cake) and cultures (where fatty foods are available). Eating a meal with fat makes people feel full, because fat delays the emptying of the stomach. This lasting feeling of fullness is called **satiety.**

Satiety—A feeling of being full after eating.

■ **MINI-SUMMARY**

Many fat cells are located just under the skin, where fat provides insulation for the body, a cushion around critical organs, and optimum body temperature in the cold. At 9 kcalories per gram, fat stores are a compact way to store lots of energy. Fat is an important part of all cell membranes. Fat also transports the fat-soluble vitamins (A, D, E, and K) throughout the body.

The essential fatty acids are needed for normal growth and development in infants and children. They are used to maintain the structural parts of cell membranes, and they have a role in the proper functioning of the immune system. From the essential fatty acids, the body makes hormonelike substances that control a number of body functions, such as blood pressure and blood clotting. Lipids also include cholesterol and lecithin. Their functions will be discussed later in this chapter. In foods, fats enhance taste, flavor, aroma, crispness, juiciness, tenderness, and texture. Fats have satiety value.

Triglycerides

Fatty acids—Major component of most lipids. Three fatty acids are present in each triglyceride.

Glycerol—A derivative of carbohydrate that is part of triglycerides.

Saturated fatty acid—A fatty acid that is filled to capacity with hydrogens.

Unsaturated fatty acid—A fatty acid with at least one double bond.

A triglyceride (Figure 4-1) is made of three **fatty acids** (*tri-* means three) attached to **glycerol,** a derivative of carbohydrate. Glycerol contains three carbon atoms, each attached to one fatty acid. You can think of glycerol as the backbone of the triglyceride.

Fatty acids in triglycerides are made of carbon atoms joined like links in a straight chain. Interestingly, the number of carbons is always an even number. Fatty acids differ from one another in two respects: the length of the carbon chain and the degree of saturation. The length of the chain may be categorized as short chain (6 carbons or less), medium chain (8 to 12 carbons), or long chain (14 to 20 carbons). Most food lipids contain long-chain fatty acids. The length of the chain influences the fat's ability to dissolve in water. Generally, triglycerides do not dissolve in water, but the short- and medium-chain fatty acids have some solubility in water, which will have implications later in our discussion on their digestion, absorption, and metabolism.

Fatty acids are referred to as **saturated** or **unsaturated.** To understand this concept, think of each carbon atom in the fatty-acid chain as having hydrogen atoms attached like charms on a bracelet, as you can see in Figure 4-2. Each C represents a carbon atom, each H represents a hydrogen atom, and each O represents an oxygen atom. Each carbon atom can have a maximum of four bonds, so it can attach to four other atoms. Typically a carbon atom has one bond each to the two carbon atoms on its sides and one bond each to two hydrogens. If each carbon atom in the chain is filled to capacity with hydrogens, it is considered a saturated fatty acid. That's how saturated fatty acid got its name: It is saturated with hydrogen atoms. When

Figure 4-1
A triglyceride

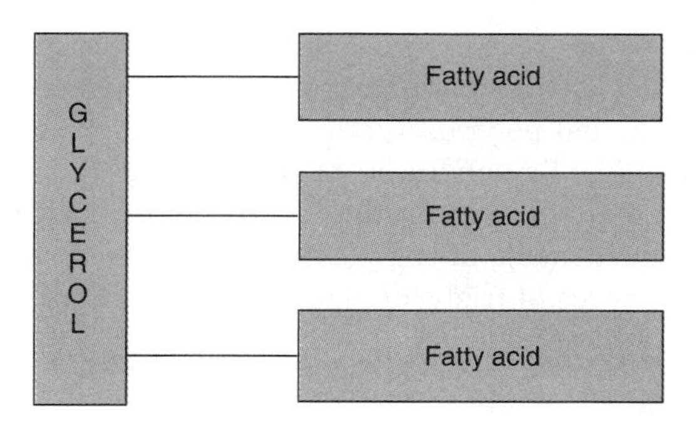

Point of unsaturation— The location of the double bond in unsaturated fatty acids.

Monounsaturated fatty acid—A fatty acid that contains only one double bond in the chain.

Polyunsaturated fatty acid—A fatty acid that contains two or more double bonds in the chain.

a hydrogen is missing from two neighboring carbons, a double bond forms between the carbon atoms, and this type of fatty acid is considered unsaturated.

If you look at Figure 4-2, the top fatty acid in the illustration is saturated: It is filled to capacity with hydrogens. By comparison, the middle and lower fatty acids are unsaturated. This is evident because there are empty spaces without hydrogens in the picture. Wherever hydrogens are missing, the carbons are joined by two lines, indicating a double bond. The spot where the double bond is located is called the **point of unsaturation.**

Now that you know what saturated and unsaturated fatty acids are, we need to look at the two types of unsaturated fatty acids. Unsaturated fatty acids are either **monounsaturated** or **polyunsaturated.** A fatty acid that

Figure 4-2
Types of fatty acids

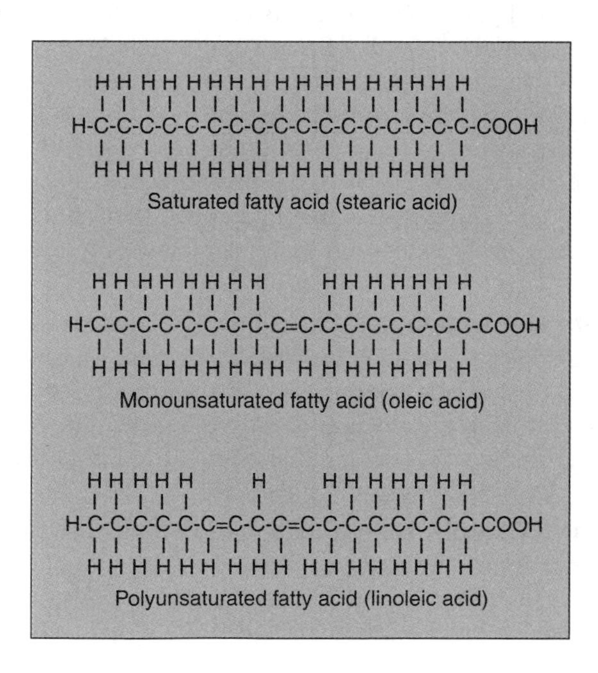

contains only one (*mono-* means one) point of unsaturation is called monounsaturated; if the chain has two or more points of unsaturation, the fatty acid is called polyunsaturated. Figure 4-2 gives an example of a monounsaturated and a polyunsaturated fatty acid.

Now that you know about the different types of fatty acids, it's time to get back to the concept of triglycerides. From the three types of fatty acids, we get three types of triglycerides, commonly called fats.

Saturated fat—A triglyceride made of mostly saturated fatty acids.

Monounsaturated fat—A triglyceride made of mostly monounsaturated fatty acids.

Polyunsaturated fat—A triglyceride made of mostly polyunsaturated fatty acids.

1. A saturated triglycercide, also called a **saturated fat,** is made of mostly saturated fatty acids.
2. A monounsaturated triglyceride, also called a **monounsaturated fat,** is made of mostly monounsaturated fatty acids.
3. A polyunsaturated triglyercide, also called a **polyunsaturated fat,** is made of mostly polyunsaturated fatty acids.

Now we are ready to see which types of foods the three different fats appear in.

Triglycerides in Food

All food fats, animal or vegetable, contain a mixture of saturated and unsaturated fats. A fat or oil is classified as saturated, monounsaturated, or polyunsaturated based on which type of fatty acid predominates.

Saturated fats are mostly found in foods of animal origin, and monounsaturated and polyunsaturated fats are mostly found in foods of plant origin and some seafoods. Foods of animal origin include meat, poultry, seafood, milk and dairy products such as butter, and eggs. Foods of plant origin include fruits, vegetables, dried beans and peas, grains, foods made with grains such as breads and cereals, nuts, seeds, and vegetable oils such as corn oil. The more unsaturated a fat is, the more liquid it is at room temperature.

Before going into more detail on foods that contain the three types of fats, let's see what each food group contributes in terms of overall fat.

1. Fruits and Vegetables. Whether fresh, canned, or frozen, most fruits and vegetables are practically fat-free. The exceptions are avocados, olives, and coconuts, which contain significant amounts of fat. Also, frozen vegetables that have butter, margarine, or sauces added are probably high in fat. Last, fried vegetables, such as french-fried potatoes, are high in fat.

2. Breads, Cereals, Rice, Pasta, and Grains. Most breads and cereals in this group are low in fat. Exceptions include granolas, croissants, biscuits, cornbread, and many crackers. Most baked goods such as cakes, pies, cookies, and quick breads are also high in fat, especially when commercially made.

3. Dry Beans and Peas, Nuts and Seeds. Dry beans and peas are very low in fat. Most nuts and seeds, however, such as peanuts and peanut butter, are quite high in fat.

4. Meat, Poultry, and Fish. Meat and poultry, and to some extent fish, contain a bit of fat. The fat content of meat tends to be higher than that of poultry, and poultry tends to have more fat than seafood does. Of course, within each group there are choices that are quite high in fat and choices that are much more moderate in fat. For example, chicken without the skin is low in fat, because most of its fat is just under the skin.

5. Dairy Foods. Most regular dairy foods are high in fat. Luckily, there are plenty of choices with no fat or reduced fat, such as skim milk, nonfat yogurt, and low-fat cheeses.

6. Fats, Oils, and Condiments. Fats, such as vegetable shortening, and oils are almost all fat. Table 4-1 lists the total calories and fat in selected fats and oils. Condiments such as regular mayonnaise and salad dressings also

TABLE 4-1 Total Calories and Fat in Selected Fats and Oils

Fat or Oil	Kcalories/ Tablespoon	Grams Fat/ Tablespoon	Grams Saturated Fat/Tablespoon
Coconut oil	120	13	12
Palm kernel oil	120	13	11
Palm oil	120	13	7
Butter, stick	108	12	7
Lard	115	13	5
Cottonseed oil	120	13	3
Olive oil	119	13	1
Canola oil	120	13	1
Peanut oil	119	13	2
Safflower oil	120	13	1
Corn oil	120	13	2
Soybean oil	120	13	2
Sunflower oil	120	13	1
Shortening	106	12	3
Margarine, stick	100	11	1–3
Margarine, soft tub	100	11	1–2
Margarine, liquid	90	10	1–2
Margarine, whipped	70	8	1–2
Margarine, spread	60	7	1–2
Margarine, diet	50	6	1

Source: U.S. Department of Agriculture and manufacturers.

contain much fat. The main contributors to fat in the American diet are beef, margarine, salad dressings, mayonnaise, cheese, milk, and baked goods. You can't see most of the fat you get in the foods you eat, except of course when you add oils and fats. The fat in meat, in milk and cheese, and in fried foods are not as obvious as the margarine you spread on bread.

In addition to understanding which foods are high in fat, let's look at which foods contain mostly saturated, monounsaturated, and polyunsaturated fat.

1. Saturated Fat. The biggest sources of saturated fat in the American diet are animal foods: cheese, beef (more than half comes from ground beef), whole milk, fats in baked goods, butter, and margarine. Saturated fat is also found in eggs, poultry skin, and other full-fat dairy products such as ice cream. Animal fat often contains at least 50 percent saturated fat. Although most vegetable oils are rich in unsaturated fats, several are high in saturated fat. These include the so-called tropical oils: coconut, palm kernel, and palm oils. They are used in some processed foods, such as baked goods and frozen whipped nondairy toppings. Table 4-2 compares the saturated fat content of selected foods.

TABLE 4-2 A Comparison of Saturated Fat in Some Foods

Food Category*	Portion	Saturated Fat (grams)
Cheese		
Regular cheddar cheese	1 oz.	6.0 grams
Low-fat cheddar cheese	1 oz.	1.2
Ground Beef		
Regular ground beef	3 oz. cooked	7.2
Extra-lean ground beef**	3 oz. cooked	5.3
Milk		
Whole milk	1 cup	5.1
Low-fat (1%) milk**	1 cup	1.6
Breads		
Croissant	1 medium	6.6
Bagel**	1 medium	0.1
Frozen Desserts		
Regular ice cream	1/2 cup	4.5
Frozen yogurt**	1/2 cup	2.5
Table Spreads		
Butter	1 tablespoon	8.0
Canola margarine**	1 tablespoon	2.0

*The food categories listed are among the major food sources of saturated fat for U.S. adults and children.

**Choice that is lower in saturated fat.

Source: Nutrition and Your Health: Dietary Guidelines for Americans. Home and Garden Bulletin No. 232. U.S. Department of Agriculture and U.S. Department of Health and Human Services, 2000.

2. Monounsaturated Fat. Good examples of monounsaturated fats include olive oil, canola oil, and peanut oil. Like other vegetable oils, these are used in cooking and in salad dressings. Canola oil is also used to make margarine.

3. Polyunsaturated Fat. Polyunsaturated fats are found in greatest amounts in safflower, corn, soybean, sesame, and sunflower oils. These oils are commonly used in salad dressings and as cooking oils. Nuts and seeds also contain polyunsaturated fats, enough to make them a rather high-calorie snack food depending on serving size.

Too much saturated fat is the primary contributor to high blood cholesterol levels, a major risk factor for heart disease. Figure 4-3 shows the fat composition of fats and oils.

Trans Fats

Trans fats—Unsaturated fatty acids that lose a natural bend or kink so they become straight (like saturated fatty acids) after being hydrogenated; they act like saturated fats in the body.

Hydrogenation—A process in which liquid vegetable oils are converted into solid fats (such as margarine) by the use of heat, hydrogen, and certain metal catalysts.

Low-trans hydrogenation—A new hydrogenation process to make shortenings and margarines with less trans fat.

Trans fats, short for **trans fatty acids,** occur naturally at low levels in meat and dairy foods. Most of the trans fats that we eat is a result of a process called **hydrogenation.** Hydrogenation, discovered at the turn of the twentieth century, converts liquid vegetable oils into solid fats by using hydrogen, heat, and certain metal catalysts. The partial hydrogenation process was quickly used to make vegetable shortening, which is simply vegetable oils that have been partially hydrogenated. Vegetable shortening is attractive because it is cheaper to make than butter or lard (pork fat) and has a longer shelf life. The hydrogenation process has also been used to make margarines (hydrogenation makes them easy to spread) and many oils used in deep-fat frying (hydrogenation gives them a high smoking point). Hydrogenation also helps these products stay fresh longer. If you check food labels, you will find partially hydrogenated oils popping up in many baked goods (cookies, cakes, muffins,etc.), fast foods, breads, snack foods, and margarines/shortenings.

During hydrogenation, some of the unsaturated fatty acids become saturated. Other unsaturated fatty acids lose their natural bend or kink and become straight, like saturated fatty acids. These are the trans fatty acids. Because they are straight, they can fit closer together, which makes them more solid. This explains why vegetable shortening is a solid. Because they are straight, they also behave like saturated fats in the body. Trans fatty acids have been shown to raise blood cholesterol levels (a major risk factor for heart disease) as much as saturated fat. This is discussed in more depth in a later section on health and lipids.

On food labels you will see the amount of trans fat per serving on its own line under "Total Fat." The amount of trans fatty acids in selected foods appears in Table 4-3.

In the future, we may see shortenings and margarines made with less trans fats due to a new processing technique called **low-trans hydrogenation.**

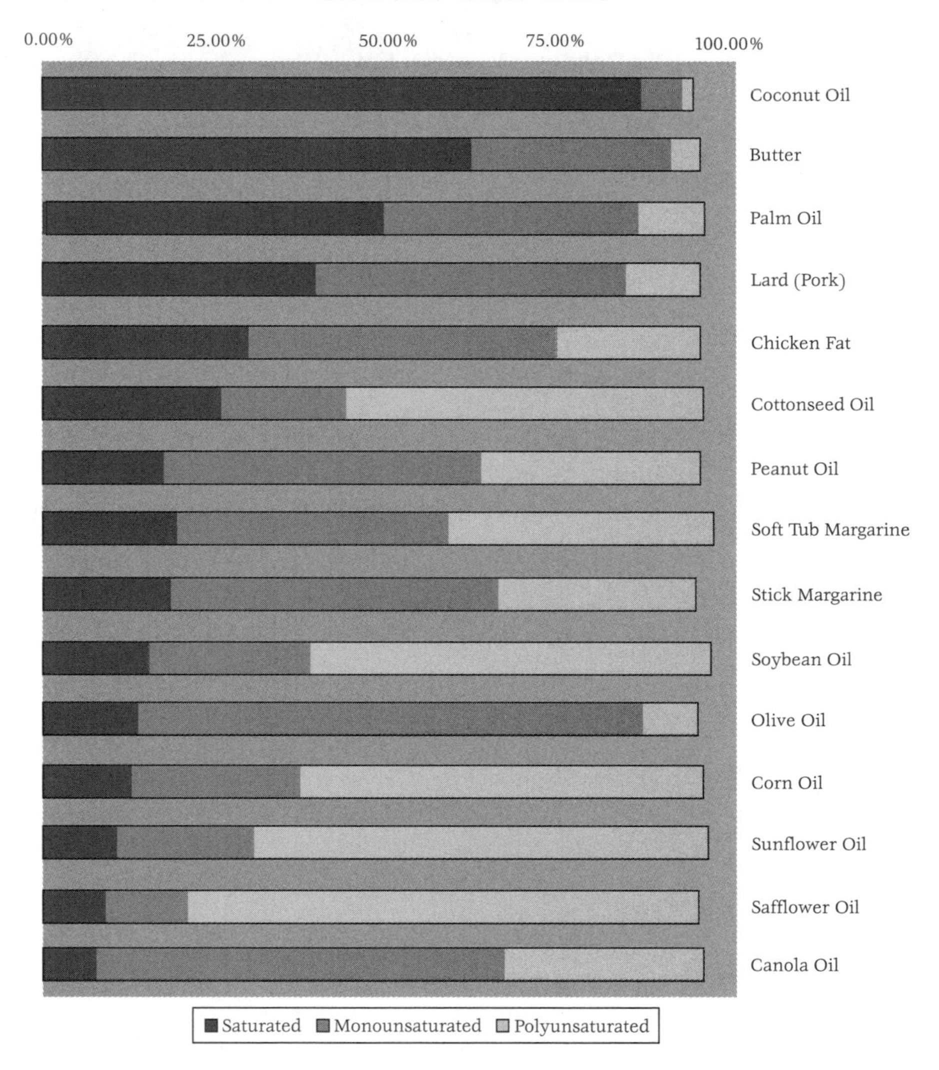

Figure 4-3
Fatty acid composition of 1 tablespoon of fats and oils

This new technique is more expensive but results in products, such as margarine, with much less trans fat.

Essential Fatty Acids

As mentioned, there are two essential fatty acids that the body can't make. The essential fatty acids are both polyunsaturated fatty acids: **linolcic acid** and **alpha-linolenic acid.** Linoleic acid is called an *omega-6 fatty acid* be-

TABLE 4-3 Trans Fatty Acids in One Serving of Selected Foods

Food	Trans Fatty Acids (grams/serving)
Vegetable shortening	1.4–4.2
Margarine, stick, 83% fat	2.4
Margarine, stick, 68% fat	1.8
Margarine, tub, 80% fat	1.1
Margarine, tub, 40% fat	0.6
Salad dressings (regular)	0.06–1.1
Vegetable oils	0.01–0.06
Pound cake	4.3
Doughnuts	0.3–3.8
Microwave popcorn (regular)	2.2
Chocolate-chip cookies	1.2–2.7
Vanilla wafers	1.3
French fries (fast food)	0.7–3.6
Snack crackers	1.8–2.5
Snack chips	0–1.2
Chocolate candies	0.04–2.8
White bread	0.06–0.7
Ready-to-eat breakfast cereals	0.05–0.5

Source: USDA food composition data, 1995.

Linoleic acid—Omega-6 fatty acid found in vegetable oils such as corn, safflower, soybean, cottonseed, and sunflower oils; this essential fatty acid is vital to growth and development, maintenance of cell membranes, and the immune system.

Alpha-linolenic acid—An omega-3 fatty acid found in several oils, notably canola, flaxseed, soybean, walnut, and wheat germ oil (or margarines made with canola or soybean oil); this essential fatty acid is vital to growth and development, maintenance of cell membranes, and the immune system, and is inadequate in many Americans' diets.

cause its double bonds appear after the sixth carbon in the chain (see Figure 4-2). The major omega-6 fatty acid in the diet is linoleic acid. Alpha-linolenic acid is the leading *omega-3 fatty acid* found in food, and its double bonds appear after the third carbon in the chain.

Linoleic acid is found in vegetable oils such as corn, safflower, soybean, cottonseed, and sunflower. Most Americans get plenty of linoleic acid from foods containing vegetable oils, such as margarine, salad dressings, and mayonnaise. Whole grains and vegetables also supply some linoleic (and linolenic) acid.

Alpha-linolenic acid is found in several oils, notably canola, flaxseed, soybean, walnut, and wheat germ oil (or margarines made with these oils). Other good sources of alpha-linolenic acid include ground flaxseed, walnuts, and soy products. The body converts alpha-linolenic acid into two other omega-3 fatty acids: docosahexaenoic acid (DHA) and eicosapentaenoic acid (EPA). Both DHA and EPA are found in fish such as salmon, mackerel, sardines, halibut, bluefish, trout, and tuna. Notice that these fish tend to be fatty fish, not lean fish. Whereas Americans generally get plenty of linoleic acid, that is not the case with alpha-linolenic acid, DHA, or EPA.

The essential fatty acids are vital to normal growth and development in infants and children. They are used to maintain the structural parts of cell membranes, and they have a role in the proper functioning of the immune system. The body uses both essential fatty acids to make substances known as *eicosanoids.* Eicosanoids affect the activities of the cells in their "local" area in a variety of ways, and sometimes their actions are opposite. For example, the eicosanoids made from linoleic acid tend to make small blood vessels smaller (which increases blood pressure) and increase blood clotting (which increases thc likclihood of a heart attack) and inflammation in the body. On the other hand, the eicosanoids made from linolenic acid dilate blood vessels and reduce blood clotting and inflammation.

The omega-3 fatty acid DHA is found in the retina of the eye and in the brain. DHA and EPA decrease blood triglyceride and cholesterol levels, and they are both used to make eicosanoids. The omega-3 fatty acids may be important in the prevention and treatment of heart disease and hypertension, and may also help in preventing and treating cancer and arthritis. Eating fatty fish rich in essential fatty acids twice a week has been shown to help prevent heart disease. As of 2002, Adequate Intakes have been set for linoleic and alpha-linolenic acids (Table 4-6).

■ MINI-SUMMARY

A triglyceride is made of three fatty acids attached to glycerol. Fatty acids can be saturated, monounsaturated, or polyunsaturated. All food fats contain a mixture of saturated and unsaturated fats. Most fruits and vegetables are fat-free, and most breads, cereals, rice, and pasta are low in fat. Regular dairy foods are generally high in fat and saturated fat. Saturated fats are mostly found in foods of animal origin, and monounsaturated and polyunsaturated fats are mostly found in foods of plant origin and some seafood. The biggest sources of saturated fat in the American diet are cheese, beef, milk, baked goods, butter, and margarine. Olive oil, canola oil, and peanut oil are rich in monounsaturated fats. Polyunsaturated fats are found in greatest amounts in safflower, corn, soybean, sesame, and sunflower oils.

Trans fats, or trans fatty acids, occur naturally in small amounts in some foods and are created by hydrogenation. Hydrogenated oils are found in many baked goods, fast foods, snack foods, and margarines/shortenings. Both saturated fat and trans fats increase blood cholesterol levels, a risk factor for heart disease.

The essential fatty acids are both polyunsaturated fatty acids: linoleic acid (an omega-6 fatty acid) and alpha-linolenic acid (an omega-3 fatty acid). Alpha-linoleic acid is found in vegetable oils such as corn, safflower, soybean, cottonseed, and sunflower. Linolenic acid is found in several oils, notably canola, flaxseed, soybean, walnut, and wheat germ oil (or margarines made with these oils). The body converts linolenic acid into two other omega-3 fatty acids: docosahexaenoic acid (DHA) and eicosapentaenoic acid (EPA). Both DHA and EPA are found in fatty fish such as salmon, mackerel, sardines, halibut, bluefish, trout, and tuna. Whereas Americans generally get plenty of linoleic acid, that is not the case with alpha-linolenic acid, DHA, or EPA. The essential fatty acids are vital to normal growth and development in infants and children. They are used to maintain the structural parts of cell membranes, and they have a role in the proper functioning of the immune system. The body uses both essential fatty acids to make substances known as eicosanoids. The omega-3 fatty acids are possibly important in the prevention and treatment of heart disease and hypertension, and may also help in preventing and treating cancer and arthritis.

Cholesterol

Cholesterol—The most abundant sterol (a category of lipids); a soft, waxy substance present only in foods of animal origin; it is present in every cell in your body.

Bile acids—A component of bile that aids in the digestion of fats in the duodenum of the small intestine.

Cholesterol is the most abundant *sterol,* a class of lipids. Pure cholesterol is an odorless, white, waxy, powdery substance. You cannot taste it or see it in the foods you eat.

Your body needs cholesterol to function normally. It is present in every cell in your body, including the brain and nervous system, muscles, skin, liver, intestines, heart, and skeleton. The body uscs cholesterol to make **bile acids,** which allow us to digest fat, and to make cell membranes and many hormones. Hormones are the chemical messengers of the body. They enter the bloodstream and travel to a target organ to influence what the organ does. For example, insulin is a hormone made in the pancreas. When it is secreted into the bloodstream, it increases the movement of glucose from the bloodstream into muscle and fat cells. Cholesterol is used to make some important hormones such as the sex hormones (estrogen and testosterone), a form of the active vitamin D hormone, and the hormones of the adrenal gland such as cortisone.

Unfortunately, high blood cholesterol is a risk factor for heart disease and is found in the plaque that clogs arteries. This will be discussed in more detail later on.

So which foods contain cholesterol? Cholesterol is found only in foods of animal origin: egg yolks (it's not in the whites), meat, poultry, milk and milk products, and fish. (See Table 4-4.) It is not found in foods of plant origin. Egg yolks and organ meats (liver, kidney, sweetbreads, brain) contain the most cholesterol—one egg yolk contains 213 milligrams of cholesterol. About 4 ounces of meat, poultry, or fish (trimmed or untrimmed) contain 100 milligrams of cholesterol, with the exception of shrimp, which is higher in cholesterol. Eggs, meat, and whole milk provide most of the cholesterol we eat, and these sources are also rich in saturated fat.

In milk products, cholesterol is mostly in the fat, so lower-fat products contain less cholesterol. For example, 1 cup of whole milk contains 33 milligrams

TABLE 4-4 Cholesterol in Foods

Food and Portion	Cholesterol (milligrams)
Liver, braised, 3 ounces	333
Egg, whole, 1	213
Beef, short ribs, braised, 3 ounces	80
Beef, ground, lean, broiled medium, 3 ounces	74
Beef, top round, broiled, 3 ounces	73
Chicken, roasted, without skin, light meat 3-1/2 ounces	75
Shrimp, cooked, moist heat, 3 ounces	167
Scallops, broiled, 3 ounces	47
Lobster, cooked, moist heat, 3 ounces	61
Haddock, baked, 3 ounces	63
Mackerel, baked, 3 ounces	64
Swordfish, baked, 3 ounces	43
Milk, whole, 8 ounces	33
Milk, 2%, 8 ounces	18
Milk, 1%, 8 ounces	10
Skim milk, 8 ounces	4
Cheddar cheese, 1 ounce	30
American processed cheese, 1 ounce	27
Cottage cheese, low-fat, 1%, 1/2 cup	5

Source: National Institutes of Health. 1994. *Step by Step: Eating to Lower Your High Blood Cholesterol.* NIH Publication No. 94–2920.

of cholesterol, whereas a cup of skim milk contains only 4 milligrams (Table 4-4).

Egg whites and foods that come from plants, such as fruits, nuts, vegetables, grains, cereals, and seeds, have *no* cholesterol.

We eat about 200 to 400 milligrams of cholesterol daily, and about half of this cholesterol is actually absorbed. The body also manufactures about 700 milligrams of cholesterol daily. The liver makes only 10 to 20 percent of this amount, and the body's cells synthesize the rest. Because the body produces plenty of cholesterol to make bile acids, hormones, and so on, cholesterol is not considered an essential nutrient.

■ MINI-SUMMARY

The body uses cholesterol to make bile acids, which allow us to digest fat, and to make cell membranes and many hormones. Cholesterol is found only in foods of animal origin, such as egg yolks, meat, poultry, fish, milk, and milk products. It is not found in foods of plant origin. The body's cells and the liver produce cholesterol, so it is not an essential nutrient.

Lecithin

Lecithin—A phospholipid and a vital component of cell membranes that acts as an emulsifier (a substance that keeps fats in solution).

Lecithin is considered a *phospholipid,* a class of lipids that are like triglycerides except that one fatty acid is replaced by a phosphate group and choline (or other nitrogen-containing group). Phospholipids are found in all living organisms. Lecithin functions as a vital component of cell membranes. It also acts as an *emulsifier.* As you may know, fats and water do not normally stay mixed together but separate into layers. An emulsifier is capable of breaking up the fat globules into small droplets, resulting in a uniform mixture that won't separate. Lecithin keeps fats in solution in the blood, and elsewhere in the body, a most important function. Lecithin is used commercially as an emulsifier in foods such as salad dressings and bakery products. Egg yolks are especially rich in lecithin and are used as emulsifiers in many baking recipes as well as in mayonnaise and hollandaise sauce.

Although the media have featured lecithin as a wonder nutrient that can burn fat, improve memory, and accomplish other similar feats, none of this is true. Since lecithin is made in the liver, it is not considered an essential nutrient.

■ **MINI-SUMMARY**

Lecithin, a phospholipid, is a vital component of cell membranes and acts as an emulsifier (keeps fats in solution). It is not an essential nutrient.

Digestion, Absorption, and Metabolism

Lingual lipase—An enzyme made in the salivary glands in the mouth that has a minor role in fat digestion in adults, and an important role in fat digestion in infants.

Gastric lipase—An enzyme in the stomach that breaks down mostly short-chain fatty acids.

Fats are difficult for the body to digest, absorb, and metabolize. The problem is simple: Fat and water do not mix. Minimal digestion of fats occurs before they reach the upper part of the small intestine. In your mouth, a salivary gland makes an enzyme, called **lingual lipase,** that has a minor role in fat digestion in adults but an important role in fat digestion in infants. Lingual lipase digests certain fatty acids in milk in infants. In the stomach, **gastric lipase** enzyme works on breaking down mostly short-chain fatty acids.

Once fats reach the small intestine, the gallbladder is stimulated to release **bile** into the intestine. Bile is made by the liver, stored in the gallbladder, and squirted into the intestinal tract when fat is present. Bile contains bile acids that emulsify fat, meaning that they split fats into small globules or pieces. In this manner, fat-splitting enzymes (including pancreatic lipase and intestinal lipase) can then do their work. The enzymes break down many triglycerides to their component parts—fatty acids and

Bile—A yellow-green liver secretion that is stored in the gallbladder and released when fat enters the small intestine because it emulsifies fat.

Monoglycerides—Triglycerides with only one fatty acid.

Lipoprotein—Protein-coated packages that carry fat and cholesterol through the bloodstream; the body makes four types classified according to their density.

Chylomicron—The lipoprotein responsible for carrying mostly triglycerides, and some cholesterol, from the intestines through the lymph system to the bloodstream.

Lipoprotein lipase—An enzyme that breaks down triglycerides from the chylomicron into fatty acids and glycerol so that they can be absorbed in the body's cells.

Very low density lipoprotein (VLDL)—Lipoproteins made by the liver to carry triglycerides and some cholesterol through the body.

Low-density lipoprotein (LDL)—Lipoproteins that contain most of the cholesterol in the blood; they carry cholesterol to body tissues.

glycerol—so that they can be absorbed across the intestinal wall. **Monoglycerides,** triglycerides with only one fatty acid, are also produced.

Once absorbed into the cells of the small intestine, triglycerides are re-formed. Both shorter-chain fatty acids and glycerol can travel freely in the blood because they are water-soluble. However, triglycerides, monoglycerides, cholesterol, and longer-chain fatty acids would float in clumps and wreak havoc in either the blood or lymph. Because of this, the body wraps them with protein and phospholipids to make them water-soluble. The resulting substance is called a **lipoprotein,** a combination of fat *(lipo-)* and protein. Lipoproteins have four components:

- Triglycerides
- Protein
- Cholesterol
- Phospholipids

The body makes four types of lipoproteins: chylomicrons, very low-density lipoproteins, low-density lipoproteins, and high-density lipoproteins. Each will now be discussed.

Chylomicron is the name of the lipoprotein responsible for carrying mostly triglycerides, and some cholesterol, from the intestines through the lymph system to the bloodstream. Lymph is similar to blood but has no red blood cells. Lymph vessels are found all around the body, and the lymph transports fat and fat-soluble vitamins to the blood as well as moves fluids found between cells to the bloodstream. The lymph system goes into areas where there are no blood vessels to feed the cells.

In the bloodstream, an enzyme—**lipoprotein lipase**—breaks down the triglycerides in the chylomicrons into fatty acids and glycerol so that they can be absorbed into the body's cells. The cells can either use the fatty acids for energy, which the muscle cells often do, or make triglycerides for storage, which fat cells often do. Once the triglycerides are broken down and taken up by the cells, what remains of the chylomicron is some protein and cholesterol that is metabolized by the liver.

The primary sites of lipid metabolism are the liver and the fat cells. The liver makes triglycerides and cholesterol that are carried through the body by **very low density lipoprotein (VLDL),** the liver's version of chylomicron. Half of the VLDL is made of triglycerides, which are broken down in the bloodstream with the help of lipoprotein lipase. Again, the body's cells absorb fatty acids and glycerol to be burned for energy or stored as triglycerides. Once the majority of triglycerides are removed, VLDL is converted in the blood into another type of lipoprotein called **low-density lipoprotein (LDL).**

LDL is the major cholesterol-carrying lipoprotein. LDL also distributes some triglycerides and phospholipids. LDL therefore supplies materials for cells to make new membranes, hormones (chemical messengers), and

High-density lipoprotein (HDL)—Lipoproteins that contain much protein and carry cholesterol away from body cells and tissues to the liver for excretion from the body.

other substances. Certain cells (especially in the liver) have the ability to absorb the entire LDL particle. The LDL not absorbed by cells is somehow involved in depositing cholesterol on the inner blood vessel wall, causing hardening and narrowing of the arteries.

A last type of lipoprotein, **high-density lipoprotein (HDL),** contains much protein and travels throughout the body picking up cholesterol. HDL carries cholesterol back to the liver for breakdown and disposal. Thus HDL helps remove cholesterol from the blood, preventing the buildup of cholesterol in the arterial walls.

■ MINI-SUMMARY

Fats are difficult for the body to digest, absorb, and metabolize because fat and water do not mix. Minimal digestion of fats occurs before they reach the small intestine. Lingual lipase works as an enzyme in the mouth, and gastric lipase acts in the stomach. Once fats reach the small intestine, the gallbladder releases bile. Bile acids emulsify fat so that pancreatic lipase and intestinal lipase can break down triglycerides into their components. Once absorbed into the cells of the small intestine, triglycerides are re-formed. Shorter-chain fatty acids and glycerol can travel in the blood, but the longer-chain fatty acids, monoglycerides, triglycerides, and cholesterol are wrapped into chylomicrons (a lipoprotein) with protein and phospholipids to make them water-soluble. Chylomicrons carry mostly triglycerides and some cholesterol from the intestines through the lymph system to the bloodstream. In the bloodstream lipoprotein lipase breaks down the triglycerides so that fatty acids and glycerol can enter the body's cell. The cells either use the fatty acids for energy (muscle cells) or make triglycerides for storage (fat cells). The primary sites of lipid metabolism are the liver and fat cells. The liver makes VLDL, which transports triglycerides around the body for use by the cells. Once the majority of triglycerides are removed, VLDL is converted in the blood into LDL (the major cholesterol-carrying lipoprotein). HDL carries cholesterol back to the liver for breakdown and disposal.

Lipids and Health

Heart Disease

Heart disease is the number-one killer of both men and women in the United States. In 1999, over 725,000 people died due to heart disease. Many Americans have elevated blood cholesterol levels, one of the key risk factors for heart disease. Anyone can develop high blood cholesterol, regardless of age, gender, race, or ethnic background.

Plaque—Deposits on arterial walls that contain cholesterol, fat, fibrous scar tissue, calcium, and other biological debris.

Atherosclerosis—The most common form of artery disease, characterized by plaque buildup along artery walls.

Angina—Symptoms of pressing, intense pain in the heart area, often due to stress or exertion when the heart muscle gets insufficient blood.

Myocardial infarction—Heart attack.

Stroke—Damage to brain cells resulting from an interruption of blood flow to the brain.

Too much circulating cholesterol can build up in the walls of your arteries, especially the heart's arteries (called the coronary arteries) that supply the heart with what it needs to keep pumping. This leads to accumulation of cholesterol-laden **plaque** in blood vessel linings, a condition called **atherosclerosis.** Atherosclerosis is a slow, complex disease that starts in childhood. Over time the buildup in the walls of the arteries causes "hardening of the arteries"—arteries become narrowed and blood flow to the heart is slowed down or blocked.

When blood flow to the heart is blocked, the heart muscle becomes starved for oxygen, causing chest pain (called **angina**). If the blood supply to a portion of the heart is completely cut off by a blockage, a heart attack (called a **myocardial infarction**) can occur. If the blood supply to part of the brain is cut off, a **stroke** can occur.

LDL cholesterol is the main source of cholesterol buildup and blockage in the arteries. The primary ways in which LDL cholesterol levels become too high is through eating too much saturated fat, trans fat, and, to a lesser extent, cholesterol. Dietary factors that lower LDL cholesterol include polyunsaturated fats, monounsaturated fatty acids and, to a lesser extent, soluble fiber and soy protein.

A variety of other things can affect cholesterol levels. These are things you can do something about.

- **Weight.** Being overweight is a risk factor for heart disease. It also tends to increase your cholesterol. Losing weight can help lower your LDL and total cholesterol levels, as well as raise your HDL and lower your triglyceride levels.
- **Physical activity.** Not being physically active is a risk factor for heart disease. Regular physical activity can help lower LDL cholesterol and raise HDL cholesterol levels. It also helps your lose weight. You should try to be physically active for 30 minutes on most, if not all, days.

Things you cannot do anything about also can affect cholesterol levels. These include:

- **Age and gender.** As women and men get older, their cholesterol levels rise. Before the age of menopause, women have lower total cholesterol levels than men of the same age. After the age of menopause, women's LDL levels tend to rise.
- **Heredity.** Your genes partly determine how much cholesterol your body makes. High blood cholesterol can run in families.

Besides high LDL cholesterol levels, there are several other risk factors for heart disease. These include cigarette smoking, high blood pressure, diabetes mellitus, obesity, and physical inactivity. If any of these is present in addition to high blood cholesterol, the risk of heart disease is even greater.

The good news is that these factors can be brought under control by changes in lifestyle—such as adopting a different diet, losing weight, beginning an exercise program, or quitting a tobacco habit. Drugs also may be necessary for some people. Sometimes one change can help bring several risk factors under control. For example, weight loss can reduce blood cholesterol levels, help control diabetes, and lower high blood pressure.

But some risk factors can't be controlled. These include age (45 or older for men, and 55 or older for women) and family history of early heart disease (father or brother stricken before age 55, mother or sister stricken before age 65).

The American Heart Association (2000) has developed guidelines to reduce the risk of cardiovascular disease by dietary and other practices. Their guidelines are summarized in Table 4-5. Heart disease is discussed in greater depth in Chapter 11.

Cancer

Cancer is the second leading cause of death in the United States following heart disease. The relationship between fat and cancer is not nearly as clear as that between fat and heart disease. Fat may be involved in certain cancers, such as prostate cancer. In other cancers, such as breast or colon cancer, obesity and eating excessive calories seem to be more related to cancer development than dietary fat alone. The American Cancer Society has written guidelines for cancer prevention that are consistent with guidelines from the American Heart Association.

TABLE 4-5 Summary of American Heart Association Dietary Guidelines (2000)

Population Goals	Major Guidelines
Overall Healthy Eating Pattern	Include a variety of fruits, vegetables, grains, low-fat or nonfat dairy products, fish, legumes, poultry, lean meats.
Appropriate Body Weight	Match energy intake to energy needs, with appropriate changes to achieve weight loss when indicated.
Desirable Cholesterol Profile	Limit foods high in saturated fat and cholesterol; substitute grains, vegetables, fish, legumes, nuts.
Desirable Blood Pressure	Limit salt and alcohol; maintain a healthy body weight and a diet with emphasis on vegetables, fruits, and low-fat or nonfat dairy products.

Source: Krauss, R.M., et al., 2000. "AHA Dietary Guidelines: A Statement for Healthcare Professionals from the Nutrition Committee of the American Heart Association." *Circulation* 101:2284.

■ **MINI-SUMMARY**

Heart disease is the number-one killer in the United States. Too much circulating cholesterol can build up in the form of plaque on the walls of your arteries, especially the coronary arteries, a condition called atherosclerosis. Atherosclerosis is a slow, complex disease that starts in childhood. Over time the buildup in the walls of the arteries causes "hardening of the arteries"—arteries become narrowed and blood flow to the heart is slowed down or blocked, causing angina, heart attacks, and strokes. The primary ways in which LDL cholesterol levels become too high is through eating too much saturated fat, trans fat, and to a lesser extent cholesterol. Dietary factors that lower LDL cholesterol include polyunsaturated fats, monounsaturated fatty acids and to a lesser extent soluble fiber and soy protein. Being overweight and inactive promotes high blood cholesterol levels. Age, gender, and heredity also can affect your blood cholesterol levels, but you can't control these factors as you can the other factors mentioned. The American Heart Association (2000) guidelines to reduce the risk of cardiovascular disease by dietary and other practices appear in Table 4-5. Fat may be involved in certain cancers, such as prostate cancer.

Dietary Recommendations

The 2002 Dietary Reference Intakes for macronutrients did not include an Adequate Intake (AI) or a Recommended Dietary Allowance (RDA) for total fat (except for infants). (The AI for infants is 31 grams fat/per day for 1 to 6 months and 30 grams fat per day for 7 to 12 months). This occurred because there wasn't enough scientific data to determine the level at which inadequacy or disease prevention occurs. However, Acceptable Macronutrient Distribution Ranges (AMDRs) were set for total fat as follows.

Age	*AMDR for Fat*
1 to 3 years old	30–40 percent of kcalories
4 to 18 years old	25–35 percent of kcalories
Over 18 years old	20–35 percent of kcalories

This range of intake is associated with reduced risk of chronic disease while providing adequate intake.

Neither an AI nor RDA were set for saturated fat or cholesterol because these substances are all made in the body and have no known role in preventing chronic diseases such as heart disease or diabetes. Likewise, no AI or RDA was determined for trans fatty acids because they are not essential and provide no known benefit to health. The Food and Nutrition Board does recommend keeping intake of saturated fat, cholesterol, and trans fatty acids as low as possible while eating a nutritionally adequate diet.

In 2002, an AI was set for the essential fatty acids: linoleic acid and alpha-linolenic acid (Table 4-6). There is insufficient evidence to set a Tolerable Upper Intake Level for either essential fatty acid.

Fat intake in the United States as a proportion of total calories is lower than it was many years ago (dropping from 45 to 34 percent), but most people still eat too much saturated fat. Although the percent of calories from fat has decreased, the number of calories eaten has increased, so we are actually eating more grams of fat.

The Dietary Guidelines for Americans (2000) and the American Heart Association's Dietary Guidelines (2000) both recommend a diet for healthy Americans that provides

- 30 percent or less of total kcalories from fat
- Less than 10 percent of total kcalories from saturated fat
- Less than 300 milligrams per day of cholesterol

Table 4-7 shows the maximum number of grams of fat and saturated fat for a variety of calorie levels using these recommendations. Keep in mind that trans fat should be considered part of your saturated fat intake. These recommendations do not apply to children age 2 and under. Children need fat in order to grow and develop properly. After age 2, children should progressively adopt these recommendations up to age of 5 years.

TABLE 4-6 Adequate Intake Values for Essential Fatty Acids

	Linoleic Acid AI (grams per day)		Alpha Linolenic Acid AI (grams per day)	
Age Group	Male	Female	Male	Female
0–6 months	4.4 grams	4.4 grams	0.5 grams	0.5 grams
7–12 months	4.6	4.6	0.5	0.5
1–3 years	7	7	0.7	0.7
4–8 years	10	10	0.9	0.9
9–13 years	12	10	1.2	1.1
14–18 years	16	11	1.6	1.1
19–30 years	17	12	1.6	1.1
31–50 years old	17	12	1.6	1.1
Over 50 years old	14	11	1.6	1.1
Pregnancy		13		1.4
Lactation		13		1.3

Source: Adapted with permission from the *Dietary References Intakes for Energy, Carbohydrates, Fiber, Fat, Protein, and Amino Acids (Macronutrients).* © 2002 by the National Academy of Sciences. Courtesy of the National Academy Press, Washington, D.C.

TABLE 4-7 Recommended Maximum Fat and Saturated Fat Intake

If Your Total Daily Kcalories Are:	Total Fat (grams)	Saturated Fat (grams)
1200	40	13
1500	50	17
1800	60	20
2000	67	22
2200	73	24
2400	80	27
2600	86	29
2800	93	31
3000	100	33

In Chapter 2, we discussed the Mediterranean diet, which includes lots of monounsaturated fat, mostly in the form of olive oil. There seems to be agreement that if fat intake goes higher than 30 percent of total calories, the diet should emphasize monounsaturated fats such as olive oil and canola oil. The American Heart Association also recommends eating fatty fish (a rich source of omega-3 fatty acids) twice a week because of increased evidence of cardiovascular benefits.

■ MINI-SUMMARY

There is no RDA or AI for fat (except for infants), saturated fat, cholesterol, or trans fatty acids. The AMDR for fat is 20 to 35 percent of kcalories for adults. An AI is set for the essential fatty acids. It is recommended to keep your intake of saturated fat, cholesterol, and trans fatty acids as low as possible while eating a nutritionally adequate diet. The Dietary Guidelines for Americans (2000) and the American Heart Association's Dietary Guidelines (2000) both recommend a diet for healthy Americans that provides 30 percent or less of total kcalories from fat, less than 10 percent of total kcalories from saturated fat, and less than 300 milligrams per day of cholesterol. The American Heart Association also recommends eating fatty fish (a rich source of omega-3 fatty acids) twice a week because of increased evidence of cardiovascular benefits.

Ingredient Focus: Meats, Poultry, and Fish

Before going into meats, poultry, and fish, two concepts need to be explained: percentage fat and rancidity. When looking at fat in food, it is important to distinguish between two different concepts: the percentage of fat by weight and the percentage of calories from fat. To explain these two concepts, let's look at an example. In a supermarket, you find sliced turkey breast that is advertised as being "96 percent fat-free." What this means is that if you weighed out a 3-ounce serving, 96 percent of the weight is lean or without fat. In other words, only 4 percent of its weight is actually fat. The statement "96 percent fat-free" does not tell you anything about how many calories come from fat.

Now, if you look at the Nutrition Facts on the label, you read that a 3-ounce serving contains 3 grams of fat, 27 calories from fat, and 140 total calories. The label also states that the percentage of calories from fat in a serving is 19 percent. To find out the percentage of calories from fat in any serving of food, simply divide the number of calories from fat by the number of total calories, then multiply the answer by 100, as follows.

$$\frac{\text{Calories from fat}}{\text{Total calories}} \times 100 = \text{Percentage of calories from fat}$$

$$\frac{\text{27 calories from fat}}{\text{140 calories}} \times 100 = \text{19 percent}$$

This percentage has become more important as many recommendations on fat consumption target 30 percent or less of total kcalories as a desirable daily total from fat. This does not mean, however, that every food you eat needs to derive only 30 percent of its kcalories from fat. If this were the case, you could not even have a teaspoon of margarine because all of its kcalories come from fat. It is your total fat intake over a few days that is important, not the percentage of fat in just one food or just one meal.

Rancidity—The deterioration of fat, resulting in undesirable flavors and odors.

Another important concept, rancidity, is related to the storing and use of fats in the kitchen. **Rancidity** is the deterioration of fat, resulting in undesirable flavors and odors. In the presence of air, fat can lose a hydrogen atom at the point of unsaturation and take on an oxygen atom. This change creates unstable compounds that start a chain reaction, quickly turning a fat rancid. You can tell whether a fat is rancid by its odd odor and taste. The greater the number of points of unsaturation, the greater the possibility that rancidity will develop. This explains why saturated fats are more resistant to rancidity than unsaturated fats. Rancidity is also quickened by heat and ultraviolet light. Luckily, vitamin E is present in plant oils, and it naturally resists deterioration of the oil.

To prevent rancidity, store fat and oils tightly sealed in cool, dark places. For butter and margarine, check the date on the packaging. When oils are refrigerated, they sometimes become cloudy and thicker. This usually clears up after they are left at room temperature again or put under warm water.

Food Facts in this chapter discusses the wide variety of oils, butter, and margarine. This section will discuss meats, poultry, and fish. Milk and dairy products, which are also often high in fat, are discussed in the next chapter.

Nutrition

Purchasing lean, fresh cuts of meat, poultry, and fish is important, but first let's compare these items nutritionally (see Table 4-8).

- Most fish is lower in fat, saturated fat, and cholesterol than are meat and poultry.
- Chicken is twice as fatty as turkey.
- Chicken and turkey breast (meaning white meat) without skin are low in fat—only about 3 grams of fat in 3 ounces of chicken, and 1 gram of fat in 3 ounces of turkey. By comparison, white meat with skin and dark meat (such as thighs and drumsticks) are much higher in fat. Also, chicken wings may be considered white meat, but they are fattier than the drumstick.
- If buying ground turkey or chicken, make sure it is made from only skinless breast meat for the least amount of fat. If the product includes skin and dark meat, it will be *much* higher in fat.
- Trimmed veal is leaner than skinless chicken.
- When choosing beef, you will get the least fat from eye of round, followed by top round and bottom round.

TABLE 4-8 Meat, Poultry, and Fish: A Comparison

Food Type (3 ounces, cooked)	Saturated Fat (grams)	Dietary Cholesterol (milligrams)	Total Fat (grams)	Kcalories
Beef, top round, broiled	3	73	8	185
Beef, whole rib, broiled	10	72	26	313
Chicken, light meat without skin, roasted	1	64	4	130
Chicken, light meat with skin, roasted	3	71	19	189
Ground turkey (breast meat only), cooked	<1	35	<2	130
Ground turkey (meat and skin), cooked	3	87	11	200
Cod, baked	<1	47	<1	89

Source: National Institutes of Health, 1994. *Step by Step: Eating to Lower Your High Blood Cholesterol.* NIH Publication No. 94-2920.

Meat is a good source of many important nutrients, including protein, iron, copper, zinc, and some of the B vitamins, such as B_6 and B_{12}. Meat is also a significant source of fat, saturated fat, and cholesterol.

In comparison to red meats, skinless white-meat chicken and turkey are comparable in cholesterol, but lower in total fat and saturated fat. The skin of chicken and turkey contains much of the bird's fat. The skin should be left on during cooking to keep in moisture but can be removed before serving. Chicken and turkey are rich in protein, niacin, and vitamin B_6. They are also good sources of vitamin B_{12}, riboflavin, iron, zinc, and magnesium. Duck and goose are quite fatty in comparison, because they contain all dark meat.

Fish and shellfish are excellent sources of protein and are relatively low in calories. Most are also low to moderate in cholesterol content and a good source of certain vitamins, such as vitamins E and K, and minerals, such as iodine and potassium. Certain fish (Tables 4-9 and 4-10) are fattier than others, such as mackerel and herring, but fatty fish are an important source of omega-3 fatty acids.

Purchasing

The first step is to select a lean cut, such as one of these:

- Beef: eye of round, inside (top) round, outside (bottom) round, sirloin tip, flank steak, top sirloin butt

TABLE 4-9 Fat Content of Fish

Low-Fat Fish (fat content less than 2.5 percent)	Medium-Fat Fish (fat content 2.5–5 percent)	High-Fat Fish (fat content over 5 percent)
Cod	Bluefish	Albacore tuna
Croaker	Swordfish	Bluefin tuna
Flounder	Yellowfin tuna	Herring
Grouper		Mackerel
Haddock		Sablefish
Pacific halibut		Salmon
Pollock		Sardines
Red snapper		Shad
Rockfish		Trout
Sea bass		Whitefish
Shark		
Sole		
Whiting		

TABLE 4-10 Seafood Nutrition Chart (based on 3-1/2-ounce portions)

Species	Kcalories	Fat (grams)	Saturated Fat (grams)*	Cholesterol (milligrams)
Finfish				
Carp, cooked, dry heat	162	7	1	84
Cod, Atlantic, cooked, dry heat	105	1	—	55
Grouper, cooked, dry heat	118	1	—	47
Haddock, cooked, dry heat	112	1	—	74
Halibut, cooked, dry heat	140	3	1	41
Herring, Atlantic, cooked, dry heat	203	12	3	77
Mackerel, Atlantic, cooked, dry heat	262	18	4	75
Perch, cooked, dry heat	117	1	—	115
Pike, northern, cooked, dry heat	113	1	—	50
Pollock, walleye, cooked, dry heat	113	1	—	96
Pompano, cooked, dry heat	211	12	5	64
Salmon, coho, cooked, moist heat	185	8	1	49
Salmon, sockeye, canned, drained solids with bone	153	7	2	44
Sea bass, cooked, dry heat	124	3	1	53
Smelt, rainbow, cooked, dry heat	124	3	1	90
Snapper, cooked, dry heat	128	2	—	47
Swordfish, cooked, dry heat	155	5	1	50
Trout, rainbow, cooked, dry heat	151	4	1	73
Tuna, bluefish, fresh, cooked, dry heat	184	6	2	49
Whiting, cooked, dry heat	115	2	—	84
Shellfish				
Clam, cooked, moist heat	148	2	—	67
Crab, Alaska king, cooked, moist heat	97	2	—	53
Crayfish, cooked, moist heat	114	1	—	178
Lobster, northern, cooked, moist heat	98	1	—	72
Oyster, eastern, cooked, moist heat	137	5	1	109
Scallops, raw	88	1	—	33
Shrimp, cooked, moist heat	99	1	—	195

* A dash (—) means less than 1 gram of saturated fat.

Source: United States Department of Agriculture.

- Veal: any trimmed cut except commercially ground and veal patties
- Pork: pork tenderloin, pork chop (sirloin), pork chop (top loin), pork chop (loin)
- Lamb: shank, sirloin
- Poultry: breast (skin removed after cooking)
- Fish: all fish and shellfish

The second step is to order a quality product, such as USDA choice for beef (avoid prime—it contains more fat), from a reputable purveyor. Check the product for quality when it is received, and reject it if necessary. Put it immediately into the refrigerator or freezer. Fresh meat should be wrapped. Fresh poultry and whole fish should be stored on shaved or crushed ice. The ice should be put in drip pans and changed daily. Whole fish should be drawn (remove entrails) as soon as possible. Fish steaks (cross-cut section), fillets (boneless lengthwise section), or other cut fish should be wrapped in moisture-proof packaging before being placed on ice. If icing is not possible, the refrigerator should maintain a temperature between 30 and 34°F. Fresh poultry and fish should be used preferably within 24 hours.

Preparation

The third step is to trim the meat of all visible fat and use other techniques that decrease calories, fat, and cholesterol. Of course, even after fat is trimmed, the meat still contains invisible fat. Studies performed on cooked poultry have shown that poultry cooked with the skin on (where most of the fat lurks) does not significantly add fat to the poultry meat itself and does help prevent the meat from drying out. So it's a good idea to cook poultry with the skin, then remove the skin before serving. Select a 4- to 5-ounce raw portion to produce a 3- to 3-1/2-ounce cooked portion.

Rubs—A dry marinade made of herbs and spices (and other seasonings), sometimes moistened with a little oil, and rubbed or patted on the surface of meat, poultry, or fish (which is then refrigerated and cooked at a later time).

Marinades—A seasoned liquid used before cooking to flavor and moisten foods; usually based on an acidic ingredient.

The fourth step is to use flavorful rubs and marinades, when appropriate, to allow new and creative flavor options. **Rubs** combine dry ground spices, such as cinnamon, and finely cut herbs, such as cilantro. Rubs may be dry or wet. Wet rubs, also called pastes, use liquid ingredients such as mustard or vinegar. Pastes produce a crust on the food. Wet or dry seasoning rubs work particularly well with beef, and can range from a mesquite barbecue seasoning rub to a Jamaican jerk rub. To make a rub, various seasonings are mixed together and spread or patted evenly on the meat just before cooking for delicate items, or up to 24 hours in advance for heartier meat cuts (see Table 4-11 for the ingredients in 13 Cajun Spice Rub). The larger the piece of meat or poultry, the longer the rub can stay on. The rub flavors the exterior of the meat as it cooks.

Marinades, seasoned liquids used for soaking a food before cooking, are useful for adding flavor as well as for tenderizing meat and poultry. Marinades bring out the strongest flavors naturally, so you don't need to drown the food in fat, cream, or sauces. Marinades allow a food to stand on its own with a light dressing, chutney, sauce, or relish (discussed more in Chapter 8). Fish can also be marinated. Although fish is already tender, a short marinating time (about 30 minutes) can develop a unique flavor. A marinade usually contains an acidic ingredient, such as wine, beer, vinegar, citrus juice, or plain yogurt, to break down the tough meat or poultry. The other ingredients add flavor. Without the acidic ingredient, you can marinate fish for a few hours to instill flavor. Oil is often used in marinades to carry flavor, but it isn't essential. Fat-free salad dressings such as Italian

TABLE 4-11

13 Cajun Spice Rub

INGREDIENTS

4 cups paprika
2 cups chili powder
4 tablespoons cayenne pepper
4 tablespoons black and white pepper
4 tablespoons garlic powder
4 tablespoons onion powder
2 cups cumin
4 tablespoons thyme
4 tablespoons oregano
4 tablespoons marjoram
4 tablespoons basil
4 tablespoons gumbo filé
4 tablespoons fennel powder

work well in marinades. To give marinated foods flavor, try minced fruits and vegetables, low-sodium soy sauce, mustard, fresh herbs, and spices. For example, fruit-juice marinades can be flavored with Asian seasonings such as ginger and lemongrass. Even a simple fat-free Italian dressing can serve as a marinade.

Cooking

The fifth step is to choose a cooking method that will produce a flavorful, moist product and that adds little or no fat to the food. Possibilities include roasting, grilling, broiling, sautéing, poaching, and braising (discussed in detail in Chapter 8). The sixth step is to think of how you want to flavor the dish (discussed in detail in Chapter 8). For example, smoking can be used to complement the taste of meat, poultry, or fish. Hardwoods or fruitwoods, such as the following, are best for producing tasty foods:

- **Fruit (apple, cherry, peach).** These woods are too strong for fish but work well with pork, chicken, or turkey.
- **Hickory and maple.** These are great for beef or pork.
- **Mesquite.** Mesquite produces an aromatic smoke that works well with beef and pork.

The seventh step is to make the portion look large and attractive on the plate. Keep in mind that a cooked 3-ounce serving of meat, poultry, or fish is about the size of a deck of cards. Slice the meat or poultry thin and fan it across the plate so it looks like more. More information on presenting foods is found in Chapter 8.

CHEF'S TIPS

- Marinate top sirloin butt or sirloin tip with tomato juice, herbs, and spices. Cut into strips and use in fajitas or stir-fries. Cut into cubes and grill them as kabobs.

- Grilling is a wonderful way to add flavor to meats, poultry, and fish without adding fat. Since it must be done just prior to serving, the food is always fresh. Marinate your foods ahead of time to add flavor and moisture.
- Remember food safety when marinating foods. Do not reuse liquid marinades. Marinate meat, poultry, and fish in covered, clean, sanitized pans in the coldest part of the refrigerator on the bottom shelf, to avoid cross-contamination of other foods. If you make a marinade and part of it is to be used in the sauce, put that portion aside *before* you marinate the meat, poultry, or fish.
- Organic chicken is very much worth the extra money for its superb sweet taste. When you butcher the whole chicken, there is also less fat under the skin.
- Fish is a very versatile and nutritious food. Anything, such as rice or beans or pasta, goes with fish. Serve fish on top of a vegetable ragout, or try salmon with couscous.
- You can marinate fish without any citrus, which ruins the texture of the fish if it is marinated for very long. By eliminating citrus, you can marinate the fish longer, for two hours or even overnight. Try a marinade that includes fish stock, chives, tarragon, thyme, and black pepper. The fish will absorb some liquid, which keeps it moister during cooking.
- Cedar-planked fish is another way to add flavor. Soak an untreated cedar plank, then put marinated fish on it and bake in the oven. The cedar plank will impart a unique flavor.
- Fish must be cooked very carefully. Fish is done when it *just* separates into flakes and turns opaque. Once cooking is completed, serve immediately for the best flavor and texture.

Check-Out Quiz

Directions: In the following columns, check off each food that is a significant source of fat and/or cholesterol.

Food	Fat	Cholesterol
1. Butter	______	______
2. Margarine	______	______
3. Split peas	______	______
4. Peanut butter	______	______
5. Porterhouse steak	______	______
6. Flounder	______	______
7. Skim milk	______	______
8. Cheddar cheese	______	______

Food	Fat	Cholesterol
9. Chocolate chip cookie made with vegetable shortening	_______	_______
10. Green beans	_______	_______

Match each statement on the left with the term on the right that it describes. The terms will be used more than once.

_______	1. Present in every cell in the body	a. Lecithin
_______	2. Emulsifies fats	b. Cholesterol
_______	3. Found only in animal foods	
_______	4. Vital component of cell membranes	
_______	5. Used to make bile	

Match each numbered statement with the lettered term it describes.

_______	1. Cottonseed oil is a good source of this fat	a. Rancidity
_______	2. Lessens possibility of blood clots	b. Chylomicron
_______	3. Liquid at room temperature	c. Monounsaturated fat
_______	4. Deterioration of fat in air and heat	d. Oil
_______	5. Olive oil is a good source of this fat	e. Emulsifier
_______	6. Breaks up fat globules into small fatty acid droplets	f. Polyunsaturated fat
_______	7. Carries triglycerides and cholesterol through the lymph to the blood	g. Long-chain omega-3 fatty acid

Activities and Applications

1. Self-Assessment

To find out if your diet is high in fat, saturated fat, and cholesterol, check *yes* or *no* for the following questions.

Do You Usually:	Yes	No
1. Put butter on popcorn?	_______	_______
2. Eat more red meats (beef, pork, lamb) than chicken and fish?	_______	_______
3. Leave the skin on chicken?	_______	_______
4. Eat whole-milk cheeses, such as cheddar, American, and Swiss, more than three times a week?	_______	_______
5. Sauté or fry foods more than once or twice a week?	_______	_______
6. Eat regular lunch meats, hot dogs, and bacon more than three times a week?	_______	_______
7. Leave visible fat on meat?	_______	_______
8. Use regular creamy salad dressings such as Russian, blue cheese, Thousand Island, and creamy French?	_______	_______

Do You Usually:	Yes	No
9. Eat potato chips, nacho chips, and/or cream dips more than twice a week?	______	______
10. Drink whole milk?	______	______
11. Eat more than four eggs a week?	______	______
12. Eat organ meats (liver, kidney, etc.) more than once a week?	______	______
13. Use mayonnaise, margarine, and/or butter often on your sandwiches?	______	______
14. Use vegetable shortening in baking or cooking?	______	______
15. Eat commercially baked goods, including cakes, pies, and cookies, more than twice a week?	______	______
Ratings: If you answered yes to:		
1–3 questions: You are probably eating a diet not too high in fat, saturated fat, and cholesterol.		
4–7 questions: You could afford to make some food substitutions, such as skim for regular milk, to reduce your fat and saturated fat intake.		
8–15 questions: Your diet is very likely to be high in fat, saturated fat, and cholesterol.		

2. Changing Eating Habits

If you are eating too much fat, you can make changes a little at a time. Check off one of these things to try (if you are not already doing it) or make up your own.

- The next time, I eat chicken, I will take the skin off.
- I will limit my daily meat and poultry servings to two 3-ounce servings a day. A 3-ounce serving is about the size of a deck of cards.
- This week, I will try a new type of fresh or frozen fish.
- I will try a low-fat cheese, such as low-fat Swiss.
- I will switch to 1-percent or skim milk.
- I will try sherbet or ice milk for dessert instead of ice cream.
- I will count the number of eggs I eat a week and see whether I meet the recommendations.
- I will try to use a lower-in-fat margarine, salad dressing, or mayonnaise.
- I will keep more fruit and vegetables in the refrigerator, so they will be handy for a snack instead of cookies or chips.
- I will buy pretzels instead of potato chips.
- For breakfast, instead of doughnuts, I will try a hot or cold cereal with skim milk and toast with jelly.
- I will top my spaghetti with stir-fried vegetables instead of a creamy sauce.

3. Reading Food Labels

Following are food labels from two brands of lasagne. One is heavy on cheese and ground beef. The other is a vegetable lasagne made with moderate amounts of cheese. Using the Nutrition Facts given, can you tell which is which? How did you tell?

Lasagne #1	Lasagne #2
NUTRITION FACTS	NUTRITION FACTS
Amount per serving	Amount per serving
Calories 230	Calories 140
Calories from Fat 140	Calories from Fat 20
Total Fat 16 g	Total Fat 2.5 g
Saturated 6 g	Saturated 0.5 g
Polyunsaturated 1 g	Polyunsaturated 1.0 g
Monounsaturated 7 g	Monounsaturated 0.5 g
Cholesterol 74 g	Cholesterol 0 g
Sodium 180 mg	Sodium 180 mg
Total Carbohydrate 0 g	Total Carbohydrate 21 g
Dietary Fiber 0 g	Dietary Fiber 5 g
Sugar 0 g	Sugar 0 g
Protein 21 g	Protein 8 g

4. Meat, Poultry, and Seafood Comparison

Pick out three meat items you eat, three poultry items you eat, and three fish/shellfish items you eat. Make a chart listing the calories, fat, saturated fat, and cholesterol of all these foods. Once the chart is done, ask yourself the following questions:

- Which food has the least/most fat?
- Which food has the least/most saturated fat?
- Which food has the least/most cholesterol?

5. Name That Fat Substitute!

Following are ingredient listings from four products made with fat substitutes. Using Hot Topics: *Fat Substitutes* on page 177 as a guide, identify the fat substitutes in these foods.

Creme-Filled Chocolate Cupcakes—0 grams fat per cupcake
Sugar, water, corn syrup, bleached flour, egg whites, nonfat milk, defatted cocoa, invert sugar, modified food starch (corn, tapioca), glycerine, fructose, calcium carbonate, natural and artificial flavors, leavening, salt, dextrose, calcium sulfate, oat fiber, soy fiber, preservatives, agar, sorbitan monostearate, mono- and diglycerides, carob bean gum, polysorbate 60, sodium stearoyl lactylate, xanthan gum, sodium phosphate, maltodextrin, guar gum, pectin, cream of tartar, sodium aluminum sulfate, artificial color.
Low-Fat Mayonnaise Dressing—1 gram fat per tablespoon
Water, corn syrup, liquid soybean oil, modified food starch, egg whites, vinegar, maltodextrin, salt, natural flavors, gums (cellulose gel and gum, xanthan), artificial colors, sodium benzoate and calcium disodium EDTA.

Lite Italian Dressing—0.5 gram fat per 2 tablespoons
Water, distilled vinegar, salt, sugar, contains less than 2% of garlic, onion, red bell pepper, spice, natural flavors, soybean oil, xanthan gum, sodium benzoate, potassium sorbate and calcium disodium EDTA, yellow 5 and red 40.
Light Cream Cheese—5 grams fat per 2 tablespoons
Pasteurized skim milk, milk, cream, contains less than 2% of cheese culture, sodium citrate, lactic acid, salt, stabilizers (xanthan and/or carob bean and/or guar gums), sorbic acid, natural flavor, vitamin A palmitate.

6. Baking with Fat Substitutes
In this activity, you will prepare three batches of brownies using mixes. Make the first batch using the instructions on the box. For the second batch, you will use applesauce or prune purée in place of all the oil called for in the recipe. For the third batch, you will substitute half the oil with applesauce or prune purée. When the brownies are all baked, put some of each type on separate paper plates and find out if your classmates can tell which product is which. You should also ask your classmates to rank the products in terms of taste, texture, appearance, and overall acceptability.

If you want to use prune purée in other baked goods, try a recipe from *www.sunsweet.com*.

Nutrition Web Explorer

The American Heart Association www.deliciousdecisions.org
Visit the nutrition site for the American Heart Association and look at the recipes in their cookbooks. Write down three cooking methods, three seasonings, and three cooking substitutions that are heart-healthy.

California Olive Oil www.olive-oil.com
Visit this website of an olive oil producer and find out what an oil mister is by clicking on "Housewares." Also write down five types of infused oils.

Low-Fat Baking www.orst.edu/dept/ehe/nutrition/puree.html
Visit this website at Oregon State to learn about substituting applesauce and prune purée for oil in cooking and baking.

Food Facts *Oils and Margarines*

There is an ever-widening variety of oils and margarines on the market. They can differ markedly in their color, flavor, uses, and nutrient makeup.

When choosing vegetable oils, pick those high in monounsaturated fats, such as olive oil, canola oil, and peanut oil. Olive oil contains from 73 to 77 percent monounsaturated fat. The color of olive oil varies from pale yellow to dark green and its flavor from subtle to a full, fruity taste. The color and flavor of olive oil depend on the olive variety, level of ripeness, and processing method. When buying olive oil, make sure you are buying the right product for your intended use.

- Extra-virgin olive oil, the most expensive form, has a rich, fruity taste that is ideal for flavoring finished dishes and in salads, vegetable dishes, marinades, and sauces. It is not usually used for cooking because it loses some flavor. It is made by putting mechanical pressure on the olives, a more expensive process than using heat and chemicals.
- Olive oil, also called pure olive oil, is golden and has a mild, classic flavor. It is an ideal, all-purpose product that is great for sautéing, stir-frying, salad dressings, pasta sauces, and marinades.
- Light olive oil refers only to color or taste. These olive oils lack the color and much of the flavor found in the other products. Light olive oil is good for sautéing or stir-frying because the oil is used mainly to transfer heat rather than to enhance flavor.

Polyunsaturated fats, such as corn oil, safflower oil, sunflower oil, and soybean oil are also good choices, but not as good as monounsaturated fats such as olive oil, canola oil, and peanut oil.

Table 4-12 gives information on various oils. Be prepared to spend more for exotic oils such as almond, hazelnut, sesame, and walnut. Because these oils tend to be cold-pressed (meaning they are processed without heat), they are not as stable as the all-purpose oils and should be purchased in small quantities. They are strong, so you don't need to use much of them. Don't purchase these oils to cook with—they burn easily.

Vegetable oils are also available in convenient sprays that can be used as nonstick spray coatings for cooking and baking with a minimal amount of fat. Vegetable-oil cooking sprays come in a variety of flavors (butter, olive, Italian, mesquite), and a quick two-second spray adds about 1 gram of fat to the product. To use, first spray the cold pan away from any open flames (the spray is flammable), heat up the pan, then add the food.

Margarine was first made in France in the late 1800s to provide an economical fat for Napoleon's army. It didn't become popular in the United States until World War II, when it was introduced as a low-cost replacement for butter. Margarine must contain vegetable oil and water and/or milk or milk solids. Flavorings, coloring, salt, emulsifiers, preservatives, and vitamins are usually added. The mixture is heated and blended, then firmed by exposure to hydrogen gas at very high temperatures (see information about hydrogenation on page 147). The firmer the margarine, the greater the degree of hydrogenation and the longer its shelf life.

TABLE 4-12

Oil	Characteristics/Uses	Smoke Point
Canola oil (monounsaturated)	Light yellow color Bland flavor Good for frying, sautéing, and in baked goods Good oil for salad dressings	420° F
Corn oil	Golden color Bland flavor Good for frying, sautéing, and in baked goods Too heavy for salad dressings	420° F
Cottonseed oil	Pale yellow color Bland flavor Good for frying, sautéing, and in baked goods Good oil for salad dressings	420° F
Hazelnut oil	Dark amber color Nutty and smoky flavor Not for frying or sautéing as it burns easily Good for flavoring finished dishes and salad dressings Use in small amounts Expensive	*
Olive oil	Varies from pale yellow with sweet flavor to greenish color and fuller flavor to full, fruity taste (color and flavor depend on olive variety, level of ripeness, and how oil was processed)	
Olive Oil	Extra virgin or virgin olive oil—good for flavoring finished dishes and in salad dressings, strong olive taste	*
	Pure olive oil—can be used for sautéing and in salad oils, not as strong an olive taste as extra virgin or virgin	280° F
	Light olive oil—the least flavorful, good for sautéing, stir-frying, or baking	
Peanut oil	Pale yellow color Mild nutty flavor Good for frying and sautéing Good oil for salad dressings	420° F
Safflower oil	Golden color Bland flavor Has a higher concentration of polyunsaturated fatty acids than any other oil Good for frying, sautéing, and in baked goods Good oil for salad dressings	470° F
Sesame oil	Light gold flavor Distinctive, strong flavor Good for sautéing Good oil for flavoring dishes and in salad dressings Use in small amounts Expensive	440° F

TABLE 4-12 (continued)

Oil	Characteristics/Uses	Smoke Point	Oil	Characteristics/Uses	Smoke Point
Soybean oil	More soybean oil is produced than any other type; used in most blended vegetable oils and margarines Light color Bland flavor Good for frying, sautéing, and in baked goods Good oil for salad dressings	420° F	Sunflower oil	Pale golden color Bland flavor Good for frying, sautéing, and in baked goods Good oil for salad dressings	340° F
			Walnut oil	Medium yellow to brown color Rich, nutty flavor For flavoring finished dishes and in salad dressings Use in small amounts Expensive	*

* Not recommended for cooking.

Standards set by the U.S. Department of Agriculture and the Food and Drug Administration require margarine and butter to contain at least 80 percent fat by weight and to be fortified with vitamin A. One tablespoon of either has approximately 11 grams of fat and 100 kcalories. You can compare the fat profile of butter and margarine in Table 4-13. Butter contains primarily saturated fat and no polyunsaturated fat, whereas margarine is low in saturated fat and rich in monounsaturated and polyunsaturated fat. Although some margarines contain more trans fat than butter does, the total of trans and saturated fat is always less than the total for butter. The total for butter is much higher because of all the saturated fat it contains.

Butter must be made from cream and milk. Salt and/or colorings may be added. Margarine must contain vegetable oil and water and/or milk. Salt, food coloring, other vitamins, emulsifying agents such as lecithin, and preservatives may be added to margarine.

If a butter or margarine product does not contain at least 80 percent fat by weight, it can't be called "butter" or "margarine," but instead is classified as a *spread.* The percent of fat (by weight) must appear on the label. Water, gums, gelatins, and various starches are used in spreads to replace some or all of the fat, or air may be whipped into the product.

Margarines basically vary along these dimensions:

- **Physical form.** Margarine comes in either sticks or in tubs. Tub margarines contain more polyunsaturated fatty acids than stick margarines, so they melt at lower temperatures and are easier to spread. Spreads come in sticks, tubs, liquids, and pumps. Liquid spreads are packaged in squeeze bottles or pump dis-

TABLE 4-13 Total Kcalories and Fat in Selected Fats and Oils

Fat or Oil	Kcalories/ Tablespoon	Grams Fat/ Tablespoon	Grams Saturated Fat/ Tablespoon	Grams Trans Fat/ Tablespoon	Grams Monounsaturated Fat/Tablespoon	Grams Polyunsaturated Fat/Tablespoon
Coconut oil	120	14	12	0	1	0
Palm kernel oil	120	14	11	0	2	0
Palm oil	120	14	7	0	5	1
Butter, stick	108	12	8	0.3	4	0
Lard	115	13	5	N/A	6	1
Cottonseed oil	120	14	4	0	2	7
Olive oil	119	14	2	0	10	1
Canola oil	120	14	1	0	8	4
Peanut oil	119	14	2	0	6	4
Safflower oil	120	14	1	0	2	10
Corn oil	120	14	2	0	3	8
Soybean oil	120	14	2	0	3	8
Sunflower oil	120	14	1	0	6	6
Shortening	106	12	3	1.4–4.2	6	3
Margarine, stick	102	11	2	2.4	5	3
Margarine, soft tub	102	11	2	1	4	5
Margarine, liquid	102	11	2	varies	3	6
Margarine, whipped	70	7	2	varies	2	3
Margarine spread	78	9	1	varies	5	3
Margarine, diet	51	6	1	varies	2	2
Margarine, fat-free	0	0	0	0	0	0

Source: U.S. Department of Agriculture Food and Drug Administration, and manufacturers.

pensers in which the margarine spread is really liquid, even in the refrigerator. They work well when drizzled on hot vegetables and other cooked dishes. Fat-free sprays also can be used to coat cooking pans.

- **Type of vegetable oil(s) used.** The vegetable oil may be mostly corn oil, safflower oil, canola oil, or others. Check the ingredients label to compare how much liquid oil and/or partially hydrogenated oil are used. If the first ingredient is liquid corn oil, then more liquid oil is used, meaning that there will be less saturated fat than there is in a product with hydrogenated corn oil as the first ingredient.
- **Percentage of fat by weight and nutrient profile.** Margarine and spreads are available with amounts of fat that vary from 0 percent to 80 percent by weight. Look on the label for the percentage of fat by weight. Also look for terms such as *light, diet,* or *fat-free.* Light margarine contains one-third fewer calories or half the fat of the regular product. Diet margarine, also called reduced-calorie margarine, has at least 25 percent fewer calories than the regular product. Fat-free margarine has less than 0.5 gram of fat per serving. If you are wondering how they make a margarine fat-free, part of the answer is gelatin. Water, rice starch, and other fillers

are used to make it taste like fat. As mentioned in this chapter, information on the trans fat content of products will appear soon on labels. A number of margarines without any trans fats are available.

Not all margarines and spreads can be used in the same way. Spreads with lots of water can make bread or toast soggy, and may spatter and evaporate quickly in hot pans, causing foods to stick. In baking, low-fat spreads are not recommended because product quality suffers.

In addition to butter and margarine, blends and butter-flavored buds are also available. Blends are part margarine and part butter (about 15 to 40 percent). They are made of vegetable oils, milk fat, and other dairy ingredients added to make the product taste like butter. Blends may have as much fat as regular margarine or butter (in other words, at least 80 percent fat) or they may be reduced in fat. Butter-flavored buds are made from carbohydrates and a small amount of dehydrated butter. They are virtually fat free and cholesterol free, and are designed to melt on hot, moist foods such as a baked potato. When mixed with water, they can make butter-flavored sauces.

Two margarinelike spreads, Take Control (made by Lipton) and Benecol (made by McNeil Consumer Products), use ingredients that lower blood cholesterol levels. Take Control uses a stanol-like ingredient from soybeans. Benecol contains a plant stanol ester that comes from pine trees.

Lipton recommends one to two servings (1 to 2 tablespoons) of Take Control every day as part of a diet low in saturated fat and cholesterol. Take Control is low in saturated fat and free of trans fats, and contains 6 grams of fat and 1.1 grams soybean extract per serving. Benecol is recommended in three daily servings (1-1/2 tablespoons total) of its regular or light spread. Both products cost considerably more than regular margarine. These spreads are examples of functional foods, a topic discussed in detail in Chapter 6.

Hot Topic Fat Substitutes

Food manufacturers are making it easier for fat-conscious consumers to eat their cake and have it too—and their cheeses, chips, chocolate, cookies, ice cream, salad dressings, and various other foods that are now available in lower-fat versions. A host of fat substitutes that replace some, or sometimes all, of the fat in a food, makes these lower-fat foods possible. Most of these fat replacers are ingredients already approved by the Food and Drug Administration for other uses in food. For instance, starches and gums, two popular fat replacers, have long since been approved as thickeners and stabilizers. Newer fat replacers, such as olestra, have undergone or will undergo close scrutiny by the FDA to assess their safety.

Fat is a difficult substance to replace because it gives taste, consistency, stability, and palatability to foods. Fat replacers can help reduce a food's fat and calorie levels while maintaining some of the desirable qualities fat brings to food, such as mouth feel, texture, and flavor.

Under FDA regulations, fat replacers usually fall into one of two categories: food additives or "generally recognized as safe" (GRAS) substances. Each has its own set of regulatory requirements.

Food additives must be evaluated for safety and approved by the FDA before they can be marketed. They include substances with no proven track record of safety; scientists just don't know that much about their use in food. Examples of food additives are olestra, polydextrose, and carrageenan, which are used as fat replacers.

GRAS substances, on the other hand, do not have to undergo rigorous testing before they are used in foods because they are generally recognized as safe by knowledgeable scientists, usually because of the substances' long history of safe use in foods. Many GRAS substances are similar to substances already in food. Examples of GRAS substances used as fat replacers are cellulose gel, dextrins, and guar gum.

In addition to water, the simplest fat substitute, fat replacers may be carbohydrate-, protein-, or fat-based. The first to hit the market used carbohydrate as the main ingredient. Avicel, for example, is a cellulose gel introduced in the mid-1960s as a food stabilizer. Carrageenan, a seaweed derivative, was approved for use as an emulsifier, stabilizer, and thickener in food in 1961. Its use as a fat replacer became popular in the early 1990s. Polydextrose (made from dextrose and small amounts of sorbitol and citric acid) came on the market in 1981 as a humectant, meaning that it helps retain moisture in a food.

Other carbohydrate-based fat replacers include starch, modified food starch, dextrins and maltodextrins (both made from starch), fiber, and gums. Gums are made from seeds, seaweed extracts, and plants. Examples include xantham gum, guar gum, alginates, and cellulose gum. Gums thicken foods and can replace some of the fat. They are popular in diet salad dressings and other foods.

Fruit purées, such as prune purée, are also useful to replace some of the fat in a product. Fruit purées have been used successfully in baked products, where they add tenderness and moisture.

Although their original purpose was often to perform certain technical functions in food that would improve overall quality, many carbohydrate-based fat replacers are now used to replace fat and reduce calories. They provide from 0 to 4 calories per gram and are used in a variety of foods including dairy-type products, sauces, frozen desserts, salad dressings, processed meat, baked goods, spreads, chewing gum, and sweets.

A carbohydrate-based fat substitute, called Z-Trim, was developed by the U.S. Department of Agriculture from a variety of low-cost agricultural by-products such as hulls of oats, soybeans, peas, and rice, or bran from corn or wheat. The bran or hulls are processed into microscopic fragments, purified, and dried and milled to an easy-flowing powder. When the fragments absorb water, they swell to provide a smooth mouth feel. Z-Trim supplies moistness, density, and smoothness in foods ranging from reduced-calorie cheese products and hamburger to baked goods, without any calories.

Protein-based fat substitutes came along in the 1990s. These (and fat-based replacers) were designed specifically to replace fat in foods. One protein-based fat substitute, known as Simplesse, is made from egg white and milk protein, which are blended and heated using a process called microparticulation. The protein is shaped into microscopic round particles that roll easily over one another. The aim of the process is to create the feel of a creamy liquid and the texture of fat. Simplesse cannot be used to fry foods but can be used in some cooking and baking. It contains 1 to 2 calories per gram. Simplesse has been used in frozen desserts. Another type of protein-based fat replacers, called protein blends, combine animal or vegetable protein, gums, food starch, and water. They are made with FDA-approved ingredients and are used in frozen desserts and baked goods.

Olestra is an example of a fat-based fat replacer, also called an engineered fat. The FDA approved olestra (its brand name is Olean, made by Procter & Gamble) in 1996 for use in preparing potato chips, crackers, tortilla chips, and other savory snacks. Olestra has properties similar to those of naturally occurring fat, but it provides zero kcalories and no fat. That's because it is indigestible. It passes through the digestive tract, but is not absorbed into the body. This is due to its unique configuration: a center unit of sucrose (sugar) with six, seven, or eight fatty acids attached. Olestra's configuration also makes it possible for the substance to be exposed to high temperatures, such as in frying—a quality most other fat replacers lack.

As promising as olestra sounds, it does have some drawbacks. Studies show that it may cause intestinal cramps and loose stool in some individuals. Also, according to clinical tests, olestra reduces the absorption of fat-soluble nutrients, such as vitamins A, D, E, and K and carotenoids, from foods eaten at the same time as olestra-containing products.

To address these concerns, the FDA required that the four vitamins be added to olestra-containing foods and that the following statement appear on products made with olestra:

> This product contains olestra. Olestra may cause abdominal cramping and loose stools. Olestra inhibits the absorption of some vitamins and other nutrients. Vitamins A, D, E and K have been added.

Another fat-based replacer, salatrim, is the generic name for a family of reduced-calorie fats that are only partially absorbed in the body. Salatrim provides 5 kcalories per gram. It is used in a brand of reduced-fat baking chips. Caprenin is a 5-kcalorie-per-gram fat substitute for cocoa butter that is used in candy bars.

In using reduced-fat foods, be aware that fat-free does *not* mean kcalorie-free. The kcalories lost in removing regular fat from a food can be regained through sugars added for palatability, as well as fat replacers, many of which provide kcalories, too. Use the Nutrition Facts panel on the product to compare kcalories and other nutrition information between fat-reduced and regular-fat foods. Many nutrition experts agree that, used properly, fat replacers can play an important role in improving adult Americans' diets. But, as with any diet or food, variety and moderation are emphasized to ensure a healthy intake.

Chapter 5
Protein

Have you ever wondered why meat, poultry, and seafood are often considered entrées, or main dishes, whereas vegetables and potatoes are side dishes? As recently as the 1950s and 1960s, the abundant protein found in meat, poultry, and seafood was considered the mainstay of a nutritious diet. You could say that these foods took center stage, or more accurately, center plate. As a child, I can remember going to visit my grandparents on Sundays and eating a roast beef dinner during our visit. Yes, we had vegetables, too, but the big deal at dinner was the roast that was carefully cooked, sliced, and served (with brown gravy, of course).

Today, protein foods continue to be an important component of a nutritious diet; however, we are much more likely to see foods such as lentils or pasta occupying the center of the plate. For adults who grew up when beef was king (and not nearly as expensive as it is today) and full-fat bologna sandwiches filled many lunchboxes, making spaghetti without meatballs takes a little getting used to, but more and more meatless meals are being served.

Protein—Major structural parts of the body's cells that are made of nitrogen-containing amino acids assembled in chains, particularly rich in animal foods.

So just what is **protein?** It is an essential part of all living cells found in animals and plants. The protein found in animal and plant foods is such an important substance that the term *protein* is derived from the Greek word meaning "first." About 16 percent of your body weight (if you're not overweight) is protein. Proteins reside in your skin, hair, nails, muscles, and tendons, to name just a few places. They function in a very broad sense to build and maintain the body. This chapter discusses protein's structure, functions, metabolism, and relationship to diet, and will help you to:

- Identify and describe the building blocks of protein
- List the functions of protein in the body
- Explain how protein is digested, absorbed, and metabolized
- Distinguish between complete and incomplete protein and list examples of foods that contain each
- Explain the potential consequences of eating too much or too little protein
- State the dietary recommendations for protein
- Discuss the purchasing, storage, cooking, and menuing of milk, dairy products, and eggs

Structure of Protein

Like carbohydrates and fats, proteins contain carbon, hydrogen, and oxygen. Unlike carbohydrates and fats, proteins contain nitrogen and provide much of the body's nitrogen. Nitrogen is necessary for the body to function; life as we know it wouldn't exist without nitrogen.

Amino acids—The building blocks of protein.

Proteins are long chains of **amino acids** strung together much like railroad cars. Amino acids are the building blocks of protein. There are 20 different

Essential or indispensable amino acids—Amino acids that either cannot be made in the body or cannot be made in the quantities needed by the body; must be obtained in foods.

Nonessential amino acids—Amino acids that can be made in the body.

Conditionally essential amino acids—Nonessential amino acids that may, under certain circumstances, become essential.

Peptide bonds—The bonds that form between adjoining amino acids.

Polypeptides—Protein fragments with 10 or more amino acids.

Primary structure—The number and sequence of the amino acids in the protein chain.

ones, each consisting of a backbone to which a side chain is attached (Figure 5-1). The amino acid backbone is the same for all amino acids, but the side chain varies. It is the side chain that makes each amino acid unique.

Of the 20 amino acids in proteins (see Table 5-1), 9 either cannot be made in thc body or cannot be made in the quantities needed. They must therefore be obtained in foods for the body to function properly. This is why we call these amino acids **essential** or **indispensable amino acids.** The remaining 11, called **nonessential amino acids,** can be made in the body. However, under certain circumstances, one or more of these amino acids may become essential, which is referred to as **conditionally essential.**

When the amino-acid backbones join end to end, a protein forms (Figure 5-2). The bonds that form between adjoining amino acids are called **peptide bonds.** Proteins often contain from 35 to several hundred or more amino acids. Protein fragments with 10 or more amino acids are called **polypeptides.**

Each of the over 100,000 different proteins in the body contains its own unique number and sequence of amino acids. In other words, each protein differs in terms of what amino acids it contains, how many it contains, and the order in which they are contained. The number and sequence of the amino acids in the protein chain is called the **primary structure.** The number of possible arrangements is as amazing as the fact that all the words in the English language are made of different sequences of 26 letters. Also, some proteins are made of more than one chain of amino acids. For example, hemoglobin contains four chains of linked amino acids.

After a protein chain has been made in the body, it does not remain a straight chain. In the instant after a new protein is created, the side chain of each amino acid in the strand either attracts or repels other side chains,

Figure 5-1
An amino acid

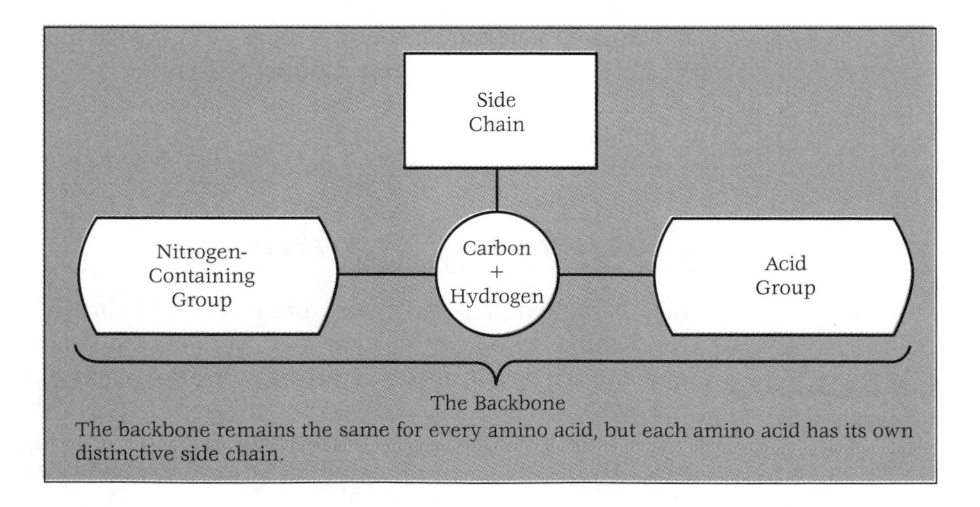

The backbone remains the same for every amino acid, but each amino acid has its own distinctive side chain.

TABLE 5-1 Amino Acids

Essential Amino Acids	Nonessential Amino Acids
Histidine	Alanine
Isoleucine	Arginine*
Leucine	Asparagine
Lysine	Aspartic acid
Methionine	Cysteine*
Phenylalanine	Glutamic acid
Threonine	Glutamine*
Tryptophan	Glycine*
Valine	Proline*
	Serine
	Tyrosine*

* Under some circumstances, these amino acids become essential.

Secondary structure—The bending and coiling of the protein chain.

Tertiary structure—The folding of the protein chain.

resulting in the protein bending and coiling (Figure 5-3). This bending and coiling is called the protein's **secondary structure.**

One more step must take place before the protein can do any work in the body. Due in part to the interaction of amino acids at some distance from each other in the chain (Figure 5-3), the protein folds and loops. This process of folding results in the protein's **tertiary structure.** In case you are wondering whether a protein's tertiary structure has any real importance, it does! A protein's tertiary structure—how it bends and folds—makes the protein able to perform its functions in the body. Proteins with two or more chains of amino acids fold and loop even more, which results in quaternary structure. So what exactly does a protein do? That's our next topic.

Figure 5-2
A part of a protein

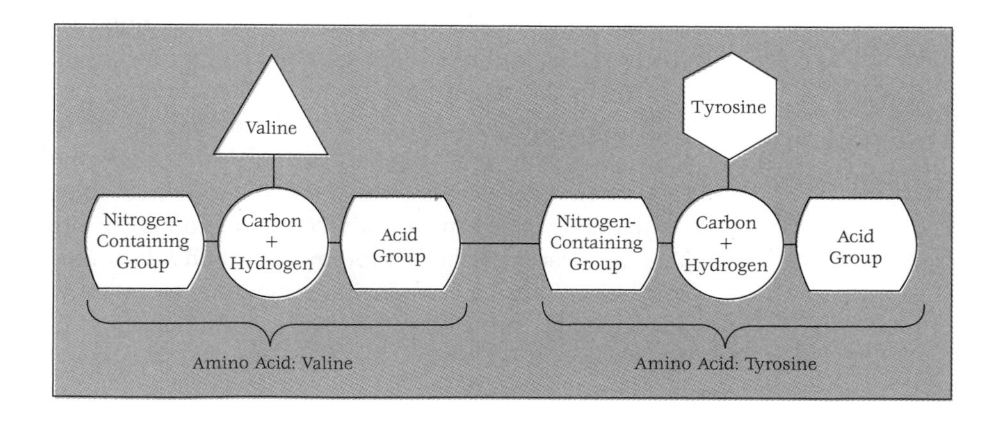

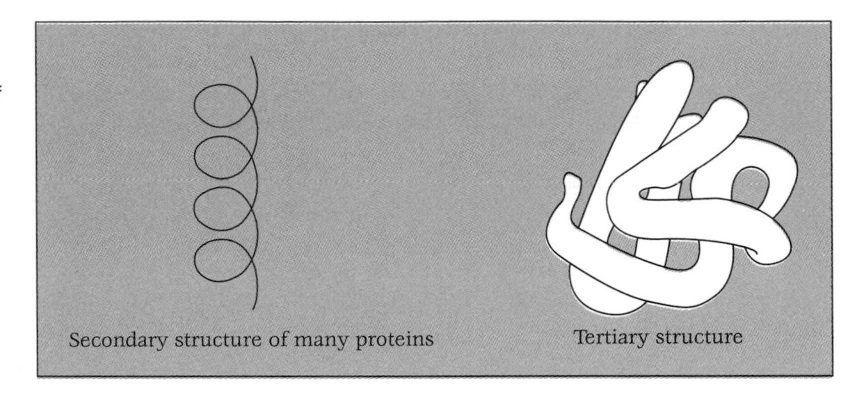

Figure 5-3
Primary structure—the number and sequence of amino acids; secondary structure—bends and coils; tertiary structure—folds and loops

■ **MINI-SUMMARY**

Proteins contain nitrogen. They are long chains of 20 different amino acids, some of which are essential, joined end to end by peptide bonds. Each protein has its own characteristic primary structure (number and sequence of amino acids), secondary structure (bending or coiling), and tertiary structure (folding and looping), which make it functional.

Functions of Protein

After reviewing all of the jobs proteins perform, you will have a greater appreciation of this nutrient. In brief, protein is part of most body structures; builds and maintains the body; is a part of many enzymes, hormones, and antibodies; transports substances around the body; maintains fluid and acid-base balance; and can provide energy for the body (Table 5-2). Now let's take a look at each function separately.

Proteins function as part of the body's structure. For example, protein can be found in skin, bones, hair, fingernails, muscles, blood vessels, the digestive tract, and blood. Protein appears in every cell.

Proteins are used for building and maintaining body tissues. Worn-out cells are replaced throughout the body at regular intervals. For instance, your skin today will not be the same skin in a few months. A skin cell only lives about one month. Skin is constantly being broken down and rebuilt or remodeled, as are most body cells, including the protein within the cells. The greatest amount of protein is needed when the body is building new tissues rapidly, such as during pregnancy or infancy. Additional protein is also needed when body protein is either lost or destroyed, as in burns, surgery, or infections.

TABLE 5-2 Functions of Protein

- Acts as a structural component of the body
- Builds and maintains the body
- Found in many enzymes and hormones, and all antibodies
- Transports iron, fats, minerals, and oxygen
- Maintains fluid and acid-base balance
- Provides energy as last resort
- Helps blood clot

Enzymes—Catalysts in the body.

Proteins are found in many **enzymes,** some hormones, and all antibodies. Thousands of enzymes have been identified. Almost all the reactions that occur in the body, such as food digestion, involve enzymes. Enzymes are catalysts, meaning that they increase the rate of these reactions, sometimes by more than a million times. They do this without being changed in the overall process. Enzymes contain a special pocket called the *active site.* You can think of the active site as a lock into which only the correct key will fit. Various substances will fit into the pocket, undergo a chemical reaction, and then exit the enzyme in a new form, leaving the enzyme to speed up other reactions. Enzymes help break down, build up, or transform substances.

Hormones—Chemical messengers in the body.
Homeostasis—A constant internal environment in the body.
Antibodies—Proteins in the blood that bind with foreign bodies or invaders.
Antigens—Foreign invaders in the body.
Immune response—The body's response to a foreign substance, such as a virus, in the body.

Hormones are chemical messengers secreted into the bloodstream by various organs, such as the liver, to travel to a target organ and influence what it does. Hormones regulate certain body activities so a constant internal environment (called **homeostasis**) is maintained. For example, the hormone insulin is released from the pancreas when your blood sugar level goes up, such as after eating lunch. Insulin pushes sugar from the blood into your cells, resulting in lower, more normal blood sugar levels. Amino acids are components of insulin as well as other hormones.

Antibodies are blood proteins whose job is to bind with foreign bodies or invaders (the scientific name is **antigens**) that do not belong in the body. Invaders could be viruses, bacteria, or toxins. Each antibody fights a specific invader. For example, there are many different viruses that cause the common cold. An antibody that binds with a certain cold virus is of no use to you if you have a different strain of the cold virus. However, exposure to a cold virus results in increased amounts of the specific type of antibody that can attack it. Next time that particular cold virus comes around, your body remembers and makes the right antibodies. This time the virus is destroyed faster, and your body's response (called the **immune response**) is enough to combat the disease.

Proteins also act as taxicabs in the body, transporting iron and other minerals, some vitamins, fats, and oxygen through the blood.

Acid-base balance—The process by which the body buffers the acids and bases normally produced in the body so the blood is neither too acidic nor too basic.
Acidosis—A dangerous condition in which the blood is too acidic.
Alkalosis—A dangerous condition in which the blood is too basic.

Protein plays a role in body fluid balance (to be discussed in Chapter 7), and the **acid-base balance** of the blood. Normal bodily processes produce acids and bases that can cause major problems, even death, if not buffered or neutralized. The blood must remain neutral; otherwise, dangerous conditions known as **acidosis** (above normal acidity) and **alkalosis** (above normal alkalinity) can occur. Some blood proteins have the chemical ability to buffer, or neutralize, both acids and bases.

In addition, amino acids can be burned to supply energy (4 kcalories per gram) if absolutely needed. Of course, burning amino acids for energy takes them away from their vital functions. Some amino acids can also be converted to glucose when necessary to maintain normal blood glucose levels.

Protein is also important for blood clotting.

■ MINI-SUMMARY

Protein is part of most body structures; builds and maintains the body; is a part of many enzymes, hormones, and antibodies; transports substances around the body; maintains fluid and acid-base balance; provides energy for the body; and is important for proper vision and blood clotting.

Denaturation

Denaturation—A process in which a protein uncoils and loses its shape, causing it to lose its ability to function; it can be caused by high temperatures, whipping, and other circumstances.

Under certain circumstances, a protein uncoils and loses its shape, causing it to lose its ability to function. This process is called **denaturation.** In most cases, the damage cannot be reversed. Denaturation can occur both to proteins in food and to proteins in our bodies.

Denaturation can be caused by high temperatures (as in cooking), ultraviolet radiation, acids and bases, agitation or whipping, and high salt concentration. For example, when you fry an egg, the proteins in the egg white become denatured and turn from clear to white. Gluten, the protein in flour, denatures during baking to give bread and other baked goods their structure. Denaturation of protein can also occur in the body whenever the blood becomes too acidic or too basic.

■ MINI-SUMMARY

When a protein is denatured, it uncoils and loses its shape so it can no longer function. Protein foods denature during cooking, and proteins in the body can denature if the blood becomes too acidic or basic.

Digestion, Absorption, and Metabolism

Pepsin—The principal digestive enzyme of the stomach.

Amino acid pool—The overall amount of amino acids distributed in the blood, organs, and body cells.

Proteins cannot be absorbed across the intestinal membranes until they are broken down into their amino acid units. Protein digestion starts in the stomach, where stomach acid uncoils the proteins (denaturation) enough to allow enzymes to enter them to do their work. The acid in the stomach, called hydrochloric acid, also converts a substance called pepsinogen to the stomach enzyme **pepsin.** Pepsin splits peptide bonds, making proteins shorter in length.

The next stop is the small intestine where protein digestion is completed. Proteases (enzymes made by the pancreas and small intestines) split up the proteins into amino acids and dipeptides. These smaller units are absorbed by the microvilli in the walls of the small intestine. Because amino acids are water soluble, they travel easily in the blood to the liver and then to the cells that require them. The dipeptides must be split into two amino acids before entering the blood.

An **amino acid pool** in the body provides the cells with a supply of amino acids for making protein. The amino acid pool refers to the overall amount of amino acids distributed in the blood, the organs (such as the liver), and the body's cells. Amino acids from foods, as well as amino acids from body proteins that have been dismantled, stock these pools. In this manner, the body recycles its own proteins. If the body is making a protein and can't find an essential amino acid for it, the protein can't be completed, and the partially completed protein is disassembled or taken apart. This is important to consider for the next section on protein quality.

■ MINI-SUMMARY

Protein digestion takes place in the stomach and small intestine, where stomach acid and enzymes help break up food proteins into amino acids to be absorbed across the wall of the small intestine. Because they are water-soluble, amino acids easily travel in the blood (as part of the amino acid pool) to the liver and cells that require them.

Protein in Food

Protein is found in animal and plant foods (Table 5-3). Animal foods, such as beef, chicken, fish, and dairy products, have the most protein. Of the plant foods, grains, legumes, and nuts usually contribute more protein than vegetables and fruits. Protein-rich animal foods are usually higher in fat

TABLE 5-3 Fat, Saturated Fat, Protein, Cholesterol, and Fiber in Animal and Plant Foods

Animal Foods	Fat (grams)	Saturated Fat (grams)	Protein (grams)	Cholesterol (milligrams)	Fiber (grams)
Beef, ground, broiled, 3 oz.	16	6	21	74	0
Chicken breast, roasted, 3 oz.	3	1	27	73	0
Cod, baked, 3 oz.	1	0	19	47	0
Milk, 2%, 8 fl. oz.	5	3	8	18	0
Cheese, American, 1 oz.	9	6	6	27	0
Egg, 1	6	2	6	274	0
Plant Foods					
Lentils, cooked, 1/2 cup	0	0	8	0	5
Peanut butter, 2 tablespoons	16	3	10	0	2
Brown rice, cooked, 1/2 cup	1	0	2	0	1
Spaghetti, whole wheat, 1 cup	1	0	7	0	3
Whole-wheat bread, 2 slices	2	0	6	0	3
Broccoli, chopped 1 cup	0	0	6	0	3
Apple, 1 medium	0.5	0	0.3	0	3

Source: United States Department of Agriculture Handbook Number 72, 8-1, 8-5, 8-13, 8-15, 8-20.

Complete proteins—Food proteins that provide all of the essential amino acids in the proportions needed.

Limiting amino acid—An essential amino acid in lowest concentration in a protein.

Incomplete proteins—Food proteins that contain at least one limiting amino acid.

and saturated fat, and always higher in cholesterol, than plant foods (plant foods have no cholesterol). Protein-rich foods also tend to be the most expensive foods on the menu.

To understand the concept of protein quality, you need to recall that 9 of the 20 amino acids either can't be made in the body or can't be made in sufficient quantity. Food proteins that provide all of the essential amino acids in the proportions needed are called high-quality or **complete proteins.** Examples of complete proteins include the animal proteins, such as meats, poultry, fish, eggs, milk, and other dairy products.

An essential amino acid in lowest concentration in a protein is referred to as a **limiting amino acid** because it limits the protein's usefulness unless another food in the diet contains it. Lower-quality or **incomplete protein** contains at least one limiting amino acid. Plant proteins, including dried beans and peas, grains, vegetables, nuts, and seeds, are incomplete. When certain plant foods, such as peanut butter and whole-wheat bread,

Complementary proteins—The ability of two protein foods to make up for the lack of certain amino acids in each other.

are eaten over the course of a day, the limiting amino acid in each of these proteins is supplied by the other. Such combinations are called **complementary proteins.**

Although plant proteins are incomplete, they are not low-quality. When plant proteins are eaten with other foods, the food combinations usually result in complete protein. This is the case, for example, when grains are consumed with legumes. Some plant proteins, such as the grains amaranth and quinoa and protein made from soybeans (isolated soy protein), are complete proteins. In adequate amounts and combinations, plant foods can supply the essential nutrients needed for growth and development and overall health. Many cultures around the world use plant proteins extensively. On average, plant protein foods contribute 65 percent of the protein for each person in the world, but for North America, plant protein foods contribute only about 32 percent of the protein for each person.

Researchers have developed various ways to score the quality of food proteins. They judge them on how much of their nitrogen the body retains, or how well the proteins support growth or maintenance of body tissue. Animal proteins tend to score higher than vegetable proteins, and animal protein is also more digestible. Protein scores have little use in countries where protein consumption is adequate, but are useful to scientists working in countries where protein intakes are low.

■ MINI-SUMMARY

Animal proteins are examples of complete proteins, and most plant proteins are examples of incomplete proteins. By eating complementary plant proteins, you can overcome the problem presented by limiting amino acids and eat a nutritionally adequate diet. In the right amounts and combinations, plant proteins can support growth and maintenance.

Protein and Health

If you eat too much or too little protein, your health may be affected. First let's look at a high-protein diet.

Eating too much protein has no benefits. It will not result in bigger muscles, stronger bones, or increased immunity. In fact, eating more protein than you need may add excessive kcalories beyond what you require. Extra protein is not stored as protein but is stored as fat if too many calories are being taken in. High-protein foods also are often rich in fat and therefore high in kcalories.

Diets high in protein can also be a concern if you are eating a lot of high-fat animal foods, such as hamburgers and cheese, and few vegetable proteins. Comparison of the fat and fiber content of animal and vegetable proteins (Table 5-3) makes clear that plant sources of protein contain less fat and more fiber. They also contain no cholesterol and are rich in vitamins and minerals. Eating too much high-fat animal food, which contains much saturated fat, raises your blood cholesterol levels, which in turn increases your risk of cardiovascular disease. By eating fewer plant foods, you are also missing out on good sources of fiber, vitamins and minerals, and phytochemicals.

Published studies show that increased protein intake leads to increased calcium loss. This does not necessarily mean that everyone who takes in too much protein is calcium-deficient, since the body will make up for this loss by absorbing more calcium in the intestine. However, if an individual has a high protein intake and a low calcium intake, the increased calcium absorption won't compensate enough for its loss. Published studies also show that high protein intakes tax the kidneys and can worsen kidney problems in patients with renal (kidney) disease.

High-protein diets have been popular for many years as a way to lose weight. People lose weight initially with these diets, but it is mostly water loss due to the restriction on eating carbohydrates (which retain water in the body). While these diets are not normally harmful for a short period of time, they are not recommended because they restrict healthful foods you need and may cause heart, bone, kidney, and liver abnormalities over time.

On the other hand, eating too little protein can cause problems too, such as slowing down the protein rebuilding and repairing process and weakening the immune system. Developing countries have the most problems with **protein-energy malnutrition (PEM).** PEM refers to a broad spectrum of malnutrition, from mild to serious cases. PEM can occur in infants, children, adolescents, and adults, although it is seen most often in infants and children. PEM develops gradually over weeks or months. In mild cases of PEM, there is weight loss, stunted or slowed growth, and more sedentary behavior.

Protein-energy malnutrition (PEM)—A broad spectrum of malnutrition from mild to serious cases.

Kwashiorkor—A type of PEM associated with children with insufficient protein intake and who have a preexisting disease.

Marasmus—A type of PEM characterized by gross underweight and severe food shortage.

In severe cases of PEM, physicians often see the clinical syndromes called kwashiorkor and marasmus. **Kwashiorkor** is usually seen in children with an existing disease who are getting totally inadequate amounts of protein and only marginal amounts of kcalories. Characterized by retarded growth and development, the child has a protruding abdomen due to edema (swelling), peeling skin, a loss of normal hair color, irritability, and sadness. **Marasmus** is characterized by severe insufficiency of kcalories and protein, which accounts for the child's gross underweight, lack of fat stores, and wasting away of muscles. There is no edema. Whereas marasmus is usually associated with severe food shortage and prolonged semistarvation, kwashiorkor is associated with poor protein intake and early weaning from mother's milk due to arrival of a new baby.

■ **MINI-SUMMARY**

There is no benefit to eating too much protein. Eating too much high-fat animal foods can increase your blood cholesterol levels, increase calcium loss from the body (a concern when calcium intake is low), and worsen kidney problems in people with renal disease. Eating too little protein is associated with protein-energy malnutrition.

Dietary Recommendations for Protein

The 2002 RDA for protein for both men and women is 0.8 gram/kilogram of body weight. For healthy adults, the RDA works out to be 0.36 gram of protein per pound of body weight. For example, if you weigh 140 pounds, you need 50 grams of protein each day.

$$140 \text{ pounds} \times 0.36 \text{ grams/pound} = 50 \text{ grams of protein}$$

This amount allows for adequate protein to make up for daily losses in urine, feces, hair, and so on. In other words, taking in enough protein each day to balance losses results in a state of protein balance, called **nitrogen balance.** The RDA for protein is generous and is based on the recommendation that proteins come from both animal and plant foods.

Nitrogen balance—The difference between the total nitrogen intake and total nitrogen loss; a healthy person has the same nitrogen intake as loss, resulting in a zero nitrogen balance.

Positive nitrogen balance—A condition in which the body excretes less protein than is taken in; this can occur during growth and pregnancy.

Negative nitrogen balance—A condition in which the body excretes more protein than is taken in; this can occur during starvation and certain illnesses.

The amount of protein needed daily is proportionally higher during periods of growth, such as during infancy, childhood, and pregnancy. Accordingly the RDA for protein is higher than 0.8 gram/kilogram of body weight during these times.

0–6 months old	9.1 grams per day (This is an AI, not RDA.)
7–12 months old	1.5 grams protein/kilogram body weight
1–3 years old	1.1 grams protein/kilogram body weight
4–8 years old	0.95 gram protein/kilogram body weight
9–13 years old	0.95 gram protein/kilogram body weight
14–18 years old	0.85 gram protein/kilogram body weight
Pregnancy and lactation (all age groups)	1.1 grams protein/kilogram body weight OR 25 grams additional protein/day

(Pregnant and lactating women need to get either 1.1 grams of protein/kilogram body weight *or* they can eat 25 grams protein above their normal RDA before they were pregnant or lactating.) During periods of growth when a person needs to eat more protein than is lost, the body is said to be in a state of **positive nitrogen balance. Negative nitrogen balance** occurs during starvation and some illnesses when the body excretes more protein than is taken in.

In the United States, meeting the RDA for protein is rarely a problem. Most Americans eat more than the RDA. According to the U.S. Department of Agriculture surveys, 14 to 18 percent of calories in the American diet come from protein, with animal protein contributing about 65 percent.

The 2002 Dietary Reference Intake report established Acceptable Macronutrient Distribution Ranges (AMDR) for protein. Adults should get from 10 to 35 percent of total kcalories from protein. The AMDR for children from 1 to 3 years old is 5 to 20 percent of kcalories, and for children 4 to 18 years old is 10 to 30 percent. Tolerable upper intake levels for protein and individual amino acids could not be set due to inadequate or conflicting data.

■ MINI-SUMMARY

The RDA for protein for both men and women is 0.8 gram/kilogram of body weight. The amount of protein needed daily is proportionally higher during periods of growth because the body is in a state of positive nitrogen balance. Negative nitrogen balance occurs during starvation and illnesses when the body excretes more protein than is taken in. The AMDR for protein for adults is 10 to 35 percent of total kcalories.

Ingredient Focus: Milk, Dairy Products, and Eggs

Milk

Pasteurized—A product, such as milk, that has been treated to kill harmful germs.

Homogenized—Milk that has had its fat particles broken up so finely that they remain uniformly dispersed throughout the milk.

The age of the long-necked glass bottle containing milk that required shaking prior to consumption was not so long ago. In the early stages of milk processing, milk was **pasteurized** (heated to kill harmful germs) but not homogenized. When this milk was allowed to stand, the lighter fat would float to the top and the denser milk would settle below. In other words, the top contained cream, and the bottom clear liquid was skim milk. By shaking the bottle prior to use, the consumer created a temporary emulsion (a combination of fat and liquid) and the milk would become smooth and blended. Since fat and liquid do not combine permanently under normal circumstances, milk would start to separate within a few minutes after shaking. Today's milk products are **homogenized** to create a permanent bond between the milk and cream. In addition, the cardboard containers help to better preserve some nutrients.

Milk is a good source of:

- High-quality protein
- Carbohydrates
- Riboflavin
- Vitamins A and D (if fortified)
- Calcium and other minerals such as phosphorus, magnesium, and zinc

Different types of fluid milk (see Table 5-4) contain varying amounts of fat. In addition to whole, reduced-fat, low-fat and fat-free, there are also the following.

- **Cultured buttermilk.** Buttermilk is made most often by adding a bacterial culture to fresh, pasteurized skim milk. The bacteria convert the sugar in milk (lactose) into lactic acid, thereby giving buttermilk a thick consistency and a tart and buttery taste. Buttermilk is enjoyed as a beverage and is used in baking when sour milk is needed. It is also used in making salad dressings and cold soups.
- **Eggnog.** Eggnog is a mixture of dairy ingredients (cream, milk, partially skimmed milk, or skim milk), eggs, and sweeteners. It may be flavored with rum extract, nutmeg, vanilla, or other flavorings. Commercial eggnog is pasteurized, so there shouldn't be any concern about the safety of the egg yolks in this product.
- **Lactase-treated milk.** Milk that has been treated with the enzyme lactase is particularly helpful for individuals who have lactose intolerance. Lactose intolerance, or lactase deficiency, is a disease caused by a lack of the enzyme lactase, which is needed by the intestine to split lactose into its two components for absorption. After drinking milk or eating a dairy product with lactose, the lactose-intolerant individual experiences

TABLE 5-4 Nutrient Composition of 1 Cup of Milk

Name	Kcalories	Fat	Saturated Fat	Calcium
Whole milk	150	8 g	5 g	300 mg
Reduced-fat milk, also called 2% milk	120	5 g	3 g	300 mg
Low-fat milk, also called 1% milk	100	2.5 g	1.5 g	300 mg
Fat-free milk, also called skim or nonfat milk	80	0 g	0 g	300 mg

Source: Step by Step: Eating to Lower Your High Blood Cholesterol. NIH Publication No. 94-2920. National Institutes of Health, 1994.

symptoms that include abdominal cramps, bloating, and diarrhea. Lactose-treated milk is otherwise nutritionally identical to regular milk.

When purchasing milk, decide on the fat content you want, and specify U.S. grade A milk that has been pasteurized, homogenized, and fortified with vitamins A and D. Fresh milk is very perishable and should be stored in the refrigerator; it can be used for up to four days after the pull date (whole milk stays fresher longer than either skim or low-fat milk).

Milk is also available with some or all of its water content removed.

- **Evaporated milk and evaporated skim milk.** These products are made by heating milk to stabilize the milk protein, then removing about 60 percent of the water. Both products are sold in cans. Evaporated milk has 7.5 percent milkfat; evaporated skim milk has no more than 0.5 percent milkfat. They are used mostly in cooking and baking. One-half cup of either product can be reconstituted with 1/2 cup of water to make 1 cup of milk.
- **Sweetened condensed milk and skimmed milk.** These products also have about 60 percent of their water removed. What makes them different is that they are heavily sweetened, usually with sugar. Sweetened condensed milk has 8 percent milkfat; sweetened condensed skimmed milk has no more than 0.5 percent milkfat. They are used mostly in cooking and baking.
- **Nonfat dry milk (powdered milk).** This product is made by removing the water from pasteurized skim milk. It can be easily reconstituted with water—3 tablespoons of the dry milk to 8 ounces of water. Although it doesn't have the taste of fluid milk, it is acceptable for use in baking and cooking.

Evaporated, condensed, and nonfat dry milk can all be stored unopened at room temperature for at least one year, or you can use the dating code marked on the container if available. Once opened, evaporated and condensed milk should be poured into another container, refrigerated, and used within five days. Once reconstituted, nonfat dry milk should be refrigerated and used within five days.

Milk products are a very important part of many recipes. The addition of milk products can add a creamy consistency to many sauces, help dissolve dry ingredients in baked products, or simply add good nutrition and flavor to a food. Substituting low-fat or skim milk for whole milk works well in many cooking and baking recipes to reduce the fat content. Evaporated skim milk can be used in place of half-and-half in some recipes such as soups.

When cooking with milk, remember a very important rule—use moderate heat, and heat the milk slowly (but not too long) to avoid curdling—a grainy appearance with a lumpy texture. From a scientific point of view, milk curdles when the casein (protein in milk) separates out of the milk. Add other food products to hot milk products slowly, stirring either with a spoon or a wire whisk, if preparing a sauce, to avoid lumps. Be especially careful when adding foods high in acid—milk has a tendency to curdle if not beaten quickly.

Another problem with milk is the creation of a "skin"—a film that forms on top of the milk during cooking. It can be prevented by keeping the pot of milk covered. A final problem with milk is scorching—when milk sticks to the bottom of the pot during cooking and eventually starts to burn, making the food you are cooking taste terrible. Because milk is very sensitive to direct heat, heat it slowly; in some cases you will need to scald the milk.

Cheese

Cheese is a versatile food and can be used in many different applications. Cheese flavors range from mild to very sharp, and cheese not only can be an important ingredient in an end product, but can be consumed as a snack.

Cheese is made from various types of milk—cow's and goat's being the most popular. In many parts of the world cheese is produced from various kinds of milk: sheep, reindeer, yak, buffalo, camel, and donkey. Cheese is produced when bacteria or rennet (or both) are added to milk and the milk then curdles. The liquid, known as the whey, is separated from the solid, known as the curd, which is the cheese. It is thought that cheese was discovered by accident thousands of years ago in the ancient Far East.

Cheese is an excellent source of nutrients such as protein and calcium. However, because most cheeses are prepared from whole milk or cream, they are also high in saturated fat and cholesterol. Ounce for ounce, meat, poultry, and most cheeses have about the same amount of cholesterol. But cheeses tend to have much more saturated fat.

Determining which cheeses are high or low in saturated fat and cholesterol can be confusing, because there are so many different kinds on the market: part-skim, low-fat, processed, and so on. Not all reduced-fat or part-skim cheeses are always low in fat; they are only lower in fat than similar natural cheeses. For instance, one reduced-fat cheddar gets 56 percent of its kcalories from fat—considerably less than the 71 percent of regular cheddar, but not super-lean, either. The trick is to read the label. Table 5-5 is a guide to fat in cheeses.

Cheese can be divided into different categories, depending on texture or whether it has been ripened (also called cured). The texture of cheese varies from soft to semisoft, firm, and hard. *Ripened cheeses* are those that have been fermented with bacteria or molds. By contrast, *unripened cheeses* are fresh and untreated. Unripened cheeses are generally more perishable

TABLE 5-5 Guide to Fat in Cheeses[†]

Lowfat 0–3 g fat/oz	Medium Fat 4–5 g fat/oz	High Fat 6–8 g fat/oz	Very High Fat 9–10 g fat/oz
Natural Cheeses			
*Cottage cheese (1/4 c) dry curd	*Mozzarella, part-skim	Blue cheese	Cheddar
Cottage cheese (1/4 c) low-fat 1%	*Ricotta (1/4 c), part-skim	*Brick	Colby
Cottage cheese (1/4 c) low-fat 2%	String cheese, part-skim	Brie	*Cream cheese (1 oz - 2 Tbsp)
Cottage cheese (1/4 c) Creamed 4%		Camembert	Fontina
Sapsago		Edam	*Gruyère
		Feta	Longhorn
		Gjetost	*Monterey Jack
		Gouda	Muenster
		*Light cream cheese (1 oz = 2 Tbsp)	Roquefort
Look for special low-fat brands of mozzarella, ricotta, cheddar, and Monterey jack.	Look for reduced-fat brands of cheddar, colby, Monterey jack, muenster, and Swiss.	Limburger	
		Mozzarella, whole-milk	
		Parmesan (1 oz = 3 Tbsp)	
		*Port du Salut	
		Provolone	
		*Ricotta (1/4 c), whole-milk	
		Romano (1 oz = 3 Tbsp)	
		*Swiss	
		Tilsit, whole milk	
Modified Cheeses			
Pasteurized process, imitation, and substitute cheeses with 3 g fat/oz or less.	Pasteurized process, imitation, and substitute cheeses with 4–5 g fat/oz.	Pasteurized process Swiss cheese	Some pasteurized process cheeses are found in this category—check the labels.
		Pasteurized process Swiss cheese food	
		Pasteurized process American cheese	
		Pasteurized process American cheese food	
		American cheese food cold pack	
		Imitation and substitute cheeses with 6–8 g fat/oz.	

[†] Check the labels for fat and sodium content. 1 serving = 1 oz. unless otherwise stated.
* These cheeses contain 160 mg or less of sodium per 1 oz.

Source: Reprinted by permission of the American Heart Association, Alameda County Chapter, 11200 Golf Links Road, Oakland, CA 94605.

than ripened cheeses. The processing time is very short and no time is needed for ripening. When the cheese is manufactured, it is packaged and sold with a relatively short expiration date compared with ripened cheeses. Popular types of unripened cheeses include ricotta cheese and cottage cheese.

Ripened cheeses need many months to age and develop into a marketable cheese. The longer they age, the sharper their flavor—and the higher their cost. Cheddar and Swiss are popular types of ripened cheese. Cheddar cheese comes in many varieties, ranging in flavor from very mild to very sharp. Cheddar is used as an appetizer, in sandwiches, and as a dessert cheese. Two kinds of imported Swiss are frequently used in the kitchen: Emmenthaler and Gruyère. Domestic Swiss is usually used in sandwiches.

When a cheese or combination of cheeses is manufactured into another type of cheese, this is referred to as a *process cheese.* Whereas natural cheeses are made directly from milk or whey, process cheese is a modified form of natural cheeses that have been ground or shredded from a variety of natural cheeses. Process cheese takes many forms.

- **Pasteurized process cheese**—a combination of cheddar and other cheeses with emulsifiers to make it smooth (for this reason, these cheeses melt better); includes American cheese
- **Pasteurized process cheese foods and spreads**—made like process cheese except for the addition of optional ingredients such as cream, milk, buttermilk, nonfat dry solids, or whey; these contain more moisture and less fat than process cheese
- **Cold pack cheese**—made by mixing ripened cheeses without heat

American cheese is probably the most well-known process cheese. It is excellent in sandwiches and melted on cheeseburgers. Table 5-6 is a guide to many of the different types of cheeses available.

Buy the best-quality cheese that you can. Some cheeses are graded by the USDA. The four grades are AA, A, B, and C. All cheddar cheese that is graded must also show the cure category:

Mild: cured for two to three months
Mellow aged: cured for four to seven months
Sharp: fully ripened—cured for eight to twelve months
Very sharp: aged over twelve months

When receiving cheese, check for mold—even when it is in airtight packaging. Mold commonly appears as white, blue, or green fuzzy spots. Also check for dryness—the cheese will have darker edges if this is the case. Many processed cheeses are dated with sell-by or use-by dates, so check for one. Make sure the cheese has been kept cold. Only four types of cheese do not require refrigeration until opened: cold pack cheese, cold pack

TABLE 5-6 A Guide to Cheeses

Cheese	Characteristics	Uses
UNRIPENED		
Cottage	Mild, slightly acid flavor; soft, open texture with tender curds of varying size; white to creamy white	Appetizers, salads, cheesecakes, dips
Cream	Delicate, slightly acid flavor; soft, smooth texture; white	Appetizers, salads, sandwiches, desserts, and snacks
Neufchâtel	Mild, acidic flavor; soft, smooth texture similar to cream cheese but lower in fat; white	Salads, sandwiches, desserts, snacks, dips
Ricotta	Mild, sweet, nutlike flavor; soft, moist texture with loose curds (fresh ricotta) or dry and suitable for grating; white	Salads, main dishes such as lasagne and ravioli, and desserts, mostly in cooked dishes
SOFT, RIPENED		
Bel Paese	Mild, sweet flavor; light, creamy-yellow interior; slate-gray surface; soft to medium-firm, creamy texture	Appetizers, sandwiches, desserts, and snacks
Brie	Mild to pungent flavor; soft, smooth texture; creamy-yellow interior; edible thin brown and white crust	Appetizers, sandwiches, desserts, snacks, salads
Camembert	Distinctive mild to pungent flavor; soft, smooth texture—almost fluid when fully ripened; creamy-yellow interior; edible thin white or gray-white crust	Appetizers, desserts, and snacks
Limburger	Highly pungent, very strong flavor and aroma; soft, smooth texture that usually contains small irregular openings; creamy-white interior; reddish-yellow surface	Appetizers, desserts, snacks, sandwiches
SEMISOFT, RIPENED		
Blue	Tangy, piquant flavor; semisoft, pasty, sometimes crumbly texture; white interior marbled or streaked with blue veins of mold; resembles Roquefort	Appetizers, salads and salad dressings, desserts, and snacks
Brick	Mild, pungent, sweet flavor; semisoft to medium-firm, elastic texture; creamy white-to-yellow interior; brownish exterior	Appetizers, sandwiches, desserts, and snacks
Gorgonzola	Tangy, rich, spicy flavor; semisoft, pasty, sometimes crumbly texture; creamy-white interior, mottled or streaked with blue-green veins of mold; clay-colored surface	Appetizers, salads, desserts, and snacks
Mozzarella (also called scamorza)	Delicate, mild flavor; slightly firm, plastic texture; creamy white	Main dishes such as pizza or lasagne, sandwiches, snacks, and salads

Muenster	Mild to mellow flavor; semisoft texture with numerous small openings; creamy-white interior; yellowish-tan or white surface	Appetizers, sandwiches, desserts, and snacks
Port du Salut	Mellow to robust flavor similar to Gouda; semisoft, smooth elastic texture; creamy white or yellow	Appetizers, desserts, and snacks
Roquefort	Sharp, peppery, piquant flavor; semisoft, pasty, sometimes crumbly texture; white interior streaked with blue-green veins of mold	Appetizers, salads and salad dressings, desserts, and snacks
Sapsago	Sharp, pungent, clover-like flavor; very hard texture suitable for grating; light green or sage green	Grated for seasoning
Stilton	Piquant flavor, milder than Gorgonzola or roquefort; open, flaky texture; creamy-white interior streaked with blue-green veins of mold; wrinkled, melon-like rind	Appetizers, salads, desserts, snacks, in cooked foods
HARD, RIPENED		
Cheddar (often called American)	Mild to very sharp flavor; smooth texture, firm to crumbly; light cream to orange	Appetizers, main dishes, sauces, soups, sandwiches, salads, desserts, and snacks
Colby	Mild to mellow flavor, similar to Cheddar; softer body and more open texture than Cheddar; light cream to orange	Sandwiches, snacks, cooked foods
Edam	Mellow, nutlike, sometimes salty flavor; rather firm, rubbery texture; creamy-yellow or medium yellow-orange interior; surface coated with red wax; usually shaped like a flattened ball	Appetizers, salads, sandwiches, sauces, desserts, and snacks
Gouda	Mellow, nutlike, often slightly acid flavor; semisoft to firm, smooth texture, often containing small holes; creamy-yellow or medium yellow-orange interior; usually has red wax coating; usually shaped like a flattened ball	Appetizers, salads, sandwiches, sauces, desserts, and snacks
Gruyère	Nutlike, salty flavor, similar to Swiss, but sharper; firm, smooth texture with small holes or eyes light yellow	Appetizers, desserts, snacks, fondue and other cooked dishes
Monterey (Jack)	Semisoft; smooth, open texture, mild flavor; cheddar-like; hard when aged	Appetizers, sandwiches, salads
Parmesan	Sharp, distinctive flavor; very hard, granular texture; yellowish white	Grated on cooked (Italian) dishes, salads, as seasoning
Provolone	Mellow to sharp flavor, smoky and salty; firm, smooth texture; cuts without crumbling; light creamy yellow; light-brown or golden-yellow surface	Appetizers, main dishes, sandwiches, desserts, and snacks
Romano	Very sharp, piquant flavor; very hard, granular texture; yellowish-white interior; greenish-black surfaces	Seasoning and general table use; when cured a year, it is suitable for grating
Swiss (also called Emmenthaler)	Mild, sweet, nutlike flavor; firm, smooth, elastic body with large round eyes; light yellow	Sandwiches, salads, snacks, fondue and other cooked dishes

Source: Reprinted from *Food Preparation for the Professional* by David Mizer. Mary Porter, Beth Sonnier, and Karen Eich Drummond. Copyright ©2000, John Wiley and Sons, Inc. *This material is used by permission of John Wiley and Sons, Inc.*

cheese food, pasteurized process cheese food, and pasteurized process cheese spread.

Because of their fat content, many cheeses readily absorb refrigerator odors, resulting in a poor-tasting product. Therefore, store all cheeses in tight plastic wrap or foil (except blue cheese, which should be wrapped loosely). Cheeses also need to be wrapped tightly to prevent them from drying out. Hard and firm cheeses keep from a week to several months in the refrigerator, while semisoft and soft cheeses are much more perishable and keep only one to two weeks. For unripened cheeses, such as cottage cheese and cream cheese, check the sell-by or use-by date. These products should last at least several days beyond the sell-by date.

There are several ways to use cheese in light and healthy cooking.

- Use a regular cheese with a strong flavor, and use less of it than called for in the recipe.
- Use less cheese.
- Substitute low-fat cheeses for regular ones.
- Use a mixture of half regular cheese and half low-fat cheese.

When cooking with cheese, observe some simple guidelines to come out with a tasty dish.

1. **Use low heat.** It is best to use as low a heat as possible when cooking with cheese. Cheese has a tendency to toughen when subjected to high heat, due to its high protein content. Avoid boiling at all costs.
2. **Use short cooking times.** Most recipes will require the addition of cheese at the end of the recipe to avoid overcooking. Remember to stir often to enhance the blend of flavors and establish a good, smooth consistency.
3. **Grate the cheese.** The best way to add a cheese to a recipe is to grate it. Grating will break the cheese into small, thin pieces that will melt and blend quickly and evenly into the end product. Grating creates an image of more cheese when melting it on top of a product (au gratin).

Cream

Cream is used in many ways in the kitchen. It is used in sauces, soups, hot and cold beverages, baked goods, and, when whipped, as a topping for desserts and hot beverages.

There are several types of cream. Most have been pasteurized. In comparison to pasteurized cream, ultrapasteurized cream has a much longer shelf life but does not whip as well.

- **Heavy whipping cream or heavy cream.** Heavy cream is a very thick, semi-fluid liquid that is at least 36 percent fat (by weight). It is often used as a topping after a short period of whipping and can be used in sauces. Remember not to overbeat heavy cream, or it will turn into butter!

- **Light whipping cream or whipping cream.** Light whipping cream contains 30 to 35 percent milkfat. Whipping cream labeled as ultrapasteurized does not whip as well as regular whipping cream but does have a longer shelf life.
- **Light cream.** Light cream is also called table cream or coffee cream. It has a fat content between 18 and 30 percent (usually it's 18 percent). It, too, has a strong, rich flavor and is often used as an ingredient in both sauces and baked products. It can't be successfully whipped.
- **Half-and-half.** Half-and-half (half milk, half cream) normally has a fat content of 10 to 12 percent. Its consistency is heavy and the flavor is still rich. Like light cream, it can't be successfully whipped.
- **Sour cream.** Sour cream is made from pasteurized cream (with 18 percent milkfat content) to which bacteria are added. The bacteria convert the milk sugar, lactose, into lactic acid, thereby giving the product its thick consistency and tangy taste.

Although cream does contain some nutrients, it mostly supplies fat (see Table 5-7).

Always purchase high-quality, very fresh cream. Be sure the container is clean, tightly sealed, and cold. Check for a date code on it as well. Cream is very perishable and should be kept refrigerated in its closed carton. Do not leave cream out on the kitchen table while you are having your morning coffee—keep it cold! Observe the use-by date on the carton. If sell-by dates are used, allow three to four days beyond the date. For sour cream, allow about ten days past the sell-by date. If sour cream gets moldy, even if it is just a few dots, throw it out. Cream that has been ultrapasteurized can be kept refrigerated (unopened) for up to six weeks. Once opened, it can keep one week.

TABLE 5-7 Kcalories and Fat in Milk and Cream

Product	Serving Size	Kcalories	Fat (grams)
Skim milk	1 cup	86	0.5
1% milk	1 cup	102	3
2% milk	1 cup	121	5
Whole milk	1 cup	150	8
Half-and-half	1 tablespoon	20	2
Light cream (coffee cream)	1 tablespoon	29	3
Light whipping cream, fluid	1 tablespoon	44	5
Heavy whipping cream, fluid	1 tablespoon	52	6
Sour cream	1 tablespoon	26	3

Source: U.S. Department of Agriculture.

If you want *real* whipped cream (in moderation anything is fine), use light whipping cream, or for a substitute try drained yogurt and whipped egg whites (described under Chef's Tips). Make sure the cream is cold, as cold cream whips better than warm cream. For best results, place the cream, bowl, and beaters into the freezer for about ten minutes before whipping. Use a bowl that is deep enough to accommodate the beaters and small enough for the beaters to maintain contact with the cream. Beat rapidly, scraping the bowl frequently for two to three minutes, until you get stiff peaks. Don't overbeat—the product will be granular and turn into butter. If you are adding sugar to the cream, do so after whipping, because it makes the cream harder to whip and less stable. Also, use confectioners' rather than granulated sugar for a smoother product. One cup of whipping cream yields two cups whipped. It is best to use whipped cream right away, but it can be put into the refrigerator, covered, for a few hours.

When cooking with sour cream, don't let it boil, or it may curdle due to the high heat. To prevent separation, you can mix in 1 tablespoon of flour per 1/2 cup of sour cream before cooking.

Ice Cream and Ice Milk

In order to be labeled as ice cream, a product must have at least 10 percent fat by weight. Premium ice creams have much more than the minimum required: about 16 percent fat. By comparison, ice milk, which is prepared from the same ingredients as ice cream, must have at least 3 percent fat. Ice milk usually has more sugar added.

Ice cream and ice milk are a source of calcium, riboflavin, and protein. Ice cream is also a significant source of fat, as seen in Table 5-8. Ice milk has much less fat.

Ice cream quality varies in three ways:

- **The amount of fat.** The more fat an ice cream contains, the richer it tastes and the more expensive it is.
- **The kind of flavoring.** Ice cream contains natural flavorings, artificial flavorings, or a combination of both. Natural flavorings make a better-quality ice cream that is then more expensive. If only natural flavorings are used, for example, the vanilla product is called vanilla ice cream; if both natural and artificial flavorings are used, it is called vanilla-flavored ice cream; if only artificial flavorings are used, it is called artificially flavored ice cream.
- **The amount of air in the ice cream.** All ice cream has air in it—less than half of the product can be air if it is to be labeled as ice cream. Air cells act as a cushion that keep the ingredients from forming into a solid, icy mass. The cranking of an ice-cream freezer whips air into the mixture. Premium ice creams have less air in them than lesser brands.

TABLE 5-8 Kcalories and Fat in Ice Cream, Ice Milk, and Frozen Yogurt

Dessert	Serving Size	Kcalories	Fat (grams)
Premium vanilla ice cream (16% fat)	1 cup	349	24
Vanilla ice cream (10% fat)	1 cup	269	14
Vanilla ice milk	1 cup	184	6
Vanilla soft-serve ice milk	1 cup	223	5
Frozen yogurt, vanilla	1 cup	240	varies

Source: U.S. Department of Agriculture.

That's why premium ice creams always weigh more than equivalent volumes of cheaper brands.

Make sure any ice cream or ice milk you receive is rock solid and the container is clean. If it feels sticky or looks frosty, it has probably thawed and refrozen. Ice cream and ice milk last about two months in the freezer. For best quality, use them within two weeks.

Yogurt

Yogurt is one of the oldest fermented milks known. Yogurt is cultured with the live active cultures *Lactobacillus bulgaricus* and *Streptococcus thermophilus.* In the modern commercial production of yogurt, the milk base for the product is pasteurized to condition it for fermentation. Only then are the culture organisms added. After approximately three hours of incubation at about 110°F, the milk acquires its custardlike texture. After cooling, the product is ready for distribution. Yogurt is not required to contain live bacterial cultures. Yogurt with live, active bacterial cultures will state this fact on the label.

Yogurt is a good source of:

- Protein
- Calcium, phosphorus, and potassium
- Riboflavin and vitamin B_{12}

Yogurt contains the same fats that are found in the milk product it was made from. Whole-milk yogurt contains at least 3.5 grams fat per 100 grams (3-1/2 ounces) of yogurt, low-fat yogurt contains between 0.5 and 2 grams fat, and nonfat yogurts contain less than 0.5 gram fat.

There are three main types of yogurt to choose from.

1. Unflavored, plain yogurt
2. Flavored, containing no fruit (such as vanilla, lemon, coffee)

3. Flavored and containing fruit, which may be of two styles:
   - **Sundae style**—fruit is at the bottom of the container with plain or flavored yogurt on top; this product is normally stirred before eating
   - **Blended style** (also called Swiss style)—fruit is blended throughout plain or flavored yogurt

The calorie content of fresh yogurt (see Table 5-9) varies dramatically, due to its fat content and whether fruit is added. Many yogurts with fruit also have a lot of sugar or other sweetener added to help preserve the fruit. Plain yogurt with fresh fruit has much less sugar and fewer kcalories.

Like fresh yogurt, frozen yogurt varies in the amount of fat it contains. On the high end are yogurts with about 6 grams of fat per 6 ounces (still lower than an equivalent amount of vanilla ice cream, with 10 grams of fat). On the low end are nonfat yogurts. Most frozen yogurt does not contain nearly the number of live bacteria (if they're alive at all) that fresh yogurt does. There are no federal standards for frozen yogurt, and many brands do not use live culture. Even when live bacteria are used, their numbers are far below those found in fresh yogurt. With or without active cultures, frozen yogurt still is a healthier option than ice cream (that is, of course, without the addition of pieces of candy bars, chocolate chips, etc.).

When yogurt is received, see that it is fresh and the containers are clean. Store fresh yogurt in the refrigerator up to the use-by date, or seven to ten days past the sell-by date. If you open the container and find some liquid sitting on top of the yogurt, don't worry, it's still safe to eat. Just drain off the liquid or stir it into the yogurt.

Plain yogurt, either nonfat or low-fat, can be substituted for many higher-fat ingredients in salad, salad dressings, soups, sauces, and desserts. Here are some examples.

- Substitute plain yogurt for mayonnaise. In situations where the taste of mayonnaise is desired, use half reduced-calorie mayonnaise and half yo-

TABLE 5-9 Kcalories and Fat in Yogurt (1-cup portion)

Type of Yogurt	Kcalories	Fat (grams)
Whole milk, plain	139	7
Low-fat, plain	144	4
Low-fat, vanilla or coffee-flavored	194	3
Low-fat, fruit-flavored	225	3
Nonfat, plain	127	0

Source: U.S. Department of Agriculture.

gurt. This works well for dishes such as potato salad, coleslaw, tuna salad, cold pasta salads, and appetizers.
- Substitute plain yogurt for sour cream in dips and salad-dressing recipes.
- In baking, substitute yogurt for sour cream in recipes for pancakes, waffles, loaf breads, and muffins.
- In cooking, substitute 1 cup yogurt for 1 cup sour cream.
- On baked potatoes, offer plain yogurt mixed with fresh herbs instead of sour cream.
- Mix plain yogurt with ricotta cheese (made from skim milk) and use as a spread on toast, bagels, and crackers.

Where you want a creamy texture, drain the yogurt first to remove some of the liquid. Because of yogurt's acidity, you may want to decrease the amount of other acidic ingredients in your recipe, such as lemon juice.

When cooking with yogurt, use only low heat. High temperatures may cause separation, evaporation of liquid, and a curdled appearance. To help keep yogurt from separating during cooking, blend 1 tablespoon of cornstarch (unless the recipe calls for flour to be mixed with it) with a few tablespoons of yogurt, then stir the mixture into the remaining yogurt to be used, and proceed according to the recipe. Yogurt might also become thin if it is overmixed, so do not overstir.

Eggs

Eggs are truly a unique food product—they provide versatility and can be used for any meal. Eggs not only are excellent traditional breakfast foods but are used in many breads, pies, cakes, custards, beverages, and entrées, to name a few.

Eggs are very nutritious and full of high-quality protein, as well as varying amounts of many vitamins and minerals. The concern with overconsumption of eggs stems from the fact that they are very high in cholesterol—215 milligrams per egg (compare that to the suggested maximum of 300 milligrams to eat daily). One egg also contributes 5 grams of fat, of which 2 grams are saturated fat.

Eggs are sold according to their grade and size. Standards for both grade and size are established by the U.S. Department of Agriculture. "Grade" refers to the quality of the egg and the shell when it is packed. Grades are AA, A, and B. The grade of an egg has nothing to do with whether it is nutritious or wholesome. Instead, grade is based on freshness, as well as the interior quality of the egg white and yolk. Lower-grade eggs have a thinner white and flatter yolk than do higher-grade eggs. U.S. Grade A is the most common grade available and has a 30-day shelf life. Grade AA is not often seen because that grade must have a 10-day expiration date.

Eggs come in various sizes: jumbo, extra large, large, medium, small, and peewee. Eggs are sized by weight, so in a given box, some eggs may be below size and others above; one dozen eggs must meet the minimum weight per dozen set for the marked size.

When eggs are received, check the carton to make sure it contains only clean, uncracked eggs. Make sure the eggs have been refrigerated.

There is no difference between brown- and white-shelled eggs. Shell color is determined by the breed of the hen, and it does not affect the grade, nutritive value, flavor, or cooking performance of the egg. Brown eggs are often more expensive because they come from larger hens that require more food.

Store eggs in the carton they came in, because the carton will help keep out odors the eggs might absorb and it also helps prevent the loss of carbon dioxide and moisture from the eggs, which causes them to age quicker. Keep eggs refrigerated, as they maintain freshness better this way. Eggs should be stored with the large end up to keep the yolk centered. As an egg ages, the yolk flattens and the thick section of the white becomes watery and thins out. Stale eggs also usually have a "rotten egg" odor that smells like sulfur.

Eggs can be cooked in many ways: baked, cooked in the shell, poached, fried, and scrambled. When making fried eggs, scrambled eggs, and omelets, the use of nonstick pans and vegetable cooking sprays is important to keep down the amount of fat. When cooking eggs, always use low to medium temperatures to prevent overcooking. Overcooked eggs are tough and rubbery.

Egg substitutes are available that are low in cholesterol but not always low in fat. They are often made from egg whites and vegetable oil. Use 1/4 cup egg substitute to equal one whole egg.

Besides using egg substitutes, there are other ways to make egg dishes with less cholesterol.

1. To make scrambled eggs and omelets, use one whole egg, and add two egg whites for each additional egg.
2. In baking, replace one whole egg with two egg whites, and two whole eggs with one whole egg and two egg whites.

CHEF'S TIPS

- To make "whipped cream," drain plain fat-free yogurt in cheesecloth to remove as much liquid as possible. Fold whipped egg whites into the yogurt and add a little honey for flavor. Use frozen pasteurized egg whites to avoid any food safety (salmonella) problem.
- To make an excellent omelet without cholesterol, whip egg whites until they foam. Add a touch of white wine, freshly ground mustard, and chives. Spray a nonstick pan with oil and add your eggs. Cook like a traditional omelet. When the omelet is close to done, put the pan under the broiler to finish. The omelet will puff up. Stuff the omelet, if desired, with vegetables or other low-fat filling, then fold over and serve.

- For color and flavor, serve an omelet with spicy salsa poured on top of it, or serve with salsa or black bean relish and blue corn tortilla chips.

Check-Out Quiz

1. Proteins contain nitrogen.
   a. True b. False
2. Essential amino acids cannot be made in the body.
   a. True b. False
3. Every protein has a unique primary structure.
   a. True b. False
4. In denaturation, the protein's shape is distorted, but the protein can still function.
   a. True b. False
5. During digestion and before absorption, proteins are broken down into their amino acid units, which can then be absorbed and transported in the blood.
   a. True b. False
6. Americans tend to eat just enough protein.
   a. True b. False
7. The stomach enzyme pepsin aids in the digestion of protein.
   a. True b. False
8. You should try to balance your intake of protein from animal and plant sources.
   a. True b. False
9. Kwashiorkor is associated with children with insufficient protein intake and who have a preexisting disease.
   a. True b. False
10. Most plant foods are examples of incomplete proteins.
   a. True b. False

Activities and Applications

1. Self-Assessment

Write the number of times per week that you eat the foods listed below in the space provided. Then think about the following questions.

- Is your protein coming mostly from animal or plant sources, or is it somewhat evenly balanced between the two?
- Are the meats, poultry, and fish usually the entrées, and the pasta, rice, vegetables, and dried beans or peas served as side dishes in smaller quantities?

■ What can you do to balance the two sides better, if necessary?

Animal Protein Sources		Plant Protein Sources	
Red meats	______	Dried beans	______
Poultry	______	Dried peas	______
Fish	______	Bread	______
Milk	______	Cereals	______
Cheese	______	Pasta	______
Yogurt	______	Rice	______
Eggs	______	Nuts and seeds	______
		Vegetables (including potatoes)	______
Total Number of Servings	______	Total Number of Servings	______

2. Reading Food Labels

Following are food labels from a beef burger and a vegetable burger. Compare and contrast their nutritional content.

Beef Burger (3 oz.)	Vegetable Burger (2.5 oz.)
NUTRITION FACTS	NUTRITION FACTS
Amount per serving	Amount per serving
Calories 230	Calories 140
Calories from Fat 140	Calories from Fat 20
Total Fat 16 g	Total Fat 2.5 g
Saturated 6 g	Saturated 0.5 g
Polyunsaturated 1 g	Polyunsaturated 1.0 g
Monounsaturated 7 g	Monounsaturated 0.5 g
Cholesterol 74 g	Cholesterol 0 g
Sodium 180 mg	Sodium 180 mg
Total Carbohydrate 0 g	Total Carbohydrate 21 g
Dietary Fiber 0 g	Dietary Fiber 5 g
Sugar 0 g	Sugar 0 g
Protein 21 g	Protein 8 g

3. How Much Protein Do You Need?

Calculate how many grams of protein you need by multiplying your weight (in pounds) times 0.36.

Example: 150 pounds × 0.36 grams protein per pound = 54 grams protein

4. How Much Protein Do You Eat?

Write down everything you ate yesterday, including approximate portion sizes. If yesterday was not a typical day, write down what you normally eat during the course of a day. Find the approximate amount of protein in each food using the following information and/or Appendix A, then total up your protein intake for the day.

1 slice bread, 1/2 English muffin or hamburger roll, 4 to 6 crackers,
1/2 cup cooked cereal or pasta, 3/4 cup dry cereal,
1/2 cup cooked beans, 1/2 cup peas or corn, 1 small potato,
1/2 cup sweet potato. . . . . 3 grams protein
Fruit, margarine, butter, salad dressing . . . . 0 grams protein
1 cup raw vegetables, 1/2 cup cooked vegetables, 1/2 cup tomato or
vegetable juice . . . . 2 grams protein
1 ounce cooked meat, poultry, or fish; 1 egg, 1 ounce cheese,
1 tablespoon peanut butter . . . . 7 grams protein
1 cup milk, 1 cup yogurt. . . . . 8 grams protein

Now you can compare how much protein you ate on one day to the RDA. Do you consume too much, too little, or just about the right amount of protein daily? If you are eating too much, what foods would you cut down on and what foods would you replace them with?

5. Meat Diet vs. Mostly Plant Diet

Using the tables in this chapter, Appendix A, and/or food labels, find the amount of protein in each food listed below and total up each list. Each list represents one day's intake.

Meat-Based Diet		Plant-Based Diet with Dairy	
2 eggs	______	1 cup oatmeal	______
2 slices white toast	______	1/2 cup raisins	______
1/2 cup orange juice	______	1 corn muffin	______
1/2 cup milk	______	1 cup milk	______
1 doughnut	______	1 apple	______
3 oz. roast beef	______	1 vegetarian burger	______
1 oz. American cheese	______	1 whole-grain bun	______
2 slices white bread	______	Lettuce and tomato slices	______
1 tablespoon mayonnaise	______	1 banana	______
1-oz. package corn chips	______	Iced tea	______
2 cupcakes	______	1 granola bar	______
2 slices pizza	______	1 cup vegetable soup	______
1 cup vegetable salad	______	1 cup meatless chili	______
1 tablespoon dressing	______	1 cup vegetable salad	______
12-oz. soft drink	______	1 tablespoon dressing	______
1 cup vanilla ice cream	______	1/2 cup milk	______
		Peach cobbler	______
Total Protein:	______	Total Protein:	______

Which diet contained more protein? Do either or both of these diets meet your protein RDA? Is it possible for you to get the protein you need without eating meat, poultry, or seafood?

6. Cooking with Egg Substitutes

Make the following recipe for a vegetable omelet using eggs, then Egg Beaters (a brand of egg substitutes). Egg Beaters is 99 percent egg whites and contains no fat or cholesterol. Some coloring and natural flavoring is added to make it look and taste like a whole egg. The product has less than half the calories of whole eggs. Compare how the two dishes look and taste.

Vegetable Omelet

Makes 2 servings

2 teaspoons vegetable oil
1/2 cup sliced fresh mushrooms
1/4 cup sliced fresh zucchini
1/4 cup fresh broccoli florets
1/4 cup sliced fresh red bell pepper
1 teaspoon Italian seasoning
4 eggs, beaten (or 1 cup Egg Beaters)

1. In a nonstick skillet, heat the oil over medium heat.
2. Add the vegetables and Italian seasoning and sauté until the vegetables are tender.
3. Remove the vegetables from the skillet and keep warm.
4. Pour the eggs (or egg substitute) into the skillet. Cook, lifting the edges to allow the uncooked eggs to flow underneath.
5. When the eggs are almost set, gently spoon the sautéed vegetables into the center of the omelet.
6. Fold the sides of the omelet over the filling and slide the omelet onto the serving plate.

Nutrition Web Explorer

Eat Ethnic www.eatethnic.com

Use this website to find out what types of high-protein foods another ethnic group eats.

Graham Kerr's Lowfat Cooking Techniques www.grahamkerr.com

On Graham Kerr's home page, click on "Cooking Techniques." Read about how he makes and uses yogurt cheese.

Food Facts *Soybeans*

The soybean plant was first domesticated in China 3,000 years ago. The Chinese call it the "yellow jewel" or the "great treasure," for several reasons. Soybeans are easy to farm, and the plants do not deplete the soil. They are inexpensive to buy, contain the most protein of all legumes (with no cholesterol), and are a very versatile food, although when merely boiled, they have a strong taste with a metallic aftertaste. Perhaps due to this problem, the soybean has been used to make a tremendous variety of products.

Soybeans are grown in abundance in this country, but most are sold as animal feed after being processed for their oil. Soybean oil is used extensively in salad dressings, in margarine, and as salad/cooking oil.

In addition to soy oil, another important soybean product, particularly for vegetarians, is tofu, or bean curd. Tofu was invented by a Chinese scholar in 164 B.C. and is the most important of the foods prepared from soybeans in the East. Tofu is made in a process similar to making cheese. Soybeans are crushed to produce soy milk, which is then coagulated, causing solid curds (the tofu) and liquid whey to form. Tofu is white in color, soft in consistency, and bland in taste. It readily picks up other flavors, making it a great choice for mixed dishes such as lasagne.

Tofu is available shaped in cakes of varying textures and packed in water, which must be changed daily to keep it fresh. Firm tofu is compressed into blocks and holds its shape during preparation and cooking. Firm tofu can be used for stir-frying, grilling, or marinating. Soft tofu contains much more water and is more delicate. Soft tofu is good to use in blenderized recipes to make dips, sauces, salad dressings, spreads, puddings, cream pies, pasta filling, and cream soups. Silken tofu is even softer and more delicate, and works well in creamy desserts. Tofu should be kept refrigerated and used within one week.

CHEF'S TIPS FOR USING TOFU

- Marinate tofu with ginger-lime sauce.
- Crumble firm tofu and sauté with chopped onions, bell peppers, and other vegetables, herbs, and seasonings to make tacos and other Mexican dishes.
- Grill tofu and serve as the "meat" in a sandwich with portobello mushrooms.
- Replace part of the cream in creamed soups with blended silken tofu.
- Use blended silken or soft tofu instead of ricotta cheese in Italian dishes and other mixed dishes, such as Indian curry or a hot Thai dish.
- Use soft tofu in place of mayonnaise in salad dressings such as green goddess.

Other soybean products include the following:

- **Soy sauce** combines fermented soy and wheat. The wheat is first roasted, and contributes both the soy sauce's brown color and its sharp, distinctive flavor.
- **Miso** is similar to soy sauce but pasty in consistency. It is made by fermenting soybeans with or without rice or other grains. A number of varieties are available, from light-colored and sweet to dark and robust. It is used in soups and gravies, as a marinade for tofu, as a seasoning, and as a spread on sandwiches and fried tofu.

- **Tempeh** is a white cake made from fermented soybeans. It is a pleasant-tasting, high-protein food that can be cooked quickly to make dishes such as barbecued or fried tempeh, or cut into pieces to add to soups. Tempeh is cultured like cheese and yogurt, and therefore must be used when fresh or it will spoil.
- **Textured vegetable protein (TVP)** is made of granules of isolated soy protein that must be rehydrated before using in recipes. TVP is actually a brand name. The generic name is textured soy protein, or TSP. It can replace up to one-quarter of the meat in a recipe without tasting unacceptable. It is a very concentrated source of protein and is almost fat-free. Because of its strong flavor, TSP is most successfully used in highly flavored dishes such as chili, spaghetti sauce, and curries.
- **Meat analogs** are imitation meat products made from soy protein without any animal products. They are offered in forms resembling meat, such as hamburgers, hot dogs, bacon, ham, and chicken patties and nuggets. They contain little or no fat and no cholesterol but are often high in sodium. Some are fortified with vitamin B_{12} and iron.

Additional soy products include soy milk, soy yogurt, soy ice cream, and soy nuts (great for salads and snacks).

Hot Topic Irradiation

Beef is one of the U.S. food industry's hottest sellers-to the tune of billions of pounds a year. In recent years, though, beef, especially ground beef, has shown a dark side: It can harbor the bacterium *E. coli* 0157:H7, a pathogen that threatens the safety of the domestic food supply. If not properly prepared, beef tainted with *E. coli* 0157:H7 can make people ill, and can, in the case of children or the elderly, kill them. In 1993, contaminated hamburgers sold by a fast-food chain were linked to the deaths of four children and hundreds of illnesses in the Pacific Northwest.

In 1997, the potential extent of *E. coli* 0157:H7 contamination came to light when Arkansas-based Hudson Foods voluntarily recalled 25 million pounds of hamburger suspected of containing *E. coli* 0157:H7. It was the largest recall of meat products in U.S. history.

Nationally, *E. coli* 0157:H7 causes about 73,000 illnesses and 61 deaths a year, according to the federal Centers for Disease Control and Prevention. Scientists have known only since 1982 that this form of *E. coli* causes human illness.

To help combat this public health problem, the Food and Drug Administration (FDA) approved in 1997 the treatment of red-meat products with a measured dose of radiation. This process, commonly called *irradiation,* has drawn praise from many food-industry and health organizations because it can control *E. coli* 0157:H7 and several other disease-causing microorganisms. Since 1963, the FDA has been allowing the irradiation of a number of foods, such as poultry, fresh fruits and vegetables, dry spices, and seasonings.

The process is similar to sending luggage through a radiation field—typically gamma rays produced from radioactive cobalt-60. That amount of energy is not strong enough to add any radioactivity to the food. The same irradiation process is used to sterilize medical products such as bandages, contact lens solutions, and hospital supplies such as gloves and gowns. Many spices solid in this country also are irradiated, which eliminates the need for chemical fumigation to control pests. American astronauts have eaten irradiated foods since 1972.

Irradiation is a "cold" process that gives off little heat, so foods can be irradiated within their packaging and remain protected against contamination until opened by users. Because a few bacteria can survive the process in poultry and meats, it's important to keep products refrigerated and to cook them properly.

Irradiation interferes with bacterial genetics, so the contaminating organism can no longer survive or multiply. Although chemicals called radiolytic products are created when food is irradiated, the FDA has found them to pose no health hazard. In fact, the same kinds of products are formed when food is cooked.

As part of its approval, the FDA requires that irradiated foods include labeling with either the statement "treated with radiation" or "treated by irradiation" and the international symbol for irradiation, the radura (Figure 5-4). Irradiation labeling requirements apply only to foods sold in stores. Irradiation labeling does not apply to restaurant foods. Dairy Queen restaurants serve hamburgers made with irradiated meat, and they indicate this fact in their advertising. The irradiated hamburgers have been well accepted.

The FDA has evaluated irradiation safety for 40 years and found the process safe and effective for many foods. Before approving red-meat irradiation, the agency reviewed numerous scientific studies conducted worldwide. These include research on the chemical effects of radiation on meat, the impact that the process has on nutrient content, and potential toxicity.

In reviews of the irradiation process, FDA scientists concluded that irradiation reduces or eliminates pathogenic bacteria, insects, and parasites. It reduces spoilage; in certain fruits and vegetables, it inhibits sprouting and delays the ripening process. Also, it does not make food radioactive, compromise nutritional quality, or noticeably change food taste, texture, or appearance, as long as it's applied properly to a suitable product.

Health experts say that in addition to reducing *E. coli* 0157:H7 contamination, irradiation can help control the potentially harmful bacteria salmonella and campylobacter, two chief causes of food borne illness. FDA officials emphasize that, though irradiation is a useful tool for reducing food-borne illness risk, it complements, but doesn't replace, proper food-handling practices by producers, processors, and consumers.

Figure 5-4
International Symbol for Irradiation.
Source: Food and Drug Administration.

Chapter 6
Vitamins

In the early 1900s, scientists thought they had found the compounds needed to prevent scurvy and pellagra, two diseases caused by vitamin deficiencies. These compounds originally were believed to belong to a class of chemical compounds called amines and were named from the Latin *vita,* or life, plus *amine—vitamine.* Later, the *e* was dropped when it was found that not all of the substances were amines. At first, no one knew what they were chemically, so vitamins were identified by letters. Later, what was thought to be one vitamin turned out to be many, and numbers were added, such as for the vitamin B complex (for example, vitamin B_6). Later on, some vitamins were found unnecessary for human needs and were removed from the list, which accounts for some of the numbering gaps. For example, vitamin B_8, adenylic acid, was later found not to be a vitamin.

Vitamins are organic substances that carry out processes in the body vital to health. This chapter will help you to:

- State the general characteristics of vitamins
- Identify the functions and food sources of each of the 13 vitamins
- List the health risks of vitamin deficiencies and outline the possible side effects of vitamin toxicity
- Describe ways to conserve vitamins when cooking fruits and vegetables
- Discuss the purchasing, storage, cooking, and menuing of fruits and vegetables
- Define functional foods and discuss their role in the diet
- Give examples of phytochemicals and the foods in which they are found

Characteristics of Vitamins

Let's start with some basic facts about vitamins.

1. Very small amounts of vitamins are needed by the human body, and very small amounts are present in foods. Some vitamins are measured in IUs (international units), a measure of biological activity; others are measured by weight, in micrograms or milligrams. To illustrate how small these amounts are, remember that 1 ounce is 28.3 grams. A milligram is 1/1000 of a gram, and a microgram is 1/1000 of a milligram.
2. Although vitamins are needed in small quantities, the roles they play in the body are enormously important, as you will see in a moment.
3. Most vitamins are obtained through food. Some are also produced by bacteria in the intestine (and are absorbed into the body), and one (vitamin D) can be produced by the skin when it is exposed to sunlight, but the body doesn't make enough.
4. There is no perfect food that contains all the vitamins in just the right amounts. The best way to ensure an adequate vitamin intake is to eat a varied and balanced diet of plant and animal foods.

Precursors—Forms of vitamins that the body changes chemically to active vitamin forms.

Fat-soluble vitamins—A group of vitamins that generally occur in foods containing fats; these include vitamins A, D, E, and K.

Water-soluble vitamins—A group of vitamins that are soluble in water and are not stored appreciably in the body; these include vitamin C, thiamin, riboflavin, niacin, vitamin B_6, folate, vitamin B_{12}, pantothenic acid, and biotin.

5. Vitamins do not contain kcalories, so they do not directly provide energy to the body. Vitamins provide energy indirectly because they are involved in extracting energy from carbohydrate, protein, and fat.
6. Some vitamins in foods are not the actual vitamin but are **precursors.** The body chemically changes the precursor to the active form of the vitamin.
7. The body can't detect whether a vitamin is made in a laboratory (called a synthetic vitamin) or is isolated from a food (called a natural vitamin). Synthetic and natural vitamins are the same except in two cases. Vitamin E is stronger in its natural form, and folic acid is stronger in its synthetic form (this is the form used when it is added to foods such as breads and cereals).

Vitamins are classified according to how soluble they are in either fat or water. **Fat-soluble vitamins** (A, D, E, and K) generally occur in foods containing fat, and they are not readily excreted (except for vitamin K) from the body. **Water-soluble vitamins** (vitamin C and the B-complex vitamins) are readily excreted from the body (except vitamin B_{12}) and therefore don't often reach toxic levels.

■ MINI-SUMMARY

Very small amounts of vitamins are needed by the human body, and very small amounts are present in foods. Although vitamins are needed in small quantities, the roles they play in the body are enormously important. Vitamins must be obtained through foods, because vitamins are either not made in the body or not made in sufficient quantities. There is no perfect food that contains all the vitamins in just the right amounts. Vitamins have no kcalories. Some vitamins in foods are not the actual vitamin but rather are precursors; the body chemically changes the precursor to the active form of the vitamin. Synthetic and natural vitamins are the same except in two cases, vitamin E and folic acid. Vitamins are classified according to how soluble they are in either fat or water.

Fat-Soluble Vitamins

Fat-soluble vitamins include vitamins A, D, E, and K. They generally occur in foods containing fats and are stored in the body either in the liver or in adipose (fatty) tissue until they are needed. Fat-soluble vitamins are absorbed and transported around the body like other fats. If anything interferes with normal fat digestion and absorption, these vitamins may not be absorbed. Many require protein carriers to be transported around the body.

Although it is convenient to be able to store these vitamins so you can survive periods of poor intake, excessive vitamin intake (higher than the

Tolerable Upper Intake Level) causes large amounts of vitamins A and D to be stored and may lead to undesirable symptoms.

Retinoids—The forms of vitamin A that are in the body: retinol, retinal, and retinoic acid.

Retinol—A form of vitamin A found in animal foods; it can be converted to retinal and retinoic acid in the body.

Carotenoids—A class of pigments that contribute red, orange, or yellow colors to fruits and vegetables; can be converted to retinol or retinal in the body.

Beta-carotene—A precursor of vitamin A that functions as an antioxidant in the body; the most abundant carotenoid.

Xerosis—A condition in which the cornea of the eye becomes dry and cloudy; often due to a deficiency of vitamin A.

Xerophthalmia—Hardening and thickening of the cornea that can lead to blindness; usually caused by a deficiency of vitamin A.

Night blindness—A condition caused by insufficient vitamin A in which it takes longer to adjust to dim lights after seeing a bright light at night; this is an early sign of vitamin A deficiency.

Vitamin A

During World War I, many children in Denmark developed eye problems. Their eyes became dry and their eyelids swollen, and eventually blindness resulted. A Danish physician read that an American scientist gave milkfat to laboratory animals to cure similar eye problems in animals. At the time Danish children were drinking skim milk, because all the milkfat was being made into butter and sold to England. When the Danish doctor gave whole milk and butter to the children, they got better. The Danish government later restricted the amount of exported dairy foods. Dr. E. V. McCollum, the American scientist, eventually found vitamin A (the first vitamin to be discovered) to be the curative substance in milkfat.

In the body, vitamin A appears in three forms: retinol, retinal, and retinoic acid. Together they are referred to as **retinoids. Retinol** is found in animal foods and can be converted to retinal and retinoic acid in the body. Some pigments in plants (called **carotenoids**), which contribute red, orange, or yellow color to fruits and vegetables, can be converted to retinol or retinal in the body. **Beta-carotene,** the most abundant carotenoid, is an orange pigment in plants such as carrots. It is split in the intestine and liver to make retinol.

Vitamin A has several roles involving the eyes. First, it is essential for the health of the cornea, the clear membrane covering your eye. Without enough vitamin A, the cornea becomes cloudy. Eventually it dries (called **xerosis**) and thickens and can result in permanent blindness **(xerophthalmia).** Vitamin A is also necessary for healthy cells in other parts of the eye such as the retina.

Vitamin A is crucial to sight for other reasons. Your eye converts light energy into nerve impulses that travel to the brain, a process that uses retinal. If you don't take in enough vitamin A, you will experience a problem seeing at night. In **night blindness,** it takes longer to adjust to dim lights after seeing a bright flash of light (such as oncoming car headlights). This is an early sign of vitamin A deficiency. If the deficiency continues, xerosis and xerophthalmia can occur.

Vitamin A is involved in other parts of the body. It is needed to produce and maintain the cells (called epithelial cells) that form the protective linings of your lungs, gastrointestinal tract, urinary tract, and other organs. Vitamin A is also essential to produce and maintain epithelial cells that make mucus. Mucus protects the cells from harmful organisms, and, in the case of the stomach, the acidic stomach juices.

Vitamin A also plays a role in reproduction, growth and development in children, as well as in proper bone growth and teeth development in chil-

Antioxidant—A compound that combines with oxygen to prevent oxygen from oxidizing or destroying important substances; antioxidants prevent the oxidation of unsaturated fatty acids in the cell membrane, DNA, and other cell parts that substances called free radicals try to destroy.

Free radical—An unstable compound that reacts quickly with other molecules in the body.

Retinol activity equivalents (RAE)—The unit for measuring vitamin A. One RAE = 1 microgram retinol, 12 micrograms beta-carotene, or 24 micrograms of other vitamin A precursor carotenoids.

dren. Vitamin A is needed for proper functioning of the immune system (so you can fight infections) and for healthy skin.

As described, the carotenoids are precursors to vitamin A. Not all beta-carotene is converted to an active form of vitamin A. Beta-carotene is an **antioxidant** and may function as an antioxidant in the body. Antioxidants combine with oxygen so that the oxygen is not available to oxidize, or destroy, important substances in the cell. Antioxidants prevent the oxidation of unsaturated fatty acids in the cell membrane, DNA (the genetic code), and other cell parts that substances called **free radicals** destroy. Free radicals are highly reactive compounds that normally result from cell metabolism and functioning of the immune system. In the absence of antioxidants, free radicals destroy cells (possibly accelerating the aging process) and alter DNA (possibly increasing the risk for cancerous cells to develop). Free radicals may also contribute to the development of cardiovascular disease. In the process of functioning as an antioxidant, beta-carotene is itself oxidized or destroyed.

The role of beta-carotene in preventing cancer or cardiovascular disease is being studied. Beta-carotene supplements have not been shown to protect against cancer or heart disease. The Institute of Medicine did not set a Tolerable Upper Intake Level for carotene or carotenoids. Instead, they concluded that beta-carotene supplements are not advisable for the general population.

Certain plant foods are excellent sources of carotenoids. These include dark green vegetables, such as spinach, and deep orange fruits and vegetables, such as apricots, carrots, and sweet potatoes. Beta-carotene has an orange color seen in many vitamin-A-rich fruits and vegetables, but in some cases its orange color is masked by dark green chlorophyll found in vegetables such as broccoli or spinach.

Sources of retinol, also called preformed vitamin A, include animal products such as liver (a very rich source), vitamin A-fortified milk, eggs, and fortified cereals. Most ready-to-eat and instant cereals are fortified with at least 25 percent of the Daily Value for vitamin A. Butter and margarine are also fortified with vitamin A. Retinol is used in fortification. Table 6-1 lists the vitamin A content of various foods.

The RDA for vitamin A is now measured in **retinol activity equivalents (RAEs).** The concept of RAE is used because the body obtains vitamin A from retinol and carotenoids. One RAE is equal to:

- 1 microgram retinol
- 12 micrograms beta-carotene
- 24 micrograms of other vitamin A precursor carotenoids

Until the RDA for vitamin A was published in 2001, vitamin A was measured in retinol equivalents (REs). Food composition tables and nutrient analysis software continue to use REs instead of RAEs. RAE and RE values

TABLE 6-1 Vitamin A in Food

Food	Micrograms Retinol Activity Equivalents
Animal Sources	
Liver, beef, cooked, 3 oz.	9107
Liver, chicken, cooked, 3 oz.	4180
Egg substitute, fortified, 1/4 cup	407
Fat-free milk, fortified with vitamin A, 1 cup	150
Cheese pizza, 1/8 of a 12-inch diameter pie	114
Cheddar cheese, 1 oz.	90
Whole egg, 1 medium	84
Swiss cheese, 1 oz.	72
Yogurt, fruit-flavored, low-fat, 1 cup	36
Plant Sources	
Carrot, 1 raw, 7-1/2 inches	1013
Carrots, boiled, sliced, 1/2 cup	894
Carrot juice, canned, 1/2 cup	646
Mango, raw, 1 fruit	389
Sweet potatoes, 1/2 cup, mashed	372
Spinach, boiled, 1/2 cup	369
Cantaloupe, raw, 1 cup cubes	258
Kale, boiled, 1/2 cup	240
Vegetable soup, prepared with water	150
Pepper, sweet, red, raw, 1/2 cup sliced	131
Apricots, without skin, 1/2 cup halves	103
Spinach, raw, 1 cup	101
Broccoli, frozen, chopped, boiled, 1/2 cup	87
Apricot nectar, canned, 1/2 cup	83
Oatmeal, instant, fortified, dry, 1 packet	53
Tomato juice, canned, 6 fl. oz.	50
Peaches, canned, 1/2 cup slices or halves	26
Orange, raw, 1 large	14
Asparagus, boiled, 4 spears	17
Tomato, red, ripe, raw, 1/2-inch-thick slice	8

Source: Facts About Dietary Supplements. Office of Dietary Supplements, National Institutes of Health. 2001. Office of Dietary Supplements homepage: http://ods.od.nih.gov.

are the same for retinol or preformed vitamin A, but the RAE value for carotenoids is about half the RE value.

Vitamin A deficiency is of most concern in developing countries where it affects the health of many children and adults, causing night blindness, blindness, poor growth, and other problems. Up to 500,000 children worldwide go blind each year because of vitamin A deficiency. Signs of deficiency include night blindness, dry skin, dry hair, broken fingernails, and decreased resistance to infections. In the United States, vitamin A deficiency is sometimes seen in the elderly, the poor, and preschool children. In children, a mild degree of vitamin A deficiency may increase their risk of developing respiratory and diarrheal infections, decrease growth rate, slow bone development, and decrease likelihood of survival from serious illness.

Hypervitaminosis A—A disease caused by prolonged use of high doses of preformed vitamin A that can cause hair loss, bone pain and damage, soreness, and other problems.

Prolonged use of high doses of preformed vitamin A (even just three times one's RDA) may cause symptoms of **hypervitaminosis A** such as hair loss, bone pain and damage, soreness, skin problems, liver damage, nausea, and diarrhea. High doses are particularly dangerous for pregnant women (they may cause birth defects) and the elderly (they can cause joint pain, nausea, muscle soreness, itching, hair loss, and liver and bone damage). Overconsumption of beta-carotene supplements can be harmful for some individuals.

Vitamin D

Vitamin D_3 (cholecalciferol)—The form of vitamin D found in animal foods.

Vitamin D differs from most other nutrients in that it can be made in the body. It also is unique in that it functions as a hormone. When ultraviolet rays shine on your skin, a cholesterol-like compound is converted into a precursor of vitamin D and absorbed into the blood. Over the next one and a half to three days, the precursor is converted to **vitamin D_3,** also called **cholecalciferol.** Of course, if you are not in the sun much or if the ultraviolet rays are cut off by heavy clothing, clouds, smog, fog, sunscreen (SPF of 8 or higher), or window glass, there will be less vitamin D produced. On the positive side, a light-skinned person needs only about 15 minutes of sun on the face, hands, and arms two to three times per week to make enough vitamin D. A dark-skinned person needs more time in the sun because melanin (dark brown to black pigments in the skin) acts as sunscreen. Several months' supply of vitamin D can be stored in the body, which is helpful during winter months when the sun is not as strong in northern climates and you need to wear more clothing. As you get older, your body makes less vitamin D.

Vitamin D_3 is converted into its active form by enzymes in the liver and then the kidney. In its active form, vitamin D functions more as a hormone than as a vitamin. Hormones are substances secreted into the bloodstream that travel to one or more organs. Once the hormone reaches what is called the target organ(s), it affects something that the organ does. The active form of vitamin D travels through the bloodstream to the intestines, kidney, and bones to increase the amount of calcium in the blood. Blood calcium levels must be kept high so enough calcium is present to build bones and teeth, contract and relax muscles, and transmit nerve impulses. The hormone form of vitamin D increases blood calcium levels in three ways: It increases calcium absorption in the intestine, decreases the amount of calcium excreted by the kidney, and pulls calcium out of the bones.

Foods of animal origin contain vitamin D_3 or cholecalciferol. Foods of plant origin contain a plant version called vitamin D_2 or ergocalciferol. Significant food sources of vitamin D include fatty fish and vitamin-D-fortified milk and cereals. If you drink 2 cups of milk each day, you will get about half the RDA of vitamin D (the rest comes from other foods and sun exposure). There is no vitamin D added to milk products such as yogurt or cheese. Liver, egg yolks, and vitamin D-fortified margarine contain some vitamin D.

Vitamin D was previously measured in international units (IU), so most nutrient composition tables use IU. The current AI for vitamin D is expressed in micrograms of cholecalciferol. The relationship between IU and micrograms of cholecalciferol is as follows.

1 IU = 0.025 micrograms cholecalciferol

The AI for vitamin D assumes that you are not getting any vitamin D from exposure to the sun.

Rickets—A childhood disease in which bones do not grow normally, resulting in bowed legs and knock knees; it is generally caused by a vitamin D deficiency.

Osteomalacia—A disease of vitamin D deficiency in adults in which the leg and spinal bones soften and may bend.

Vitamin D deficiency in children causes **rickets,** a disease in which bones do not grow normally, resulting in soft bones and bowed legs. Vitamin D deficiency in adults causes **osteomalacia,** a disease in which leg and spinal bones soften and may bend and break. Rickets is rarely seen, but osteomalacia may be seen in elderly individuals with poor milk intake and little sun exposure, and in individuals with diseases such as renal disease or alcoholism.

Osteomalacia is not the same disease as osteoporosis. It is estimated that over 25 million adults in the Untied States have, or are at risk of developing, osteoporosis. Osteoporosis is a disease characterized by fragile bones. It results in increased risk of bone fractures. Having normal storage levels of vitamin D in your body helps keep your bones strong and may help prevent osteoporosis in elderly individuals who are non-ambulatory and in postmenopausal women.

Vitamin D supplements are often recommended for exclusively breast-fed infants because human milk may not contain adequate supplies. Mothers of infants who are exclusively breast-fed and have limited sun exposure should consult with a pediatrician on this issue. Since infant formulas are routinely fortified with vitamin D, formula-fed infants usually have adequate dietary intake of vitamin D.

Vitamin D, when taken in excess of the AI, is the most toxic of all the vitamins. All you need is about four to five times the AI to start feeling symptoms of nausea, vomiting, diarrhea, fatigue, confusion, and thirst. It can lead to calcium deposits in the heart, blood vessels, and kidneys that can cause severe health problems and even death. Young children and infants are especially susceptible to the toxic effects of too much vitamin D, and megadoses can cause growth failure.

Vitamin E

Vitamin E has an important function in the body as an antioxidant in the cell membrane and other parts of the cell. Vitamin E is of particular importance to cell membranes at the highest risk of oxidation, which includes cells in the lungs, red blood cells, and brain. Vitamin E even protects vitamin A from oxidation. Vitamin E is also important for the metabolism of iron and healthy immune system and nervous tissues. Vitamin E may have a role in preventing cardiovascular disease and cancer, but research is still in the early stages.

Vitamin E is widely distributed in plant foods. Rich sources include vegetable oils, margarine and shortening made from vegetable oils, salad dressings made from vegetable oils, seeds, and nuts. In oils, vitamin E acts like an antioxidant, thereby preventing the oil from going rancid or bad. Other good sources include whole-grain breads and cereals such as oatmeal. Animal foods are poor sources of vitamin E.

The RDA for vitamin E is expressed in milligrams of alpha-tocopherol (the most active form of vitamin E). Food composition tables usually measure vitamin E in milligrams of tocopherol equivalents, which includes forms of vitamin E in addition to alpha-tocopherol. If you know the number of tocopherol equivalents in a food, multiply that number by 0.8 to come up with the alpha-tocopherol content.

Deficiency is rare, except in infants born prematurely, in part because of vitamin E's wide distribution in plant foods and because the body has significant storage capacity. Toxicity is also rare and can cause bleeding problems.

Vitamin K

Vitamin K has an essential role in the production of a number of blood-clotting factors, such as prothrombin. Blood clotting prevents excessive blood loss when the skin is broken. Vitamin K is also needed to make an important protein used to form bone.

Vitamin K is found in certain foods and is also produced in the body. There are billions of bacteria that normally live in your intestines, and some of them make a form of vitamin K. It is thought that the amount of vitamin K produced by the bacteria is significant and may meet about half of your needs. (An infant is normally given this vitamin after birth to prevent bleeding because the intestine does not yet have the bacteria to produce vitamin K.) Food sources of vitamin K provide the balance needed. Excellent sources of vitamin K include liver, green leafy vegetables such as spinach and cabbage, broccoli, and vegetable oils.

Vitamin K deficiency is rare, but it can occur when you are taking antibiotics (because the medication destroys the bacteria in the intestines) or have problems absorbing fat. Toxicity is normally not a problem because vitamin K is readily excreted. However, excessive supplementation of a synthetic version of vitamin K (water-soluble menadione) can be toxic, so supplements of vitamin K alone are not available unless prescribed by a physician. No Tolerable Upper Intake Level has been set for vitamin K.

■ **MINI-SUMMARY**

Table 6-2 summarizes the functions and sources of the fat-soluble vitamins: A, D, E, and K.

Water-Soluble Vitamins

Water-soluble vitamins include vitamin C and the B-complex vitamins. The B vitamins work in every body cell, where they function as coenzymes. A coenzyme combines with an enzyme to make it active. Without the coenzyme, the enzyme is useless. The body stores only limited amounts of water-soluble vitamins (except vitamin B_{12}). Due to their limited storage, these vitamins need to be taken in daily. Excesses are excreted in the urine. Even though excesses are excreted, excessive supplementation of certain water-soluble vitamins can cause toxic side effects.

Vitamin C

Scurvy—A vitamin C deficiency disease marked by bleeding gums, weakness, loose teeth, and broken capillaries under the skin.

Collagen—The most abundant protein in the body; a fibrous protein that is a component of skin, bone, teeth, ligaments, tendons, and other connective structures.

Scurvy, the name for vitamin C deficiency disease, has been known since biblical times. It was most common on ships, where sailors developed bleeding gums, weakness, loose teeth, broken capillaries (small blood vessels) under the skin, and eventually death. Because sailors' diets included fresh fruits and vegetables only for the first part of a voyage, longer voyages resulted in more cases of scurvy. Once it was discovered that citrus fruits prevented scurvy, British sailors were given daily portions of lemon juice. In those days, lemons were called limes; hence British sailors got the nickname "limeys."

Vitamin C (its chemical name is ascorbic acid, meaning "no-scurvy acid") is important in forming **collagen,** a protein substance that provides strength and support to bones, teeth, skin, cartilage, blood vessels, and healing wounds. It has been said that vitamin C acts like cement, holding together our cells and tissues.

Vitamin C is also important in helping to make some hormones, such as thyroxine, which regulates your metabolic rate. While vitamin C has received much publicity as a cure for the common cold, the most it seems to do is shorten the length of the cold by a day. However, vitamin C does have a very important connection to your body's immune system. White blood

TABLE 6-2 Vitamins

Vitamin	Recommended Intake	Functions	Sources
Fat-Soluble Vitamins			
Vitamin A	(2001 RDA): Males: 900 micrograms RAE (14–70+) Females: 700 micrograms RAE (14–70+) UL: 3000 micrograms/day of preformed vitamin A	Health of eye (especially cornea and retina), vision. Epithelial cells that form protective linings of lungs, gastrointestinal tract, and urinary tract. Reproduction. Growth and development. Bone and teeth development. Immune system function. Healthy skin.	Preformed: Liver, fortified milk, fortified cereals, eggs. Provitamin: Dark green vegetables, deep orange fruits and vegetables.
Vitamin D	(1998 AI): 5 micrograms cholecalciferol (31–50) 10 micrograms cholecalciferol (51–70+) UL: 50 micrograms cholecalciferol	Maintenance of blood calcium levels. Building bones and teeth (calcium). Muscle contraction and nerve impulses (calcium).	Fortified milk, fatty fish, fortified cereals, egg yolk, sunshine.
Vitamin E	(2000 RDA): 15 milligrams alpha-tocopherol (14–70+) UL: 1,000 milligrams alpha-tocopherol (synthetic forms from supplements and/or fortified foods)	Antioxidant—especially helps red blood cells, lungs, and brain. Iron metabolism. Immune system function. Nervous system function.	Vegetable oils, margarine, shortening, salad dressings, seeds, nuts, whole-grain breads and cereals.
Vitamin K	(2001 AI): Males: 120 micrograms vitamin K (19–70+) Females: 90 micrograms vitamin K (19–70+)	Blood clotting. Healthy bones.	Liver, green leafy vegetables, broccoli, vegetable oils. Made in intestine.
Water-Soluble Vitamins			
Vitamin C	(2000 RDA): Males: 90 milligrams vitamin C (19–70+) Females: 75 milligrams vitamin C (19–70+) UL: 2,000 milligrams vitamin C	Collagen formation. Wound healing. Synthesis of some hormones. Healthy immune system. Antioxidant. Absorption of iron and vitamin A.	Citrus fruits, bell peppers, kiwi fruit, strawberries, tomatoes, broccoli, potatoes; fortified juices, drinks, and cereals.
Thiamin	(1998 RDA): Males: 1.2 milligrams thiamin (14–70+) Females: 1.1 milligrams thiamin (19–70+)	Part of coenzyme in energy metabolism. Normal growth. Nerve function.	Pork, sunflower seeds, wheat germ, peanuts, acorn squash, dry beans, whole-grain and enriched/fortified breads and cereals.

TABLE 6-2 *(continued)*			
Vitamin	Recommended Intake	Functions	Sources
Riboflavin	(1998 RDA): Males: 1.3 milligrams riboflavin (14–70+) Females: 1.1 milligrams riboflavin (19–70+)	Part of coenzymes in energy metabolism. Normal growth. Formation of vitamin B_6 coenzyme and niacin. Normal vision.	Milk, milk products, organ meats, whole-grain and enriched/fortified breads and cereals, eggs, some meats.
Niacin	(1998 RDA): Males: 16 milligrams niacin equivalent (14–70+) Females: 14 milligrams niacin equivalents (14–70+) UL: 35 milligrams niacin equivalents (synthetic forms from supplements and/or fortified foods)	Part of coenzymes in energy metabolism. Normal growth.	Meat, poultry, fish, organ meats, whole-grain and enriched/fortified breads and cereals, peanut butter, milk, eggs.
Vitamin B_6	(1998 RDA): Males: 1.3 milligrams vitamin B_6 (14–50) 1.7 milligrams vitamin B_6 (51–70+) Females: 1.3 milligrams vitamin B_6 (19–50) 1.5 milligrams vitamin B_6 (51–70+) UL: 100 milligrams vitamin B_6	Part of coenzyme involved in carbohydrate, fat, and especially protein metabolism. Synthesis of red blood cells, white blood cells, and neurotransmitters.	Meat, poultry, fish, potatoes, fruits such as bananas, some leafy green vegetables, fortified cereals.
Folate	(1998 RDA): 400 micrograms dietary folate equivalent (14–70+) UL: 1000 micrograms dietary folate equivalents (synthetic forms from supplements and/or fortified foods)	Part of coenzyme required to make DNA and new cells. Formation of neurotransmitters in the brain. Amino acid metabolism.	Green leafy vegetables, legumes, orange juice, fortified breads and cereals.
Vitamin B_{12}	(1998 RDA): 2.4 micrograms vitamin B_{12} (14–70+)	Conversion of folate into active forms. Normal nervous system function. Healthy bones.	Animal foods such as meat, poultry, fish, shellfish, eggs, milk, and milk products.
Pantothenic Acid	(1998 AI): 5 milligrams pantothenic acid (14–70+)	Part of coenzyme in energy metabolism.	Widespread.
Biotin	(1998 AI): 30 micrograms biotin (19–70+)	Part of coenzymes involved in metabolism of carbohydrates, fats, and proteins.	Widespread. Made in intestine.
Choline	(1998 AI): Males: 550 milligrams choline (14–70+) Females: 425 milligrams choline (19–70+) UL: 3.5 grams choline	Synthesis of neurotransmitter. Synthesis of lecithin (a phospholipid) found in cell membranes.	Widespread; milk, eggs, peanuts.

cells, which defend your body against undesirable invaders, have the highest concentration of vitamin C in the body.

Like beta-carotene and vitamin E, vitamin C is an important antioxidant in the body. For example, it prevents the oxidation of vitamin A and iron in the intestine so these nutrients can be absorbed. Its antioxidant properties have made variations of vitamin C widely used as a food additive. It may appear on food labels as sodium ascorbate or calcium ascorbate. Neither of these substances has vitamin C activity.

Foods rich in vitamin C include citrus fruits (oranges, grapefruits, limes, and lemons), bell peppers, kiwi fruit, strawberries, tomatoes, broccoli, and potatoes. Only foods from the fruit and vegetable groups contribute vitamin C. There is little or no vitamin C in the meat group (except in liver) or the dairy group. Some juices are fortified with vitamin C (if not already rich in vitamin C), as are most ready-to-eat cereals. Many people meet their needs for vitamin C simply by drinking orange juice. This is a good choice because vitamin C is easily destroyed in food preparation and cooking. Table 6-3 lists the vitamin C content of selected foods.

Certain situations require additional vitamin C. These include pregnancy and nursing, growth, fevers and infections, burns, fractures, surgery, and cigarette smoking. Smoking produces oxidants, which depletes vitamin C. The RDA for smokers is 35 milligrams of vitamin C daily in addition to the normal RDA (75 mg for women, 90 mg for men).

Deficiencies resulting in scurvy are rare. Scurvy might occur in elderly people with poor intake of fruits and vegetables, and in alcoholics.

The Tolerable Upper Intake Level for vitamin C is 2000 mg or 2 grams per day. At that level, gastrointestinal symptoms such as stomach cramps and diarrhea may occur. High doses can interfere with certain medical tests.

Thiamin, Riboflavin, and Niacin

Coenzyme—A molecule that combines with an enzyme and makes the enzyme functional.

Thiamin, riboflavin, and niacin all play key roles as part of **coenzymes** in energy metabolism. They are essential in the release of energy from carbohydrates, fats, and proteins. They are also needed for normal growth.

Thiamin also plays a vital role in the normal functioning of the nerves. Riboflavin is part of coenzymes that help form the vitamin B_6 coenzyme and make niacin in the body from the amino acid tryptophan. Riboflavin also promotes normal vision.

Because thiamin, riboflavin, and niacin all help to release food energy, the needs for these vitamins increase as caloric needs increase.

Thiamin is widely distributed in foods, but mostly in moderate amounts. Pork is an excellent source of thiamin. Other sources include sunflower seeds, wheat germ, peanuts, acorn squash, dry beans, and whole-grain and enriched/fortified breads and cereals.

TABLE 6-3 Vitamin C in Foods

Food	Milligrams Vitamin C
Fruits	
Orange, 1	80
Kiwi, 1 medium	75
Cranberry juice cocktail, 6 ounces	67
Orange juice, from concentrate, 1/2 cup	48
Papaya, 1/2 cup cubes	43
Strawberries, 1/2 cup	42
Grapefruit, 1/2	41
Grapefruit juice, canned, 1/2 cup	36
Cantaloupe, 1/2 cup cubes	34
Tangerine, 1	26
Mango, 1/2 cup, slices	23
Honeydew melon, 1/2 cup cubes	21
Banana, 1	10
Apple, 1	8
Nectarine, 1	7
Vegetables	
Broccoli, chopped, cooked, 1/2 cup	49
Brussels sprouts, cooked, 1/2 cup	48
Cauliflower, cooked, 1/2 cup	34
Sweet potato, baked, 1	28
Kale, cooked, chopped, 1/2 cup	27
White potato, baked, 1	26
Tomato, 1 fresh	22
Tomato juice, 1/2 cup	22
Cereals	
Cornflakes, 1 cup	15

Source: U.S. Department of Agriculture.

Milk and milk products are the major source of riboflavin in the American diet. Other sources include organ meats such as liver (very high in riboflavin), whole-grain and enriched/fortified breads and cereals, eggs, and some meats.

The main sources of niacin are meat, poultry, and fish. Organ meats, again, are quite high in niacin. Whole-grain and enriched/fortified breads and cereals, as well as peanut butter, are also important sources of niacin. All foods containing complete protein, such as those just mentioned and

Tryptophan—An amino acid present in protein foods that can be converted to niacin in the body.

Niacin equivalents (NE)—The unit for measuring niacin. One niacin equivalent is equal to one milligram of niacin or 60 milligrams of tryptophan.

also milk and eggs, are good sources of the precursor of niacin, tryptophan. **Tryptophan,** an amino acid present in some of these foods, can be converted to niacin in the body. This is why the RDA for niacin is stated in **niacin equivalents.** One niacin equivalent is equal to one milligram of niacin or 60 milligrams of tryptophan. Less than half the niacin we use is made from tryptophan.

Deficiencies in thiamin, riboflavin, and niacin are rare in the United States, in large part because breads and cereals are enriched with all three nutrients. General symptoms for B-vitamin deficiencies include fatigue, decreased appetite, and depression. Alcoholism can create deficiencies in these three vitamins due in part to limited food intake.

Toxicity is not a problem except in the case of niacin. Nicotinic acid, a form of niacin, is often prescribed by physicians to lower elevated blood cholesterol levels. Unfortunately, it has some undesirable side effects. Starting at doses of 100 milligrams, typical symptoms include flushing, tingling, itching, rashes, hives, nausea, diarrhea, and abdominal discomfort. Flushing of the face, neck, and chest lasts for about 20 minutes after taking a large dose. More serious side effects of large doses include liver damage and high blood sugar levels.

Vitamin B_6

Vitamin B_6 is a water-soluble vitamin that exists in three major chemical forms: pyridoxine, pyridoxal, and pyridoxamine. Vitamin B_6 plays an important role as part of a coenzyme involved in carbohydrate, fat, and protein metabolism. In its coenzyme form, it is particularly crucial to protein metabolism. Vitamin B_6 is used to make red blood cells (which transport oxygen around the body) and white blood cells (which are crucial for the immune system to function properly). Vitamin B_6 is also used to break down glycogen to glucose, keeping your blood sugar level steady, and to convert tryptophan to niacin.

Vitamin B_6 is needed for the synthesis of neurotransmitters such as serotonin and dopamine. These neurotransmitters are required for normal nerve cell communication. Lower levels of serotonin have been found in individuals suffering from depression and migraine headaches. So far, however, vitamin B_6 supplements have not proven effective for relieving these symptoms. Researchers have also been investigating the relationship between vitamin B_6 status and a wide variety of neurological conditions such as seizures, chronic pain, and Parkinson's disease.

Good sources for vitamin B_6 include meat, poultry, and fish. Vitamin B_6 also appears in plant foods; however, it is not as well absorbed from these sources. Good plant sources include potatoes, some fruits (such as bananas and watermelon), and some leafy green vegetables (such as broccoli and spinach). Fortified ready-to-eat cereals are also good sources of vitamin B_6. Table 6-4 gives the amount of vitamin B_6 in many foods.

Deficiency of vitamin B_6, which may occur in some women and older adults, can cause symptoms such as fatigue, depression, and irritability. Symptoms can become much more serious if the deficiency continues.

Excessive use of vitamin B_6 (more than 2 grams daily for two months or more than 200 milligrams daily for a longer period of time) can cause irreversible nerve damage and symptoms such as numbness in hands and feet and difficulty walking. The Tolerable Upper Intake Level for B_6 is 100 milligrams. The problem with B_6 is that, unlike other water-soluble vitamins, it is stored in the muscles. Vitamin B_6 supplementation became popular when it appeared that the vitamin may relieve some of the symptoms of premenstrual syndrome and carpal tunnel syndrome, a condition in which a compressed nerve in the wrist causes much pain. There is no scientific research showing that B_6 helps in either condition.

TABLE 6-4 Vitamin B_6 in Food

Food and Serving Size	Milligrams Vitamin B_6	% DV
Ready-to-eat cereal, fortified with 100% of the DV, 3/4 cup	2.00	100
Potato, baked, flesh and skin, 1 medium	0.70	35
Banana, raw, 1 medium	0.68	34
Garbanzo beans, canned, 1/2 cup	0.57	30
Chicken breast, meat only, cooked, 1/2 breast	0.52	25
Ready-to-eat cereal, fortified with 25% of the DV, 3/4 cup	0.50	25
Oatmeal, instant, fortified, 1 packet	0.42	20
Pork loin, lean only, cooked, 3 oz.	0.42	20
Roast beef, eye of round, lean only, 3 oz.	0.32	15
Trout, rainbow, cooked, 3 oz.	0.29	15
Sunflower seeds, kernels, dry-roasted, 1 oz.	0.23	10
Tomato juice, canned, 6 fl. oz.	0.20	10
Avocado, raw, sliced, 1/2 cup	0.20	10
Salmon, sockeye, cooked, 3 oz.	0.19	10
Tuna, canned in water, drained, 3 oz.	0.18	10
Wheat bran, crude or unprocessed, 1/4 cup	0.18	10
Peanut butter, smooth, 2 tablespoons	0.15	8
Walnuts, English/Persian, 1 oz.	0.15	8
Spinach, frozen, cooked, 1/2 cup	0.14	8
Lima beans, frozen, cooked, drained, 1/2 cup	0.10	6
Soybeans, green, boiled, drained, 1/2 cup	0.05	2

Source: Facts About Dietary Supplements. Office of Dietary Supplements, National Institutes of Health. 2001. Office of Dietary Supplements homepage: http://ods.od.nih.gov.

Folate and Vitamin B_{12}

Folate and vitamin B_{12} often work together in the body. Folate is a component of coenzymes required to form DNA, the genetic material contained in every body cell. Folate is therefore needed to make all new cells. Much folate is used to produce adequate numbers of red blood cells, white blood cells, and digestive tract cells, since these cells divide frequently. Folate also is involved in forming neurotransmitters in the brain and in amino acid metabolism.

Excellent sources of folate include green leafy vegetables (the word *folate* comes from the word *foliage,* meaning leaves), legumes, orange juice, and fortified breads and ready-to-eat cereals. Meats and dairy products contain little folate. Much folate is lost during food preparation and cooking, so fresh and lightly cooked foods are more likely to contain more folate. Table 6-5 gives the folate content of selected foods.

Dietary folate equivalents (DFEs)—The unit for measuring folate that takes into account the amount of folate that is absorbed from natural and synthetic sources.

Megaloblastic (macrocytic) anemia—A form of anemia caused by a deficiency of vitamin B_{12} or folate and characterized by large, immature red blood cells.

Neural tube defects—Diseases in which the brain and/or the spinal cord form improperly in early pregnancy.

The RDA for folate is stated in micrograms of **dietary folate equivalents (DFEs).** DFEs take into account the amount of folate that is absorbed from natural and synthetic sources. Synthetic folate, used in supplements and fortified foods such as breads, is absorbed at 1.7 times the rate of folate that naturally occurs in foods such as leafy green vegetables. Whereas 100 micrograms of naturally occurring folate is counted as 100 micrograms DFE, 100 micrograms of synthetic folate is counted as 170 micrograms DFE. The Tolerable Upper Intake Limit for folate is 1000 micrograms of the synthetic form found in supplements and fortified foods.

A folate deficiency can cause **megaloblastic (macrocytic) anemia,** a condition in which the red blood cells are larger than normal and function poorly. The red blood cells are large because they have not matured normally due to the fact that DNA synthesis has slowed down. Other symptoms may include digestive tract problems such as diarrhea, and mental confusion and depression. Groups particularly at risk for folate deficiency are pregnant women, low-birth-weight infants, and the elderly.

The need for folate is critical during the earliest weeks of pregnancy—when most women don't know they are pregnant. A folate deficiency may cause **neural tube defects.** The neural tube is the tissue in the embryo that develops into the brain and spinal cord. The neural tube closes within the first month of pregnancy. Neural tube defects are diseases in which the brain and spinal cord form improperly in early pregnancy. They affect one to two of every 1,000 babies born each year. Neural tube defects include anencephaly, in which most of the brain is missing, and spina bifida. In one form of spina bifida, a piece of the spinal cord protrudes from the spinal column, causing paralysis of parts of the lower body.

Because of the importance of folate during pregnancy and the difficulty most women encounter trying to get enough folate in the diet, the Food and Drug Administration requires manufacturers of enriched bread, flour, pasta, cornmeal, rice, and some other foods to fortify their products with a very absorbable form of folate. Women eating folate-fortified foods should not assume that these foods will meet all their folate needs. They should still seek out folate-rich foods (Table 6-5). Folate requirements increase from 400 to 600 micrograms DFE during pregnancy.

A number of frequently used medications interfere with the normal use of folate in the body. Frequent users of aspirin, antacids, anticonvulsant medications, or oral contraceptives should speak with their physicians about obtaining adequate folate.

Vitamin B_{12}, a group of related compounds that contain the mineral cobalt, is present in all body cells. One of its important functions is to convert folate into its active forms so that it can make DNA. It also helps in the normal functioning of the nervous system by maintaining the protective cover around nerve fibers. Bone cells also depend on vitamin B_{12}.

Vitamin B_{12} differs from other vitamins in that it is found only in animal foods such as meat, poultry, fish, shellfish, eggs, milk, and milk products. Plant foods do not naturally contain any vitamin B_{12}. Unlike the other B vitamins, vitamin B_{12} is easily destroyed when foods containing it are microwaved.

Intrinsic factor—A protein-like substance secreted by stomach cells that is necessary for the absorption of vitamin B_{12}.

Pernicious anemia—A type of anemia caused by a deficiency of vitamin B_{12} and characterized by macrocytic anemia and deterioration in the functioning of the nervous system.

Vitamin B_{12} also differs from other vitamins in that it requires **intrinsic factor** (a proteinlike substance produced in the stomach) to be absorbed. Vitamin B_{12} must be separated from protein in food before it can bind with intrinsic factor and be absorbed. Hydrochloric acid in the stomach helps separate the vitamin. As it enters the small intestine, vitamin B_{12} attaches to the intrinsic factor and is then carried to the ileum (the last portion of the small intestine), where it is absorbed. Vitamin B_{12} is stored in the liver.

A vitamin B_{12} deficiency in the body is usually due not to poor intake, but rather to a problem with absorption. Absorption problems are often due to a lack of intrinsic factor and/or a lack of hydrochloric acid. Both conditions are more likely as you age. In fact, 10 to 30 percent of older people may be unable to absorb vitamin B_{12} in food. Individuals with disorders of the stomach or small intestine may also not absorb enough vitamin B_{12}.

When vitamin B_{12} is not properly absorbed, **pernicious anemia** develops. Pernicious means ruinous or harmful, and this type of anemia is marked by a megaloblastic anemia (as with folate deficiency), also called macrocytic anemia, in which there are too many large, immature red blood cells. Symptoms include extreme weakness and fatigue. Nervous system problems also erupt. The cover surrounding the nerves in the body becomes damaged, making it difficult for impulses to travel along them. This causes a poor sense of balance, numbness and tingling sensations in the arms and legs, and mental confusion. Pernicious anemia can result in paralysis and death if not treated.

Because deficiency in either folate or vitamin B_{12} causes macrocytic anemia, a physician may mistakenly administer folate when the problem is really a vitamin B_{12} deficiency. The folate would treat the anemia, but not the deterioriation of the nervous system due to a lack of vitamin B_{12}. If untreated, this damage can be significant and sometimes irreversible, although it takes many years to occur. When vitamin B_{12} is deficient due to an absorption problem, injections of the vitamin must be given.

Vegetarians who do not eat meats, fish, eggs, milk or milk products, or B_{12}-fortified foods consume no vitamin B_{12} and are at high risk of developing a

TABLE 6-5 Folate in Food

Food	Micrograms Dietary Folate Equivalents	% Daily Value
Ready-to-eat cereal, fortified with 100% of the DV, 3/4 cup	400	100%
Beef liver, cooked, braised, 3 oz.	185	45
Cowpeas (black-eyed peas) immature, cooked, boiled, 1/2 cup	105	25
Breakfast cereals, fortified with 25% of the DV, 3/4 cup	100	25
Spinach, frozen, boiled, 1/2 cup	100	25
Great Northern beans, boiled, 1/2 cup	90	20
Asparagus, boiled, 4 spears	85	20
Wheat germ, toasted, 1/4 cup	80	20
Orange juice, chilled, 3/4 cup	70	20
Turnip greens, frozen, boiled, 1/2 cup	65	15
Vegetarian baked beans, canned, 1 cup	60	15
Spinach, raw, 1 cup	60	15
Green peas, boiled, 1/2 cup	50	15
Broccoli, chopped, frozen, cooked, 1/2 cup	50	15
Egg noodles, cooked, enriched, 1/2 cup	50	15
Rice, white, long-grain, parboiled, cooked, enriched, 1/2 cup	45	10
Avocado, raw, all varieties, sliced, 1/2 cup	45	10
Peanuts, dry roasted, 1 oz.	40	10
Lettuce, romaine, shredded, 1/2 cup	40	10
Tomato juice, canned, 6 fl. oz.	35	10
Orange, fresh, 1 small	30	8
Bread, white, enriched, 1 slice	25	6
Egg, whole, raw, fresh, 1 large	25	6
Cantaloupe, raw, 1/4 medium	25	6
Papaya, raw, 1/2 cup cubes	25	6
Banana, raw, 1 medium	20	6
Broccoli, raw, 1 spear	20	6
Lettuce, iceberg, shredded, 1/2 cup	15	4
Bread, whole-wheat, 1 slice	15	4

Source: Facts About Dietary Supplements. Office of Dietary Supplements, National Institutes of Health. 2001. Office of Dietary Supplements homepage: http://ods.od.nih.gov.

deficiency of vitamin B_{12}. When adults adopt a vegetarian diet, deficiency symptoms can be slow to appear because it usually takes years to deplete normal body stores. However, severe symptoms of B_{12} deficiency, most often featuring poor neurological development, can show up quickly in children and breast-fed infants of women who follow a strict vegetarian diet. Fortified cereals are one of the few plant food sources of vitamin B_{12} and are an important dietary source of B_{12} for vegetarians who consume no eggs, milk, or milk products. Vegetarian adults who do not eat vitamin-B_{12}-fortified foods need to consider taking a supplement. Likewise, vegetarian mothers should consult with a pediatrician regarding appropriate vitamin B_{12} supplementation for their infants and children.

A deficiency of folate, vitamin B_{12}, or vitamin B_6 may increase your level of homocysteine, an amino acid normally found in your blood. There is evidence that an elevated homocysteine level is an independent risk factor for heart disease and stroke. The evidence suggests that high levels of homocysteine may damage coronary arteries or make it easier for blood clotting cells called platelets to clump together and form a clot. It is very important to eat good sources of each of these nutrients. There is currently no evidence to suggest that lowering homocysteine levels with vitamin supplements will reduce your risk of heart disease. Clinical intervention trials are needed to determine whether supplementation with folate, vitamin B_{12}, or vitamin B_6 can help protect you against developing coronary heart disease.

Pantothenic Acid and Biotin

Both pantothenic acid and biotin are parts of coenzymes involved in energy metabolism. Pantothenic acid is needed to release energy from carbohydrates, fats, and protein, and to make fatty acids. Biotin is involved in the metabolism of carbohydrates, fats, and proteins.

Both pantothenic acid and biotin are widespread in foods. Good sources of pantothenic acid include peanuts, eggs, meat, chicken, milk, some vegetables, and legumes. Good sources of biotin include peanuts, egg yolks, cheese, and soybeans. Intestinal bacteria make considerable amounts of biotin. Deficiency and toxicity concerns are not known.

Choline and Vitaminlike Substances

Choline can be made in the body in small amounts. Choline is needed to make the neurotransmitter acetylcholine and the phospholipid lecithin, the major component of cell membranes. Lecithin is also a required component of VLDL, the lipoprotein that carries triglycerides and other lipids made in the liver to the body cells. Without enough lecithin, fat and cholesterol accumulate in the liver. Choline can be considered an essential nutrient, because when the diet contains no choline, the body can't make enough of it and liver damage can result. It is rare that the diet would contain no choline because it is so widespread in foods (rich sources include milk, eggs, and peanuts). Although the Food and Nutrition Board has not yet declared choline to be a vitamin (not all scientists agree with the Food and Nutrition Board on this), they have set an Adequate Intake and Tolerable Upper Intake Level for choline based on gender and age.

Vitaminlike substances, such as carnitine, lipoic acid, inositol, and taurine, are necessary for normal metabolism, but the body makes enough, so they are therefore not vitamins. They are added to formula for infants.

Other substances are promoted as being vitamins, or at least important to human nutrition, that are clearly not vitamins or will ever even be vitamins. Examples include para-aminobenzoic acid (PABA), bioflavonoids (incorrectly called vitamin P), pangamic acid (incorrectly called vitamin B_{15}), and laetrile (incorrectly called vitamin B_{17}—a supposed cancer "cure" that is, in fact, harmful).

■ **MINI-SUMMARY**

Table 6-2 summarizes the functions and sources of the water-soluble vitamins.

Ingredient Focus: Fruits and Vegetables

Nutrition

Vegetables and fruits, in general, have the following characteristics:

- Low in kcalories
- Low or no fat (except avocados)
- No cholesterol
- Good sources of fiber
- Excellent sources of vitamins and minerals, particularly vitamins A and C (see Table 6-6)
- Low in sodium (except for canned vegetables)

Also, dried fruits, such as raisins and apricots, provide some iron. The nutritional content of many fruits and vegetables appear in Tables 6-6 and 6-7.

Purchasing and Receiving

Always purchase quality fruits and vegetables, preferably fancy or no. 1 grade. Seriously consider organic produce, for its fuller flavor, if your budget will allow it. Figure 6-1 shows some specialty vegetables and fruits that you may want to consider purchasing for variety.

TABLE 6-6 Produce Nutrition

"Eating five fruits and vegetables a day is one of the most important choices you can make to help maintain your health." —National Cancer Institute

	Household Serving Sizes	Serving Size (g)	Serving Size (oz)	Kcalories (Kcal)	Protein (g)	Carbohydrate (g)	Fat (g)	Sodium (mg)	Dietary Fiber (g)	Vitamin A (% of U.S. RDA)	Vitamin C (% of U.S. RDA)	Calcium (% of U.S. RDA)	Iron (% of U.S. RDA)
Apple	1 med. apple	154	5.5	80	0	18	1	0	5	*	6	*	*
Asparagus	5 spears	93	3.5	18	2	2	0	0	2	10	10	*	*
Avocado	1/3 med. avocado	55	2	120	1	3	12	5	2	*	5	*	*
Banana	1 med. banana	126	4.5	120	1	28	1	0	3	*	15	*	2
Bell pepper	1 med. pepper	148	5.5	25	1	5	1	0	2	2	130	*	*
Broccoli	1 med. stalk	148	5.5	40	5	4	1	75	5	10	240	6	4
Cabbage	1/12 med. head	84	3	18	1	3	0	30	2	*	70	4	*
Cantaloupe	1/4 med. melon	134	5	50	1	11	0	35	0	80	90	2	2
Carrot	1 med., 7" long, 1-1/4" diameter	78	3	40	1	8	1	40	1	330	8	2	*
Cauliflower	1/6 med. head	99	3	18	2	3	0	45	2	*	110	2	2
Celery	2 med. stalks	110	4	20	1	4	0	140	2	*	15	4	*
Cherry	21 cherries; 1 cup	140	5	90	1	19	1	0	3	*	10	2	*
Cucumber	1/3 med. cucumber	99	3.5	18	1	3	0	0	0	4	6	2	2
Grape	1-1/2 cups grapes	138	5	85	1	24	0	3	2	3	9	2	2
Grapefruit	1/2 med. grapefruit	154	5.5	50	1	14	0	0	6	6	90	4	*
Green bean	3/4 cup cut beans	83	3	14	1	2	0	0	3	2	8	4	*
Green onion	1/4 cup chopped	25	1	7	0	1	0	0	0	3	20	*	5
Honeydew	1/10 med. melon	134	5	50	1	12	0	50	1	*	40	*	2
Iceberg lettuce	1/6 med. head	89	3	20	1	4	0	10	1	2	4	*	*
Kiwi fruit	2 med. kiwi fruit	148	5.5	90	1	18	1	0	4	2	230	4	4
Leaf lettuce	1-1/2 cups shredded	85	3	12	1	1	0	40	1	20	4	4	*
Lemon	1 med. lemon	58	2	18	0	4	0	10	0	*	35	2	*
Lime	1 med. lime	67	2.5	20	0	7	0	1	3	*	35	2	2
Mushrooms	5 med. mushrooms	84	3	25	3	3	0	0	0	*	2	*	*
Nectarine	1 med. nectarine	140	5	70	1	16	1	0	3	20	10	*	*
Onion	1 med. onion	148	5.5	60	1	14	0	10	3	*	20	4	*
Oranges	1 med. orange	154	5.5	50	1	13	0	0	6	*	120	4	*
Peach	2 med. peaches	174	6	70	1	19	0	0	1	20	20	*	*
Pear	1 med. pear	166	6	100	1	25	1	1	4	*	10	2	2

TABLE 6-6 ***(continued)***

"Eating five fruits and vegetables a day is one of the most important choices you can make to help maintain your health." —National Cancer Institute

	Household Serving Sizes	Serving Size (g)	Serving Size (oz)	Kcalories (Kcal)	Protein (g)	Carbohydrate (g)	Fat (g)	Sodium (mg)	Dietary Fiber (g)	Vitamin A (% of U.S. RDA)	Vitamin C	Calcium	Iron
Pineapple	2 slices, 3" diameter 3/4" thick	112	4	90	1	21	1	10	2	*	35	*	*
Plum	2 med. plums	132	4.5	70	1	17	1	0	1	9	20	*	*
Potato	1 med. potato	148	5.5	110	3	23	0	10	3	*	50	*	8
Radishes	7 radishes	85	3	20	0	3	0	35	0	*	30	*	*
Strawberries	8 med. berries	147	5.5	50	1	13	0	0	3	*	140	2	2
Summer squash	1/2 med. squash	98	3.5	20	1	3	0	0	1	4	25	2	2
Sweet corn	kernels from 1 med. ear	90	3	75	3	17	1	15	1	5	10	*	3
Sweet potato	1 med., 5" long, 2" diameter	130	4.5	140	2	32	0	15	3	520	50	3	4
Tangerine	2 med., 2-3/8" diameter	168	6	70	1	19	0	2	2	30	85	2	*
Tomato	1 med. tomato	148	5.5	35	1	6	1	10	1	20	40	*	2
Watermelon	1/18 med. melon; 2 cups diced	280	10	80	1	19	0	10	1	8	25	*	2

* Contains less than 2% of the U.S. RDA of this nutrient.

Source: U.S. Food and Drug Administration.

When receiving fresh produce, randomly check for the following.

1. Characteristic color—should be bright and lively
2. Characteristic shape—misshapen produce is often inferior in taste and texture and harder to prepare
3. Appropriate size—small or immature vegetables lack full flavor, and overgrown vegetables are often coarse
4. Freshness, which may be evident by crispness, lack of wilting, etc., depending on the item
5. Bruises, blemishes, decay, cracks, frostbite, etc.

Frozen vegetables and fruits have much the same flavor as fresh fruit, but the texture often changes during freezing. Vegetables and fruits are frozen at the peak of ripeness and are very convenient to use. Look for

TABLE 6-7 Vitamin A and C and Fiber in Fruits and Vegetables

In selecting your daily intake of fruits and vegetables, the National Cancer Institute recommends choosing:

- At least one serving of a vitamin A-rich fruit or vegetable a day.
- At least one serving of a vitamin C-rich fruit or vegetable a day.
- At least one serving of a high-fiber fruit or vegetable a day.
- Several servings of cruciferous vegetables a week. Studies suggest that these vegetables may offer additional protection against certain cancers, although further research is needed.

High in Vitamin A*	High in Vitamin C*	High in Fiber or Good Source of Fiber*	Cruciferous Vegetables
Apricots	Apricots	Apple	Bok choy
Cantaloupe	Broccoli	Banana	Broccoli
Carrots	Brussels sprouts	Blackberries	Brussels sprouts
Kale, collards	Cabbage	Blueberries	Cabbage
Leaf lettuce	Cantaloupe	Brussels sprouts	Cauliflower
Mango	Cauliflower	Carrots	
Mustard greens	Chili peppers	Cherries	
Pumpkin	Collards	Cooked beans and peas (kidney, navy, lima, and pinto beans, lentils, black-eyed peas)	
Romaine lettuce	Grapefruit	Dates	
Spinach	Honeydew melon	Figs	
Sweet potato	Kiwi fruit	Grapefruit	
Winter squash (acorn, hubbard)	Mango	Kiwi fruit	
	Mustard greens	Orange	
	Orange	Pear	
	Orange juice	Prunes	
	Pineapple	Raspberries	
	Plum	Spinach	
	Potato with skin	Strawberries	
	Spinach	Sweet potato	
	Strawberries		
	Bell peppers		
	Tangerine		
	Tomatoes		
	Watermelon		

* Based on FDA's food labeling regulations.

Source: National Cancer Institute.

frozen fruit that is thoroughly frozen and shows no signs of having thawed and refrozen.

Canned fruits are available in heavy syrup, in light syrup, juice-packed, and water-packed. Syrup-packed fruit contains much sugar. Canned fruits and vegetables are mostly available as grade A or grade B—each offers the same nutrition, but the grade A product will have a better appearance, texture, and flavor. Look for clean cans without dents, swelling, or rust.

SPECIALTY FRUITS & VEGETABLES

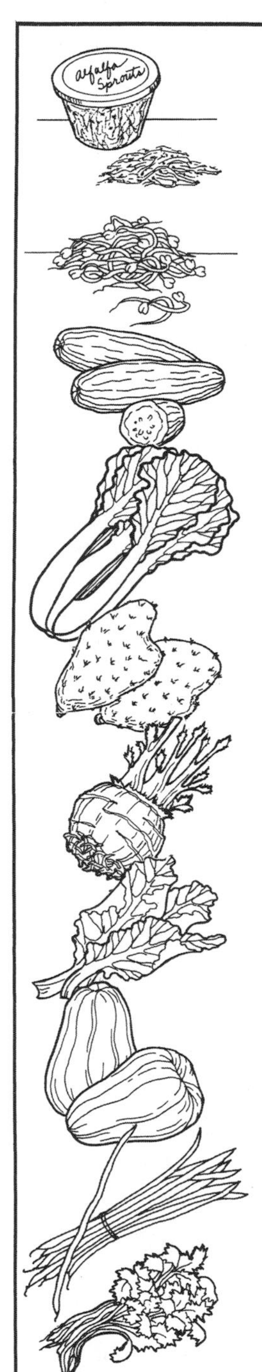

ALFALFA SPROUTS are the tender young sprouts produced by alfalfa seeds and harvested only six days after they begin growing. They have a crunchy, nutty flavor.

BEAN SPROUTS grow from mung beans and are harvested and sold within six days from the time they first appear. The shorter the bean sprouts, the crispier and more tender they are.

BITTER MELON (Balsam Pear) is shaped like a cucumber, 6–8 inches long. The outer surface is clear green and wrinkled; inside it contains a layer of white or pink spongy pulp and seeds.

BOK CHOY (Chinese Chard, White Mustard Cabbage) is a combination of chard and celery with long and broad crisp white stalks with shiny, deep green leaves. The flavor is sweet but mild.

CACTUS LEAVES (Nopales) grow on a cactus plant and are light green and crisp, but also tender. Served like a vegetable, they have the texture and flavor of green beans.

CELERIAC (Celery Root) looks like celery, but only the bulb-type root is edible. The small young root knob is more tender and less woody than the larger, mature roots.

CHARD (Swiss Chard) is a lush green, leafy vegetable resembling spinach in appearance, with flavor and texture similar to asparagus. The leaves and ribs are excellent as fresh salad greens.

CHAYOTE (Vegetable Pear) has a very dark green hard surface and can be either pear shaped or round. It is 3–5 inches long, looks like an acorn squash but has a more delicate flavor, and has flesh the color of honeydew melon.

CHINESE LONG BEANS (Dow Kwok) are pencil-thin beans, 12–25 inches long, with light green tender pods. The tiny beans inside resemble immature blackeyed peas.

CORIANDER (Cilantro) resembles parsley with small leafy sprigs, but is more tender. The flavor is stronger than parsley and lingers on the tongue. Therefore, as a parsley substitute, less is used.

CRABAPPLES are tiny apples with a strong, pleasing flavor for cooking and use in jellies and jams. They are too tart to eat out of hand. They can be yellow, red or green.

DAIKON (Japanese White Radish) is a large, tapered white radish about 8–10 inches long. It has a crisp texture and a flavor similar to an ordinary radish, but sharper and hotter.

FENNEL (Anise) has broad leaf stalks which overlap at the base of the stem and form a firm, rounded, white bulb. The leaves are green and featherlike and the plant has a licorice aroma and flavor. The bulb as well as the inside stalks are edible.

GINGER ROOT is the underground stem or root of the tropical ginger plant. It is light reddish brown and knobby, and is used as a tangy flavoring when shredded or mashed.

HORSERADISH ROOT (German Mustard) has a potent flavor and is used in small quantities. The edible part of the plant is the white root, which should be grated or shredded.

JERUSALEM ARTICHOKES (Sunchokes) are root vegetables of a gnarled, knotty appearance with a crisp, crunchy texture and delicious flavor. Eaten cooked or raw, they are very versatile and can be used as a substitute for potatoes.

JICAMA is a large turnip-shaped root vegetable with a light brown peel. Inside, it is white and crisp with a delicate sweet flavor similar to water chestnuts. The peel is not usually eaten.

KIWIFRUIT has an unattractive brown fuzzy surface. The inside, however, is smooth, creamy, bright green flesh with a strawberry-melon flavor and tiny edible black seeds. Kiwifruit is about the size of a large egg, 3 inches long and 2 inches in diameter.

KOHLRABI is a green leafy vegetable with a stem that thickens above ground and looks like a bulb. The whole plant is edible. The leaves are used like spinach and the bulb (about 2–3 inches in diameter) is served cooked or raw.

Figure 6-1

Specialty fruits and vegetables. Courtesy United Fresh Fruit and Vegetable Association

KUMQUATS are small, orange-gold citrus fruits with sweet-flavored skins and tart, tangy insides. Eaten whole, the flavors mix nicely. Remove seeds before eating.

MANGOS are subtropical fruits of varied size, shape and color. Their smooth skins can be green, yellow or red, and their shapes vary from round to oval. Sizes can be from 2 to 10 inches and 2 to 5 pounds.

NAPPA (Sui Choy, Chow Choy, Won Bok, Chinese Cabbage) has broad-ribbed stalks varying in color from white to light green, resembling romaine, with crinkly leaves forming a long, slender head. The flavor is mild.

PAPAYA is an oblong melonlike tropical fruit, green to yellow on the outside with golden yellow to orange flesh inside. A medium papaya weighs about 1 pound. The small black seeds inside are not to be eaten.

PERSIMMONS are a brilliant orange in color and soft to the touch (like a tomato). They are about the size of an apple. Enjoy their flavorful taste by eating out of hand.

PLANTAINS look like large green bananas with rough, mottled peels. They are used as a vegetable and must be cooked. Never eat them raw. Black skins indicate that they are sweet and fully ripe.

POMEGRANATES are about the size and shape of apples, golden yellow to deep red, with hundreds of kernels or seeds which are juicy and sweet. Inside, the pulp is crimson and too bitter to eat. Only the seeds are edible.

PRICKLY PEARS (Cactus Pear, Indian Fig, Barberry Fig, Tuna) are pear-shaped with yellow to crimson skins covered with spines. Usually the spines have been removed before marketing. The flesh is yellow and sweet with a taste of watermelon; the skin is bright red when ripe. Peel back the skin to eat.

QUINCE looks like an apple with an odd-shaped stem end and skin that is pale yellow when the fruit is fully ripe. Quince has a distinctive biting flavor and should be served cooked in jams, jellies, sauces, puddings, etc., not eaten raw.

SALSIFY (Oyster Plant) looks like parsnip with a full, grassy top, a firm gray-white root, and the juicy flavor of fresh oysters. The edible part is the carrot-shaped root, which should be peeled and then cooked.

SNOW PEAS (Sugar Peas, China Peas) are an oriental vegetable looking like regular peas, except that the pods are much flatter and translucent. They have a delicate flavor and add crunch and texture to a dish. Serve hot or cold after blanching 2–3 minutes.

SPAGHETTI SQUASH (Cucuzzi, Calabash, Suzza Melon) is a large edible gourd that is round to oval, about 2 feet long and 3–4 inches in diameter. The skin is pale yellow and edible; the inside forms translucent strands similar to spaghetti when cooked. Use it like pasta.

SUGAR CANE is woody sugar stalks cut into 4–6 inch lengths. Eat by pulling away the outer covering and chewing the cane. Boil the cane for a sweet syrup topping.

TARO ROOT (Dasheen) is a highly digestible starchy root with a mild flavor, used in many different ways like a potato. It should be peeled and cooked before eating.

TOFU (Soybean Curd) is similar to a very soft cheese or custard, with a bland flavor and the ability to pick up the flavor of the foods with which it is cooked.

TOMATILLOS (Ground Tomatoes, Husk Tomatoes) are small green vegetables that grow on ground vines and look like tiny tomatoes. They turn from bright green to yellow when fully ripe but are often used unripe, raw or cooked.

UGLI FRUIT is an unattractive, juicy citrus hybird, a cross between a grapefruit and a tangerine. It is the size and shape of a grapefruit, with a mottled rough peel that looks loose. The rind has light-green blemishes that turn orange when ripe. The fruit has a mild, orangelike flavor.

WATER CHESTNUTS (Chinese Water Chestnuts, Waternuts) are the small edible roots of the tropical Water Chestnut plant. The deep brown skin of the roots is removed and only the white, crunchy inside is used (mainly in salads or stir-fry dishes) after blanching 2–4 minutes.

YUCCA ROOT (Manioc, Cassava, Casava) is the large, starchy root of the Yucca plant. Used like a potato, it must be peeled before cooking and cooked before eating. Its flavor is mild but very pleasing.

Figure 6-1 *(continued)*

Storage

All vegetables and fruits require careful handling and storage to conserve quality. Before storing, discard any that are damaged. Date all containers to make sure produce is used on a first-in, first-out basis. Stack containers so that air can circulate and produce can continue to breathe. Most produce needs refrigeration, except green bananas (the enzymes responsible for ripening bananas are more active at warm temperatures), avocados, onions, and potatoes.

The saying about a bad apple spoiling the whole bunch is true: One damaged apple (or any fruit) can accelerate the decay of fruit stored with it. The same is true for potatoes. Also, never store tomatoes near lettuce, because tomatoes cause the lettuce to brown.

Most produce need not be washed before storing. In particular, if you wash the following vegetables and fruits, it may cause mold or wilting: strawberries, raspberries, blueberries, mushrooms, plums, lettuce, and grapes.

Some fruits are picked when they are ripe and therefore don't ripen after being packed. These include grapes, citrus fruit, berries, and apples. Other fruits benefit from further ripening or softening. These include avocados, bananas, kiwi fruit, melons, peaches, nectarines, and tomatoes.

Store frozen vegetables and fruits at 0°F or below for 10 to 12 months. Canned vegetables and fruits can be stored in a cool, dry place for as long as a year. When stored too long or at too high a temperature, canned vegetables and fruits lose quality (but are still safe to eat). Once opened, store in another container and use within two to three days. Dried fruits should be stored in tightly closed plastic bags or containers. They can last up to one month at room temperature or six months in the refrigerator.

Preparation and Cooking

When preparing and cooking vegetables and fruits, an important consideration is doing so without the loss of valuable nutrients. The Food Facts section in this chapter explains how to retain nutrients.

Vegetables can be cooked using many different cooking methods, including (but not limited to) steaming, baking, broiling, stir-frying, grilling, and sautéing. Following are guidelines for cooking and serving vegetables with optimum color, flavor, texture, and maximum nutrient retention.

1. Cook vegetables for as short a time as possible. Do not overcook in order to retain an appropriate texture and the most nutrition, color, and flavor.
2. Steam vegetables when possible, because steaming is an excellent cooking method for retaining nutrients, color, flavor, and texture.
3. Be very careful about timing when a vegetable is ready to come out of the steamer, to avoid overcooking. Vegetables such as broccoli, cauliflower,

Brussels sprouts, and cabbage develop an unpleasant taste and smell when overcooked.

4. When boiling vegetables, start with boiling water to decrease the cooking time and use enough water to cover them (to retain color and flavor), except for green and strong-flavored vegetables. Green vegetables actually need more water to keep their color, and strong-flavored vegetables need more water to help release their strong flavors. Boil green and strong-flavored vegetables uncovered for the same reasons. An exception is spinach, which can be cooked covered in a small amount of water.
5. Do not add baking soda to vegetables, because it causes a mushy texture and nutrient loss.
6. Reheat canned vegetables by bringing half their liquid to a boil, then adding the vegetables and bringing to serving temperature. Canned vegetables are already cooked.
7. When microwaving single servings of vegetables, cover them and be careful not to overcook.

Fruits are not cooked as frequently as vegetables, but they can be easily baked (baked apples), broiled (broiled bananas), sautéed (sautéed peaches), and simmered (applesauce).

On a menu, vegetables have endless possibilities—in appetizers, salads, soups, stir-fries, ragouts, pasta, pizza, and even breads (zucchini bread), cakes (carrot cake), and pies (pumpkin pie). Like vegetables, fruits offer wonderful colors and a variety of fresh flavors and textures to be incorporated into any menu category. Fruits are excellent raw or in salads, poached in a sweetened liquid, roasted, puréed in fruit soup or batter for baked products, chopped in salsas and chutneys, cooked in sauces, and baked in cobblers and pies. Salsas are chunky mixtures of fruits and/or vegetables and flavorings that are served cold as sauces for meat, poultry, and fish. Chutney is a sweet-and-sour condiment made of fruits and/or vegetables cooked in vinegar with sugar and spices. Recipes for chutney and salsa appear in Chapter 8.

CHEF'S TIPS

- Fruits work especially well with breakfast. For example, there is fruit compote (fresh or dried fruit, cooked in syrup flavored with spices or liqueur), glazed grapefruit, baked apple wrapped with phyllo dough, or French crêpes with chopped kiwi salsa.
- Fruits are a natural in salads with vegetables: combine pineapple, raisins, and carrots, or grapes with cucumbers. Feature colorful fruits such as kiwi with carrots in a chef's salad. Cooked fruits and vegetables have many possibilities, too: sweet potatoes with apples or lemon-glazed baby carrots.
- Roasted fruits are a wonderful base to place entrées on. For example, place a chicken paillard (cutlet pounded and grilled or sautéed) on a bed

of roasted peaches and mango, or monkfish on a bed of roasted pears and fennel. Fruits can also be used to make relishes.

- Fruits have always been a natural for dessert—fresh, roasted, or perhaps baked into a cobbler with a granola crust or into a strudel.
- Fruits such as pineapple, kiwi, mango, and papaya work well in salsa and relishes.
- Berries, such as blueberries or strawberries, are wonderful when you want vibrant colors.
- Vegetables allow you to serve what appears to be a sumptuous portion without the dish being high in kcalories, fat, or cholesterol. What's also wonderful about vegetables is that they are not very expensive when in season.
- When using vegetables, you need to *think*—about what's in season for maximum flavor, about how the dish will look, and about how the dish will taste. Think flavor and color. For example, halibut works well with beets, haricots verts, and a grilled tomato.
- Also, think variety. Serving vegetables doesn't mean switching from broccoli to cauliflower and then back to broccoli. There are many, many varieties of vegetables to choose from. Be adventurous.
- Most of your meal's eye appeal comes from vegetables—use them to your advantage.

Check-Out Quiz

1. Vitamin E deficiency in adults causes osteomalacia.
   a. True b. False
2. Water-soluble vitamins are not toxic when taken in excess of the RDA because they are fully excreted.
   a. True b. False
3. Vitamin E is a fat-soluble vitamin.
   a. True b. False
4. Vitamins are needed in very small amounts.
   a. True b. False
5. Vitamins supply calories and energy to the body.
   a. True b. False
6. Beta carotene is an active form of the vitamin.
   a. True b. False
7. Water-soluble vitamins are more likely than fat-soluble vitamins to be needed in the diet on a daily basis.
   a. True b. False
8. Vitamin B_6 requires intrinsic factor to be absorbed.
   a. True b. False
9. Thiamin, riboflavin, and niacin all play key roles as coenzymes in energy metabolism.
   a. True b. False

10. Good sources of folate include green leafy vegetables, legumes, and orange juice.
    a. True b. False
11. Name the vitamins described in the following.
    A. Which vitamin(s) is only present in animal foods?
    B. Which vitamin(s) is found in high amounts in pork and ham?
    C. Which vitamin(s) is found mostly in fruits and vegetables?
    D. Which vitamin(s) needs a compound made in the stomach in order to be absorbed?
    E. Which vitamin, when deficient, causes osteomalacia?
    F. Which vitamin is made from tryptophan?
    G. Which vitamin(s) is made by intestinal bacteria?
    H. Which vitamin(s) do you need more of if you eat more protein?
    I. Which vitamin(s) is needed for clotting?
    J. Which vitamin(s) is increased during pregnancy?
    K. Which vitamin(s) is known for forming a cellular cement?
    L. Which vitamin has a precursor called beta-carotene?
    M. Which vitamin is made in your skin?
    N. Which vitamin(s) is an antioxidant?
    O. Which vitamin(s) is needed for bone growth and maintenance?
    P. Which vitamin(s), when deficient, causes night blindness?
    Q. Which vitamin(s) is purposely put into milk because there are no other good sources of it available?

Activities and Applications

1. Your Eating Style

Using Table 6-2, circle any foods that you do not eat at all, or that you eat infrequently, such as dairy products or green vegetables. Do you eat most of the foods containing vitamins, or do you hate vegetables and maybe fruits, too? In terms of frequency, how often do you eat vitamin-rich foods? The answers to these questions should help you assess whether your diet is adequately balanced and varied, which is necessary to ensure adequate vitamin intake.

2. Supermarket Sleuth

Check at your local pharmacy or supermarket to view the selection of supplements available. How many supply only 100 percent of the RDA? How many supply 500 percent and more of the RDA? Are any "nonvitamins" being sold? Name three.

3. Vitamin Salad Bar

You are to set up a salad bar using Figure 6-2. You may use any foods you like in your salad bar as long as you have a good source of each of the 13

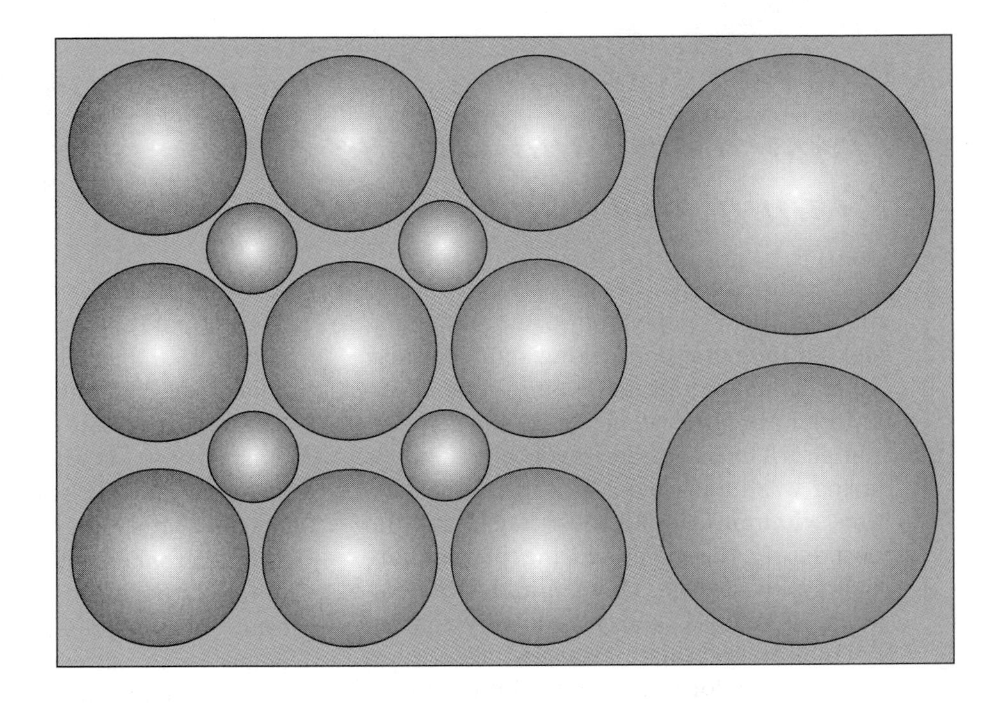

Figure 6-2
Vitamin salad bar

vitamins and you fill each of the circles. In each circle, write down the name of the food and which vitamin(s) it is rich in.

4. Rainbow Dinner

Along with your classmates, you are to write a dinner menu of high-vitamin foods that, once prepared, will include a rainbow of colors and be a good source of all the vitamins in Table 6-2. Once you have written the menu, check that you have a good source of all the vitamins. If possible, prepare your menu and enjoy the meal with your classmates and others.

Nutrition Web Explorer

Food and Drug Administration http://vm.cfsan.fda.gov/~dms/supplmnt.html

This is a special FDA site related to supplements. In the "Topics" box on their home page, click on "GBC, GHB, and BD" and learn about these nonvitamins.

Vitamins www.bookman.com.au/vitamins

Use this site to read about the results of two scientific studies on vitamins.

5 a Day www.5aday.com

On this home page for the Eat 5 Fruits and Vegetables a Day campaign, click on "5 a Day Better Health Cookbook." Find a recipe you like and then use a nutrient composition table to count up how many vitamins are in the recipe.

Food Facts *How Food Processing, Storage, Preparation, and Cooking Affect Vitamin and Mineral Content*

Five factors are responsible for most nutrient loss: heat, exposure to air and light, cooking in water, and baking soda. Because the fat-soluble vitamins are insoluble in water, they are fairly stable in cooking. Water-soluble vitamins easily leach out of foods during washing or cooking. Of all the vitamins, vitamin C is the most fragile and the most easily destroyed during preparation, cooking, or storage. Oxygen and high temperatures readily oxidize or destroy vitamin C. Thiamin and folate are also fragile. Here are some tips for retaining food nutrients.

1. Buy fresh, high-quality food.
2. Examine fresh fruits and vegetables thoroughly for appropriate color, size, and shape.
3. Store fresh fruits and vegetables in the refrigerator (except green bananas, avocados, potatoes, and onions) to inhibit enzymes that make fruits and vegetables age and lose nutrients. The enzymes are more active at warm temperatures. Refrigerated goods should be maintained at a temperature of 41°F or lower, freezer goods at 0°F or lower. Thermometers should be kept in the refrigerator and freezer to monitor temperatures.
4. Foods should not be stored for too long, as lengthy storage causes nutrient loss. Store canned goods in a cool place.
5. When storing fresh fruits and vegetables, close up wrapping tightly to decrease exposure to the air, which pulls out water and decreases the nutrient content. Refrigerated goods should be maintained at a temperature of 41°F or lower, freezer goods at 0°F or lower. Thermometers should be kept in the refrigerator and freezer to monitor temperatures.
6. Wash fruits and vegetables when ready to use, and avoid soaking them.
7. Potatoes and other vegetables that are boiled or baked without being peeled retain many more nutrients than peeled and cut vegetables. In general, the smaller you cut vegetables before cooking, the higher the vitamin loss, because leaching and oxidation are increased with more surface area. In order to conserve vitamins A, C, E, and some B vitamins, cut vegetables no more than necessary.
8. Keep skins of fruits as much as possible, because more vitamins and minerals reside under the skin than in the center of the fruit.
9. Steaming, microwaving, and stir-frying are good choices to retain nutrients when cooking. Each method is fast and uses little or no water. For boiled vegetables, the longer the cooking time and the more water used, the higher the nutrient loss. Cook quickly and use as little water as possible. Also, add the vegetables to the water after it is boiling to conserve vitamin C.
10. Frying temperatures can destroy vitamins such as A, C, E, and K. For instance, french-fried potatoes lose much of their vitamin C.

11. Never use baking soda with green vegetables to improve appearance, as it causes nutrient loss.
12. Broiled or roasted meats retain more B vitamins than meats that are braised or stewed.
13. Use the cooking water from vegetables and the drippings from meats (after skimming off the fat) to prepare soup and gravy.
14. Prepare foods close to the time they will be served.
15. Don't keep milk in clear glass containers, as light destroys the riboflavin. Factors that destroy vitamins often spoil the color, flavor, and texture of food as well.

Hot Topic Functional Foods and Phytochemicals

Foods such as bread and salt were originally enriched and fortified to prevent nutritional deficiency diseases. Today, the fastest-growing category of foods is made up of so-called *functional foods.* These are foods supplemented with ingredients thought to help prevent diseases, such as cancer and heart disease, or to improve health. In other words, functional foods, also called designer foods or nutraceuticals, go beyond basic nutrition. At the supermarket you will find margarine with an ingredient to lower cholesterol, drinks that contain medicinal herbs such as ginseng, and genetically engineered foods such as canola oil with beta-carotene, a *carotenoid.* In some stores, you'll even find memory boosters in chewing gum.

Many of the health-promoting ingredients used in functional foods are vitamins, minerals, herbs, or *phytochemicals*—substances such as beta-carotene that are found largely in fruits and vegetables and that seem to be helpful in preventing cancer and/or heart disease when consumed regularly. How phytochemicals work and optimum amounts for human consumption are still being investigated. Some act as antioxidants, such as *lycopene* in tomatoes or *phenols* in tea, or they may act in other ways.

For example, broccoli contains the chemical *sulforaphane,* which seems to initiate increased production of cancer-fighting enzymes in the cells. *Isoflavonoids,* found mostly in soy foods, are known as plant estrogens or *phytoestrogens* because they are similar to estrogen and interfere with its actions (estrogen seems to promote breast tumors). Members of the cabbage family (cabbage, broccoli, cauliflower, mustard greens, kale), also called *cruciferous vegetables,* contain phytochemicals such as *indoles* and *dithiolthiones.* They activate enzymes that destroy cancer-causing substances. *Flavonoids,* which are found in citrus fruits, onions, apples, grapes, wine, and tea, are thought to be helpful in preventing heart disease and cancer.

Various phytochemicals may reduce heart disease risk by reducing blood cholesterol levels or preventing blood clotting. Examples of these phytochemicals occur in grapes and garlic.

There are concerns about how effective and safe functional foods are. When manufacturers add herbs to foods, they don't have to disclose how much is added. Is it safe? Do herbs or other added ingredients really work? Will making tea with St. John's wort (a herb) really help your depression? Will beta-carotene in a food it normally doesn't appear in still work like the beta-carotene in a carrot? Will any of the added ingredients interact with a medication or dietary supplement you take? Is the fortified food a healthy one, or is it a fortified candy bar?

There are many questions about functional foods, and phytochemicals as well. Although it may be possible to isolate specific components of food that may reduce the risk of diseases like cancer, it is unclear whether phytochemicals added to foods have the same health benefits as whole foods because compounds in foods may act synergistically to impart health benefits. Like dietary supplements, functional foods will not compensate for a poor diet. Whole foods still contain the right amount and balance of nutrients and phytochemicals to promote health.

Chapter 7

Water and Minerals

Water

Major Minerals

Calcium and Phosphorus

Magnesium

Sodium

Potassium

Chloride

Other Major Minerals

Trace Minerals

Iron

Zinc

Iodine

Selenium

Fluoride

Chromium

Copper

Other Trace Minerals

Osteoporosis

Ingredient Focus: Nuts and Seeds

Chef's Tips

Food Facts: The Dangers of Lead

Hot Topic: Dietary Supplements

Major minerals—Minerals needed in relatively large amounts in the diet—over 100 milligrams daily.

Trace minerals—Minerals needed in smaller amounts in the diet—less than 100 milligrams daily.

Bioavailability—The degree to which a nutrient is absorbed and available to be used in the body.

If you were to weigh all the minerals in your body, they would amount to only 4 or 5 pounds. You need only small amounts of minerals in your diet, but they perform enormously important jobs in your body—building bones and teeth, regulating your heartbeat, and transporting oxygen from the lungs to tissues, to name a few.

Some minerals are needed in relatively large amounts in the diet—over 100 milligrams daily. These minerals, called **major minerals,** include calcium, chloride, magnesium, phosphorus, potassium, sodium, and sulfur. Other minerals, called **trace minerals** or trace elements, are needed in smaller amounts—less than 100 milligrams daily. Iron, fluoride, and zinc are examples of trace minerals. Table 7-1 lists the major and trace minerals.

Minerals have some distinctive properties not shared by other nutrients. For example, whereas over 90 percent of dietary carbohydrates, fats, and proteins are absorbed into the body, the percentage of minerals that is absorbed varies tremendously. Only 15 percent of the iron in your diet is normally absorbed, about 30 percent of calcium is absorbed, and almost all the sodium you eat is absorbed. Minerals in animal foods tend to be absorbed better than those in plant foods, because plant foods contain fiber and other substances that bind minerals, preventing them from being absorbed. The degree to which a nutrient is absorbed and available to be used in the body is called **bioavailability.** Sometimes minerals even compete with each other for absorption.

Unlike vitamins, minerals are organic elements that are not destroyed in food storage or preparation. They are, however, water-soluble, so there is some loss in cooking liquids. Like vitamins, minerals can be toxic when consumed in excessive amounts and may interfere with the absorption and metabolism of other minerals. Like vitamins, minerals interact with one another. For example, high phosphorus intake limits the absorption of magnesium.

Deny someone food and he or she can still live for weeks. But death comes quickly, in a matter of a few days, if you deprive a person of water. Nothing survives without water, and virtually nothing takes place in the body without water playing a vital role.

This chapter will help you to:

- Identify the percentage of body weight made up of water
- List the functions of water in the body
- Identify the functions and food sources of the major minerals (calcium, phosphorus, sodium, potassium, chloride, magnesium, and sulfur) and the trace minerals (chromium, copper, fluoride, iodine, iron, manganese, molybdenum, selenium, and zinc)
- List the health risks of mineral deficiencies and outline possible side effects of mineral toxicity
- Discuss the storage and use of nuts and seeds in cooking
- Identify instances when mineral supplements may be necessary

TABLE 7-1 Major and Trace Minerals

Major Minerals	Trace Minerals
Calcium	Chromium
Chloride	Copper
Magnesium	Fluoride
Phosphorus	Iodine
Potassium	Iron
Sodium	Manganese
Sulfur	Molybdenum
	Selenium
	Zinc

Water

The average adult's body weight is generally 50 to 60 percent water—enough, if it were bottled, to fill 40 to 50 quarts. For example, in a 150-pound man, water accounts for about 90 pounds and fat about 30 pounds, with protein, carbohydrates, vitamins, and minerals making up the balance. Men generally have more water than women, a lean person more than an obese person. Some parts of the body have more water than others. Human blood is about 92 percent water, muscle and brain tissue about 75 percent, and bone 22 percent.

The body uses water for virtually all its functions: digestion, absorption, circulation, excretion, transporting nutrients, building tissue, and maintaining temperature. Almost all body cells need and depend on water to perform their functions. Water carries nutrients to the cells and carries away waste materials to the kidneys.

Water is needed in each step of the process of converting food into energy and tissue. Water in the digestive secretions softens, dilutes, and liquefies the food to facilitate digestion. It also helps move food along the gastrointestinal tract. Differences in the fluid concentration on either side of the intestinal wall enhance the absorption process.

Water serves as an important part of body lubricants, helping to cushion the joints and internal organs; keeping tissues in the eyes, lungs, and air passages moist; and surrounding and protecting the fetus during pregnancy.

Many adults take in and excrete between 8 and 10 cups of fluid daily. Nearly all foods have some water. Milk, for example, is about 87 percent water, eggs about 75 percent, meat between 40 and 75 percent, vegetables from 70 to 95 percent, cereals from 8 to 20 percent, and bread around 35 percent.

The body gets rid of the water it doesn't need through the kidneys and skin and, to a lesser degree, from the lungs and gastrointestinal tract. Water is also excreted as urine by the kidneys along with waste materials carried from the cells. About 4 to 6 cups a day are excreted as urine. The amount of urine reflects, to some extent, the amount of an individual's fluid intake, although despite the amount consumed, the kidneys will always excrete a certain amount each day (about 2 cups) to eliminate waste products generated by the body's metabolic actions. In addition to the urine, air released from the lungs contains some water, and evaporation that occurs on the skin (when sweating or not sweating) contains water as well.

If normal and healthy, the body maintains water at a constant level. A number of mechanisms, including the sensation of thirst, operate to keep body water content within narrow limits. You feel thirsty when the blood starts to become too concentrated. Unfortunately, by the time you feel thirsty, you are already much in need of extra fluid. It is therefore very important not to ignore feelings of thirst, a concern that is particularly appropriate for the elderly, whose thirst mechanism is compromised. The well-known recommendation to drink 8 cups of fluid daily is too much for some, like many elderly, and too little for others, like athletes.

There are, of course, conditions in which the various body mechanisms for regulating water balance do not work, such as severe vomiting, diarrhea, excessive bleeding, high fever, burns, and excessive perspiration. In these situations, large amounts of fluids and minerals are lost. These conditions are medical problems to be managed by a physician.

■ **MINI-SUMMARY**

The average adult's body weight is generally 50 to 60 percent water. The body uses water for virtually all its functions: digestion, absorption, circulation, excretion, transporting nutrients, building tissue, and maintaining temperature. Almost all body cells need and depend on water to perform their functions. Water carries nutrients to the cells and carries away waste materials to the kidneys. A number of mechanisms, including the sensation of thirst, operate to keep body water content within narrow limits. You feel thirsty when the blood starts to become too concentrated.

Major Minerals

Calcium and Phosphorus

Calcium and phosphorus are used for building bones and teeth. Over 80 percent of the body's calcium and phosphorus is found in the bones and teeth, where they give rigidity to the structures. Bone is being rebuilt every day, with new bone being formed and old bone being taken apart. There is little turnover of calcium in teeth.

Calcium also circulates in the blood, where a constant level is maintained so it is always available for use. Calcium helps blood to clot, muscles to contract (including the heart muscle), and nerves to transmit impulses. Calcium may help in lowering blood pressure and reducing the risk of colon cancer. In cases of inadequate dietary intake, calcium is taken out of the bones to maintain adequate blood levels.

Like calcium, phosphorus circulates in the blood. Phosphorus is involved in the metabolic release of energy from fat, protein, and carbohydrates. It is also a part of DNA (genetic material) and is therefore needed for growth. Normal body processes produce acids and bases that can cause major blood and body problems, such as coma and death, if not buffered (or neutralized) somehow. In each of the body's cells, phosphorus has a role to buffer both acids and bases. Many enzymes become active when a phosphate group is attached.

The major sources of calcium are milk and milk products. Not all milk products are as rich in calcium as milk (see Table 7-2). As a matter of fact, butter, cream, and cream cheese contain little calcium. One cup of milk or yogurt or 1-1/2 ounces of cheese each have a little less than one-third of the AI (Adequate Intake) for most adults.

Without milk or milk products in your diet, it may be difficult to get enough calcium. Other good sources of calcium include tofu made with calcium carbonate, calcium-fortified foods such as orange juice, and several greens such as broccoli, collards, kale, mustard greens, and turnip greens. Other greens such as spinach, beet greens, Swiss chard, and parsley are calcium-rich but also contain a binder **(oxalic acid)** that prevents some calcium from being absorbed. Dried beans and peas, whole-wheat bread, and certain shellfish contain moderate amounts of calcium but are usually not eaten in sufficient quantities to make a significant contribution. **Phytic acid,** a binder found in wheat bran and whole grains, also prevents some calcium from being absorbed.

Oxalic acid—An organic acid found in spinach and other leafy green vegetables that can decrease the absorption of certain minerals such as calcium.

Phytic acid—A binder found in wheat bran and whole grains that can decrease the absorption of certain nutrients such as calcium and iron.

About 25 to 30 percent of the calcium you eat is absorbed. The body absorbs more calcium (up to 60 percent) during growth and pregnancy, when additional calcium is needed. Absorption is higher for younger people than for older people. Postmenopausal women, who are at higher risk of developing osteoporosis, often absorb the least calcium unless they are taking estrogen. Sufficient vitamin D helps calcium absorption and is added to milk.

Calcium deficiency is much more common in women than men and is a major contributing factor in a disease called osteoporosis, discussed later in this chapter. Calcium can be toxic when large doses of supplements are taken. The Tolerable Upper Intake Level for calcium is 2,500 milligrams per day; amounts above that can contribute to the development of calcium deposits in the kidneys and other organs, kidney failure, and other problems.

Phosphorus is widely distributed in foods and is rarely lacking in the diet. Milk and milk products are excellent sources of phosphorus, as they are for calcium. Other good sources of phosphorus are meat, poultry, fish,

TABLE 7-2 Calcium in Selected Foods

Food	Calcium Content (milligrams)
Dairy Group	
Milk, skim, 8 ounces	302 milligrams
Milk, 2%, 8 ounces	297
Milk, whole, 8 ounces	291
Yogurt, low-fat, 8 ounces	415
Yogurt, low-fat with fruit, 8 ounces	345
Yogurt, frozen, 1 cup	200
Ice cream, vanilla, 1 cup low-fat	176
Cottage cheese, low-fat, 1 cup	155
Swiss cheese, 1 ounces	272
Parmesan, grated, 2 tablespoons	138
Cheddar, 1 ounce	204
Mozzarella, 1 ounce	147
American cheese, 1 ounce	174
Cheese pizza, 1/4 of 14-inch pie	332
Macaroni and cheese, 1/2 cup	181
Meat, Fish, and Alternatives	
Sardines, drained, 2 ounces	175
Oysters, cooked, 3 ounces	76
Shrimp, cooked, 3 ounces	33
Tofu, calcium set, 3-1/2 ounces	128
Dried navy beans, cooked, 1 cup	95
Vegetables	
Turnip greens, frozen and cooked, 1 cup	249
Kale, frozen and cooked, 1 cup	179
Mustard greens, 1 cup	104
Broccoli, cooked, 1/2 cup	35
Other Foods	
Oatmeal, instant, fortified, 1 packet	160
Pancakes, from mix, one 4-inch pancake	30
Wheat bread, 2 slices	40
Orange juice, calcium fortified, 8 ounces	330

Source: U.S. Department of Agriculture Handbook 8, and Home and Garden Bulletin No. 72, and manufacturers.

eggs, and legumes. Fruits and vegetables are generally low in this mineral. Compounds made with phosphorus are used in processed foods, especially soft drinks (phosphoric acid).

Magnesium

Magnesium is found in all body tissues, with about 60 percent in the bones and the remainder in the soft tissues, such as muscles, and in the blood.

Your body works very hard to keep blood levels of magnesium constant. It is essential to many enzyme systems responsible for energy metabolism, protein synthesis, and other functions. Magnesium is used in building bones and maintaining teeth. Like calcium, magnesium is involved in muscle relaxation, blood clotting, and nerve transmission. Magnesium also helps the immune system to work properly.

Magnesium is important to carbohydrate metabolism. It may influence the release and activity of insulin, the hormone that helps control blood glucose levels. Elevated blood glucose levels, which are seen in poorly controlled diabetes, causes increased losses of magnesium in the urine, which in turn lowers blood levels of magnesium.

Evidence suggests that magnesium may play an important role in regulating blood pressure. Diets that provide plenty of fruits and vegetables, which are good sources of magnesium and potassium, are consistently associated with lower blood pressure. The Joint National Committee on Prevention, Detection, Evaluation, and Treatment of High Blood Pressure recommends maintaining an adequate magnesium intake, as well as potassium and calcium, as a positive lifestyle modification for preventing and managing high blood pressure.

Magnesium is a part of chlorophyll, the green pigment found in plants, so good sources include green leafy vegetables, potatoes, nuts (especially almonds and cashews), seeds, legumes, and whole-grain cereals. Seafood is also a good source. Meat and dairy products supply small amounts. Absorption is only 30 to 40 percent.

The magnesium content of refined foods is usually low. Whole-wheat bread, for example, has twice as much magnesium as white bread because the magnesium-rich germ and bran are removed when white flour is processed.

Although magnesium is present in many foods (see Table 7-3), it usually occurs in small amounts. As with most nutrients, daily needs for magnesium cannot be met from a single food. Eating a wide variety of foods, including five servings of fruits and vegetables daily and plenty of whole grains, helps to ensure an adequate intake of magnesium.

Even though dietary surveys suggest that many American do not consume magnesium in recommended amounts, deficiency is rarely seen in the United States in adults. When magnesium deficiency does occur, it is usually due to excessive loss of magnesium in urine, gastrointestinal disorders that cause a loss of magnesium or limit magnesium absorption, or a chronically low intake of magnesium. Treatment with diuretics (water pills), some antibiotics, and some medicines used to treat cancer can increase the loss of magnesium in urine. Poorly controlled diabetes and a high alcohol intake also increase excretion of magnesium. Signs of magnesium deficiency include confusion, disorientation, loss of appetite, depres-

TABLE 7-3 Food Sources of Magnesium

Foods	Milligrams	% Daily Value
Avocado, Florida, 1/2 medium	103	26
Wheat germ, toasted, 1 oz.	90	22
Almonds, dry-roasted, 1 oz.	86	21
Cereal, shredded wheat, 2 rectangular biscuits	80	20
Cashews, dry-roasted, 1 oz.	73	18
Nuts, mixed, dry-roasted, 1 oz.	66	17
Spinach, cooked, 1/2 cup	65	16
Bran flakes, 1/2 cup	60	15
Cereal, oats, instant/fortified, 1 cup	56	14
Potato, baked with skin, 1 medium	55	14
Soybeans, cooked, 1/2 cup	54	14
Peanuts, dry-roasted, 1 oz.	50	13
Peanut butter, 2 tablespoons	50	13
Chocolate bar, 1.45 oz	45	11
100 percent bran, 2 tablespoons	44	11
Vegetarian baked beans, 1/2 cup	40	10
Potato, baked without skin, 1 medium	40	10
Avocado, California, 1/2 medium	35	9
Lentils, cooked, 1/2 cup	35	9
Banana, raw, 1 medium	34	9
Shrimp, mixed species, raw, 12 large, 3 oz.	29	7
Tahini, 2 tablespoons	28	7
Raisins, golden seedless, 1/2 cup	28	7
Cocoa powder, unsweetened, 1 tablespoon	27	7
Bread, whole-wheat, 1 slice	24	6
Spinach, raw, 1 cup	24	6
Kiwi fruit, raw, 1 medium	23	6
Hummus, 2 tablespoons	20	5
Broccoli, chopped, boiled, 1/2 cup	19	5

Source: Facts About Dietary Supplements. Office of Dietary Supplements, National Institutes of Health. 2001. Office of Dietary Supplements homepage: http://ods.od.nih.gov.

sion, muscle contractions and cramps, tingling, numbness, abnormal heart rhythms, coronary spasm, and seizures.

Very high doses of magnesium supplements can cause diarrhea. Especially in the elderly, it can also cause problems with the kidneys because the kidneys are trying to remove excess magnesium. The elderly are at risk of magnesium toxicity because kidney function declines with age and they are more likely to take magnesium-containing laxatives and antacids. When excess magnesium impacts the kidneys, symptoms include nausea, diarrhea, appetite loss, muscle weakness, difficulty breathing, and irregular heartbeat.

Sodium

Electrolytes—Chemical elements or compounds that ionize in solution and can carry an electric current; they include sodium, potassium, and chloride.

Ion—An atom or group of atoms carrying a positive or electric charge.

Water balance—The process of maintaining the proper amount of water in each of three body "compartments": inside the cells, outside the cells, and in the blood vessels.

Acid-base balance—The process by which the body buffers the acids and bases normally produced in the body so that the blood is neither too acidic nor basic.

Sodium, potassium, and chloride are collectively referred to as **electrolytes** because when dissolved in body fluids, they separate into positively or negatively charged particles called **ions.** Potassium, which is positively charged, is found mainly within the cells. Sodium (positively charged) and chloride (negatively charged) are found mostly in the fluid outside the cells.

The electrolytes maintain two critical balancing acts in the body: **water balance** and **acid-base balance.** Water balance means maintaining the proper amount of water in each of the body's three "compartments": inside the cells, outside the cells, and in the blood vessels. Electrolytes maintain the water balance by moving the water around in the body. Electrolytes are also able to buffer, or neutralize, various acids and bases in the body. In addition to its roles in water and acid-base balance, sodium is needed for muscle contraction and transmission of nerve impulses.

The major source of sodium in the diet is salt—a compound made of sodium and chloride. Salt by weight is 39 percent sodium, and 1 teaspoon contains 2300 milligrams (a little more than 2 grams) of sodium. Many processed foods have high amounts of sodium added during processing and manufacturing, and it is estimated that these foods provide fully 75 percent of the sodium in most people's diets. The following is a list of processed foods high in sodium:

- Canned, cured, and/or smoked meats and fish, such as bacon, salt pork, sausage, scrapple, ham, bologna, corned beef, frankfurters, luncheon meats, canned tuna fish and salmon, and smoked salmon
- Many cheeses, especially processed cheeses such as processed American cheese
- Salted snack foods, such as potato chips, pretzels, popcorn, nuts, and crackers
- Food prepared in brine, such as pickles, olives, and sauerkraut
- Canned vegetables, tomato products, soups, and vegetable juices
- Prepared mixes for stuffings, rice dishes, and breading
- Dried soup mixes and bouillon cubes
- Certain seasonings such as salt, sea salt, garlic salt, onion salt, celery salt, seasoned salt, soy sauce, Worcestershire sauce, horseradish, ketchup, and mustard

Table 7-4 illustrates the salt content of food in the various food groups. Salt is also used in food preparation and at the table for seasoning.

In addition to the sodium in salt, sodium appears in monosodium glutamate (MSG), baking powder, and baking soda. Other possible sources of dietary sodium include the sodium found in some local water systems and in medications, such as some antacids. Unprocessed foods also contain nat-

TABLE 7-4 Where's the Salt?

Food Groups	Sodium, mg	Food Groups	Sodium, mg
Bread, Cereal, Rice, and Pasta		Natural cheeses, 1-1/2 oz.	110–450
Cooked cereal, rice, pasta, unsalted, 1/2 cup	Trace	Process cheeses, 2 oz.	800
Ready-to-eat cereal, 1 oz.	100–360	*Meat, Poultry, Fish, Dry Beans, Eggs, and Nuts*	
Bread, 1 slice	110–175	Fresh meat, poultry, fish, 3 oz.	Less than 90
Vegetable		Tuna, canned, water pack, 3 oz.	300
Vegetables, fresh or frozen, cooked without salt, 1/2 cup	Less than 70	Bologna, 2 oz.	580
		Ham, lean, roasted, 3 oz.	1020
Vegetables, canned or frozen with sauce, 1/2 cup	140–460	*Other*	
Tomato juice, canned, 3/4 cup	660	Salad dressing, 1 tbsp.	75–220
Vegetable soup, canned, 1 cup	820	Ketchup, mustard, steak sauce, 1 tbsp.	130–230
Fruit		Soy sauce, 1 tbsp.	1030
Fruit, fresh, frozen, canned 1/2 cup	Trace	Salt, 1 tsp.	2000
		Dill pickle, 1 medium	930
Milk, Yogurt, and Cheese		Potato chips, salted, 1 oz.	130
Milk, 1 cup	120	Corn chips, salted, 1 oz.	235
Yogurt, 8 oz.	160	Peanuts, roasted in oil, salted, 1 oz.	120

Source: U.S. Department of Agriculture.

ural sodium, but in small amounts (with the exception of milk and some milk products).

The Nutrition Facts Label lists a Daily Value of 2400 mg per day for sodium. Overconsumption of sodium, particularly as salt, is common and is a concern because a high sodium intake aggravates high blood pressure in individuals who are salt sensitive.

Potassium

Potassium, an electrolyte found mainly in the fluid inside individual body cells, helps maintain water balance and acid-base balance along with sodium. In the blood, potassium assists in muscle contraction, including maintaining a normal heartbeat, and sending nerve impulses.

Potassium is distributed widely in foods, both plant and animal. Unprocessed, whole foods such as fruits and vegetables (especially winter squash, potatoes, oranges, grapefruits, and bananas), milk, yogurt, legumes, and meats are excellent sources of potassium.

A potassium deficiency is uncommon in healthy people but may result from dehydration, certain diseases, or drugs. Diuretics, a certain class of blood pressure drugs, cause increased urine output, and some cause an increased excretion of potassium as well. Symptoms of deficiency include muscle cramps, weakness, nausea, and abnormal heart rhythms that can be very dangerous, even fatal.

Chloride

Chloride, another important electrolyte, helps maintain water balance and acid-base balance. It is also a part of hydrochloric acid, which is quite highly concentrated in the stomach juices. Hydrochloric acid aids in protein digestion, destroys harmful bacteria, and increases the absorption of calcium and iron. Chloride also helps remove carbon dioxide (a waste product) from the cells so the red blood cells can carry carbon dioxide to the lungs for disposal (by being exhaled). The most important source of dietary chloride is sodium chloride, or salt. If sodium intake is adequate, there will be ample chloride as well.

Other Major Minerals

The body doesn't use the mineral sulfur by itself, but uses the nutrients it is found in, such as protein, thiamin, and biotin. The protein in hair, skin, and nails is particularly rich in sulfur. As part of sulfate, sulfur helps maintain acid-base balance in the body. There is no RDA for sulfur. Protein foods supply plentiful amounts of sulfur, and deficiencies are unknown.

■ MINI-SUMMARY

The functions and sources of the major minerals are listed in Table 7-5.

Trace Minerals

Trace minerals (Table 7-1) represent an exciting area for research, because our understanding of many trace minerals is still emerging. Like vitamins, minerals can be toxic at high doses. Unlike vitamins, many trace minerals are toxic at levels only several times higher than recommendations. Also, trace minerals are highly interactive with each other. For example, taking extra zinc can cause a copper deficiency.

Hemoglobin—A protein in red blood cells that carries oxygen to the body's cells.

Iron

Iron, one of the most abundant metals in the universe and one of the most important in the body, is a key component of **hemoglobin,** a part of red

TABLE 7-5 Minerals

Mineral	Adequate Intake/ Recommended Dietary Allowance*	Functions	Sources
Calcium (AI)	1000 milligrams (19–50) 1200 milligrams (51+) UL: 2500 milligrams	Building bones and teeth. Blood clotting. Muscle contraction. Nerve transmission. May lower blood pressure.	Milk and many milk products, calcium-fortified foods, tofu made with calcium carbonate, several greens (broccoli, collards), legumes, whole wheat bread, some shellfish.
Phosphorus (RDA)	700 milligrams (19+) UL: 4000 milligrams (19–70) 3000 milligrams (71+)	Metabolic release of energy from fat, protein, and carbohydrates. Part of DNA. Buffer of acids and bases. Activation of enzymes.	Milk and milk products, meat, poultry, fish, eggs, legumes.
Magnesium (RDA)	Males: 400 milligrams (19–30) 420 milligrams (31+) Females: 310 milligrams (19–30) 320 milligrams (31+) UL: 350 milligrams	Energy metabolism Protein synthesis Bones and teeth. Muscle contraction. Blood clotting. Nerve transmission. Immune system. Release and activity of insulin.	Green leafy vegetables, potatoes, nuts, seeds, legumes, whole-grain cereals, seafood.
Sodium	*	Water balance. Acid-base balance. Muscle contraction. Nerve transmission.	Salt, processed foods (luncheon meats, American cheese, salted snack foods, canned foods, etc.), milk and dairy products.
Potassium	*	Water balance. Acid-base balance. Muscle relaxation. Nerve transmission.	Fruits and vegetables (winter squash, potatoes, oranges, grapefruits, banana), milk, yogurt, legumes, meats.
Chloride	*	Water balance. Acid-base balance. Part of hydrochloric acid in stomach. Removal of carbon dioxide from cells.	Salt
Sulfur	None	Part of protein, thiamin, and biotin. Part of hair, skin, and nails. Acid-base balance.	Protein foods
Iron (RDA)	Males: 8 milligrams (19+) Females: 18 milligrams (19–50) 8 milligrams (51+) UL: 45 milligrams	Component of hemoglobin and myoglobin. Energy metabolism. Synthesis of amino acids and certain hormones and neurotransmitters. Immune system.	Beef, enriched breads and other baked goods, shellfish, legumes, fortified cereals, green leafy vegetables, poultry.

TABLE 7-5 (continued)

Mineral	Adequate Intake/ Recommended Dietary Allowance*	Functions	Sources
Zinc (RDA)	Males: 11 milligrams (19+) Females: 8 milligrams (19+) UL: 40 milligrams	Wound healing. Bone formation. DNA synthesis. Protein, carbohydrate, and fat metabolism. Development of sexual organs. General tissue growth and maintenance. Taste perception. Acid-base balance. Vitamin A activity. Protection of cell membranes from free-radical attacks. Storage and release of insulin.	Shellfish, meat, poultry, legumes, dairy foods, whole grades, fortified cereals.
Iodine (RDA)	150 micrograms (19+) UL: 1100 micrograms	Normal functioning of thyroid gland. Normal metabolic rate. Growth and development.	Iodized salt, saltwater fish, grains grown in iodine-rich soil.
Selenium (RDA)	55 micrograms (19+) UL: 400 micrograms	Antioxidant. Immune system. Thyroid gland.	Plant foods when grown in selenium-rich soil, meat, seafood, bread, nuts.
Fluoride (AI)	Males: 4 milligrams (19+) Females: 3 milligrams (19+) UL: 1100 micrograms	Strong teeth and bones.	Fluoridated water, fish, tea.
Chromium ((AI)	Males: 35 micrograms (19–50) Females: 25 micrograms (19–50) 20 micrograms (51+) UL: Not established	Works with insulin. Lipid metabolism.	Unprocessed foods such as whole grains, wheat germ, nuts.
Copper (RDA)	900 micrograms (19+) UL: 10,000 micrograms	Works with iron to form hemoglobin. Synthesis of collagen. Nervous system. Energy metabolism. Immune system.	Organ meats, seafood, nuts, seeds, milk, tea, chicken.
Manganese (AI)	Males: 2.3 milligrams (19+) Females: 1.8 milligrams (19+) UL: 11 milligrams	Bone and skin. Metabolism of carbohydrate, fat, and protein. Part of enzyme that acts as antioxidant.	Whole grains, dried beans, nuts, leafy vegetables, tea.
Molybdenum (RDA)	45 micrograms (19+) UL: 2,000 micrograms	Cofactor for several enzymes.	Legumes, whole grains, nuts.

*The values for sodium, potassium, and chloride will be set in 2003.

Myoglobin—A muscle protein that stores and carries oxygen that the muscles will use to contract.

blood cells that carries oxygen to body cells. Cells require oxygen to break down glucose and produce energy. Iron is also part of **myoglobin,** a muscle protein that stores and carries oxygen that the muscles use to contract. Iron works with many enzymes in energy metabolism and is needed to make amino acids as well as certain hormones and neurotransmitters. Iron also is a part of enzymes found in leukocytes (white blood cells—part of the immune system). About 15 percent of your body's iron is stored for future needs (in the bone marrow, spleen, and liver) and mobilized when dietary intake is inadequate.

About one-third of iron in the American diet comes from beef and enriched breads, rolls, and crackers. Other sources include shellfish, legumes, fortified cereals, green leafy vegetables, and poultry (see Table 7-6).

Only about 15 percent of dietary iron is absorbed in healthy individuals. The greatest influence on iron absorption is the amount stored in your body. Iron absorption significantly increases when body stores are low. When iron stores are high, absorption decreases to help protect against iron overload. The body also absorbs iron more efficiently when there is a high need for red blood cells, such as during growth spurts or pregnancy or due to blood loss.

Heme iron—The predominant form of iron in animal foods, it is absorbed and used more readily as iron in plant foods.

Nonheme iron—A form of iron found in all plant sources of iron and also as part of the iron in animal food sources.

Iron-deficiency anemia—A condition in which the size and number of red blood cells are reduced; may result from inadequate iron intake or from blood loss; symptoms include fatigue, pallor, and irritability.

The ability of the body to absorb and use iron from different foods varies. The predominant form of iron in animal foods, called **heme iron,** is absorbed and used twice as readily as iron in plant foods, called **nonheme iron.** Animal foods also contain some nonheme iron. The presence of vitamin C in a meal increases nonheme iron absorption, as does consuming meat, poultry, and fish. Calcium, substances found in tea and coffee, oxalic acid (in some vegetables such as spinach), and phytic acid (in grain fiber) can decrease the absorption of nonheme iron. Some proteins found in soybeans also inhibit nonheme iron absorption.

Iron deficiency is the most common nutritional deficiency. Iron deficiency results in feelings of being tired, irritable, or depressed. If severe enough, it results in **iron-deficiency anemia,** a condition in which the size and number of red blood cells are reduced. The blood hemoglobin concentration also falls. This condition may result from inadequate intake of iron, inadequate intestinal absorption, excessive blood loss (from heavy menses, ulcers, or types of cancer), and/or increased needs. Women of childbearing age, pregnant women, older infants and toddlers, and teenage girls are at greatest risk of developing iron deficiency anemia because they have the greatest needs. Premenopausal women have higher needs for iron than men because of monthly blood loss during menstruation.

Signs of iron-deficiency anemia include feeling tired and weak, decreased work and school performance, slow cognitive and social development during childhood, difficulty maintaining body temperature, and decreased immune function. During pregnancy, iron deficiency is associated with increased risk of premature delivery, giving birth to infants with low

TABLE 7-6 Iron in Foods

Foods	Milligrams Iron
Meat and Poultry	
Beef liver, 3 ounces	5.3
Sirloin steak, 3 ounces	2.6
Hamburger, 3 ounces, lean	1.8
Chicken breast, 3 ounces	0.9
Shellfish	
Oysters, breaded, fried, 1	3.0
Clams, raw, 3 ounces	2.6
Vegetables	
Spinach, 1 cup, frozen, cooked	2.9
Legumes	
Great Northern beans, 1 cup	4.9
Black beans, 1 cup	2.9
Tofu, 2-1/2 inch × 2-3/4 inch × 1 inch cube	2.3
Eggs	
Egg yolk, 1	0.9
Dried Fruits	
Apricots, 1/4 cup	1.5
Raisins, 1/4 cup	0.7
Breads and Cereals	
Corn flakes cereal, 1 ounce	1.8
Whole-wheat bread, 1 slice	1.0
White bread, 1 slice	0.7

Source: Nutritive Value of Foods, U.S. Department of Agriculture Home and Garden Bulletin No. 72, 1989.

Iron overload (hemochromatosis)—A common genetic disease in which individuals absorb about twice as much iron from their food and supplements as other people.

birth weight, and maternal complications. Iron supplementation, such as ferrous sulfate pills, authorized by a physician are generally helpful.

Although the body generally avoids absorbing huge amounts of iron, some people can absorb large amounts. The problem with iron is that once it is in the body, it is hard to get rid of. For individuals who can absorb much iron, large doses of iron supplements can damage the liver and do other damage, a condition called **iron overload** or **hemochromatosis.** This condition is usually caused by a genetic disorder.

It is especially important to keep iron supplements away from children, because they are so toxic they can kill. Consuming 1 to 3 grams of iron can be fatal to children under 6, and lower doses can cause severe symptoms such as vomiting and diarrhea. It is important to keep iron supplements tightly capped and away from children's reach.

Zinc

Zinc is in every cell in the body. It is a cofactor (a substance that binds with an enzyme and allows the enzyme to do its job) for nearly 100 enzymes. Zinc is involved in:

- Wound healing
- Bone formation
- DNA synthesis
- Protein, carbohydrate, and fat metabolism
- Development of sexual organs
- General tissue growth and maintenance
- Taste perception
- Acid-base balance
- Vitamin A activity
- Protection of cell membranes from free-radical attacks
- Storage and release of insulin

The effect of zinc treatments on the severity or duration of cold symptoms is inconclusive. Recent research suggests that the effect of zinc may be influenced by the ability of the specific supplement formula to deliver zinc ions to the oral mucosa. Additional research is needed to determine whether zinc compounds have any effect on the common cold.

Protein-containing foods are all good sources of zinc, particularly shellfish (especially oysters), meat, and poultry (see Table 7-7). Legumes, dairy foods, whole grains, and fortified cereals are good sources as well. Zinc is much more readily available, or absorbed better, from animal foods. Like iron, zinc is more likely to be absorbed when animal sources are eaten and when the body needs it. Only about 40 percent of the zinc we eat is absorbed into the body. Phytates, which are found in whole-grain breads, cereals, legumes, and other foods, can decrease zinc absorption.

Deficiencies are more likely to show up in pregnant women, the young, and the elderly. Adults deficient in zinc may have symptoms such as poor appetite, diarrhea, skin rash, and hair loss. Signs of severe deficiency in children include growth retardation, delayed sexual maturation, decreased sense of taste, poor appetite, delayed wound healing, and immune deficiencies. Marginal deficiencies do occur in the United States.

Zinc toxicity has been shown in both acute and chronic forms. Intakes of 150 to 450 milligrams of zinc per day have been associated with low copper status, altered iron function, reduced immune function, and reduced levels of high-density lipoproteins (the good cholesterol). The Tolerable Upper Intake Level for adults is 40 milligrams. Since zinc supplements can

TABLE 7-7 Food Sources of Zinc

Food	Milligrams	% Daily Value
Oysters, battered, fried, 6 pieces	15.0	100
Breakfast cereal, fortified with 100% of the DV for zinc per serving, 3/4 cup	15.0	100
Beef shank, lean only, cooked, 3 oz.	8.9	60
Beef chuck, arm pot roast, lean only, cooked, 3 oz.	7.4	45
Beef tenderloin, lean only, cooked, 3 oz.	4.8	30
Pork shoulder, arm picnic, lean only, cooked, 3 oz.	4.2	30
Beef, eye of round, lean only, cooked, 3 oz.	4.0	25
Breakfast cereal, fortified with 25% of the DV for zinc, 3/4 cup	3.7	25
Breakfast cereal, complete wheat bran flakes, 3/4 cup	3.7	25
Chicken leg, meat only, roasted, 1 leg	2.7	0
Pork tenderloin, lean only, cooked, 3 oz.	2.5	15
Pork loin, sirloin roast, lean only, cooked, 3 oz.	2.2	15
Yogurt, plain, low fat, 1 cup	2.2	15
Baked beans, 1/2 cup	1.7	10
Cashews, dry-roasted without salt, 1 oz.	1.6	10
Yogurt, fruit, low fat, 1 cup	1.6	10
Pecans, dry-roasted without salt, 1 oz.	1.4	10
Raisin bran, 3/4 cup	1.3	8
Chickpeas, canned, 1/2 cup	1.3	8
Mixed nuts, dry-roasted with peanuts, no salt, 1 oz.	1.1	8
Cheese, Swiss, 1 oz.	1.1	8
Almonds, dry-roasted, without salt, 1 oz.	1.0	6
Walnuts, black, dried, 1 oz.	1.0	6
Milk, fluid, any kind, 1 cup	1.0	6
Chicken breast, meat only, roasted, 1/2 breast	0.9	6
Cheese, cheddar, 1 oz.	0.9	6
Cheese, mozzarella, part-skim, low-moisture, 1 oz.	0.9	6
Beans, kidney, cooked, 1/2 cup	0.8	6
Peas, green, frozen, boiled, 1/2 cup	0.8	6
Oatmeal, instant, low-sodium, 1 packet	0.8	6
Flounder/sole, cooked, 3 oz.	0.5	4

Source: Facts About Dietary Supplements. Office of Dietary Supplements, National Institutes of Health. 2001. Office of Dietary Supplements homepage: http://ods.od.nih.gov.

be fatal at lower levels than many of the other trace minerals, zinc supplements should be avoided unless a physician prescribes them.

Thyroid gland—A gland found on either side of the trachea that produces and secretes two important hormones that regulate the level of metabolism.

Iodine

Iodine, the form in which iodine is found in food, is required in extremely small amounts for normal **thyroid gland** functioning. Once in the body, iodine is chemically changed to iodide. The thyroid gland, located in the neck, is responsible for producing two important hormones that maintain a normal level of metabolism and are essential for normal growth and development, body temperature, protein synthesis, and much more. Iodide is a part of both these hormones.

Iodine is not found in many foods: mostly saltwater fish and grains grown in iodine-rich soil (once covered by the oceans, soil in the central states contains little iodine). Iodized salt was introduced in 1924 to combat iodine deficiencies. Iodine also finds its way accidentally into milk (cows receive iodine-containing drugs, and dairy equipment is sterilized with iodine-containing compounds), and into baked goods through iodine salts used as dough conditioners. Processed foods in the United States do not use iodized salt.

Average intake in the U.S. is more than recommended but less than toxic. A deficiency can cause **hypothyroidism,** a condition in which less thyroid hormone is made, leading to low metabolic rate, tendency to weight gain, and drowsiness. A deficiency can also cause **simple goiter,** in which the thyroid gland becomes very large and the affected person feels lethargic, gains weight, and has a decreased body temperature. If a woman has an iodine deficiency during pregnancy (and possibly also a selenium deficiency), the development of the fetus will be harmed, and it could cause **cretinism,** a condition of mental and physical retardation.

Hypothyroidism—A condition in which there is less production of thyroid hormones; this leads to symptoms such as low metabolic rate, fatigue, and weight gain.

Simple goiter—Thyroid enlargement caused by inadequate dietary intake of iodine.

Cretinism (congenital hypothyroidism)—Lack of thyroid secretion; causes mental and physical retardation during fetal and later development.

Selenium

Until 1979, it was not known that selenium was an essential mineral. The first RDA for selenium was announced in 1989. Selenium is an important part of antioxidant enzymes that protect cells against the effects of free radicals that are produced during normal oxygen metabolism. Antioxidants help control levels of free radicals, which can damage cells and contribute to the development of some chronic diseases. Selenium is also essential for normal functioning of the immune system and thyroid gland.

Plant foods are the major dietary sources of selenium. The amount of selenium in soil, which varies by region, determines the amount of selenium in the plant foods that are grown in that soil. Researchers know that soils in the high plains of northern Nebraska and the Dakotas have very high levels of selenium. People living in those regions generally have the highest selenium intakes in the United States.

Selenium can also be found in some meats and seafood (see Table 7-8). Animals that eat grains or plants that were grown in selenium-rich soil have higher levels of selenium in their muscle. In the United States, meat and bread are common sources of dietary selenium. Some nuts, in particular Brazil nuts and walnuts, are also very good sources of selenium.

Selenium deficiency is most commonly seen in parts of China where the selenium content in the soil, and therefore selenium intake, is very low. Selenium deficiency is linked to Keshan disease (it is named after the

TABLE 7-8 Selenium Content of Foods

Food	Micrograms	% Daily Value
Brazil nuts, unblanched, 1 oz.	840	1200
Tuna, canned in oil, drained, 3-1/2 oz.	78	111
Beef/calf liver, 3 oz.	48	69
Cod, cooked with dry heat, 3 oz.	40	57
Noodles, enriched, boiled, 1 cup	35	50
Macaroni and cheese (box mix), 1 cup	32	46
Turkey breast, oven roasted, 3-1/2 oz.	31	44
Macaroni, elbow, enriched, boiled, 1 cup	30	43
Spaghetti with meat sauce, 1 cup	25	36
Chicken, meat only, 1/2 breast	24	34
Beef chuck roast, lean only, oven-roasted, 3 oz.	23	33
Bread, enriched, whole-wheat, 2 slices	14	20
Rice, enriched, long-grain, cooked, 1 cup	14	20
Cottage cheese, lowfat (2%), 1/2 cup	11	16
Walnuts, black, 1 oz.	5	7
Cheddar cheese, 1 oz.	4	6

Source: Facts About Dietary Supplements. Office of Dietary Supplements, National Institutes of Health. 2001. Office of Dietary Supplements homepage: http://ods.od.nih.gov.

province in China where it was studied), in which the heart becomes enlarged and does not function properly. Selenium deficiency also may affect thyroid function because selenium is essential for the synthesis of active thyroid hormone. Researchers also believe selenium deficiency may worsen the effects of iodine deficiency on thyroid function, and that adequate selenium nutritional status may help protect against some of the effects of iodine deficiency. Selenium deficiency is rare in the United States.

There is a moderate to high health risk associated with too much selenium. High blood levels of selenium can result in a condition called selenosis. Symptoms include gastrointestinal upset, hair loss, white blotchy nails, and mild nerve damage. Selenium toxicity is rare in the United States.

Fluoride

Fluoride—The form of fluorine that appears in drinking water and in the body.

Fluoride is the term used for the form of fluorine that appears in drinking water and in the body. In children, fluoride strengthens the mineral composition of the developing teeth, so they resist the formation of dental cavities, and it also strengthens bone. In adults, fluoride helps teeth by decreasing the activity of the bacteria in your mouth that cause dental caries.

The major source of fluoride is drinking water, although fish and most teas contain fluoride as well. Some water supplies are naturally fluoridated, and many supplies have fluoride added, usually at a concentration of one part fluoride to a million parts water. Fluoride levels in water are stated in concentrations of parts per million (ppm). About 1 ppm is ideal. Less than 0.7 ppm isn't adequate to protect developing teeth. More than about 1.5 to 2.0 ppm can lead to mild **fluorosis,** a condition that causes small, white, virtually invisible opaque areas on teeth. In its most severe form, fluorosis causes a distinct brownish mottling or discoloring. Fluorosis can occur only during tooth development. To prevent fluorosis, children should be advised to use small amounts of fluoride toothpaste and not to swallow it. Also, monitor the fluoride content of your local water supply and use fluoride supplements as directed by your doctor.

Fluorosis—A condition in which the teeth become mottled and disordered due to high fluoride ingestion.

Only fluoride taken internally, whether in drinking water or dietary supplements, can strengthen babies' and children's developing teeth to resist decay. Once the teeth have erupted, they are beyond help from ingested fluoride. Supplements are often prescribed for the approximately 40 percent of Americans who do not have adequately fluoridated water supplies.

For both children and adults, fluoride applied to the surface of the teeth can nonetheless add protection, at least to the outer layer of enamel, where it plays a role in reducing decay. The most familiar form, of course, is fluoride-containing toothpaste, introduced in the early 1960s. Fluoride rinses are also available, as are applications by dental professionals. They are considered effective adjuncts to ingested fluoride and are the only useful sources of tooth-strengthening fluoride for teenagers and adults.

Chromium

Chromium works with insulin to transfer glucose and other nutrients from the bloodstream into the body's cells. Chromium deficiency results in a condition much like diabetes, in which the blood glucose level is abnormally high. Chromium also plays an important role in the body's metabolism of lipids. Although chromium has been advertised as helping you lose weight and put on muscle, well-designed research studies have not shown these effects.

Chromium is widely distributed throughout the food supply, but many foods contribute only small amounts of this nutrient. Good sources of chromium are whole, unprocessed foods, such as whole grains and breads and cereals made with whole grains, wheat germ, and nuts. The richest source is brewer's (nutritional) yeast.

Because it is difficult for scientists to identify who is deficient in chromium, it is not known if chromium deficiency is a concern in otherwise healthy people. It is also not known whether chromium supplements, such as chromium picolinate, are harmful for humans. Studies have shown large doses can be harmful to animals.

Copper

Copper works as an important part of many enzymes—for example, it acts with iron to form hemoglobin. It also aids in forming collagen, a protein that gives strength and support to bones, teeth, muscle, cartilage, and blood vessels. As part of many important enzymes, it is involved in the nervous system, energy release, and immune system.

Copper occurs mostly in unprocessed foods. Good sources include organ meats, seafood, nuts, and seeds. Some foods that are consumed in substantial quantities, such as milk, tea, chicken, and potatoes, also contain the nutrient, but at lower levels.

Copper deficiency is rare, but marginal deficiencies do occur. Single doses of copper only four times the recommended level can cause vomiting and nervous system disorders.

Other Trace Minerals

Manganese is needed to form bone and skin, and as a cofactor for many enzymes involved in metabolism of carbohydrate, fat, and protein as well as other metabolic processes. Manganese is also part of an enzyme that acts as an antioxidant. It is found in many foods, especially whole grains, dried beans, nuts, leafy vegetables, and tea. Too much or too little manganese is rare.

Molybdenum is a cofactor for several enzymes. It appears in legumes, whole grains, and nuts. Deficiency does not seem to be a problem. Too much molybdenum is rare, but may decrease copper absorption and damage the kidneys.

As time goes on, more trace minerals will be recognized as essential to human health. There are currently several trace minerals essential to animals that are likely to be essential to humans as well. Possible candidates for nutrient status include arsenic, boron, nickel, silicon, and vanadium. Based on adverse effects noted in animal studies, Tolerable Upper Intake Levels have been set for boron, nickel, and vanadium.

■ MINI-SUMMARY

The functions and sources of the trace minerals are listed in Table 7-5.

Osteoporosis

Osteoporosis is the most common bone disease. Characterized by loss of bone density and strength, osteoporosis is associated with debilitating fractures, especially in people age 45 and older. Bone loss develops over a span

Osteoporosis—The most common bone disease, characterized by loss of bone density and strength; it is associated with debilitating fractures, especially in people 45 and older, due to a tremendous loss of bone tissue in midlife.

of many years and is largely symptomless, although some women may experience chronic spinal pain or muscle spasms in the back. Often, the first sign of osteoporosis is a wrist or hip fracture or a compression fracture that causes the vertebrae in the upper back to collapse, curving the spine into the "dowager's hump" that has come to symbolize osteoporosis.

As many as 10 million Americans, 80 percent of them women, now suffer from the condition, with more than 1.5 million osteoporosis-related fractures occurring annually. A little less than half of all women over 50 will experience an osteoporosis-related fracture in their lifetime. Men are also at risk for the bone-thinning disease. Osteoporosis is seen less often in men because men generally have larger, stronger bones, and because men don't usually experience the abrupt and substantial hormonal changes that women do following menopause. However, the National Institutes of Health says that the problem of osteoporosis in men recently has been recognized as an important public health issue, especially in light of estimates that the number of men above age 70 will double between 1993 and 2050. Today, more than 2 million American men have osteoporosis, and another 3 million are at risk. Each year, men suffer one-third of all hip fractures. In addition to hip fractures, men most often experience fractures of the spine and wrist due to osteoporosis.

Luckily, osteoporosis can be prevented, detected, and treated, and it is never too late to do something about it.

Although bones seem to be as lifeless as rocks, they are in fact composed of living tissue that is continually being broken down and rebuilt in a process called *remodeling*. It takes about 90 days for old bone to be broken down and replaced by new bone; then the cycle begins anew. Bones continue to grow in strength and size until a person's early thirties, when peak bone mass is attained. Men achieve more peak bone mass than women. Optimum bone mass and size will be attained only if there has been enough calcium in the diet. After that, bone is broken down faster than it is deposited, resulting in decreased bone mass (about 1 percent per year). In men, bone loss is slow but constant. For women, bone loss speeds up during the five years following menopause due to decreased production of estrogen, and then slows to about the same rate as before menopause.

Besides being influenced by age, sex, and estrogen levels, bone health is also influenced by diet and exercise. For maximum bone health, adequate amounts of calcium and vitamin D need to be taken in. Unfortunately, most women (especially teenagers) do not consume the AI for calcium. During the five to ten years after the beginning of menopause, optimum calcium intake is important. Although some bone loss in inevitable, it can be kept to its programmed minimum with adequate calcium from the diet. The Adequate Intake for women and men over 50 is 1200 milligrams of calcium.

Exercise also influences bone health, and participation in sports and exercise increases bone density in children. For older adults, exercise helps improve strength and balance, making it less likely they will fall. To benefit bone health, exercise must be weight-bearing or involve strength training. Also, exercise benefits only the bones used, such as the leg bones when biking or walking.

The best approach to osteoporosis is prevention-the reason why calcium intake is so important. Starting in childhood through young adulthood (when bones are forming), adequate calcium intake is vital to having more bone mass at maturity. Adequate intake of calcium is also important after early adulthood.

Individuals who are aware of the problems of osteoporosis sometimes take calcium supplements. Many calcium supplements include calcium carbonate, a good source of calcium. There are also powdered forms of calcium-rich substances, such as bone meal and dolomite (a rock mineral). These are dangerous because they may contain lead and other elements in amounts that constitute a risk. Choose a calcium supplement with the USP seal of approval. Excessive intake of calcium can cause problems such as urinary stone formation, constipation, and decreased absorption of iron and other nutrients.

Other ways to prevent osteoporosis include regular weight-bearing or strength-training exercise (as already mentioned), consumption of adequate milk for vitamin D, exposure to the sun (for more vitamin D), estrogen therapy for women, moderate consumption of alcohol and caffeine, and avoiding smoking. Although estrogen decreases the risk of osteoporosis, with long-term use it may increase the risk of breast and endometrial cancers and blood clotting.

Although it cannot be cured, osteoporosis can be slowed down. A special kind of X-ray, the bone mineral density test, is a safe, accurate, quick, painless, and noninvasive way to detect low bone density, monitor the effectiveness of treatments, and predict the risk for future fractures. Some patients who have low bone density are prescribed a medication called Fosamax (alendronate). Fosamax is approved for use in men and women to increase bone mass. It works by reducing the activity of the cells that cause bone loss.

■ MINI-SUMMARY

The best approach to osteoporosis is prevention—taking in the AI for calcium, regular exercise, consuming milk for adequate vitamin D, consuming moderate amounts of alcohol, and avoiding smoking. Medications, such as Fosamax, can increase bone density.

Ingredient Focus: Nuts and Seeds

Nuts and seeds pack quite a few vitamins (such as folate and vitamin E) and minerals, along with fiber and protein, in their small sizes. Nuts, in particular, also contain quite a bit of fat. Luckily, most of the fat (except in walnuts) is monounsaturated. Walnuts and flaxseed are rich in the omega-3 fatty acid linolenic acid, an essential fatty acid. One ounce of many nuts contains from 13 to 18 grams of fat, making them also a relatively high-kcalorie food. By comparison, seeds contain less fat and more fiber but still quite a few kcalories. Nuts and seeds also contain many phytochemicals.

Nuts usually grow on trees and are characterized by a hard, removable outer shell. Some commonly used nuts include the following:

- **Almonds** were common ingredients in the cuisines of ancient China, Greece, Turkey, and the Middle East. Today much of the world's supply of almonds is grown in California. Almonds are sweet with a delicate butterlike flavor.
- **Brazil nuts** are the firm but tender fruit of a South American tree. They have a clean, slightly oily taste. They are high in calories.
- **Cashews** form on the bottom of a pear-shaped fruit. Because of the process needed to remove the shell, they are not readily available in the shell.
- **Macadamia nuts** are grown in Hawaii, Australia, and Central America. They are high in cost and very high in fat (store in the refrigerator).
- **Peanuts** probably originated in Brazil and are grown in the southern United States, among other places. Three types of peanuts are most commonly grown: Virginias and runners, which have red skins, and Spanish, which are smaller and have a skin that is more tan.
- **Pecans** grow on huge trees native to the Mississippi River Valley. Georgia is the main source of pecans, which are wonderful all-purpose nuts. Kernels are best stored in the refrigerator.
- **Pine nuts,** also known as pignoli, are popular in Italian cuisine, where they are used in rice, sauces, and cakes. They are also used in Turkish, Middle Eastern, and Mexican cooking. There are two varieties: the Mediterranean or Italian pine nut, with a light flavor, and the Chinese pine nut, with a stronger flavor.
- **Pistachios,** originally from the Middle East and Asia, are now grown in California. The pistachio shell splits naturally as part of the ripening process. Pistachios were originally dyed red by importers to cover stains in imported nuts. Most California pistachios are sold with their natural ivory shell.
- **Walnuts** were introduced to California by the Franciscan fathers in the 1700s. The mellow flavor of the walnut works well with a variety of foods. Most walnuts marketed in the United States are the English variety. The black walnut is sweet and has a deeper flavor. Walnuts are a good source of omega-3 fatty acids.

Nuts in the shell can be stored at room temperature in a cool, dry location. Once shelled, most nuts need to be refrigerated.

Nuts are used in all their forms and styles (whole, sliced, pieces, ground, butters, oils) in baked goods, in stews and ragouts, and as toppings for salads, cooked vegetables, and entrées. Nuts often add eye appeal and an unexpected change in texture. In part due to their high calorie and fat content, nuts are often used in small amounts. By toasting or roasting nuts, you can bring out a more intense flavor and use less. Small amounts of flavorful nuts and seeds can often replace fats such as butter or margarine. Other foods, such as pumpkin seeds or roasted chickpeas, can also be used to replace part or all of the nuts in a dish and still provide a crunchy texture.

Seeds are versatile as well.

- **Pumpkin seeds** are common in the cuisines of Austria and parts of Mexico, where people like their zesty flavor. Pumpkin seeds can be coated with olive oil and roasted to bring out their nutty flavor, then tossed on salads. Pumpkin seeds can also be pulverized into a thick powder or paste and used as a thickener, or toasted and used as a crust. In Austria, pumpkin seed oil, which has a very strong flavor, is used in small amounts in salad dressings. It is used in the United States now as well by some chefs.
- **Sunflower seeds** are large compared to seeds such as sesame and caraway. They can be used in casseroles, stews, vegetables, stuffings, or salads.
- **Sesame seeds** and **caraway seeds** are often used in baking. Toasted sesame seeds can be sprinkled on soups, fish, and cooked vegetables for flavor and texture.

Seeds should be stored in a tightly covered container in a cool, dry, dark area.

CHEF'S TIPS

- To toast nuts such as almonds, spread them in a single layer in a shallow pan. Bake at 325°F for 10 to 15 minutes, or until the almonds are lightly colored. Toss occasionally. They will continue to brown slightly after you remove them from the oven.
- To roast nuts, lightly coat the kernels with oil (about 2 tablespoons per pound of nuts) and proceed as in toasting.
- Nuts are wonderful in muffins, such as honey-almond muffins or walnut-strawberry muffins.
- Nuts and seeds work well in granolas, give crunch and flavor to casseroles, and add interest to salads, such as fennel, orange, watercress, and walnut salad.
- Nuts and seeds turn rancid easily due to their fat content. Store in airtight containers away from heat and light.

Check-Out Quiz

1. Our understanding of many trace minerals is still emerging.
   a. True b. False
2. Two cups of yogurt contain about as much calcium as 1 cup of milk.
   a. True b. False

3. Sodium and potassium are involved in muscle contraction relaxation and transmission of nerve impulses.
   a. True b. False
4. Canned soft drinks are high in sodium.
   a. True b. False
5. Iodine is needed to maintain a normal metabolic rate.
   a. True b. False
6. Few minerals are toxic in excess.
   a. True b. False
7. Nearly all foods contain water.
   a. True b. False
8. The kidneys will always excrete a certain amount each day to eliminate waste products.
   a. True b. False
9. Sodium, potassium, and calcium are referred to as electrolytes.
   a. True b. False
10. Name the mineral(s) referred to by the following descriptions:
    a. Involved in bone formation.
    b. Found mostly in milk and milk products.
    c. Helps maintain water balance and acid-base balance.
    d. Some diuretics deplete the body of this mineral.
    e. Found in the stomach juices.
    f. Important for a healthy heart.
    g. Found in certain water supplies.
    h. Found in salt.
    i. Found in heme.
    j. Causes a form of anemia.
    k. Occurs in the soil.

Activities and Applications

1. Your Eating Style

Using Table 7-5, circle any food sources of minerals that you do not eat at all, or eat infrequently, such as dairy products or green vegetables. Do you eat many of the foods containing minerals, or only a fair amount of them? How often do you eat mineral-rich foods? The answers to these questions should help you assess whether your diet is adequately balanced and varied to ensure adequate mineral intake.

2. How Much Do You Drink?

Keep a diary of how many fluid ounces you drink in one day, including beverages such as coffee and soft drinks, but excluding alcoholic beverages. Convert the number of fluid ounces to the number of cups by dividing the total ounces by 8. Compare this number of cups to the 6 to 10 cups you need daily. Did you drink enough fluid? When might you need over 10 cups daily?

3. Mineral Salad Bar

You are to set up a salad bar using Figure 7-1. You may use any foods you like in your salad bar as long as you have a good source of each of the minerals listed in Table 7-4 and you fill each of the circles. In each circle, write down the name of the food and which mineral(s) it is rich in.

4. Sodium Countdown

Using Appendix A, list the sodium content of 10 of your favorite foods. How much would each contribute to the recommendation of 2400 milligrams sodium (maximum)?

5. One-Day Food Record and Nutrient Analysis

Now that you have learned about all the nutrients, you can see how many nutrients you take in during a typical day. To do so, write down everything you eat and drink (except water) for one day. Include a description of each food and the portion size, such as "1 large apple." Also include on your food record any supplements that you take, including the name of each nutrient and how much is in the supplement. To do the nutrient analysis, use the program your teacher requests, or go to the following website and use NAT Version 2.0. www.nat.uiuc.edu/mynat

When entering the website, create a username and password (this is a free website). Follow the directions provided to do your nutrient analysis.

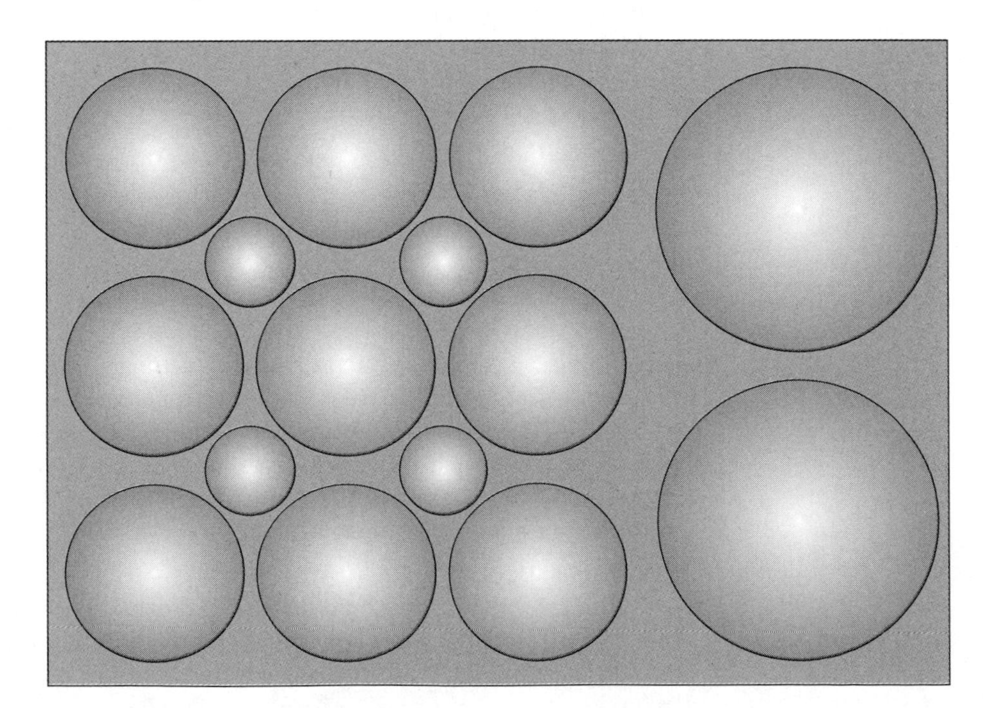

Figure 7-1
Minerals salad bar

Once completed, use your nutrient analysis to see which nutrients you get too much or too little of. Do you need to change some of your food choices to improve the results? If so, describe.

Nutrition Web Explorer

National Institutes of Health Osteoporosis Resource Center www.osteo.org.
At this site, click on "Fact Sheets," then click on "Osteoporosis." Pick a topic on this page and write a summary of what you read.

Iron Overload Diseases Association www.ironoverload.org
Click on "Two Important Dietary Cautions." What are the two important dietary cautions they suggest?

Medline Plus Health Information on Supplements
www.nlm.nih.gov/medlineplus/vitaminandmineralsupplements.html
Click on a topic under "Latest News" and write a summary of what you read.

National Center for Complementary and Alternative Medicine www.nccam.nih.gov
The National Center for Complementary and Alternative Medicine is dedicated to exploring complementary and alternative healing practices in the context of rigorous science. Read "Using Complementary and Alternative Medicine" to learn about this new area.

Food Facts *The Dangers of Lead*

Lead has no known functions or health benefits for humans. In fact, it is a highly toxic metal that can damage the nervous, cardiovascular, renal, immune, and gastrointestinal systems and is particularly dangerous for children. In children, lead has a particularly damaging effect on intellectual development. In addition, lead interferes with the manufacture of heme, the oxygen-carrying part of hemoglobin in red blood cells. Lead consumption in childhood can mean stunted growth and a lower IQ. Damage to the child's nervous system is permanent.

New research on lead shows that it may be dangerous at low levels in adults. Low levels of lead may contribute to hypertension and harm the kidneys. Significant sources of lead include lead-based paint, the overwhelming source, and also drinking water carried in lead pipes, lead-soldered cans, and some ceramic dishes coated with lead-containing glaze. Lead can be in any home's or business's water, so it is advisable to have your water tested. In addition to the problem with lead pipes, some faucets contain lead and leach lead into the water.

Although the number of cans produced in the United States that use lead solder has decreased to under 4 percent, the number of imported cans with lead solder is unknown. The only way to make sure that a can is not soldered with lead is to choose one-piece aluminum cans (such as those for soft drinks) or cans with welded seams that have shiny metal around the seam. Cans that are soldered are not so shiny around the seam because some solder is usually obvious.

Most ceramic glazes contain lead; if properly fired and sealed, they do not cause any concern. However, some ceramic cookware (and dishware), from both outside and inside the United States, has been found to leach lead in dangerous amounts into food. Ceramic items may include fine china, stoneware, earthenware, and ironstone. Unfortunately, there is no way of telling whether a ceramic piece has an unsafe amount of lead unless you use a lead-testing kit. Lead also leaches from lead crystal, and it can leach into any liquid, not just alcohol. Here are precautions to follow to lessen your chances of lead poisoning.

1. Don't store foods in ceramic cookware or dishes unless you are sure they are lead-free.
2. Avoid using ceramic dishes to serve acidic foods and beverages, which cause more lead to be leached out. Examples of acidic foods include citrus juices, apple juice, tomato products, cola-flavored soft drinks, coffee, or tea.
3. Heat also causes more lead to be leached out, so avoid cooking and microwaving with ceramic cookware or dishes that you are not sure of.
4. Do not use, and perhaps dispose of, any china on which the glaze is corroded or has a chalky gray residue when dry.
5. Be cautious about using very old china and highly decorated hand-crafted china.
6. If buying new ceramic cookware or dishes, select a manufacturer, such as Corning, that produces lead-free glazes.
7. Do not use lead crystal every day, and never use it to store food or beverages. Also, never let children use it.
8. If in doubt, check it out with a lead-testing kit. They cost $25 and up in hardware stores.

Hot Topic Dietary Supplements

Surveys show that more than half of the U.S. adult population uses dietary supplements. Traditionally, the term "dietary supplements" referred to products made of one or more of the essential nutrients, such as vitamins, minerals, and protein. But the 1994 Dietary Supplement Health and Education Act (DSHEA) broadened the definition to include, with some exceptions, any product intended for ingestion as a supplement to the diet. In addition to vitamins, minerals, and proteins, dietary supplements may include herbs, botanicals, and other plant-derived substances, as well as amino acids, concentrates, metabolites, constituents, and extracts of these substances.

It's easy to spot a supplement, because DSHEA requires manufacturers to include the words "dietary supplement" on product labels. Also, a Supplement Facts panel (Figure 7-2) is required on the labels of most dietary supplements.

Dietary supplements come in many forms, including tablets, capsules, powders, softgels, gelcaps, and liquids. Though commonly associated with health-food stores, dietary supplements also are sold in grocery, drug, and national discount chain stores, as well as through mail-order catalogs, TV programs, the Internet, and direct sales.

One thing dietary supplements are not is drugs. A drug, which sometimes can be derived from plants used as traditional medicines, is intended to diagnose, cure, relieve, treat, or prevent disease. Before marketing, a drug must undergo clinical studies to determine its effectiveness, safety, possible interactions with other substances, and appropriate dosages, and the FDA must review these data and authorize the drug's use before it is marketed. The FDA does not authorize or test dietary supplements.

A product sold as a dietary supplement and touted in its labeling as a new treatment or cure for a specific disease or condition would be considered an unauthorized—and thus illegal—drug. Labeling changes consistent with the provisions in DSHEA would be required to maintain the product's status as a dietary supplement.

Another thing dietary supplements are not is replacements for conventional diets. Supplements do not provide all the known—and perhaps unknown—nutritional benefits of conventional food.

As with food, federal law requires manufacturers of dietary supplements to ensure that the products they put on the market are safe. But supplement manufacturers do not have to provide information to the Food and Drug Administration (FDA) to get a product on the market. FDA review and approval of supplement ingredients and products is not

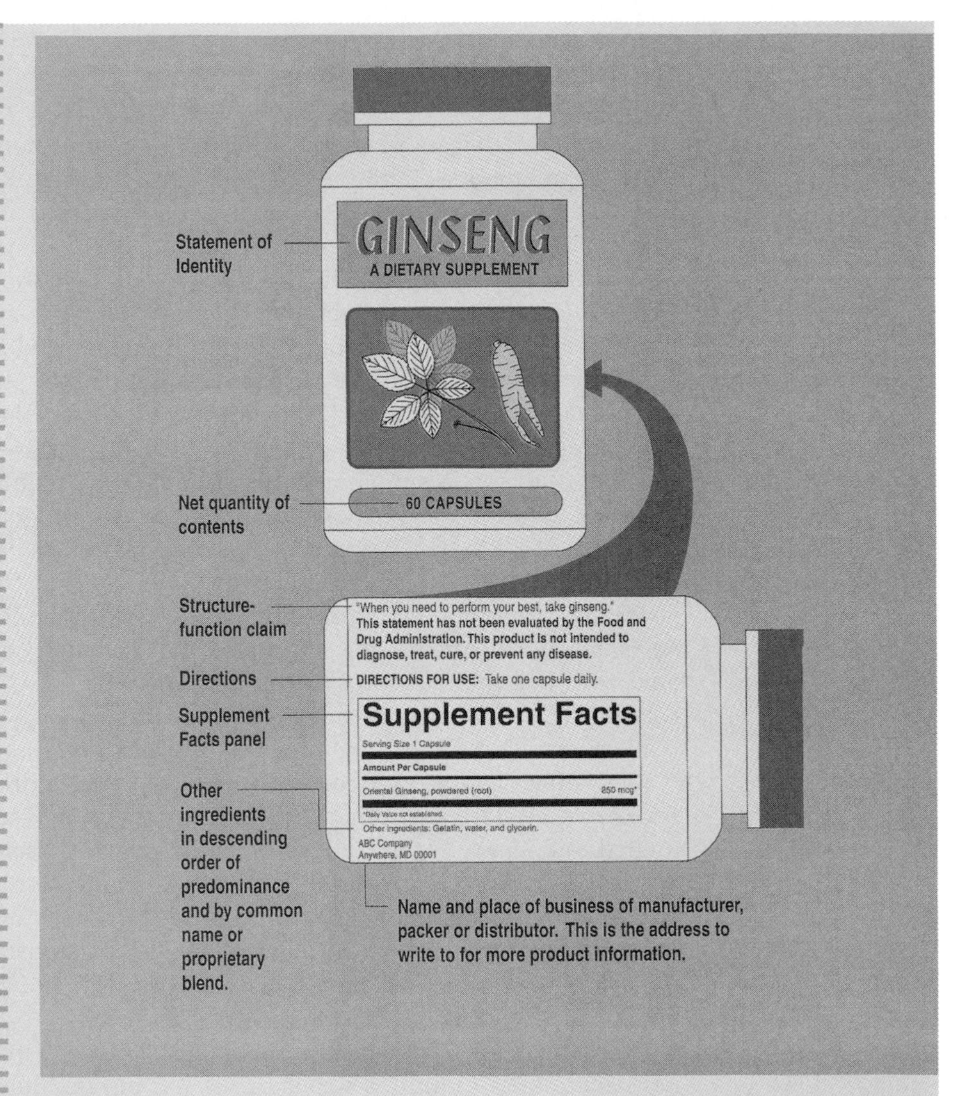

Figure 7-2
Anatomy of the new requirements for dietary supplement labels (effective March 1999).
Source: Food and Drug Administration.

required before selling. The FDA oversees safety and manufacturing and product information, such as claims, in a product's labeling, package inserts, and accompanying literature. The Federal Trade Commission regulates the advertising of dietary supplements.

Under DSHEA, once a dietary supplement is marketed, the FDA has the responsibility for showing that it is unsafe before it can take action to restrict the product's use or take it off the market.

Claims that tout a supplement's health benefits have always been a controversial feature of dietary supplements. Manufacturers often rely on them to sell their products. But consumers often wonder whether they can trust the claims.

Under DSHEA and previous food-labeling laws, supplement manufacturers are allowed to use, when appropriate, three types of claims: nutrient claims, health claims, and nutrition support claims, which include "structure-function claims."

Nutrient claims describe the level of a nutrient in a food or dietary supplement, and are discussed in Chapter 2. For example, a supplement containing at least 200 milligrams of calcium per serving could carry the claim "high in calcium."

Health claims show a link between a food or substance and a disease or health-related condition. The FDA authorizes these claims based on a review of the scientific evidence. For example, a claim may show a link between folate in the product and a decreased risk of neural-tube defects in pregnancy if the supplement contains enough folate. Table 2-10 lists claims currently allowed.

Nutrition support claims can describe a link between a nutrient and the deficiency disease that can result if the nutrient is lacking in the diet. For example, the label of a vitamin C supplement could state that vitamin C prevents scurvy. When these types of claims are used, the label must mention the prevalence of the nutrient-deficiency disease in the United States.

These claims also can refer to the supplement's effect on the body's structure or function, including its overall effect on a person's well-being. These are known as *structure-function claims.* Examples include:

- Calcium builds strong bones.
- Antioxidants maintain cell integrity.
- Fiber maintains bowel regularity.

Manufacturers can use structure-function claims without FDA approval. They base their claims on their review and interpretation of the scientific literature. Like all label claims, they must be true and not misleading. They must also be accompanied by the disclaimer "This statement has not been evaluated by the Food and Drug Administration. This product is not intended to diagnose, treat, cure, or prevent any disease."

Manufacturers who plan to use a structure-function claim must inform the FDA of the use of the claim no later than 30 days after the product is first marketed. While the manufacturer must be able to substantiate its claim, it does not have to share the substantiation with the FDA or make it publicly available. If the submitted claim promotes the product as a drug instead of a supplement, the FDA can advise the manufacturer to change or delete the claim.

To help protect themselves, consumers should:

- Look for ingredients in products with the USP notation, which indicates that the manufacturer followed standards established by the U.S. Pharmacopoeia.
- Avoid substances that are not known nutrients.
- Be aware that the term *natural* doesn't guarantee that a product is safe.
- Consider the name of the manufacturer or distributor. Supplements made by a nationally known food and drug manufacturer have probably been made under tight controls, because these companies already have in place manufacturing standards for their other products.
- Write to the supplement manufacturer for more information.
- Be sure to consult your doctor before purchasing or taking any supplement.
- Ask the pharmacist about possible interactions between supplements and prescription and over-the-counter (OTC) medicines. Taking a combination of supplements or using these products together with prescription or OTC drugs could, under certain circumstances, produce adverse effects, some of which could be life-threatening. For example, St. John's worst may reduce the effectiveness of prescription drugs for heart disease, depression, seizures, or certain cancers; it may also diminish the effectiveness of oral contraceptives.

Consumers should also be sure to tell their health care providers about the supplements they take.

Although most Americans can get needed vitamins and minerals through food, situations do occur when supplements may be needed:

- Women in their childbearing years and pregnant or lactating women, who may need iron and/or folate
- People with known nutrient deficiencies, such as an iron-deficient woman
- Elderly people who are eating poorly, have problems chewing, or have other concerns
- Drug addicts or alcoholics
- People eating less than 1,200 kcalories a day—such as dieters—who may need supplements because it is hard to get enough nutrients in such low-calorie diets
- People on certain medications or with certain diseases

If you really feel you need additional nutrients, your best bet is to buy a multivitamin-and-mineral supplement that supplies 100 percent of the RDA or AI. It can't hurt, and it may act as a safety net for individuals who eat haphazardly. But keep in mind that more is not always better, and that no supplement can adequately take the place of food and serve as a permanent substitute for improving a poor diet. In other words, use these products to supplement a good diet, not to substitute for a poor diet.

Part Two

Developing and Marketing Healthy Recipes and Menus

Chapter 8

Developing Healthy Menus and Recipes

In 2000, the typical person over 8 years of age ate 16.8 meals each week at home and 4.2 meals prepared away from home. With 858,000 restaurant and foodservice locations across the country, the restaurant industry presents consumers with more menu choices than ever before that can be part of a healthy diet. The far majority of operators promote healthy choices, from adding more fruits and vegetables to substituting a lower in sodium sauce for a customer on a salt-restricted diet. In fact, restaurants are reporting that over 70 percent of customers are more interested in customizing their food choices than they were two years ago. From healthy salads to decadent desserts, taste and presentation are important for all menu items. This chapter will help you to:

- Describe the characteristics of a balanced meal
- Explain the process for developing and evaluating healthy menu items
- Define seasoning, flavoring, herbs, and spices
- Suggest ingredients and methods to develop flavor
- Describe techniques and cooking methods that are healthy
- List the elements to consider when presenting foods
- List examples of healthy dishes for each section of the menu
- Identify healthier substitutions for common ingredients

Healthy Menus

The healthy menu provides choice: Nutritious dishes are available for guests who desire them. So what is a healthy menu item or meal? Can it be defined? Yes, it can be defined, although in different ways. You may want to define a healthy menu item simply as one that is moderate in the amount of kcalories, fat, cholesterol, and sodium that it contains, or you may want to be more precise and use the following guidelines.

A balanced and moderate *meal* will generally have no more than:

- 30 percent of its total kcalories from fat, including 10 percent of its total calories from saturated and trans fat
- 150 milligrams of cholesterol
- 1000 milligrams or less of sodium
- 15 percent or less of its total kcalories from protein
- 55 percent of its total kcalories from carbohydrates (10 percent or less from simple carbohydrates)

This does not mean that each menu item should follow these guidelines. These guidelines are more appropriate for a meal. It is possible to include a cheeseburger, which has more than 30 percent of its calories from fat, in a meal. As long as the other components of the meal are lower-fat foods, such as fruits, vegetables, and grains, the percent of total calories from fat in the entire meal will average out to meet the goal.

A menu may simply highlight two or more entrées and one or two appetizers and desserts. To develop some healthy menu items, the first step is to look seriously at your existing menu while engaging in some old-fashioned menu planning. You may go in one of three directions.

1. **Use existing items on your menu.** Certain menu selections, such as fresh vegetable salads or grilled skinless chicken, may already meet your needs.
2. **Modify existing items to make them more nutritious.** For example, pan-sear fish instead of pan-frying it with a butter sauce. In general, modification centers on ingredients, preparation, and cooking techniques. Modifying an existing item may simply mean offering a half portion.
3. **Create new selections.** Many resources are available to obtain healthy recipes: cookbooks, magazines, websites. Or draw on your culinary skills and creativity, and make your own recipes.

Whenever you are involved in menu planning, keep in mind the following considerations.

1. Is the menu item tasty? Taste is the key to customer acceptance and the successful marketing of these items. If the food does not taste delicious and have a creative presentation, then no matter how nutritious it may be, it is not going to sell.
2. Does the menu item blend with and complement the rest of the menu?
3. Does the menu item meet the food habits and preferences of the guests?
4. Is the food cost appropriate for the price that can be charged?
5. Does each menu item require a reasonable amount of preparation time?
6. Is there a balance of color in the foods themselves and in the garnishes?
7. Is there a balance of textures, such as coarse, smooth, solid, and soft?
8. Is there a balance of shape, with different-sized pieces and shapes of food?
9. Are flavors varied?
10. Are the food combinations acceptable?
11. Are cooking methods varied?
12. Can each menu item be prepared properly by the cooking staff?

To develop healthy menu items that sell and satisfy customers, you need to use the following three steps.

Step 1: Foundations: Flavor

A solid foundation in foods and cooking is necessary in order to develop healthy menus and recipes. You are expected to know basic culinary terminology and techniques and have a working knowledge of ingredients, from almonds to zucchini. A basic culinary skill that needs some further refinement when cooking healthy is that of flavoring. Because you can't rely on more than moderate amounts of fat, salt, or sugar for taste and flavor, you will need to develop excellent flavor-building skills.

Seasonings— Substances used in cooking to bring out a flavor already present.
Flavorings—Substances used in cooking to add a new flavor or modify the original flavor.

Seasonings and **flavorings** are very important in healthy cooking, because they help replace missing ingredients such as fat and salt. Seasonings are used to bring out flavor already present in a dish, whereas flavorings add a new flavor or modify the original one. The difference between them is one of degree.

Herbs and Spices

Herbs and spices are key flavoring ingredients in nutritional menu planning and execution, and are the backbone of most menu items, lending themselves to cultural and regional food styles. Good sound nutritional cooking can be virtually equal to classical cooking in terms of technique, creative seasoning, flavor blending, and presentation. It's helpful when moderating fat, cholesterol, and sugars to enhance recipes with an abundance of seasonings such as cinnamon, nutmeg, mace, vanilla beans, ginger, fennel, star anise, juniper, and cardamom. These spices give you a sense of sweet satisfaction as well as bold character to your recipes.

The use of herbs in recipes changes the flavor direction to whichever herb is prominent. For instance, the use of basil, oregano, and thyme in a tomato vinaigrette points the dish to an Italian flavor. Take that same dish and add cilantro and lime juice and you move south of the border; add fresh chopped tarragon with shallots and you're in France. There is no end to your creative abilities once you understand the basic format to healthy cooking.

Herbs—The leafy parts of certain plants that grow in temperate climates; they are used to season and flavor foods.
Spices—The roots, bark, seeds, flowers, buds, and fruits of certain tropical plants; they are used to season and flavor foods.

Herbs are the leafy parts of certain plants that grow in temperate climates. **Spices** are the roots, bark, seeds, flowers, buds, and fruits of certain tropical plants. Figure 8-1 shows a number of herbs and spices. Herbs are generally available fresh and dried. Spices are mostly available in dried form.

Fresh herbs, as opposed to dry, are far superior and more versatile when creating recipes. Herbs commonly available fresh include parsley, cilantro, basil, dill, chives, tarragon, thyme, and oregano. Fresh herbs are great when you need a crisp clean taste and maximum flavor. They can withstand only about 30 minutes of cooking, so they work best finishing dishes.

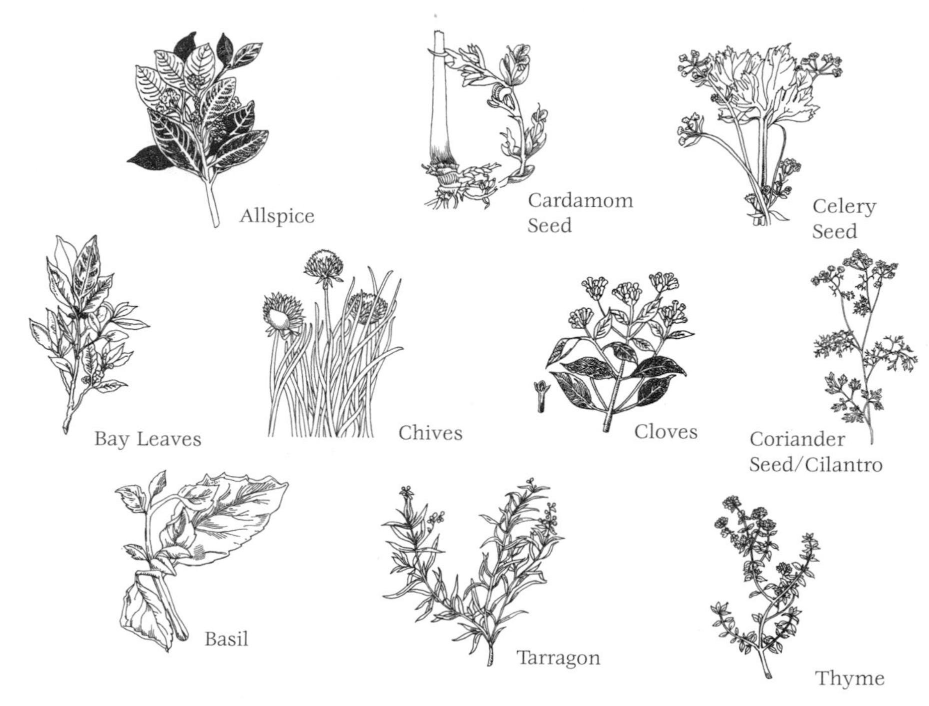

Figure 8-1
Herbs and spices.
Courtesy American Spice Trade Association

Although the use of fresh herbs is not always possible, dry herbs can be substituted with better-than-average results. Dried herbs work well in longer cooking—such as in stocks, stews, and sauces. You can use dried herbs along with fresh herbs toward the end of cooking to get a richer and cleaner flavor.

The real purpose of herbs and spices is not to rescue, remedy, flavor, or season, but to build. Spices and herbs are basically flavor builders. This is their proper use in cooking; they should be cooked with the dish as it is being made so that their flavors blend smoothly, giving character and depth to the dish.

Learning to identify the innumerable different herbs and spices requires a keen sense of taste and smell. Simply looking at them is not enough. Taste them, smell them, feel them, use them. The key to most of them is aroma, for in their aroma is about 60 percent of their flavor. Their aromatic quality not only adds flavor to the food as it is eaten, but heightens the anticipation of the diner as the food is being cooked and served.

There are many, many herbs and spices. Let's look at those most likely to be found in the kitchen. To help you understand them, we'll sort them into groups, but first let's look at the many forms of pepper.

Pepper comes in three forms: black, white, and green. White and black pepper both come from the oriental pepper plant. Black pepper is the dried unripe berry; white pepper is the kernel of the ripe berry. Green peppercorns are picked before ripeness and preserved.

Black pepper comes in four forms: whole black peppercorns, crushed, butcher's grind, and table-ground pepper. Whole pepper is used as a flavor builder during cooking, as in making stocks. Crushed black pepper can function as a flavor builder during cooking as well, or it can be added as flavoring to a finished dish. Many Americans enjoy the flavor contrast of fresh crushed peppercorns straight from the pepper mill on a crisp green salad. The flavor of ground black pepper is characteristic of certain cuisines and certain parts of the country. Cooks catering to these clienteles are likely to add this flavor as they season the food.

As a seasoning, black pepper is used only in dark-colored foods; it spoils the appearance of light-colored foods. *White pepper* is used in light-colored foods because its presence is concealed. White pepper comes in two forms: whole peppercorns and ground white pepper. White peppercorns are used in the same ways as black.

Ground white pepper is good for all-around seasoning. It blends imperceptibly into white dishes both in appearance and in flavor, and it has the strength necessary to season dark dishes. Ground white pepper is chosen by most good cooks as the true seasoning pepper. It is seldom used as a table pepper, as it is expensive.

Green peppercorns are preserved either by packing them in liquid (such as vinegar or brine) or by drying them. They are used in white-tablecloth (luxury) restaurants to complete certain recipes.

Pink peppercorns are not true peppercorns, but they look like peppercorns and have a sweet, slightly peppery taste. They are native to South America, but they are sometimes mistakenly called Japanese peppers because they are one of the few spices used in Japanese cooking. Possible adverse reactions to pink peppercorns have been reported when added generously to dishes, so use in small amounts.

Red pepper, also called *cayenne,* is completely unrelated to white or black pepper. It comes from dried pepper pods. It is quite hot and easily overused. Added with restraint in soups and sauces, it can lend a spicy hotness. When used without as much restraint, it creates the hot flavor of many foods from Mexico, South America, and India.

Let's look at nine herbs and spices that are used as often for their distinctive flavors as for general flavor enrichment. As a flavor builder, each goes beyond the subtlety of the stock herbs and spices—even though you can't single it out from the flavor of the dish as a whole. Used in quantities large enough to taste, they become major flavors rather than flavor builders.

Basil, oregano, and tarragon are available fresh and also come in the form of crushed dried leaves. They look somewhat alike, but their tastes are very different.

Basil has a warm, sweet flavor that is welcome in many soups, sauces, entrées, relishes, salsas, and dressings, as well as with vegetables such as tomatoes, peppers, eggplant, and squash. It blends especially well with tomato, lemons, and oranges. Like many other herbs, it has symbolisms: In India it expresses reverence for the dead; in Italy it is a symbol of love.

Oregano belongs to the same herb family as basil, but it makes a very different contribution to a dish-a strong bittersweet taste and aroma you may have met in spaghetti sauce. It is used in many Italian, Mediterranean, Spanish, South American, and Mexican dishes.

Tarragon has a flavor that is somehow light and strong at the same time. It tastes something like licorice. It is used in poultry and fish dishes, as well as in salads, sauces, and salad dressings.

Rosemary, like bay leaf, is used in dishes where a liquid is involved—soups, stocks, sauces, stews, and braised foods. The leaf of an evergreen shrub of the mint family, it has a pungent, hardy flavor and fragrance. Fresh or dried, it looks and feels like pine needles. It is used mostly with meats, game, poultry, mushrooms, and flavorful ragouts.

Dill and **mustard** have flavors that will be very familiar to you: dill as in pickle and mustard as in the hot dog condiment. Fresh or dried dill leaves, often called dill weed, are used in soups, fish dishes, stews, salads, and butters. Whole dill seed is used in some soups and sauerkraut. Dry mustard, a powdered spice made from the seed of the mustard plant, comes in three varieties: white, yellow, and brown. The brown has the sharper and more pungent flavor. All are used to flavor sauces, dips, dressings, and entrées. Prepared mustards are also made from all varieties and serve as excellent flavor enhancers.

Paprika is another powdered spice that comes in two flavors, mild and hot. Both kinds are made from dried pods of the same pepper family as red pepper and cayenne, and they look something like the seasoning peppers, but they do not do the work of seasonings. Hungarian paprika is the hot spicy one; Spanish paprika is mild in flavor and its red color has lots of eye appeal. Hungarian paprika is used to make goulash and other braised meats and poultry. Spanish is used for coloring, blending rubs, and mild seasoning. Paprikas are sensitive to heat and will turn brown if exposed to direct heat.

Still another branch of this same pepper pod family gives us chili peppers, the crushed or dried pods of several kinds of Mexican peppers and Asian dried red chilis. Colors range from red to green and flavors from mild to hot. Chili peppers are used in Mexican, Asian, Thai, Peruvian, Indian, Cuban, and other South American cuisines.

Several spice blends are available. Two of them are standards in any kitchen. **Chili powder** is one, which is a combination of toasted ground dried chili peppers. Chili powder varies from mild to very hot. It is used, of course, in chili, where it functions as a major flavor, and in many Mexican, South American, Cuban, and Southwestern dishes.

Curry powder is a blend of up to 20 spices. In India, where it originated, cooks blend their own curry powders, which may vary considerably. In the United States curry powder comes premixed in various blends from mild to hot. Curry powders usually include cloves, black and red peppers, cumin, garlic, ginger, cinnamon, coriander, cardamom, fenugreek, mustard, turmeric (which provides the characteristic yellow color), and sometimes other spices.

A group of powdered sweet aromatic spices from the tropics are used frequently in baking, in dessert cookery, and occasionally in sauces, vegetables, and entrées. Among these are cinnamon, nutmeg and its counterpart mace, and ginger. **Cinnamon** comes from the dried bark of the cinnamon or cassia tree, **nutmeg** and **mace** from the seed of the nutmeg tree, and **ginger** from the dried root of the ginger plant. In hot foods, the nutmeg flavor goes well with potatoes, dumplings, spinach, quiche, and some soups and entrées. Mace, a somewhat paler alternative to nutmeg, has a similar flavor and is used in bratwurst, savory dishes, baked goods, and pâtés. Ground cinnamon and ginger are used in a variety of cuisines, both sweet and savory. Cinnamon is also available in sticks.

Mint is a sweet herb, with the familiar flavor you meet in toothpaste and chewing gum. Mint is available in many varieties. The most popular are spearmint and peppermint. Others include chocolate, licorice, orange, and pineapple. The flavor of a mint sauce offers a refreshing complement to lamb. Fresh mint makes a good flavoring and garnish for fruits, vegetables, salsas, relishes, salads, dressings, iced tea, desserts, and sorbets.

Many herbs and spices can be combined to produce blends with distinctive flavors. For example, a cattleman's blend with paprika, peppers, chilis, and other dried herbs can be used as a steak seasoning. A seed blend, such as ground cardamom, fennel, anise, star anise, cumin, and coriander, is excellent in soups, marinades, vegetables, and chutneys. Ethnic blends, such as the following, also present many flavoring possibilities.

- Italian: basil, oregano, garlic, onion
- Asian: ginger, five spices, garlic, scallion
- French: tarragon, mustard, chive, chervil, shallot
- South American: chili powder, lime juice, cilantro
- Indian: ground nutmeg, fennel, coriander, cinnamon, fenugreek, curry
- Mediterranean: oregano, thyme, pepper, coriander, onion, garlic

Blends such as these are foundations for starting or finishing a dish, and are wonderful to add depth of flavor. Always check the salt and sugar content of premade blends.

Another way to get maximum flavor out of herbs and spices is to toast them. Table 8-1 lists whole spices that can be toasted. Toast them in a hot nonstick pan, then grind them and use to season marinades, salad dressings, rubs, soups, stews, ragouts, salsas, and relishes.

Table 8-2 is a reference chart for many herbs and spices.

TABLE 8-1 Examples of Whole Toasted Spices
■ Mustard seed
■ Fennel
■ Coriander
■ Star anise
■ Cardamom
■ Caraway
■ Cumin
■ Juniper
■ Allspice

Juices

Juice can be used as is for added flavor, or it can be reduced (boiled or simmered down to a smaller volume) to get a more intense flavor, vibrant color, and syrupy texture. Reduced juices make excellent sauces and flavorings. Use a good-quality juicer or buy quality premade juice.

For example, orange juice can be reduced (simmered) to orange oil, which is excellent in salad dressings, marinades, and sauces. Also, freshly made beet juice can be used to enhance stocks, glazes, and sauces. In a squirt bottle, reduced beet juice can be used lightly on plates for decoration and flavor, especially with salads, appetizers, and entrées. Other juices that are useful in the kitchen are carrot, fennel, celeriac, pomegranate, ginger, celery, asparagus, bell pepper (yellow, red, orange), herb (watercress, cilantro, parsley, basil, chive), leek, and radish.

Lemon and lime juice are seldom called seasonings, and yet their use as seasonings is not unusual. Many recipes call for small amounts of lemon rind or juice. When used with restraint to spark the flavor of the dish itself and the citrus flavor cannot be perceived, lemon and lime juice are seasonings.

Vinegars and Oils

Various types of vinegars can add flavor to a wide variety of dishes, from salads to sauces. They have a light, tangy taste and add flavor without fat. Popular vinegars include wine vinegars (made from white wine, red wine, rosé wine, rice wine, champagne, or sherry), cider vinegar (made from apples), and balsamic vinegar. True balsamic vinegar, a dark brown vinegar with a rich sweet-sour flavor, is made from the juice of a very sweet white grape and is aged for at least ten years. Vinegars can also be infused, or flavored, with all sorts of ingredients, such as chili peppers, roasted garlic, or any herbs, vegetables, and fruits. For example, lemon vinegar works well in salad dressings and cold sauces.

TABLE 8-2 Herb and Spice Reference Chart

Product	Market Forms	Description	Uses
Allspice (spice)	Whole, ground	Dried, dark brown berries of an evergreen tree indigenous to the West Indies and Central and South America. Smells of cloves, nutmeg, and cinnamon.	Braised meats, curries, baked goods, puddings, cooked fruit
Anise seed (spice)	Whole, ground	Tiny dried seed from a plant native to Eastern Mediterranean. Licorice (sweet) flavor.	Baked goods, fish, shellfish, soups, sauces
Basil (herb)	Fresh, dried: crushed leaves	Bright green tender leaves of an herb in the mint family. Sweet, slightly peppery flavor.	Tomatoes, eggplant, squash, carrots, peas, soups, stews, poultry, red sauces
Bay leaves (herb)	Whole, ground	Long, dark green, brittle leaves from the bay tree, a small tree from Asia. Pungent, warm flavor when leaves are broken. Always remove bay leaves before serving (due to toughness).	Stocks, sauces, soups, braised dishes, stews, marinades
Caraway seed (spice)	Whole, ground	Crescent-shaped brown seed of a European plant. Slightly peppery flavor.	German and Eastern European cooking (such as sauerkraut and coleslaw), rye bread, pork
Cardamom (spice)	Whole pod, ground seeds	Tiny seeds inside green or white pods that grow on a bush of the ginger family. Sweet and spicy flavor. Very expensive.	Poultry, curries and other Indian cooking, Scandinavian breads and pastries, puddings, fruits
Cayenne, red pepper (spice)	Ground	Finely ground powder from several hot types of dried red chili peppers. Very hot. Use in small amounts.	In small amounts in meats, poultry, seafood, sauces, and egg and cheese dishes
Celery seed (spice)	Whole, ground, ground mixed with salt or pepper	Small, gray-brown seeds produced by celery plant. Distinctive celery flavor.	Dressings, sauces, soups, salads, tomatoes, fish
Chervil (herb)	Fresh, dried: crushed leaves	Fernlike leaves of a plant in the parsley family. Like parsley with slight pepper taste, smells like anise.	Stocks, soups, sauces, salads, egg and cheese dishes, French cooking
Chives (herb)	Fresh, dried	Thin grasslike leaves of a plant in the onion family. Mild onion flavor.	Poultry, seafood, potatoes, salads, soups, egg and cheese dishes

Cinnamon (spice)	Stick, ground	Aromatic bark of the cinnamon tree, a small evergreen tree of the laurel family. Sweet, warm flavor.	Baked goods, desserts, fruits, lamb, ham, rice, carrots, sweet potatoes, beverages
Cloves (spice)	Whole, ground	Dried flower buds of a tropical evergreen tree. Sweetly pungent and very aromatic flavor.	Stocks, marinades, sauces, braised meats, ham, baked goods, fruits
Coriander seeds (spice)	Whole, ground	Small seeds from the cilantro plant. Mild, slightly sweet and musty flavor.	Pork, pickling, soups, sauces, chutney, casseroles, Indian cooking
Cumin seed (spice)	Whole, ground	Seed of a small plant in the parsley family. Looks like caraway seed but lighter in color. Pungent, strong, earthy flavor.	Used to make curry and chili powders. Cooking of India, Middle East, North Africa, and Mexico; sausage, Muenster cheese, sauerkraut
Curry powder (spice blend)	Ground blend	A blend of up to twenty spices. Often includes black pepper, cloves, coriander, cumin, ginger, mace, and turmeric. Depending on brand, flavor and hotness can vary tremendously.	Indian cooking, eggs, beans, soups, rice
Dill weed (herb)	Fresh, dried: crushed	Delicate, green leaves of dill plant. Dill pickle flavor.	Salads, dressings, sauces, vegetables, fish, dips
Dill seed (spice)	Whole, ground	Small, brown seed of dill plant. Bitter flavor—much stronger than dill weed.	Pickling, sour dishes, sauerkraut, fish
Epazote (herb)	Fresh	Coarse leaves of a wild plant that grows in the Americas. Strong, exotic flavor.	Mexican and Southwestern cooking
Fennel seed (spice)	Whole, ground	Oval, green-brown seeds of a plant in the parsley family. Mild licorice flavor.	Fish, pork, Italian sausage, tomato sauce, pickles, pastries
Fenugreek seed (spice)	Whole, ground	Small beige seeds of a plant in the pea family. Bittersweet flavor. Smells like curry.	Indian cooking such as curries and chutneys
Ginger (spice)	Fresh whole, dried whole, dried ground (also candied or crystallized)	Dried tan root of the tropical ginger plant. Hot but sweet flavor.	Asian dishes such as curries, baked goods, fruits, beverages
Juniper berries (spice)	Whole	Purple berries of an evergreen bush. Pinelike flavor.	Venison and other game dishes, pork, lamb, marinades

TABLE 8-2 *(continued)*

Product	Market Forms	Description	Uses
Lemongrass (herb)	Fresh stalks, dried: chopped	Tropical and subtropical scented grass. White leaf stalks and lower part are used. Bright, lemon flavor.	Soups, marinades, stir-fries, curries, salads, Southeast Asian cooking
Mace (spice)	Whole blades or ground	Lacy orange covering on nutmeg. Like nutmeg in flavor but less sweet.	Baked products, fruits, pork, poultry, some vegetables
Marjoram (herb)	Fresh, dried: crushed leaves	Leaves from a plant in the mint family. Mild flavor similar to oregano with a hint of mint.	Lamb, poultry, stuffing, sauces, vegetables, soups, stews
Mint (herb)	Fresh, dried: crushed leaves	A family of plants that include many species and flavors such as spearmint, peppermint, and chocolate. Cool, minty flavor.	Lamb, fruits, some vegetables, tea and other vegetables
Mustard seed (spice)	Whole, ground (prepared mustard)	Tiny seeds of various mustard plants, seed may be white or yellow, brown, or black. The darker the seed, the sharper and more pungent the flavor.	Meats, sauces, dreasings, pickling spices, prepared mustard
Nutmeg (spice)		Large brown seed of the fruit from the nutmeg tree. Sweet, warm flavor.	Baked products, puddings, drinks, soups, sauces, many vegetables
Oregano (herb)	Fresh leaves, dried: crushed	Dark green leaves of oregano plant. Pungent, pepperlike flavor.	Italian foods such as tomato sauce and pizza sauce, meats, sauces, Mexican cooking
Paprika (spice)	Ground	Fine powder from mild varieties of red peppers. Two varieties: Spanish and Hungarian. Hungarian is darker in color and much stronger in flavor.	Spanish used mostly as garnish; Hungarian used in braised meats, sauces, gravies, some vegetables
Parsley (herb)	Fresh, dried flakes	Green leaves and stalks of several varieties of parsley plant. Mild, sweet flavor.	Bouquet garni, fines herbes, almost any food
Peppercorns, black and white (spice)	Whole, crushed, ground	Dried, black or white hard berries from same tropical vine that are picked and handled in different ways. Black: pungent earthy flavor. White: similar but more mild than black.	Almost any food

Pink peppercorns (spice)	Whole	Dried or pickled red berries of an evergreen. Not related to black pepper. Bitter flavor, not as spicy as black pepper.	Use in small quantities in meat, poultry, and fish dishes, and in whole pepper mixtures
Poppy seed (spice)	Whole	Tiny cream-colored or deep blue seeds from the poppy plant. Nutty flavor.	On breads and rolls, in salads and noodles, ground poppy seed in pastries
Rosemary (herb)	Fresh, dried leaves	Stiff green leaves that look like pine needles of a shrub. Strong flavor like pine.	Roasted and grilled meats such as lamb, sauces such as tomato, soup
Saffron (herb)	Whole (threads), ground	Dried flower stigmas of a member of the crocus family. Used in very small amounts, has a bitter yet sweet taste, and colors foods yellow. Most expensive spice. Mix with hot liquid before using.	Paella, risotto, bouillabaisse, seafood, poultry, baked products
Sage (herb)	Whole (fresh or dried), rubbed (chopped), ground	Gray-green leaves and blue flowers of a member of the mint family. Strong, musty flavor.	Pork, sausage, stuffing, salads, beans
Savory (herb)	Crushed leaves	Small, narrow leaves of plant in mint family. Summer savory is preferable to winter savory. Bitter flavor.	Meat, poultry, sausage, fish, vegetables, beans
Sesame seeds (spice)	Whole	Creamy oval seeds of tall, tropical sesame plant. Nutty flavor.	On breads and rolls; ground seeds used to make tahini
Star anise (spice)	Whole, ground	Dried, star-shaped fruit of an evergreen native to China. Dark red. Licoricelike flavor.	Chinese cooking
Tarragon (herb)	Fresh, dried: crushed	Small plant with long narrow leaves and gray flowers. Delicate sweet flavor with hint of licorice.	Bearnaise sauce, vinegars, dressings, poultry, fish, salads
Thyme (herb)	Fresh, dried: crushed or ground	Tiny leaves and purple flowers of a short, bushy plant. Spicy, slightly pungent flavor.	Bouquet garni, meat, poultry, fish, soups, sauces, tomatoes
Turmeric (spice)	Ground	Orange-yellow root of member of ginger family. Musky, peppery flavor. Colors foods yellow.	Curry powder, curries, chutney

Like vinegar, oils can be infused with ingredients such as ground spices, fresh herbs, juices, and fresh roots. Small amounts of flavored oils can add much flavor to finish sauces, dressings, marinades, relishes, or salads, or they can be used alone to drizzle over foods ready to be served. Use a neutral oil such as canola, safflower, grapeseed, or corn oil.

To make a ground-spice oil, mix the spice first with water (three tablespoons spice to one tablespoon water) to wake up the flavor. Then mix with about two cups of oil. Let sit for about four to six days at room temperature, shaking several times a day to fuse flavors. Filter the oil and reserve for use.

To make a tender-herb oil, blanch the herbs first and then shock in ice water. Drain and dry the herbs, then purée the herbs with oil and strain through a fine filter. Keep in the refrigerator for about one week. Herbs with a harder texture can be chopped and mixed in a food processor with oil. Let sit several days and then strain. Keep in the refrigerator for several weeks.

To make an oil with vegetable and fruit juices, reduce the juice to a syrup, then blend in a food processor with a little oil, Dijon mustard, and a touch of honey. The oil is ready to use.

Stock

Stock, a flavored liquid used in making soups, sauces, stews, sautés, and braised foods, functions as the body of many foods as well as a flavor builder. *Body* refers to the amount of flavor—its strength or richness. The French call stock *fond de cuisine,* the base of cooking, which describes its role exactly.

Stocks are made by simmering water, bones, regular mirepoix (onion, carrot, celery, and sometimes tomatoes or leek) or white mirepoix (onion, celery, leek, fennel), herbs, and spices. As they simmer, the flavor-producing substances are extracted from the bones and flavor builders and dissolve in the water. Gelatin is also drawn from the bones, providing a major source of body. Though it may be imperceptible in a hot stock, gelatin causes the stock to thicken or jell when chilled.

Chicken stock is made using chicken bones. If the mirepoix and bones are caramelized (browned) first before being added to the stock, the result will be a brown chicken stock. Without this browning process, the result is a white chicken stock.

Fish stock is made like a white stock, simply using cleaned fish bones. Fish stock cooks very quickly, in less than an hour at a rolling simmer. Depending upon the types of bones/shells used, you can make clam stock, lobster stock, or other types. Crustacean stock is made usually by caramelizing the shells and mirepoix with tomato and paprika for a rich brown-red color. Fish or chicken stock is added. A blend of both stocks creates most-favorable results.

Brown stock is made like white stocks, except that the bones (beef or veal) and mirepoix are browned first. Another method is to simmer the bones with onion brulée, bouquet garni, tomato paste, whole carrots, and red wine.

Vegetable stock is made without any animal products. Vegetables are sweated in a touch of oil or stock, then herbs and spices are added. The vegetables are covered with water and simmered for 1-1/2 hours. Wine may be added. Vegetable stock is normally considered a white stock unless tomatoes are added.

In building a good stock, you need to know the nature of the product you are aiming for. Here are the principal measures of stock quality.

- A good stock is fat-free.
- A good stock is clear—translucent and free of solid matter.
- A good stock is pleasant to the senses of smell and taste.
- A good stock is flavorful, but the flavor is neutral. The flavor of the main ingredient, though predominant, is not overpowering. No one flavor builder is identifiable over the flavor of the main ingredient.

And here are some very important guidelines to observe in making stock.

- Use good raw bones—bones that smell pleasant and fresh. Wash chicken and fish bones.
- Remove excess fat from the bones. Fat will produce grease in the stock, spoiling its flavor and appearance.
- Start with cold liquid. Wash chicken and fish bones. They are naturally filled with blood and other impurities. These impurities will dissolve in cold water, and then as the stock is heated, they will become solid and rise to the surface, where you can skim them off. This is especially true with beef and veal stocks. Therefore, a cold-water start will produce a clear stock, whereas starting with hot water will produce a cloudy one.
- Use a tall, narrow pot to minimize evaporation. A certain amount of flavor is lost in evaporation, and the rate of evaporation depends on the surface area of the liquid.
- After bringing the cold water to a boil, reduce the heat to a simmer (about 185°F or 85°C). Keep the cooking temperature below the boil. It takes long, slow simmering to extract the flavors you want from the bones and flavor builders. Too high a temperature will increase evaporation and loss of desirable flavors. It will also break down vegetable textures, producing undesirable flavors and a cloudy stock.
- Skim occasionally—that is, remove the impurities that rise to the surface, using a skimmer or ladle.
- Do not salt the stock because it will usually be further cooked—in a sauce or soup, for example. As a stock is cooked further, it reduces, meaning its volume decreases due to evaporation. A stock that tastes lightly salted when prepared will taste much saltier as it is cooked further and reduces in volume.

- If you add kitchen scraps to stock, make sure they are clean and wholesome.
- Degrease the finished stock—that is, remove the fat from the surface. The most effective method is to chill the stock and remove the layer of fat that congeals on top. If you must use the stock immediately, you can skim the hot fat off the top with a ladle.
- Cool the stock quickly and store it properly.
- A stock's shelf life is no more than three to five days in the cooler. Stocks can be frozen without loss of quality.

To reduce the amount of fat in stocks, use only a small amount of oil to sauté the mirepoix, or sweat it in stock, wine, or other liquid.

Stocks are a low-kcalorie way to support flavor for the recipes they are used in. One cup of stock is about 40 kcalories at most, or only 5 kcalories per fluid ounce. Through the use of herbs, spices, and aromatic vegetables, as well as the flavor from caramelized bones, stocks can be made quite flavorful without adding any kcalories. Also, stocks can be thickened without high-fat roux. Arrowroot or cornstarch does an admirable job of thickening without fat. Puréed vegetables or potatoes can be used to give body, or simply reduce the stock to a glaze.

Glazes are basic preparations in classical cookery and are the forerunners of today's convenience products. They are simply stocks reduced to a thick, gelatinous consistency with flavoring and seasonings. Meat glaze, or *glace de viande,* is made from brown stock. *Glace de volaille* is made from chicken stock, and *glace de poisson* is made from fish stock. To prepare a glaze, you reduce stock over moderate heat, frequently skimming off the foam and impurities that rise to the top. When the stock has reduced by about half, strain it through cheesecloth into a smaller heavy pan. Place it over low heat and continue to reduce until the glaze will form an even coating on a spoon. Cool, cover, and refrigerate or freeze.

Small amounts of glazes (remember, they are very concentrated) are used to flavor sauces and other items. They make sauces cleaner and fresher tasting than sauces made with thickeners. They can also be added to soups to improve and intensify flavor. However, they cannot be used to re-create the stock from which they were made; the flavor is not the same after the prolonged cooking at higher temperatures.

Concentrated convenience bases are widely used. The results vary widely, partly because of the bewildering variety of products on the market and partly because they are often misused. Few of them can function as instant stocks. Compare the taste of a convenience base prepared according to instructions with the taste of a stock made from scratch, and you will see why. Many convenience bases have a high salt content and other seasonings and preservatives. This gives them strong and definite tastes that are difficult to work with in building subtle flavors for soups and sauces.

Among the many bases available, take care to choose the highest-quality products, though these are expensive. Look for those that list beef, chicken, or fish extract as the first ingredient (not salt). Avoid those with lots of chemical additives. If you must use a base as a stock, fortify it using these steps.

1. Dry-sauté a white or regular mirepoix (small dice) until translucent for white stock or carmelized for brown stock.
2. Add base, water, and sachet with fresh herbs, garlic, shallots, bay leaf, and peppercorns.
3. Simmer for 30 minutes.
4. Strain and reserve for use.

Cooking without fresh stock means using your head. If a recipe calls for stock, you will have to analyze the role the stock is to play in that recipe and choose the type of convenience item accordingly. Use with caution, and taste the product as you go. Remember that salt is the major ingredient in nearly every base, and adjust the amount of salt in your recipe.

Rubs and Marinades

Rubs—A dry marinade made of herbs and spices (and other seasonings), sometimes moistened with a little oil, and rubbed or patted on the surface of meat, poultry, or fish (which is then refrigerated and cooked at a later time).

Marinades—A seasoned liquid used before cooking to flavor and moisten foods; usually based on an acidic ingredient.

Rubs combine dry ground spices, such as coriander, paprika, and chili powder, and finely cut herbs, such as thyme, cilantro, and rosemary. Rubs may be dry or wet. Wet rubs, also called pastes, use liquid ingredients such as mustard or vinegar. Pastes produce a crust on the food. Wet or dry seasoning rubs work particularly well with beef and pork and can range from a smoked paprika barbecue seasoning rub to a Jamaican jerk rub. To make a rub, mix various seasonings together and spread or pat evenly on the meat, poultry, or fish—just before cooking for delicate items or at least 24 hours before for large cuts of meat. The larger the piece of meat or poultry, the longer the rub can stay on. The rub flavors the exterior of the meat as it cooks.

Marinades, seasoned liquids in which foods are soaked before cooking, are useful for adding flavor as well as for tenderizing meat and poultry. Marinades bring out the biggest flavors naturally, so you don't need to drown the food in fat, cream, or sauces. Marinades allow a food to stand on its own with a light dressing, chutney, sauce, or relish. Fish can also be marinated. Although fish is already tender, a short marinating time (about 30 minutes) can develop a unique flavor. A marinade usually contains an acidic ingredient, such as wine, beer, vinegar, citrus juice, or plain yogurt, to break down the tough meat or poultry. The other ingredients add flavor. Without the acidic ingredient, you can marinate fish for a few hours to instill flavor. Oil is often used in marinades to carry flavor, but it isn't essential. Fat-free salad dressings such as Italian work well in marinades. A simple fish marinade is fish stock, lemon rind, white wine, tarragon, thyme, dill, black pepper, shallots, Dijon mustard, and a few drops of oil.

To give marinated foods flavor, try citrus zest, diced vegetables, fresh herbs, shallots, garlic, low-sodium soy sauce, mustard, and toasted spices. For example, citrus and pineapple marinades can be flavored with Asian seasonings such as ginger and lemongrass. Tomato juice with allspice, Worcestershire sauce, garlic, cracked black pepper, mustard, fresh herbs, and coriander is great for flank steak. Adding chopped kiwi fruit to a marinade helps to soften tough meats because kiwi contains an enzyme that breaks down tough muscle.

Aromatic Vegetables

Onions and their cousins garlic, scallions, leeks, shallots, and chives are a special category of flavorings that add strong and distinctive flavors and aromas to both cooked and uncooked foods. These bulbous plants of the lily family arrive in the kitchen whole and fresh rather than dried and powdered, though some are available in dried forms. We use them in greater quantity, except for garlic, than the "pinch" or the "few" that is our limit on most herbs and spices.

The onion is the scaly bulb of an herb used since ancient times and grown the world over, the commonest and most versatile flavor builder in the kitchen. Raw onion adds a pungent flavor to salads and cold sauces. Cooked, it has a sweet, mellow flavor that blends with almost anything. Cooked onions are also served as a vegetable.

Garlic, the bulb of a plant of the same family, is available as cloves (bulblets), chopped, powdered, granulated, or in juice form, with the fresh clove having by far the best flavor. Garlic is used as a flavor builder in many preparations, including stocks, stews, sauces, salads, and salad dressings. Roasted garlic has a strong flavor that can successfully replace salt in some recipes. When puréed, it gives a binding, creamy texture to sauces, dressings, beans, grains, and soups, as well as a pleasant flavor.

Chives are another bulbous herb of the onion family, the only one whose leaves rather than the bulb are eaten. Chives are usually used raw, since most of their flavor is lost if they are cooked. They are clipped from the plant and added, usually finely sliced, to many foods.

The leek, a mild-flavored relative of the onion, has a cylindrical bulb. It is the partner of the onion in the mirepoix. Some dark leaves may be used in stock—too many will cause a gray-green color. For most other preparations, the green is cut off because it can be bitter. Braised leeks are very popular as a vegetable in France, where they are known as the poor man's asparagus. The leek's triumph is the cold soup known as vichyssoise.

The scallion is a young onion, also known as a green onion or spring onion. It has a mild flavor as onions go. Minced or sliced, it is added to salads, marinades, dressings, salsas, and relishes with all of the white bulb and some green top. It can pinch-hit in cooking for the full-grown onion, and its green top, minced, can substitute for chives in an emergency.

The shallot is a cluster of brown-skinned bulbets similar to garlic. It is somewhere between garlic and onion in both size and flavor but is milder and more delicate than either. One thinks of shallots with wine cookery and with mushrooms in a marvelous stuffing called duxelles, but they are useful to build flavor in a wide variety of dishes.

Sauce Alternatives: Vegetable Purées, Coulis, Salsas, Relishes, Chutneys, Compotes, and Mojos

Alternatives to classic sauces, many of which are high in fat, take many forms. Puréeing of vegetables or starchy foods is commonly done as a means to thicken soups, stews, sauces, and other foods without any fat, or to create a simple sauce. Sauces made with vegetables are light and low in fat and kcalories. With the addition of herbs and spices, they make a colorful, tasty complement to entrées and side dishes. For example, purée roasted yellow peppers with ginger to make a sauce for grilled swordfish or tuna.

Before puréeing, vegetables should be cooked just until tender using methods such as steaming, roasting, grilling, or sautéing. Roasting caramelizes the vegetables' natural sugars, making them taste rich and sweet. Flour, seasoned bread crumbs, and egg whites are used to give the vegetables a coating to provide crispness, or oil can simply be used. Vegetables that roast well include peppers, eggplant, and zucchini. More delicate vegetables, such as mushrooms, can be sautéed. Other attractive vegetables that make excellent purées are beets, asparagus, peas, and carrots.

Coulis—A sauce made of a purée of vegetables or fruits.

Coulis, a French term, refers to a sauce made of a purée of vegetables or fruits. A vegetable coulis, such as bell pepper coulis, can be served hot or cold to accompany entrées and side dishes. Fruit coulis is usually served cold as a dessert sauce.

A vegetable coulis is often made by cooking the main ingredient, such as tomatoes, with typical flavoring ingredients, such as onions and herbs, in a liquid such as stock. The vegetable and flavorings are then puréed, and the consistency and flavoring of the product are adjusted. A vegetable coulis can also be made by cooking the vegetable with potato or rice (2 ounces per gallon) to give the sauce a silky texture and a smooth consistency once puréed and strained.

The texture of a vegetable or fruit coulis is quite variable, depending on the ingredients and how it is to be used. A typical coulis is about the consistency and texture of a thin tomato sauce. The color and flavor of the main ingredient should stand out.

Salsas—Chunky mixtures of vegetables and/or fruits and flavor ingredients.

Salsas and relishes are versatile, colorful, low-fat sauces. Salsas and relishes are chunky mixtures of vegetables and/or fruits and flavor ingredients.

Salsa is a Latino word, and traditional salsa is made with tomatoes. Most raw salsas are a chopped or almost puréed vegetable (most often tomatoes), fruit, or herb, to which a strong flavor is added, such as onion, garlic, or lime juice. Many include several herbs, spices, and chilies. Chilies are frequently cooked to tame their flavor.

Chutney—A sauce from India that is made with fruits, vegetables, and herbs.

Compote—A dish of fruit, fresh or dried, cooked in syrup flavored with spices or liqueur; it is often served as an accompaniment or dessert.

Mojo—A spicy Caribbean sauce; it is a mixture of garlic, citrus juice, oil, and fresh herbs.

Relishes are often spicy and made from pickling foods. Served cold, they are excellent as sauces for meat, poultry, and seafood. Since salsas and relishes contain little or no fat, they rely on intense flavor ingredients, such as cilantro, jalapeño peppers, lime juice, lemon juice, garlic, dill, pickling spice, coriander, and mustard.

Chutney, such as tomato-papaya chutney, is made from fruits, vegetables, and herbs, and comes from India originally. Recipe 8-21 explains how to make Papaya and White Raisin Chutney. A **compote** is a dish of fruit, fresh or dried, cooked in syrup flavored with spices or liqueur. It is often served as dessert or as an accompaniment.

A **mojo** is a spicy Caribbean sauce. It is a mixture of garlic, citrus juice, oil, and fresh herbs.

With their many colors, flavors, and textures, each of these items is a thoroughly contemporary addition to sauces.

Alcoholic Beverages

Wines, liqueurs, brandy, cognac, and other spirits are often added as flavorings at the end of cooking. Sherry is a popular American flavoring for sauces. Brandy is often poured over a dish and flamed—set afire—at the time of service. This adds some flavor, but is done more for show.

In such dishes as sauces, wines and spirits may be added during cooking to become part of the total flavor. They are then flavor builders rather than flavorings. The same product can play one role in one dish and a different role in another dish.

Extracts and Oils

Extracts and oils from aromatic plants are used in small quantities primarily in the bake shop-extracts of vanilla, lemon, and almond; oils such as peppermint and wintergreen.

■ MINI-SUMMARY

Table 8-3 lists the powerhouses of flavor described in this section, as well as other possibilities. Flavoring adds a complementary flavor to a dish at the end of its preparation. It creates a blend in which both the original flavor and the added flavor are identifiable, as in the addition of black pepper to a green salad. Most flavorings are products with distinctive tastes, capable of holding their own in a dish.

TABLE 8-3 Powerhouses of Flavor

- Fresh herbs
- Toasted spices
- Herb and spice blends
- Freshly ground pepper
- Citrus juices, citrus juice reductions
- Strong-flavored vinegars and vinaigrettes
- Wines
- Strong-flavored oils such as extra-virgin olive oil or walnut oil
- Infused vinegars and oils
- Reduced stock (glazes)
- Rubs and marinades
- Raw, roasted, or sautéed garlic
- Caramelized onions
- Roasted bell peppers
- Chili peppers
- Grilled or oven-roasted vegetables
- Coulis, salsas, relishes, chutneys, mojos
- Dried foods: tomatoes, cherries, cranberries, raisins
- Fruit and vegetable purées
- Horseradish and Dijon mustard
- Extracts

Step 2: Healthy Cooking Methods and Techniques

Healthy cooking methods and techniques either do not add fat or add moderate amounts of primarily monounsaturated or polyunsaturated fat. Before looking at cooking methods, let's review some techniques that are used often to develop flavor in healthy recipes.

Reduction—Boiling or simmering a liquid down to a smaller volume.

Searing—Exposing meat's surfaces to a high heat before cooking at a lower temperature; this process adds color and flavor to the meat.

- **Reduction** means boiling or simmering a liquid down to a smaller volume. In reducing, the simmering or boiling action causes some of the liquid to evaporate. The purpose may be to thicken the product or to concentrate the flavor, or both. A soup or sauce is often simmered for one or both reasons. The use of reduction eliminates thickeners, and intensifies and increases flavor so you can use a smaller portion.
- **Searing** means exposing the surfaces of a piece of meat to high heat before cooking at a lower temperature. Searing, sometimes called browning, can be done in a hot pan in a little oil or in a hot oven. Searing is done to give color and to produce a distinctive flavor. Dry searing can be done over high heat in a nonstick pan using vegetable oil spray or a spray of olive oil.

Deglazing—Adding cold liquid to the hot pan used in making sauces and meat dishes; any browned bits of food sticking to the pan are scraped up and added to the liquid.
Sweating—Cooking slowly in a small amount of fat over low or moderate heat without browning.
Puréeing—Mashing or straining a food to a smooth pulp.

- **Deglazing** means adding cold liquid to the hot pan used in making sauces and meat dishes. Any browned bits of food sticking to the pan are scraped up and added to the liquid.
- **Sweating** means cooking slowly in a small amount of fat over low or moderate heat without browning. You can sweat vegetables and other foods without fat. Instead, sweat in stock or wine.
- **Puréeing** of vegetables or starchy foods is commonly done as a means to thicken soups, stews, sauces, and other foods without any fat. While the food processor is useful for this process because you can slowly pulse the mixture, emulsion blenders or high-speed table blenders can make a very smooth purée with a silky texture and smoothness.

Dry-Heat Cooking Methods

Dry-heat cooking methods are acceptable for healthy cooking when heat is transferred with little or no fat, and excess fat is allowed to drip away from the food being cooked. Both pan frying and deep frying add varying amounts of fat, kcalories, and perhaps cholesterol, depending on the source of the fat, and frying is therefore not an acceptable cooking method. Sautéing can be made acceptable by using nonstick pans and little or no vegetable oil.

Roasting. Roasting, cooking with heated, dry air, is an excellent method for cooking larger, tender cuts of meat, poultry, and fish that will provide multiple servings. When roasting, always place meat, poultry, or seafood on a rack so the drippings fall to the bottom of the pan and the meat therefore doesn't cook in its own juices. Also, cooking on a rack allows for air circulation and more even cooking. In addition to meat, poultry, and seafood, vegetables can be oven-roasted to bring out their flavor. The browning that occurs during roasting adds rich flavors to meats, poultry, seafood, and vegetables. For example, potato wedges or slices can be seasoned and roasted. Vegetables and potatoes don't need to be roasted on a rack. Season with herbs, spices, pepper, stock, garlic, onion, and a few sprays of oil.

For an accompaniment to meat or poultry, you may want to simply use the jus. To give the jus some additional flavor, add a mirepoix to the roasting pan during the last 30 to 40 minutes. To remove most of the fat from the jus, you can use a fat-separator pitcher or fat-off ladle. If time permits, you can refrigerate the jus and the fat will congeal at the top.

If you prefer to thicken the natural jus to make jus lié, first remove the fat from the jus. Because you will be using the roasting pan to make this product, also pat out the fat from the bottom of the pan. Add some of the jus and a little wine to deglaze the pan. Then add some vegetables and cook on a moderately high heat so they brown or caramelize. At this point, add more jus so the vegetables don't burn. Stir the ingredients to release the food from the pan and get its flavor. Continue to add jus, deglaze the pan, and reduce the jus until the color is appropriate. If there is not enough time to reduce the jus down to the proper consistency, you can thicken it with a cornstarch slurry.

Another way to thicken the natural jus is to add starchy vegetables such as beans or potatoes to the jus during the last 30 to 45 minutes of cooking. These vegetables can then be puréed to naturally bind the jus.

To develop flavor, rubs and marinades for meat, poultry, and fish are two possibilities that have been discussed. Smoking can also be used to complement the taste of meat, poultry, or fish. Hardwoods or fruitwoods, such as the following, are best for producing quality results:

- Fruit (apple, cherry, peach) woods work well with light entrées such as poultry or fish.
- Hickory and sugar maple are more flavorful and work better with beef, pork, sausage, and salmon.
- Mesquite produces an aromatic smoke that also works well with beef and pork.

To use hardwoods, you need to soak them in water first for 30 to 40 minutes, then drain. This way the wood does not burn, but smolders.

Smoke-roasting, also called pan-smoking, is done not in the oven, but on top of the stove. It works best with smaller, tender items such as chicken breast or fish fillet, but it can be used to add flavor to larger pieces. Place about a half-inch of soaked wood chips in the bottom of a roasting pan or hotel pan lined with aluminum foil. Next, place the seasoned food on the rack and reserve. Heat the pan over a moderate-high heat until the wood starts to smoke, then lower the heat. Place the rack over the wood and cover the pan. Smoke until the food has the desired smoke flavor, then complete the cooking in the oven if the food is not yet done. Smoking for too long can cause undesirable flavors. Also, be very careful when opening the lid, due to the heat and smoke.

CHEF'S TIPS FOR ROASTING

- Trim excess fat before cooking.
- Roast on a rack and uncovered (otherwise you are steaming the food). Cook at an appropriate temperature to reduce drying out. Basting during cooking will also reduce drying. Use a meat thermometer and allow for carryover cooking when you remove the dish from the oven. Lastly, don't slice the meat until it has rested sufficiently so you don't lose valuable juices, and slice across the grain to maximize tenderness.
- To develop flavor in meats, poultry, and fish, use rubs and marinades. You can also stuff the food (with vegetables and grains, for example), sear and/or season the food before cooking, or smoke the food over hardwood chips.

- When smoking, add dried pineapple skins, dried grapevines, or dried rosemary sticks to the wood chips for added flavor possibilities.
- Vegetables can be roasted in the oven, which results in wonderful texture and flavor, due largely to caramelization of the natural sugars. Root vegetables, such as celery root or sweet potatoes, are hardier and take longer to oven roast than vegetables such as peppers, radishes, or patty pan squash. Flavorings such as shallots, thyme, and pepper add flavor. Recipe 8-25 features roasted summer vegetables.
- Other accompaniments that add flavor with moderate or no fat are vegetable coulis, chutney, vinaigrettes, salsas, compotes, and mojos.

Broiling and Grilling. Broiling, cooking with radiant heat from above, is wonderful for single servings of steak, chicken breast, or fish with a little more fat such as salmon, tuna, or swordfish that can be served immediately. The more well done you want the product, or the thicker it is, the longer the cooking time and the farther from the heat source it should be. Otherwise, the outside of the food will be cooked but the inside will not be done.

Grilling, cooking with radiant heat from below, is also an excellent method for cooking meat, poultry, seafood, and vegetables. Like broiling, grilling browns foods and the resulting caramelization adds flavor.

Once considered a tasty way to prepare hamburgers, steaks, and fish, grilling is now used to prepare a wide variety of dishes from around the world. For example, chicken is grilled to make fajitas (Tex-Mex style of cooking) or to make jerk barbecue (Jamaican style of cooking). Grilling is also an excellent method to bring out the flavors of many vegetables. Grill them with a little oil, vinegar, or lemon juice, and selected seasonings.

Grilling foods properly requires much cooking experience to get the grilling temperature and timing just right. Some general rules to follow:

1. Cook meat and poultry at higher temperatures than seafood and vegetables.
2. For speedy cooking, use boneless chicken that has been pounded flat and meats that are no more than a half inch thick. Fat should be trimmed off meats.
3. Don't try to grill thin fish fillets, such as striped bass, because they will fall apart. Firm-fleshed, thicker pieces of fish, such as swordfish, salmon, and tuna, do much better on the grill. Don't turn foods too quickly-they'll stick and tear. Mark foods by turning the food around 90 degrees without turning it over.

To flavor foods that will be broiled or grilled, consider marinades, rubs, herbs, and spices. Lean fish can be marinated, sprinkled with Japanese crumbs, and glazed in a broiler. If grilling, consider placing soaked hardwood directly on the coals to smoke the food.

During the grilling of beef, poultry, and fish at high temperatures, substances can form on the surface of the food that cause cancerous tumors in animals. The substances, called heterocyclic amines (HCAs), are made from the amino acids in protein. Also, when fat from the grilled food falls on the hot coals or lava/ceramic bricks, another cancer-causing substance is produced (polycyclic aromatic hydrocarbon or PAH) and the smoke carries it to the food. Neither HCAs or PAHs have been shown to produce cancer in people. Tips to reduce problems with these substances are as follows.

1. Trim the fat off meat and poultry so it doesn't drip onto the coals. This will reduce development of PAHs.
2. Marinate meats, poultry, and fish before grilling. Do not use too much fat in the marinade to avoid the problem with PAHs and flare-ups. For some reason, marinades reduce the amount of HCAs formed during cooking.
3. Another way to reduce HCA formation is to turn foods often.
4. Do not serve blackened or charred foods.

CHEF'S TIPS FOR BROILING AND GRILLING

- Keep the grill clean and properly seasoned to prevent sticking. The broiler must also be kept clean and free of fat buildup; otherwise it will smoke.
- Marinating tender or delicately textured foods for broiling or grilling will firm up their texture, so that they are less likely to fall apart during the cooking process.
- Use marinades, rubs, herbs, spices, crumbs, and smoking to add flavor. When grilling, consider using hardwood chips.
- During broiling, butter has traditionally been used to prevent food from drying out from the intense direct heat. In its place, you may spray the food with olive oil, or baste it with marinade, stock, wine, or reduced-fat vinaigrette during cooking. To finish, sprinkle a thin layer of Japanese bread crumbs and glaze under the broiler.
- To retain flavor during cooking, turn foods on the broiler or grill with tongs—not forks, which cause loss of juices.
- If making kabobs, soak the wooden skewers ahead of time so they can endure the cooking heat without excessive drying or burning.
- Prepare these foods to order, and serve immediately.
- Serve with flavorful sauces such as chutneys, relishes, or salsas.

Sauté and Dry Sauté. Sautéing, to cook food quickly in a small amount of fat over high heat, can be used to cook tender foods that are in either single portions or small pieces. Sautéing can also be used as a step in a recipe to add flavor to foods such as vegetables, by either cooking or reheating. Sautéing adds flavor in large part from caramelization (browning) that occurs during cooking at relatively high temperatures.

When sautéing, use a shallow pan to allow moisture to escape, and allow space between the food items in the pan. Use a well-seasoned or nonstick pan and add about half a teaspoon or two sprays of oil per serving after preheating the pan. Instead of oil, you can use vegetable oil cooking sprays, which come in a variety of flavors, such as butter, olive, Asian, Italian, or mesquite. A quick two-second spray adds about 1 gram of fat (9 kcalories) to the product. To use these sprays, spray the preheated pan away from any open flame (the spray is flammable) and then add the food. New pump spray bottles are available that allow you to fill them with the oil of your choice.

For an even lower-in-fat cooking method, use the dry sauté technique. Using this technique, heat a nonstick pan, spray with vegetable-oil cooking spray, then wipe out the excess with a paper towel. Heat the pan again, then add the food.

If browning is not important, you can simmer the ingredient in a small amount of fat-free liquid, such as wine, vermouth, flavored vinegar, or defatted stock, to bring out the flavor. Vegetables naturally high in water content, such as tomatoes or mushrooms, can be cooked with little or no added fluid, at a very high quick heat.

When sautéing or dry sautéing, you can deglaze the pan after cooking with stock, wine, or other low-fat liquid. Then add shallots, garlic, or other seasonings to the sauce.

CHEF'S TIPS FOR SAUTÉING AND DRY SAUTÉING

- To add flavor, use marinades, herbs, or spices
- Use high heat or moderately high heat and a well-seasoned or nonstick pan used only for dry sautéing.
- Pound pieces of meat or poultry flat to increase the surface area for cooking and thereby reduce the cooking time.
- To sauté vegetables, simmer them with a liquid such as stock, juice or wine in a nonstick pan, and add seasonings such as cardamom, coriander, or fennel. Cook over a moderately high heat. As the vegetables start to brown and stick to the pan, add more liquid and deglaze the pan. Once the vegetables are ready, add a little bit of butter or flavorful oil, perhaps 1 teaspoon for four servings, to give the product a rich flavor without much fat, and a shiny texture.
- Another sauté method is to blanch vegetables in boiling water to desired doneness, then shock in ice water. Drain and dry-sauté in a hot nonstick pan with stock, wine, fresh herbs, garlic, and shallots (chopped), and finish with fresh black pepper and extra virgin olive oil, butter, or nut oil (1 teaspoon for four servings).
- Once prepared, serve sautéed foods immediately. They do not hold well.
- A sautéed entrée goes well with portions of stews, ragouts, pilafs, or risotto.

Stir-frying. Stir-frying, cooking small-sized foods over high heat in a small amount of oil, preserves the crisp texture and bright color of vegetables and cooks strips of poultry, meat, or fish quickly. Typically, stir-frying is done in a wok, but a nonstick pan can be used. Steam-jacketed kettles and tilt frying pans can also be used to make quantity stir-fry menu items. Cut up ingredients as appropriate into small pieces, thin strips, or diced portions.

CHEF'S TIPS FOR STIR-FRYING

- Coat the cooking surface with a thin layer of oil. Peanut oil works well because it has a strong flavor (so you can use just a little) and a high smoking point. You can also use vegetable cooking spray, and wipe away any excess.
- Have your ingredients ready next to you, because this process is fast.
- Partially blanch the thick vegetables first (such as carrots, or broccoli), so that they will cook completely without excessive browning.
- Preheat the equipment to a high temperature.
- Foods that require the longest cooking times, usually the meat or poultry, should be the first ingredients that you start to cook.
- Stir the food rapidly during cooking and don't overfill the pan.
- Use garlic, scallion, ginger, rice wine vinegar, low-sodium soy sauce, and chicken/vegetable stock for flavor.
- Add a little sesame oil at the end for taste (about 1 teaspoon per four servings).

Moist-Heat Cooking Methods

Moist-heat cooking methods involve water or a water-based liquid as the vehicle of heat transfer, and are often used with secondary cuts of meat and fowl, or legs and thighs of poultry. When moist-heat cooking meat or poultry, the danger is that the fat in the meat or poultry, although leaner, stays in the cooking liquid. This problem can be resolved to a large extent by chilling the cooking liquid so the fat separates and is then removed before the liquid is used. If the liquid needs to be chilled quickly, place in an ice bath for quickest results.

Compared to most dry-heat cooking methods, these methods do not add the flavor that dry-heat cooked foods get from browning, deglazing, or reduction. In order for foods that use moist-heat cooking to be successful, you will need:

- Very fresh ingredients
- Seasoned cooking liquids using fresh stock, wine, fresh herbs, spices, aromatic vegetables, and other ingredients
- Strongly flavored sauces or accompaniments to achieve flavor and balance

Steaming has been the traditional method of cooking vegetables in many quantity kitchens because it is quick and retains flavor, moisture, and nutrients. It's healthy too, because it requires no fat. The best candidates for steaming include foods with a delicate texture, such as fish, shellfish, chicken breasts, vegetables, and fruits. Be sure to use absolutely fresh ingredients to come out with a quality product.

Fish is great when steamed en papillote (in parchment) or in grape, spinach, or cabbage leaves. The covering also helps retain moisture, flavor, and nutrients. Consider marinating fish beforehand to add moisture and flavor.

It is possible to introduce flavor into steamed foods (as well as poached foods) by adding herbs, spices, citrus juices, and other flavorful ingredients to the water. For example, steam halibut over a carrot-lobster broth. Steamed foods also continue to cook after they come out of the steamer, so allow for this in your cooking time.

Poaching, cooking a food submerged in liquid at a temperature of 160 to 180°F (71 to 82°C), is used to cook fish, tender pieces of poultry, eggs, and some fruits and vegetables. To add flavor, you can poach the foods in liquids such as chicken stock, fish stock, or wine flavored with fresh herbs, spices, ginger, mirepoix, vegetables such as garlic or shallots, or citrus juices. For sweet poaching liquids, pick fruit juices, wine, honey, whole spices, and berries. Serve poached foods with flavorful sauces or accompaniments.

Fish is often poached in a flavored liquid known as court bouillon. The liquid is simmered with vegetables (such as onions, carrots, and celery), seasonings (such as herbs), and an acidic product (such as wine, lemon juice, or vinegar). It may be used to cook fish to be served either hot or cold, but it is not generally used in making sauce. Cook the court bouillon for about 20 minutes, strain out vegetables and herbs, then add the fish to infuse these flavors. Place the fish to be poached on a rack or wrap it up in cheesecloth.

Braising, or stewing (stewing usually refers to smaller pieces of meat or poultry), involves two steps: searing or browning the food (usually meat) in a small amount of oil or its own fat, then adding liquid and simmering until done. Foods to be braised are also often marinated before searing to develop flavor and tenderize the meat. When browning meat for braising, sear it in as little fat as possible without scorching, and then place it in a covered braising pan to simmer in a small amount of liquid. To add flavor, place roasted vegetables, herbs, spices, and other flavoring ingredients in the bottom of the braising pan before adding the liquid.

Following are the steps for braising meat or poultry.

1. Season the meat or poultry. Trim fat.
2. In a small amount of oil in a brazier, sear the meat or poultry to brown it and develop some flavor.
3. Remove the meat and any excess oil from the pan.
4. Put the pan back on the heat. Add vegetables and caramelize (brown) them.
5. Deglaze the pan with wine and stock, being sure to scrape the fond.

6. Add tomato paste, wine, stock, and other aromatic vegetables. Reduce.
7. Return the meat to the brazier, cover, and put into the oven, covered, to simmer. Make sure the meat is three-quarters covered with liquid. This allows the meat to cook more evenly and prevents the bottom from getting scorched.
8. When the meat is done (fork tender), strain the juices. Skim off fat and reduce juices. Purée the vegetables from the juices and use them (or you can use cornstarch) to thicken the jus to make a light gravy.

Vegetables, beans, and fish can also be braised.

Microwaving is another wonderful method for cooking vegetables, because no fat is necessary and the vegetables' color, flavor, texture, and nutrients are retained. Boiling or simmering vegetables is not nearly as desirable as steaming or microwaving, because nutrients are lost in the cooking water and more time is required.

■ MINI-SUMMARY

Healthy cooking methods include roasting, broiling, grilling, stir-frying, sautéing with little or no oil, steaming, boiling, simmering, poaching, microwaving, and braising, when the fat is removed from the cooking liquid. Cooking techniques that are especially useful in preparing healthy items include reduction, searing, deglazing, sweating, and puréeing.

Step 3: Presentation

Here are the main considerations that underlie the art of presentation.

- **Height** gives a plate interest and importance. A raised surface or high point calls attention to itself. A flat and level surface is monotonous. When actual height is difficult to attain, implied height, or an illusion of height, can often be achieved by causing the eye to focus on a particular point. This can be done in several ways. One is by arranging ingredients in a pattern that guides the eye to that point. Another is to use an eye-catching ingredient or garnish to establish a focal point.
- **Color** is very, very important. Too many colors tend to confuse the eye and dissipate the attention, so don't overdo it.
- **Shape** is important too. Vary the shapes on one plate.
- Match the **layout** of the menu item with the shape of the plate. For example, salads are usually presented on round plates. This means that the lines, forms, and shapes of the salad ingredients must be arranged in a pattern that fits harmoniously into a circle. The pattern may repeat the curve

of the plate's edge, or echo its roundness on a smaller scale, or complement it with balance and symmetry. The flow of your food presentation should be tight, meaning having food fairly close together to retain heat. The flow from left to right should curve inward to guide the eye back to the middle of the plate. The pattern begins with the rim of the plate, so never place anything on the rim. It is the frame of your design.

- The most effective **garnish** is something bright, eye-catching, contrasting in color, pleasing in shape, and simple in design. It should enhance the plate and not be the focus. At times, sauces may act as the garnish.

One of the tricks of presentation with healthy dishes is to make less look like more. When you are serving smaller portions of meat, poultry, or seafood, various techniques can be used to make the portion size look larger. By slicing meat or poultry thin, you can fan out the slices on the plate to make an attractive arrangement, and also one that looks plentiful. You can also arrange a piece of meat, poultry, or seafood on a bed of grains, vegetables, and/or fruits, or cover it with sautéed vegetables. In addition, serving larger portions of side dishes with the entrée also helps to make the plate look full. Sauces such as vegetable coulis, salsas, and relishes also help cover the plate, and they provide eye appeal and color.

A common problem that crops up when plating healthy foods is that many dishes lose heat quickly and dry out fast. High-fat sauces help keep a dish hot. When meat is sliced for presentation, it loses more juice and heat, so it dries out quickly. To overcome this problem, chefs often place foods close together on the plate, putting the densest food in the center to keep the other foods warm. When slicing meats for plating, you can slice just part of the meat for appearance and leave the remaining piece whole for the guest to cut.

Keep all garnishes simple. Some dishes, such as angel-food cake with a fruit sauce, has a natural appeal, although you may top them with a slice or two of fresh fruit.

■ MINI-SUMMARY

Pay special attention to height, color, shapes, unity, and garnishes for maximum plate presentation.

Healthy Recipes and Chef's Tips

You can modify recipes for many reasons, such as to reduce the amount of kcalories, fat, saturated fat, cholesterol, sodium, or sugar. You may also modify a recipe to get more of a nutrient, such as fiber or vitamin A.

Whether modifying a recipe to get more or less, there are four basic ways to go about it.

1. Change/add healthy preparation techniques.
2. Change/add healthy cooking techniques.
3. Change an ingredient by reducing it, eliminating it, or replacing it.
4. Add a new ingredient(s), particularly to build flavor.

If you do decide to modify a recipe, follow these steps.

1. Examine the nutritional analysis of the product and decide how and how much you want to change the product's nutrient profile. For example, in a meat loaf recipe, you may decide to decrease its fat content to less than 40 percent and increase its complex carbohydrate content to 10 grams per serving.
2. Next, you need to consider flavor. What can you do to the recipe to keep maximum flavor? Should you try to mimic the taste of its original version, or will you have to introduce new flavors? Will you be able to produce a tasty dish?
3. Next, modify the recipe using any of the methods just discussed. When modifying ingredients, think about what functions each ingredient performs in the recipe. Is it there for appearance, flavor, texture? What will happen if less of an ingredient is used, or a new ingredient is substituted? You also have to consider adding flavoring ingredients in many cases.
4. Evaluate your product to see whether it is acceptable. This step often leads to further modification and testing. Be prepared to test the recipe a number of times, and also be prepared for the fact that some modified recipes will never be acceptable.

Of course, if you don't want to go through the trouble of modifying current recipes, you can select and test recipes from healthy cookbooks or other sources, or create your own recipes. When developing new recipes, be sure to choose fresh ingredients and cooking methods and techniques that are low in fat. Also, pay attention to developing flavor, and cook foods to order as much as possible.

Appetizers

Appetizers are a very creative part of the menu. Ingredients might include fresh fruits and vegetables, fresh seafood and poultry, fresh herbs, spices, infused oils, vinegars, and pasta.

Recipes for Roast Chicken and Shredded Mozzarella Tortellinis (Recipe 8-1), Scallop Rolls in Rice Paper (Recipe 8-2), Crab Cakes (Recipe 8-3), Eggplant Rollatini with Spinach and Ricotta (Recipe 8-4), and Mussels Steamed in Saffron and White Wine (Recipe 8-5) are found in this chapter.

Additional ideas for appetizers include:

- Ricotta Cheese and Basil Tortellonis with Salsa Cruda and Arugula
- Napoleon of Grilled Vegetables and Wild Mushrooms with Roasted-Pepper Sauce
- Potato and Onion Tart with Forelli Pear Vinaigrette
- Pan-Seared Scallops with Butter Potato Pancake, Sunsprouts, and Blood-Orange Dressing
- Spicy Chicken and Jack Cheese Quesadillas with Tomatilla Salsa
- Maine Crab Cakes with Smoked-Pepper Sauce and Baby Lettuces
- Red Lentil Chili with Baked Whole-Wheat Tortilla Chips

CHEF'S TIPS FOR APPETIZERS

- Appetizers can often be sized-down entrées. For example, Maine Crab Cakes with Smoked-Pepper Sauce and Baby Lettuces can be made in larger or smaller portions to appear as entrée or appetizer.
- There are certain ingredients with which you can make a wide variety of appetizers. Consider using wonton skins and rice paper as wrappers, and stuff them with fillings such as white beans and artichokes with roasted garlic, or spiced butternut squash.
- Dried beet chips are a wonderful way to add color to appetizers. To make, slice red beets thin, and dip in simple syrup. Dry on a silk mat. Put in 275°F oven until crisp, about 15–20 minutes.
- Creative sauces and relishes can help sell appetizers. For example, serve smoked shrimp with mango sauce and cucumber and black bean relish.

Soups

Soups make up some of the most nutritious meals, from the hearty minestrone to robust butternut squash soup, creamed with nonfat yogurt and spiced with nutmeg. Soups can be balanced into a meal as an appetizer or given first billing as an entrée.

Soups are a wonderful place to spotlight more than just vegetables. Beans, lentils, split peas, and grains such as rice are also healthy ingredients that work well with soups. Some of these ingredients are starchy enough to be used to thicken soups, instead of using traditional roux (a thickening agent of fat and flour in a one-to-one ratio by weight), which is high in kcalories and fat. Examples of starchy foods that work well as thickeners include beans, lentils, rice, other grains, and puréed vegetables such as potatoes and squash. These foods can be used to make soups such as black bean, lentil, split pea, Pasta e Fagioli (Recipe 8-6), and vegetable chili soup.

To make a cream soup such as cream of broccoli without using cream or roux, start by dry-sautéing broccoli, onion, herbs, garlic, and shallots in chicken stock and white wine. Deglaze the pan, then add potato and cover with vegetable or chicken stock. Once the potato is done, purée the ingredients to the proper consistency. Garnish the soup with small, steamed broccoli florets.

Rice also is an excellent thickener and lends a creamy texture to the soup. Rice can be used successfully to thicken corn, carrot, and squash soups. Use about six ounces rice to one gallon of stock. See Recipe 8-7, Butternut Squash Bisque, for an example.

CHEF'S TIPS FOR SOUPS

- Strain soups such as broccoli, celery, and asparagus through a large-holed china cap to remove fibers.
- Also purée bean soups such as black bean or split pea to get a homogeneous product. Next, strain to remove skins.
- Rice and potatoes work well as thickeners in many soups.
- Replace ham in bean soups with smoked chilies, or your own smoked turkey, or veal bacon.
- Garnish soups with an ingredient of the soup whenever possible. For example, put pieces of baked tortillas on top of Mexican soup. The use of fresh vegetables, fruits, or herbs as a garnish adds interest, color, and flavor. Also consider garnishing some soups, as appropriate, with a small amount (such as 1 teaspoon) of cream, nuts, croutons, sour cream, smoked chicken, or avocado.

Salads and Dressings

Components of salads go way beyond simple raw vegetables. Consider using grains, beans, lentils, pasta, fresh fruits and juices, oven-dried vegetables, fresh poultry or seafood, game, and herbs and spices such as ginger, Kafir lime leaves, star anise, cardamom, curry, lavender, lemon balm, and fresh cinnamon. Salads are a wonderful place to feature high-fiber, low-fat ingredients.

Recipes for Wild Mushroom Salad (Recipe 8-8) and Baby Mixed Greens with Shaved Fennel and Orange Sections (Recipe 8-9) appear in this chapter. Other possible salad combinations include the following.

- Baby Lentil and Roasted Vegetables
- Yellow and Red Tomato Salad with Fresh Basil, Oregano, and Double-Roasted Garlic
- Haricots Verts with Trio of Roasted Peppers
- Carrot, Celery, Golden Pineapple, and Dried Pear Salad with Sweet-and-Sour Dressing
- Bow-tie Pasta with Fresh Tuna and Chives in a Lemon Basil Dressing
- Orzo pasta with Tomato, Cilantro, Scallion, and Cucumbers
- Yogurt Chicken with Fresh and Dried Fruits
- Wheat Berry Salad with Wild Mushrooms and Rosemary Thyme Dressing

- Organic Baby Lettuces with Marinated Cucumber and Tomato with Classic French Dressing

Dressings are used in much more than salads. They can often be used as an ingredient of entrées, appetizers, relishes, vegetables, and marinades. There are many categories of dressings. The best place to start is the basic vinaigrette, because it is simple and you can use ingredients such as herbs and spices to make many variations. The best ingredients to use include good-quality vinegars, first-pressing olive and nut oils, and fresh herbs, because you need the strongest flavor with the least amount of fat. Other good ingredients that add flavor without fat include Dijon mustard, shallots or garlic (which may be roasted for a robust flavor), a touch of honey, reduced vinegars, and lemon or lime juice.

For examples of vinaigrette recipes, see recipes for Basic Herb Vinaigrette (Recipe 8-10), Orange Vinaigrette (Recipe 8-11), and Sherry Wine Vinaigrette (Recipe 8-12). If you look at Basic Herb Vinaigrette, you will notice that instead of a ratio of 3 parts oils to 1 part vinegar, this recipe uses 1 part oil, 1 part vinegar, and about 2 parts thickened chicken or vegetable stock. This results in a satisfactory product whose flavor profile is boosted with fresh herbs and garlic and high-quality olive oil and vinegar.

Other salad dressings often fit into one of one of these categories.

1. **Creamy dressings**—Tofu can be processed to produce a creamy dressing, as in Green Goddess Dressing (Recipe 8-14). Other creamy ingredients are nonfat or low-fat yogurt, nonfat sour cream, or ricotta cheese. Puréed fruits and vegetables are creamy and can be used as emulsifiers in salad dressings.
2. **Puréed dressings**—Examples of puréed dressings include potato vinaigrette, hummus, and smoked-pepper coulis. Some of these dressings, such as hummus, work well as dips.
3. **Reduction dressings**—Examples include orange (see Recipe 8-11), beet, carrot-balsamic, and apple cider. These dressings can be made simply and are powerhouses of flavor.

CHEF'S TIPS FOR SALADS AND DRESSINGS

- As elsewhere in the kitchen, use fresh, high-quality ingredients. Choose ingredients for compatibility of flavors, textures, and colors.
- Vegetables, both raw and cooked, go well with dressings having an acid taste such as vinegar or lemon.
- Legumes make wonderful salads. For example, black-eyed peas go well with flageolets and red adzuki beans. To add a little more color and develop the flavor, you might add chopped tomatoes, fresh cilantro (Chinese parsley), haricots verts (green beans), and roasted peppers.
- The reduction dressings, such as reduced beet juice, can be put into a squirt bottle and used to decorate the plate for a salad, appetizer, or entrée.

- Plan your presentation carefully in terms of height, color, and composition. Keep it simple. Do not overcolor or overgarnish.

Entrées

Developing healthy entrées will draw upon your entire knowledge of ingredients, cooking techniques and methods, and nutrition. You have a wide variety of ingredients, cooking techniques, and methods at your disposal to create numerous delicious entrées, ranging from traditional beef to meatless dishes, such as the following.

- Classic Beef Stew with Horseradish Mashed Potatoes
- Grilled Pork Chop Adobo with Spicy Apple Chutney (Recipe 8-16)
- Sautéed Veal Loin with Barley Risotto and Roasted Peppers
- Pan-Seared Louisiana Spiced Breast of Chicken
- Grilled Chicken Breast and Quinoa Salad with Cucumber, Tomato, Corn, and Peppers
- Mediterranean Chicken Breast with Basmati Rice
- Grilled Chicken/Turkey Burger with Oven-Baked Blue Corn Tortillas and Salsa Roja
- Stir-Fried Chicken and Garden Vegetables with Brown Rice
- Cedar-Planked Wild Striped Bass with Ratatouille and Red Bean Salad
- Lemon-Scented Tuna with Charred Tomato Salsa

Entrée recipes in this chapter include Potato Lasagna (Recipe 8-17), Chicken Sausage (Recipe 8-18), Slates of Salmon (Recipe 8-19), and Braised Lamb (Recipe 8-20).

After reading up to this point, you may have noticed the lack of classical sauces. Rather than foods covered with rich, heavy sauces, the emphasis in this book is on the taste and appearance of the food itself. In sauce making, this has meant a great change in the techniques of thickening. Followers of the new style often use purées and reductions instead of roux to make sauces, as you have seen in vegetable coulis and meat juices that have been thickened with puréed vegetables. Also, stock can be flavored and reduced to make a quality sauce that can be used in many dishes on the menu. This chapter features the following alternatives to traditional sauces.

- Papaya and White Raisin Chutney (Recipe 8-21)
- Papaya-Plantain Salsa (Recipe 8-22)
- Red Pepper Coulis (Recipe 8-23)
- Hot and Sour Sauce (Recipe 8-24)

CHEF'S TIPS FOR ENTRÉES

- For meat, poultry, and fish entrées, a 3- to 4-ounce cooked portion is adequate. See Chapter 4 for a list of lean meat, poultry, and fish.
- Use bulgur to extend ground meat. For every pound of meat, add 1/2 cup of cooked bulgur.

- Fish is a very versatile and nutritious food. Almost any food, such as rice or beans or pasta, goes with fish. Serve fish on top of a vegetable ragout, or salmon with vegetable curried couscous.
- When choosing legumes for a dish, think color and flavor. Make sure the colors you pick will look good when the dish is complete. Also think of other ingredients you will use for flavor.
- Bigger beans, such as gigante white beans, hold their shape well and lend a hearty flavor to stews, ragouts, and salads.
- When using cheese in an entrée, use a small amount of a strong cheese such as gorgonzola or goat cheese. Also, instead of using cheese throughout the entrée, just use some on the top. Choose cheese varieties that are low in fat, such as skim milk mozzarella or ricotta.
- Create new fillings for pasta that don't rely totally on cheese. For example, sweat puréed butternut squash and potato. Add fresh thyme, roasted shallots, and perhaps a little roasted duck for flavor. Or purée together cooked artichokes, white beans, roasted garlic, and carrots or ratatouille for flavor and color.
- Top casseroles and baked pasta dishes with reduced-fat cheese near the end of the cooking time and heat just until melted. The lower the fat content, the longer the melting time. Too much heat and/or direct heat may toughen the cheese, so cook reduced-fat cheeses at lower temperatures and for as short a period of time as possible.

Side Dishes

There is no end to the variety of substitutions and side dishes for every entrée. The same dish can take on a new face simply by changing the starch or the vegetable. Besides the traditional side dish of vegetables and potatoes, consider using grains such as wheat berries or barley; try legumes such as black beans or lentils, tofu, or fresh fruits. Also consider techniques such as the following:

- **Puréeing.** For example, purée sweet potatoes, butternut squash, and carrots, flavored with cinnamon, honey, and thyme.
- **Roasting.** For example, roast onions with cinnamon, ginger, vinegar, and sugar. Recipe 8-25 shows how to prepare Roasted Summer Vegetables.
- **Grilling.** For example, grill portobello mushrooms with polenta, garden tomatoes, and roasted elephant garlic.
- **Stir-fry.** Try Hot and Sour Stir-Fry with Seared Tofu and Fresh Vegetables.

Additional examples of side dishes include these:

- Roasted Garlic and Yogurt Red Mashed Potatoes with Dill
- Couscous with Dried Fruit, Cucumber, and Mint

- Seven-Vegetable Stir-Fried Rice
- Ratatouille Strudel with Oven-Dried Tomatoes Wrapped in Phyllo
- Oven-Baked French Fries with Cajun Rub
- Swiss-Style Marinated Red Cabbage
- Sautéed Wild Mushrooms
- Risotto with Spring Onions and Pesto

Grains, such as rice, are versatile and make excellent side dish ingredients. See Recipe 8-26, which shows how to make Mixed-Grain Pilaf.

CHEF'S TIPS FOR SIDE DISHES

- When using vegetables, you need to *think*—what's in season for maximum flavor, how the dish will look (its colors), and how the dish will taste. Think flavor and color.
- Also, think variety. Serving vegetables doesn't mean switching from broccoli to cauliflower and then back to broccoli. There are many, many varieties of vegetables to choose from. Be adventurous.
- Add grains to vegetable dishes, such as brown rice with stir-fried vegetables.
- Serve grains and beans. For example, rice and beans is a very popular and versatile dish using grains and legumes. Mix purple rice with white beans, or wild rice with cranberry beans for appearance.
- Salads can often be used as side dishes.

Desserts

You can make a wide variety of desserts without compromising health. Think of fruits (either fresh or as a key ingredient), quick breads, cobblers, puddings, phyllo strudels, and even some cakes and cookies. The following recipes show how fruit can be used in many forms.

- Oatmeal-Crusted Peach Pie
- Apple Strudel with Caramel Sauce
- Fruit Sorbets
- Spiced Carrot Cake with Orange Custard Sauce
- Fresh Fruit Compote with Buttermilk Dumplings
- Chocolate Raspberry Meringue Torte
- Banana Ginger Pudding
- Poached Sickle Pears with Merlot Syrup and Almond Tuille
- Grilled Peppered Pineapple with Vanilla Ice Cream and Crunchy Cookies
- Yogurt Fruited Cheesecake

Recipe 8-28 highlights a way to use cocoa and bittersweet chocolate to make a chocolate pudding cake that is baked in sugar molds in a bain-marie (hot-water bath). Recipe 8-29 uses soft tofu, egg whites, raspberry purée, and sugar to make ice cream that, with the help of an ice cream machine, can be made right in the kitchen. This raspberry creamed ice goes well with Recipe 8-30, Angel Food Savarin. Recipe 8-31, Heart-Healthy Oatmeal Raisin Cookies, uses mostly applesauce (with a tiny amount of canola oil) to replace the usual fat. This is an excellent cookie for use of a fat substitute, because it is naturally sweet (due to the raisins) and spicy.

CHEF'S TIPS FOR DESSERTS

- To make sorbet without sugar, simply purée and strain the fruits. Make sure the fruits are at the peak of ripeness.
- Use angel food cake as a base to build a dessert. Serve it with fresh fruit sorbet, warm sautéed apples and cranberries, pear and ginger compote, or mango, mint, and papaya salsa.
- Likewise, use phyllo in many ways. For example, stuff it with strawberries and nonfat granola, or bake in a muffin pan and fill with sautéed apples garnished with dried apple chips to give it some crunch.
- Compote is an additional way to serve fruit. Compote is a dish of fruit, fresh or dried, cooked in syrup flavored with spices or liqueur. When you consider how many fruits you have at your disposal, as well as spices and flavorings, the possibilities are endless.

Breakfasts

Hot and cold cereals have been the foundation of breakfast for generations in many ethnic cultures, from Swiss birchenmuesli (oats with fresh and dried fruits soaked in milk or cream) to hot English oatmeals to contemporary American granolas. There are numerous variations from these classic foundations. For instance, the liquid used to make hot cereals like oatmeal can be a variety of fruit juices such as apple, pineapple, or orange. They can also be spiced with cinnamon, nutmeg, ginger, allspice, or cloves. For more adventurous customers, you can use jalapeño jack cheese, star anise, lavender, lemon balm, or any fresh herb combination.

The traditional Swiss birchenmuesli is made with rolled oats, heavy cream, sugar, nuts, and dried and fresh fruits. To modify this recipe, more fresh fruits are added to the cereal mixture, skim milk replaces heavy cream, and nonfat yogurt and spices complete the taste needed to make this Old World classic a modern hit.

Pancakes, French toast, and toppings are quite honestly the apple pie of breakfast. It's hard to imagine a breakfast menu without blueberry pancakes or thick crispy French toast with syrup. A typical pancake batter contains whole eggs, oil, and sugar. To use less of these ingredients, you will have to add other ingredients that you can sink your teeth into. For example, you can use wheat germ, rolled oats, stone-ground wheat, or millet to create a hearty texture. You also need to include spices and fruit flavorings. By putting leftover berries into batter or using overripe fruit to make syrups, you utilize your inventory while creating a quality product. You can also fine-cut the fruits and toss them with fresh mint or lemon balm to create a sweet salsa or compote.

Quick breads, muffins, and scones are also staple items for breakfast menus as well as buffets. Fruit juices, concentrates, and purées are a wonderful source of flavor when baking quick breads, muffins, and scones, as well as in hot cereals and pancake batters. Low-fat spreads might include flavored nonfat ricotta cheese or flavored low-fat cream cheeses.

Some menu possibilities for breakfast include the following:

- Glazed Grapefruit and Orange Slices with Maple Vanilla Sauce (Recipe 8-35)
- Crunchy Cinnamon Granola (Recipe 8-33)
- Pineapple Ginger Cream of Wheat
- Jalapeño Jack Cheese Grits
- Spiced Oatmeal with Dried Cranberries
- Stuffed French Toast Layered with Light Cream Cheese and Bananas (Recipe 8-34), topped with strawberry syrup
- Wheat-Berry Pancakes with Fresh Fruits
- German Pancakes with Orange Syrup
- Whole-Wheat Peach Chimichangas
- Banana Ginger Raisin Bread
- Dried-Cherry Scones

Breakfast is a wonderful time for a buffet. It is a time saver for guests to give them wide variety. Consider a buffet with platters of sliced fresh fruits with Maple Vanilla Yogurt, Dried-Cherry Scones, Jalapeño Jack Grits, Banana Pancakes, Classic Vegetable Frittata, Crunchy Cinnamon Granola, and High-Five Fruit Muffins.

CHEF'S TIPS FOR BREAKFASTS

- To make an excellent omelet without cholesterol, whip egg whites until they foam. Add a touch of white wine, Dijon mustard, and chives. Whip to a loose meringue. Spray a hot nonstick pan with oil and add your eggs. Cook the way you do a whole-egg omelet. When the omelet is close to done, put the pan under the broiler to finish. The omelet will puff up. Stuff the omelet, if desired, with something like grilled, roasted, or sautéed vegetables, then fold over and serve.
- For color and flavor, serve an omelet with spicy vegetable relish poured on top of it, or place the omelet on a grilled blue corn tortilla and serve with salsa roja.
- When writing breakfast menus, make sure you provide balanced, healthful, and flavorful breakfasts, such as Glazed Grapefruit and Orange Slices with Maple Vanilla Sauce, Blueberry Wheat Pancakes with Strawberry Syrup, and a side dish of Chicken Hash.

- Breakfast is probably the best time of day to offer freshly squeezed juices. Make sure they are fresh and that you offer a good variety.

Check-Out Quiz

1. Sweating vegetables can be done in stock.
   a. True b. False
2. Flavorings are used to bring out flavor that is already in a dish.
   a. True b. False
3. Ginger is an example of a fresh herb.
   a. True b. False
4. Basil, oregano, and garlic are often used in Indian dishes.
   a. True b. False
5. Toasted spices can be ground and used in marinades, salad dressings, rubs, and soups.
   a. True b. False
6. Gelatin is what gives stock body.
   a. True b. False
7. A good stock is brimming with flavor.
   a. True b. False
8. The shallot is a green onion.
   a. True b. False
9. Chutney refers to a sauce made of a purée of vegetables or fruits.
   a. True b. False
10. Portable vertical blenders are a great tool for puréeing foods.
    a. True b. False

Activities and Applications

1. Recipe Modification

Use one of the substitutions listed in Table 8-4 to prepare a recipe. Compare the flavor, texture, shape, and color of both products. Ask someone (who doesn't know which is the modified version) to test the products and determine which one tastes better.

2. Nutrient Content of Modified Recipes

Following is a recipe for Monte Cristo Sandwiches that has been modified to make it more nutritious. Calculate and compare the nutrient content per serving before and after modification.

TABLE 8-4 Recipe Substitutions

In Place of	Use
Butter	Margarine
Whole milk	Skim or low-fat milk (Add some nonfat dry milk, if desired, to make it creamy)
1 cup shortening or lard	3/4 cup vegetable oil
1 cup heavy cream	1 cup evaporated skim milk
1 cup light cream	1 cup evaporated skim milk
1 cup sour cream	1 cup nonfat plain yogurt or low-fat cottage cheese blended with 1 to 2 tablespoons buttermilk or lemon until creamy
Cream cheese	Neufchâtel or reduced-calorie cream cheese
1 ounce baking chocolate	3 tablespoons cocoa and 1 tablespoon vegetable oil
1 egg	1/4 cup cholesterol-free egg substitute or 2 egg whites
Whipped cream	Whip 1 cup low-fat ricotta cheese with 3 tablespoons nonfat plain yogurt and 2 tablespoons sugar. Use honey, vanilla, or other extracts to avor.
Whole-milk mozzarella	Part-skim mozzarella, low moisture
Whole-milk ricotta	Part-skim ricotta
Creamed cottage cheese	Low-fat cottage cheese or pot cheese
Cheddar cheese	Low-fat cheddar cheese
Swiss cheese	Low-fat Swiss cheese, such as Swiss Lorraine
Ice cream	Ice milk or frozen yogurt
Mayonnaise	Nonfat plain yogurt

Original Monte Cristo Sandwiches

Yield: 50 portions

Ham, cooked, boneless	50 1-oz. slices
Turkey, cooked, boneless	50 1-oz. slices
Swiss cheese	50 1-oz. slices
White bread	100 slices
Whole milk	3 cups
Salt	1 teaspoon
Eggs, whole, slightly beaten	1 quart (24 eggs)
Shortening, melted	2 cups

Place one slice each of ham, turkey, and cheese on one slice of bread and top with a second slice. Blend milk, salt, and egg. Dip each side of the sandwich into the egg and milk mixture; drain. Grill each sandwich on a well-greased griddle about 2 minutes on each side, or until golden brown and the cheese is melted.

Modified Monte Cristo Sandwiches

Yield: 50 portions

Low-fat or fat-free honey ham	50 1-oz. slices
Turkey, fresh roasted	50 1-oz. slices
Swiss cheese	50 0.5-oz. slices
Rye bread	100 slices
Skim milk	3 cups
Egg whites, lightly beaten	1 quart
Butter-flavored cooking spray	As needed

Place one slice each of ham, turkey, and cheese on one slice of bread and top with a second slice. Blend the milk and egg whites. Dip each side of the sandwich into the egg and milk mixture; drain. Grill each sandwich on a grill sprayed with butter-flavored cooking spray. Cook about 2 minutes on each side, or until golden brown and the cheese is melted.

Using nutrient composition information, figure out which ingredient in the recipe for Original Monte Cristo Sandwiches contributes the most fat. Which ingredient contributes the most kcalories? Which ingredient contributes the most carbohydrates?

3. Flavorings

Pick out five recipes from a traditional cookbook and five recipes from a healthy cookbook. Determine the sources of flavor for each recipe and record them. Compare the ingredients used to flavor each set of recipes. Does either set of recipes tend to use more fat or more herbs and spices?

4. Computerized Nutrient Analysis

Do a nutrient analysis of a recipe from a cookbook that has all the recipes analyzed. Compare your results to the book's nutrient analysis. Are they close? Why or why not?

5. Menu-Planning Exercise I

Using a menu from a restaurant or foodservice, recommend two healthy entrées and two healthy desserts that would fit well on this menu. Be ready to explain why you selected these menu items and how they fit with the rest of the menu and the clientele.

6. Menu-Planning Exercise II

In a small group, plan a meal using the recipes in the chapter. The meal should consist of an appetizer or soup, salad with dressing, entrée, side dish, and dessert. Use the nutrient analysis information to determine the calories, fat, sodium, and cholesterol. Also, calculate the percent of calories from fat. Limit fat to 30 percent or less of total kcalories, sodium to 1000 milligrams, and cholesterol to 150 milligrams. Each group will write up their menu and nutrient analysis to present and compare to the other groups.

Nutrition Web Explorer

Cooking Light magazine http://www.cookinglight.com

On the home page of this popular magazine on healthy cooking and living, click on "Cooking 101." Find out "Today's Cooking 101 Tip." Then click on "Techniques" and find out how to roast bell peppers. Lastly, click on "Ingredient Glossary" under "Resource" and get definitions for adobo sauce, ghee, plantain, and saffron.

The Culinary Institute of America's Professional Chef site
http://www.ciaprofchef.com

Visit this site to check out one of their free online courses, such as "Contemporary Flavors with California Raisins" or "Beef and the Global Bistro."

Recipes

8-1 Roast Chicken and Shredded Mozzarella Tortellinis
8-2 Scallop Rolls in Rice Paper with Vegetable Spaghetti
8-3 Crab Cakes
8-4 Eggplant Rollatini with Spinach and Ricotta
8-5 Mussels Steamed in Saffron and White Wine
8-6 Pasta e Fagioli
8-7 Butternut Squash Bisque
8-8 Wild Mushroom Salad
8-9 Baby Mixed Greens with Shaved Fennel and Orange Sections
8-10 Basic Herb Vinaigrette
8-11 Orange Vinaigrette
8-12 Sherry Wine Vinaigrette
8-13 Ginger Lime Dressing
8-14 Green Goddess Dressing
8-15 Capistrano Spice Rub
8-16 Grilled Pork Chop Adobo with Spicy Apple Chutney
8-17 Potato Lasagne
8-18 Chicken Sausage
8-19 Slates of Salmon
8-20 Braised Lamb
8-21 Papaya and White Raisin Chutney
8-22 Papaya-Plantain Salsa
8-23 Red Pepper Coulis
8-24 Hot-and-Sour Sauce
8-25 Roasted Summer Vegetables
8-26 Mixed-Grain Pilaf
8-27 Sautéed Cabbage
8-28 Warm Chocolate Pudding Cake with Almond Cookie and Raspberry Sauce
8-29 Raspberry Creamed Ice
8-30 Angel Food Savarin
8-31 Fresh Berry Phyllo Cones
8-32 Heart-Healthy Oatmeal Raisin Cookies
8-33 Crunchy Cinnamon Granola
8-34 Stuffed French Toast Layered with Light Cream Cheese and Bananas
8-35 Glazed Grapefruit and Orange Slices with Maple Vanilla Sauce

Appetizers

Recipe 8-1 *Roast Chicken and Shredded Mozzarella Tortellinis*

Category: Appetizer Yield: 6 portions

INGREDIENTS

4 oz. chicken breast
1/2 teaspoon canola oil
Pinch fresh oregano, chopped
Pinch fresh basil, chopped
Pinch paprika
1/2 cup reduced-fat shredded mozzarella
1/2 cup skim-milk ricotta
1/4 teaspoon fresh garlic, chopped
2 pinches freshly ground black pepper
1 teaspoon fresh basil, chopped
1/2 teaspoon fresh parsley, chopped
18 wonton skins
1 cup water
18 arugula leaves
1 tomato, chopped
1 recipe Red Pepper Coulis (Recipe 8-23)

STEPS

1. Preheat oven to 350°F.
2. Coat chicken breast with oil, herbs, and paprika. Roast about 5 minutes, or grill over hardwood for 1 minute. Let cool and fine-julienne. Reserve.
3. In bowl, mix cheeses, garlic, pepper, and herbs. Add chicken and mix well.
4. Lay out wonton skins and paint with water.
5. Place 1 full teaspoon of chicken mixture per skin, or more, to use up all the filling.
6. Fold skin to make a triangle, then connect ends in opposite direction to make tortellini shape.
7. Poach tortellinis for 3 minutes in boiling water or steam for 2 minutes.
8. While pasta is cooking, chop tomato and toss with black pepper. Let juice accumulate. Paint leaves of arugula lightly with juice of tomato for flavor.
9. Arrange each plate with 3 arugula leaves facing out like spokes of a wheel, with a tortellini between each 2 leaves. Place red pepper coulis (see Recipe 8-23) and a spoonful of chopped tomato in the middle.

NUTRITIONAL ANALYSIS:

Kcalories	Protein (gm)	Fat (gm)	Carbo (gm)	Sodium (mg)	Chol (mg)
150	12	4	15	221	24

Recipe 8-2 *Scallop Rolls in Rice Paper with Vegetable Spaghetti*

Category: Appetizer Yield: 15 portions

INGREDIENTS

1 large carrot
1 large daikon
2 large zucchini
4 tablespoons Ginger Lime Dressing (Recipe 8-13)
1 pound tofu
3 cups bean sprouts
2 whole red peppers, julienne
2 cups spinach leaves
1/2 cup snow peas, julienne
1/2 small head of white cabbage, shredded
1/4 cup Ginger Lime Dressing (Recipe 8-13)
22 ounces sea scallops (30 scallops), sliced in half
15 sheets rice paper
2 tablespoons cilantro, whole leaves
2 ounces Hot-and-Sour Sauce (Recipe 8-24)

STEPS

1. Make carrot and daikon spaghetti with oriental spaghetti machine.
2. Make zucchini noodles with mandoline.
3. Toss spaghetti and noodles with 2 tablespoons Ginger Lime Dressing.
4. Grill tofu and marinate with 2 tablespoons Ginger Lime Dressing. Dice and reserve for service.
5. Marinate bean sprouts, red pepper, spinach leaves, snow peas, and white cabbage in 1/4 cup Ginger Lime Dressing while you do the next step.
6. Grill scallops on one side for color.
7. Soak sheets of rice paper in warm water one at a time for next step.
8. Layer scallops, vegetables, and cilantro leaves on moistened rice paper and fold like an eggroll. Steam about 3 minutes.
9. In a soup bowl, rest the rice paper roll against the marinated vegetable noodles. Sprinkle tofu around the vegetable noodles. Pour hot-and-sour sauce over mixture.

NUTRITIONAL ANALYSIS:

Kcalories	Protein (gm)	Fat (gm)	Carbo (gm)	Sodium (mg)	Chol (mg)
107	10	3	11	237	13

Recipe 8-3 *Crab Cakes*

Category: Appetizer   Yield: 10 servings

INGREDIENTS

1-1/4 pounds jumbo lump crabmeat
1-1/2 teaspoons chives, chopped fine
1-1/2 teaspoons parsley, chopped fine
1 tablespoon Old Bay seasoning
3/4 teaspoon Dijon mustard
3/4 teaspoon thyme, chopped
5 ounces potatoes, cooked and mashed
1-1/2 egg whites
3/4 teaspoon lemon juice
1-1/2 teaspoons white wine
1/2 egg white
Japanese bread crumbs, as needed
3/4 teaspoon chives, chopped fine
Vegetable oil cooking spray

STEPS

1. Pick crabmeat to remove bits of shell.
2. Add in chives, parsley, Old Bay seasoning, mustard, thyme, and riced potatoes. Mix gently.
3. In a stainless-steel bowl, add 1-1/2 egg whites, lemon juice, and white wine. Whip to form stiff peaks.
4. Fold whipped egg whites into crab mixture and mold into 3-ounce crab cakes.
5. In a metal pan, place 1/2 egg white. Dip the crab cakes in the egg white, then in the crumbs mixed with chives.
6. Sauté the crab cakes to a crisp golden brown in a nonstick pan sprayed with vegetable spray. Serve with salsa or mojo of your choice.

NUTRITIONAL ANALYSIS:

Kcalories	Protein (gm)	Fat (gm)	Carbo (gm)	Sodium (mg)	Chol (mg)
97	12	2	7	266	24

Recipe 8-4 *Eggplant Rollatini with Spinach and Ricotta*

Category: Appetizer **Yield: 20 servings**

INGREDIENTS

3 lbs. fresh spinach
1-1/2 pounds eggplant, peeled and cut lengthwise
1/4 cup balsamic vinegar
Olive oil spray
Vegetable oil cooking spray
4 ounces onions, finely chopped
2 teaspoons garlic, chopped
1 teaspoon cracked black pepper
1 teaspoon garlic herb seasoning
1/4 cup Italian parsley, chopped
2 tablespoons fresh basil, chopped
16 ounces skim milk ricotta
2 ounces feta cheese, crumbled
2 tablespoons grated Parmesan cheese
3/4 cup whole-wheat bread crumbs
1 egg
1 egg white, slightly beaten
1 quart tomato sauce
1 teaspoon chives, chopped

STEPS

1. Steam spinach. Drain well and rough chop. Reserve.
2. Paint eggplant with balsamic vinegar on both sides, and spray with olive oil. Place on sheet pan sprayed with vegetable-oil cooking spray.
3. Bake in 400°F oven for 10 minutes. Remove and flip over. Finish baking until tender. Reserve.
4. Spray nonstick skillet with vegetable oil cooking spray. Quickly sauté onions and garlic. Add drained spinach. Remove from heat and let cool. Add in herb seasoning, herbs, cheeses, and bread crumbs. Mix in egg and beaten egg white. Chill mixture.

5. Place two heaping tablespoons of spinach mixture on each eggplant slice. Roll up and place open end down in casserole dish. Cover with tomato sauce. Bake in 350° oven for 30 minutes.
6. To serve, place 1 rollatini on plate. Top with tomato sauce and sprinkle with chives.

NUTRITIONAL ANALYSIS:

Kcalories	Protein (gm)	Fat (gm)	Carbo (gm)	Sodium (mg)	Chol (mg)
121	7	3	18	154	21

Recipe 8-5 *Mussels Steamed in Saffron and White Wine*

Category: Appetizer **Yield: 10 servings**

INGREDIENTS

Broth

1 tablespoon garlic, chopped
1 teaspoon olive oil
7 ounces white wine
1 gram saffron
1 ounce fresh lemon juice
2 ounces chicken stock
1/2 teaspoon pepper, fresh ground
1 teaspoon fresh thyme, chopped

Mussels

80 fresh New Zealand mussels

Garlic Rouille

2 teaspoons roasted garlic
1 cup Yukon Gold potatoes, cooked and riced
1 tablespoon chicken stock
1/2 teaspoon fresh lemon juice
1 teaspoon red chili paste
1 teaspoon chives
10 slices of whole-wheat bread, cut into 2-1/2-inch rounds
1 cup carrots, blanched, julienne
1 cup celery, blanched, julienne
2 tablespoons chives

STEPS

1. For the broth, sauté garlic in olive oil. Add wine and saffron. Let reduce for 2 minutes.
2. Add lemon juice, stock, black pepper, and fresh thyme. Cook over low heat for 1 minute. Cool and store for service.
3. Scrub mussels and remove beard.
4. Make garlic rouille by puréeing garlic and mixing with potatoes, chicken stock, lemon juice, and red chili paste. Fold in chives, spread on whole-wheat bread, and toast.
5. At service, steam carrots, celery, and mussels in saffron broth. Place mussels in a large bowl. Finish with chives and broth. Garnish with toasted whole-wheat crouton spread with garlic rouille.

NUTRITIONAL ANALYSIS:

Kcalories	Protein (gm)	Fat (gm)	Carbo (gm)	Sodium (mg)	Chol (mg)
139	5	2	22	448	3

Soups

Recipe 8-6 *Pasta e Fagioli*

Category: Soup   Yield: 16 servings (approximately 1 cup)

INGREDIENTS

1 large yellow onion, chopped
6 cloves garlic, chopped
2 tablespoons olive oil, extra virgin
2 pounds pinto, kidney, and black beans, dried and soaked overnight
1/2 cup tomato paste
2 quarts vegetable or chicken stock, defatted
1-1/2 pounds plum tomatoes, peeled, seeded
2 bay leaves
1 tablespoon red pepper flakes
1 tablespoon fresh cracked black pepper
2 tablespoons fresh oregano, chopped
2 tablespoons fresh thyme, chopped
1 tablespoon fresh rosemary, chopped

STEPS

1. In a large pot, sauté chopped onion and garlic in olive oil.
2. Drain the beans and add to the pot. Add in tomato paste to blend. Sauté.
3. Add 2 quarts of stock and bring to a boil. Reduce to simmer and stir occasionally.
4. Add chopped plum tomatoes, bay leaves, red pepper flakes, and black pepper.
5. Simmer for 1 to 1-1/2 hours until beans are soft. Add herbs during last 30 minutes.
6. Remove 1 cup of beans and purée. Return to pot. Purée more beans to achieve desired thickness. Remove bay leaves.
7. Adjust flavor, if necessary, by adding more herbs or balsamic vinegar.

NUTRITIONAL ANALYSIS:

Kcalories	Protein (gm)	Fat (gm)	Carbo (gm)	Sodium (mg)	Chol (mg)
242	13	3	42	281	0

Recipe 8-7 *Butternut Squash Bisque*

Category: Soup   Yield: 24 portions

INGREDIENTS

1 ounce sweet butter
2 teaspoons shallots, chopped
1 cup celery, chopped
1 cup onion, chopped
2 teaspoons garlic, chopped
6 ounces rice
2 quarts chicken stock
1/2 teaspoon cinnamon
1/2 teaspoon nutmeg
1 bay leaf
Cinnamon, as needed for garnish

1 tablespoon fresh thyme
4 pounds butternut squash, cleaned and diced
Nonfat plain yogurt, as needed for garnish
Chives, as needed for garnish

STEPS

1. In soup pot, melt butter. Sauté shallots, celery, onion, garlic, and thyme.
2. Add butternut squash, rice, chicken stock, cinnamon, nutmeg, and bay leaf. Cook until rice is tender.
3. Purée with blender and strain.
4. Garnish with cinnamon, nonfat plain yogurt, and chives.

NUTRITIONAL ANALYSIS:

Kcalories	Protein (gm)	Fat (gm)	Carbo (gm)	Sodium (mg)	Chol (mg)
101	2	1	15	207	3

Salads and Dressings

Recipe 8-8 *Wild Mushroom Salad*

Category: Salad or Appetizer **Yield: 1 portion**

INGREDIENTS

Dressing
1 tablespoon sherry vinegar
3 twists black pepper
1 teaspoon Dijon mustard
1 teaspoon white wine
1/2 teaspoon shallots
1/4 teaspoon fresh rosemary, chopped
1 teaspoon apple juice
1/2 teaspoon fresh thyme, chopped

Salad
1 cup of greens: red oak leaf, frisée, mustard greens, mesclun mix
1/2 cup of mushrooms: shiitake mushrooms, oyster mushrooms, domestic mushrooms
1 tablespoon diced peppers
1 teaspoon fresh chives, chopped

STEPS

1. Incorporate all dressing ingredients together in a bowl.
2. Clean and wash lettuce, and arrange on plate.
3. Toss mushrooms and peppers in bowl with dressing. Sear on flat-top or hot sauté pan.
4. Place warm (or chilled, if preferred) mushrooms in center.
5. Top with fresh chives.

NUTRITIONAL ANALYSIS:

Kcalories	Protein (gm)	Fat (gm)	Carbo (gm)	Sodium (mg)	Chol (mg)
29	1	1	5	34	0

Recipe 8-9 *Baby Mixed Greens with Shaved Fennel and Orange Sections*

Category: Salad or Appetizer Yield: 10 servings

INGREDIENTS

14 cups mixed greens such as baby Bibb, baby romaine, frisée, lolla rosa, tatsoi
3 heads fresh fennel
5 fresh oranges
20 ounces sherry vinaigrette

STEPS

1. Prepare salad greens and reserve for service.
2. Cut fennel in half and shave paper-thin on a meat slicer. Place in ice water to crisp. Drain and reserve for service.
3. Dry some fennel slices in a low oven to use as garnish (about 30 minutes).
4. Section oranges.
5. In mixing bowl, toss fennel lightly in sherry vinaigrette.
6. Toss mixed greens with dressing. Use only 2 ounces of dressing per person.
7. Coat orange sections lightly with dressing.
8. Arrange orange sections on outside of plate. Make a tower of lettuce and fennel. In the center of the plate top with fennel slices that have been dried in the oven. Serve immediately.

NUTRITIONAL ANALYSIS:

Kcalories	Protein (gm)	Fat (gm)	Carbo (gm)	Sodium (mg)	Chol (mg)
63	3	0.5	13	35	0

Recipe 8-10 *Basic Herb Vinaigrette*

Category: Dressing Yield: Approximately 1 gallon, 2-ounce portion size

INGREDIENTS

9 cups chicken stock or vegetable stock
4 cups balsamic vinegar
1/2 ounce cornstarch slurry (cornstarch with cold water)
2 tablespoons fresh chopped garlic
2 teaspoons black pepper, coarsely ground
2 tablespoons fresh thyme, chopped
2 ounces fresh parsley, chopped
2 tablespoons, fresh oregano, chopped
1-1/2 ounces fresh basil, chopped
2 ounces chives, chopped
4 cups extra-virgin olive oil

STEPS

1. Use stock or a quality low-sodium vegetable base. Heat to a rolling boil.
2. Thicken with cornstarch slurry, to a nappe (to coat) consistency.

3. Cool, and add garlic, pepper, herbs, and vinegar.
4. Whisk olive oil into the mixture to emulsify.
5. Cool and store.

You can substitute different flavored vinegars or virgin nut oil to create alternative dressing options.

NUTRITIONAL ANALYSIS:

Kcalories	Protein (gm)	Fat (gm)	Carbo (gm)	Sodium (mg)	Chol (mg)
149	1	14	5	39	4

Recipe 8-11 *Orange Vinaigrette*

Category: Dressing **Yield: 10 2-ounce servings**

INGREDIENTS

1 pint orange juice
1 tablespoon Dijon mustard
1 tablespoon honey
1 tablespoon shallots, finely diced
1/2 teaspoon coriander, ground
1/2 teaspoon black pepper, ground
2 ounces white wine vinegar
2 ounces extra-virgin olive oil
1 teaspoon thyme, fresh, chopped
1 teaspoon chives, fresh, chopped
1 teaspoon basil, fresh, chopped

STEPS

1. In a saucepan, reduce orange juice to 1/2 cup, or use 1/2 cup concentrate.
2. Place cooled orange syrup in food processor. Add mustard, honey, shallots, coriander, black pepper, and white-wine vinegar.
3. Start food processor and slowly add oil to emulsify.
4. Add fresh herbs last, but do not purée.
5. Reserve in refrigerator until needed.

NUTRITIONAL ANALYSIS:

Kcalories	Protein (gm)	Fat (gm)	Carbo (gm)	Sodium (mg)	Chol (mg)
68	0.4	6	7	10	0

Recipe 8-12 *Sherry Wine Vinaigrette*

Category: Dressing **Yield: 10 1-1/2-ounce servings**

INGREDIENTS

1/2 cup sherry vinegar
10 twists fresh ground black pepper
3-1/3 tablespoons Dijon mustard
3-1/3 tablespoons white wine
1-2/3 tablespoons shallots
2-1/2 teaspoons rosemary, chopped
2 ounces apple juice
1-2/3 tablespoons fresh thyme, chopped
3-1/3 tablespoons chives, finely chopped
1/2 cup peppers, tricolored, diced

STEPS

1. Mix all ingredients together in a bowl. Hold for service. Lasts 1 week in refrigerator.

NUTRITIONAL ANALYSIS:

Kcalories	Protein (gm)	Fat (gm)	Carbo (gm)	Sodium (mg)	Chol (mg)
21	0.5	0.5	3	30	0

Recipe 8-13 *Ginger Lime Dressing*

Category: Dressing **Yield: 8 2-ounce servings**

INGREDIENTS

1 cup soy sauce, low sodium
1/2 cup lime juice
Grated rind from 2 limes
1/4 cup scallions, finely sliced
1/2 cup water
1 teaspoon garlic, chopped
1 teaspoon ginger, chopped

STEPS

1. Mix all ingredients together in a bowl. Let sit overnight. Lasts 2 weeks in refrigerator.

NUTRITIONAL ANALYSIS:

Kcalories	Protein (gm)	Fat (gm)	Carbo (gm)	Sodium (mg)	Chol (mg)
22	2	0	5	800	0

Recipe 8-14 *Green Goddess Dressing*

Category: Dressing **Yield: 6 portions**

INGREDIENTS

4 ounces tofu (firm), well drained
1/2 cup cider vinegar
2 stalks celery
1/2 cup spinach leaves, washed and dried
1/3 cup fresh parsley (no stems)
1 tablespoon lemon juice
2 scallions
Fresh tarragon leaves
Fresh ground pepper

STEPS

1. Blend together in food processor until smooth. Use just enough tarragon and pepper to taste. This dressing will last 2 days in the refrigerator.

NUTRITIONAL ANALYSIS:

Kcalories	Protein (gm)	Fat (gm)	Carbo (gm)	Sodium (mg)	Chol (mg)
23	2	1	3	19	0

Recipe 8-15 *Capistrano Spice Rub*

Category: Rub **Yield:** 5 cups

INGREDIENTS

1 cup dried oregano
1 cup dried basil
1/2 cup dried thyme
1 cup fresh rosemary, chopped
1/2 cup black pepper, butcher's grind
1/2 cup garlic powder

STEPS

1. Blend all ingredients together. Store in an airtight container to preserve freshness.
2. Coat meat, fish, or poultry generously with spice rub prior to cooking. Can also be used to spice up soups, dressings, and vegetable dishes.

Entrées

Recipe 8-16 *Grilled Pork Chop Adobo with Spicy Apple Chutney*

Category: Entrée **Yield:** 20 portions

INGREDIENTS

Adobo Spice Rub
3 fresh green chiles (poblano)
2 fresh jalapeño peppers
10 garlic cloves
3 tablespoons fresh oregano
2 tablespoons ground cumin
2 tablespoons freshly ground black pepper
1 teaspoon ground cinnamon
1/2 pound tomatillos, husks removed
1 cup red wine vinegar

Pork
5 pounds pork loin, center cut, boneless

Chutney
1 chipotle pepper
1 onion, diced
1/2 tablespoon garlic, chopped
8 apples, cored and diced
1/2 cup raisins
1 ounce lemon juice
zest from 1 lemon
1 tablespoon ground cardamon
3 tablespoons sugar
1 cup cider vinegar
1/2 tablespoon ground fennel seed
1/4 teaspoon mace
1/2 bunch fresh cilantro, chopped

STEPS

1. Combine all the adobo spice rub ingredients in a food processor and blend until a paste forms.
2. Generously rub the pork loin with the paste and let marinate overnight in refrigerator.
3. Rehydrate the chipotle pepper in hot water until softened. Cut in half, remove seeds, and chop.

4. In a sauté pot, combine all chutney ingredients except cilantro.
5. Let simmer for 20 minutes until slightly thickened. Finish with cilantro. Cool and serve at room temperature. (The chutney can be made in advance and refrigerated.)
6. Sear pork loin until nicely caramelized. Set up smoking station with wood chips soaked in water. Smoke with medium smoke for about 15 minutes.
7. Remove and finish in a slow-roasting oven at 300°F until pork temperature is 155°F.
8. Let rest 15 minutes before slicing. Serve with apple chutney.

NUTRITIONAL ANALYSIS:

Kcalories	Protein (gm)	Fat (gm)	Carbo (gm)	Sodium (mg)	Chol (mg)
144	17	3	12	34	48

Recipe 8-17 *Potato Lasagne*

Category: Entrée (or Appetizer) **Yield: 6 servings**

INGREDIENTS

16 ounces skim-milk ricotta
1/4 teaspoon fresh garlic, chopped
1/2 teaspoon fresh basil, chopped
1 teaspoon fresh parsley, chopped
1/4 teaspoon oregano leaves, dried
1/8 teaspoon black pepper, freshly ground
3 large Idaho potatoes, peeled
3 to 4 cups tomato sauce
8 ounces skim-milk mozzarella
Fresh basil leaves, for garnish

STEPS

1. Combine ricotta cheese, garlic, basil, parsley, oregano, and black pepper.
2. Coat baking dish with nonstick vegetable-oil spray.
3. Cut potatoes lengthwise into 1/8-inch slices.
4. Heat oven to 350°F.
5. To assemble lasagne, spread about 1/3 of the potatoes evenly over bottom of baking dish. Top with 1/2 of the ricotta cheese mixture, 1/3 of the tomato sauce, and 1/2 of the mozzarella cheese. Repeat the layers. Place remaining potatoes evenly over top, and spoon on remaining tomato sauce.
6. Cover with foil. Bake 1 hour or until potatoes are tender.
7. Uncover and allow to cool 10 minutes before serving.
8. Cut lasagne into 6 servings. Garnish each serving with basil leaves.

This recipe can be made a day in advance.

NUTRITIONAL ANALYSIS:

Kcalories	Protein (gm)	Fat (gm)	Carbo (gm)	Sodium (mg)	Chol (mg)
321	21	12	34	460	45

Recipe 8-18 *Chicken Sausage*

Category: Entrée Yield: 5 servings

INGREDIENTS

2-1/2 ounces wheat bread, diced
2 egg whites
2 ounces nonfat plain yogurt
1 pound chicken breast, cleaned and diced
Freshly ground black pepper to taste
1 teaspoon fresh oregano, chopped
1/2 teaspoon fresh thyme, chopped
1 teaspoon fresh basil, chopped
2 teaspoons roasted garlic, sliced
2 tablespoons dried tomatoes, diced

STEPS

1. In a bowl, soak wheat bread with egg whites and nonfat yogurt to make *panada*.
2. In a food processor, pulse chicken meat to purée. Add panada, pepper, and herbs. Pulse to a smooth consistency, adding about 2 tablespoons of crushed ice to keep temperature down.
3. Fold in roasted garlic and tomatoes.
4. Pipe into sausage casing or plastic wrap. Tie and poach in 160 to 170°F water until an internal temperature of 160°F is reached. Shock in ice water. Cool and unwrap for service.
5. Grill to reheat and finish cooking.
6. Serve with greens, tomato salad, roasted onions, German-style potato salad, or grilled vegetables.

NUTRITIONAL ANALYSIS:

Kcalories	Protein (gm)	Fat (gm)	Carbo (gm)	Sodium (mg)	Chol (mg)
208	32	4	9	372	77

Recipe 8-19 *Slates of Salmon*

Category: Entrée Yield: 1 serving

INGREDIENTS

4 ounces salmon steak, cut very thin on bias
1 teaspoon olive oil
Pinch black pepper
1 cup arugula
1/2 cup endive
1 tablespoon plum tomatoes, seeded and diced
1 tablespoon cucumbers, peeled, seeded, and diced
2 ounces Green Goddess Dressing (Recipe 8-14)

STEPS

1. Paint the salmon with olive oil and black pepper. Grill to desired temperature.

2. Toss greens with Green Goddess Dressing and place in middle of plate. Place salmon against greens and sprinkle with tomato and cucumbers.

NUTRITIONAL ANALYSIS:

Kcalories	Protein (gm)	Fat (gm)	Carbo (gm)	Sodium (mg)	Chol (mg)
187	24	9	3	89	59

Recipe 8-20 *Braised Lamb*

Category: Entrée **Yield:** 12 portions

INGREDIENTS

12 lamb hind shanks
1 tablespoon olive oil
4 tablespoons vegetable seasoning
5 cloves garlic, sliced
2 onions, diced
4 carrots, sliced
6 celery stalks, diced
3 bay leaves
2 whole thyme sprigs
2 whole rosemary sprigs
8 ounces red wine
8 ounces fresh tomatoes
1/2 cup tomato paste
1 gallon lamb stock, defatted
3 tablespoons Worcestershire sauce
1 tablespoon cracked black pepper
36 large shiitake mushroom caps
24 ounces cannellini, cooked
3 tablespoons garlic herb seasoning

STEPS

1. Trim all fat from lamb shanks.
2. Place olive oil into large heated braising pan. Season shanks with vegetable seasoning. Sear in the large pot until brown. Remove shanks.
3. Quickly sauté garlic, onions, carrots, celery, bay leaves, thyme, and rosemary in same pot. Deglaze with red wine, fresh tomatoes, and tomato paste, and reduce.
4. Return shanks to pot with defatted lamb stock, Worcestershire sauce, and pepper. Bring to boil, reduce heat, and simmer gently for 90 minutes. Take out shanks. Strain and put sauce through a food mill. Reduce to proper consistency and skim. Pour over shanks. Cool.
5. For each order, cook 3 shiitake caps with 2 ounces cannellini beans in 3 ounces of lamb stock and 1 teaspoon garlic herb seasoning. Cook lightly to reduce. Heat each shank in sauce slowly.
6. Place 1 piece of shank on plate with vegetables and sauce. Top with shiitakes and beans.

This dish can be served with a variety of vegetables, beans, and grains.

NUTRITIONAL ANALYSIS:

Kcalories	Protein (gm)	Fat (gm)	Carbo (gm)	Sodium (mg)	Chol (mg)
391	40	10	33	471	107

Relishes, Salsas, Coulis, Chutneys, and Sauces

Recipe 8-21 *Papaya and White Raisin Chutney*

Category: Relishes, Salsas, Coulis, and Chutneys **Yield:** 16 servings

INGREDIENTS

6 pounds very ripe papaya, diced
2 cups onion, diced
1 tablespoon garlic, chopped
1/2 cup brown sugar
1/2 cup white sugar
1 cup raisins, seedless
1 cup white wine vinegar
1 teaspoon cardamom
1 teaspoon cinnamon
2 bay leaves
2 teaspoons fresh thyme, chopped

STEPS

1. In a small saucepot, mix all ingredients except thyme.
2. Reduce to a thick paste. Add thyme while chutney is still hot.
3. Let cool; remove bay leaves. Can be stored up to 2 weeks.

NUTRITIONAL ANALYSIS:

Kcalories	Protein (gm)	Fat (gm)	Carbo (gm)	Sodium (mg)	Chol (mg)
126	1	0	31	8	0

Recipe 8-22 *Papaya–Plantain Salsa*

Category: Relishes, Salsas, Coulis, and Chutneys **Yield:** 10 servings

INGREDIENTS

1 plantain, ripe, finely diced
1 teaspoon extra-virgin olive oil
1 papaya, finely diced
1/2 peppers, red and orange, finely diced
1/2 red onion, finely diced
2 teaspoons cilantro, chopped
2 teaspoons chives, finely sliced
1 cup white wine vinegar
2 teaspoons honey
Lime juice from 2 limes
5 teaspoons extra-virgin olive oil

STEPS

1. In a nonstick sauté pan, toast the diced plantain in 1 teaspoon of extra-virgin olive oil until crisp outside and tender inside.
2. In a stainless-steel mixing bowl, place half the papaya, peppers, and red onion. Add the toasted plantain, cilantro, and chives. Reserve.
3. In a food processor, place the remaining papaya, vinegar, honey, and lime juice. Purée until smooth, adding 5 teaspoons olive oil.
4. Add more vinegar if too thick.
5. Strain through a fine sieve into the reserved plantain-papaya mixture.
6. Reserve in refrigerator for use. Lasts about 5 days.

NUTRITIONAL ANALYSIS:

Kcalories	Protein (gm)	Fat (gm)	Carbo (gm)	Sodium (mg)	Chol (mg)
75	0.6	3	11	4	0

Recipe 8-23 *Red Pepper Coulis*

Category: Relishes, Salsas, Coulis, and Chutneys Yield: 36 ounces, or 18 2-ounce servings

INGREDIENTS

4 pounds red peppers (or substitute other vegetables)
1 ounce minced shallots
1 tablespoon minced garlic
1/2 teaspoon minced jalapeño pepper
2 ounces olive oil
2 ounces tomato paste
2 90-count potatoes, peeled and diced
18 ounces chicken or vegetable stock
2 tablespoons fresh basil, chopped
2 teaspoons fresh thyme, chopped
2 teaspoons fresh oregano, chopped
1-3/4 ounces balsamic vinegar

STEPS

1. Cut peppers in half and remove seeds. Place on oiled sheet pans and roast in hot oven, or grill the peppers. Weigh out 3-1/4 pounds of grilled red peppers.
2. Sauté shallots, garlic, and peppers in oil.
3. Add tomato paste and sauté. Do not brown.
4. Add red peppers, potatoes, and stock. Simmer until peppers and potatoes are tender.
5. Add basil, thyme, and oregano. Cook 5 more minutes.
6. Purée in a blender.
7. Finish with vinegar and strain through large-hole china cap.

NUTRITIONAL ANALYSIS:

Kcalories	Protein (gm)	Fat (gm)	Carbo (gm)	Sodium (mg)	Chol (mg)
67	2	3	8	26	0

Recipe 8-24 *Hot-and-Sour Sauce*

Category: Sauce Yield: 16 2-ounce portions

INGREDIENTS

24 ounces chicken stock, defatted
2 ounces sesame oil
4 ounces white wine
3 ounces rice wine vinegar
1-1/2 tablespoons cornstarch
1 tablespoon lime juice, fresh-squeezed
1/3 teaspoon lime rind
2 ounces scallions, chopped
2 teaspoons ginger, chopped
1/2 teaspoon jalapeño pepper
1/4 teaspoon cumin
1 teaspoon cilantro

STEPS

1. Heat chicken stock and wine. Thicken with cornstarch slurry made with vinegar.
2. Take off heat and cool. Add remaining ingredients except oil. Whip in oil at the end. Refrigerate for later service.

NUTRITIONAL ANALYSIS:

Kcalories	Protein (gm)	Fat (gm)	Carbo (gm)	Sodium (mg)	Chol (mg)
54	1	4	3	101	5

Side Dishes

Recipe 8-25 *Roasted Summer Vegetables*

Category: Side Dish **Yield:** 10 servings

INGREDIENTS

2 pounds carrots, peeled and cut oblique
1 pound red bell peppers
1 tablespoon olive oil
2 tablespoons shallots, chopped
4 cloves garlic, chopped
1 pound corn, cleaned
1-1/2 pounds assorted radishes, sliced 1/4-inch thick
1/2 pound pattypan squash
1/2 tablespoon thyme
2 bay leaves
Black pepper to taste
4 ounces chicken stock
1 tablespoon basil leaves, chopped
1/2 tablespoon rub of your choice

STEPS

1. Blanch carrots for 2 minutes.
2. Cut peppers in half and remove seeds. Place on oiled sheet pans and roast in hot oven or grill the peppers. Remove skins.
3. Mix all ingredients except basil and rub together and place in hot roasting pan in a 350°F oven. Let roast, stirring occasionally, for 30 minutes. Add more stock if needed to prevent burning. If vegetables are getting too brown, cover with foil.
4. Add basil and rub at end for more flavoring.

NUTRITIONAL ANALYSIS:

Kcalories	Protein (gm)	Fat (gm)	Carbo (gm)	Sodium (mg)	Chol (mg)
123	3	2	26	52	0

Recipe 8-26 *Mixed-Grain Pilaf*

Category: Side Dish **Yield:** 10 servings

INGREDIENTS

1 cup wheat berries
2 cups water
2 thyme stems
1 bay leaf
1 cup quinoa
1 cup hot water
1 bay leaf
1/2 ounce chicken stock, defatted
1 ounce onions, diced
3 ounces brown rice
1 pint chicken stock, defatted
1 bay leaf
1 head roasted garlic
2 teaspoons olive oil
2 ounces onions, chopped
2 tablespoons chopped basil, oregano, and chives
Chicken stock, as needed

STEPS

1. In a sauce pot of boiling water, add wheat berries with thyme stems and bay leaf and cook for about 1 hour. Drain and cool.
2. Soak quinoa in cold water to take out any impurities. Steam quinoa in water with bay leaf. Cover and let steam for about 20 minutes.
3. Heat 1/2 ounce of chicken stock in a stockpot. Add onions and sweat until they are translucent.
4. Add rice, 1 pint stock, thyme, and bay leaf. Simmer and cover pilaf style in a 350°F oven for about 20 minutes, or until done.
5. Purée roasted garlic.
6. Heat oil in stockpot. Sweat onions until translucent. Add puréed garlic, wheat berries, quinoa, brown rice, and herbs.
7. Let simmer for 5 minutes. Adjust consistency with stock if needed.

NUTRITIONAL ANALYSIS:

Kcalories	Protein (gm)	Fat (gm)	Carbo (gm)	Sodium (mg)	Chol (mg)
211	7	3	39	31	0

Recipe 8-27 *Sautéed Cabbage*

Category: Side Dish **Yield: 8 portions**

INGREDIENTS

32 ounces shredded cabbage
1 quart chicken stock
1 quart fish stock
2 tablespoons ground caraway seeds
1 tablespoon juniper berries
1 tablespoon peppercorns
2 sprigs fresh thyme
3 bay leaves
2 ounces Dijon mustard
3 ounces sugar
1/2 cup white wine vinegar
4 twists fresh black pepper

STEPS

1. In saucepan, mix cabbage, stocks, and ground caraway seeds.
2. Place juniper berries, peppercorns, thyme, and bay leaves in a sachet bag. Place in saucepan with cabbage.

3. Mix mustard, sugar, vinegar, and pepper. Add to cabbage. Simmer over low heat until tender, about 15 to 20 minutes.

NUTRITIONAL ANALYSIS:

Kcalories	Protein (gm)	Fat (gm)	Carbo (gm)	Sodium (mg)	Chol (mg)
79	99	2	7	271	25

Desserts

Recipe 8-28 *Warm Chocolate Pudding Cake with Almond Cookie and Raspberry Sauce*

Category: Dessert   Yield: 10 portions

INGREDIENTS

1 tablespoon orange zest
3 cups skim milk
3-1/2 ounces sugar
1 ounce cocoa powder
3 ounces cornmeal
10 egg whites
3-3/4 ounces sugar
3 ounces bittersweet chocolate
1 pint raspberries
1 ounce kirschwasser
4 ounces white wine
2-3/4 ounces honey
1-1/2 ounces sugar
1 ounce almond paste
1/2 ounce bread flour
1/4 teaspoon cinnamon
1 egg white
Pinch salt
2 teaspoons cream
2 teaspoons skim milk
Confectioners' sugar, about 1 tablespoon for dusting cookies

STEPS

1. Steep orange zest in milk and bring to a boil. Simmer.
2. Combine sugar, cocoa, and cornmeal. Pour in steady stream into simmering milk. Stir until thick and cornmeal is cooked. Allow to cool.
3. Whip egg whites and 3-3/4 ounces sugar. Fold into base. Then fold in bittersweet chocolate.
4. Pour mixture into sugared molds and bake in bain-marie about 20 to 24 minutes at 400°F. Keep warm.
5. Purée 3/4 pint raspberries (reserve others for garnish) in food processor with the kirschwasser, wine, and honey. Strain and reserve.
6. Cream sugar, almond paste, flour, and cinnamon. Add remaining ingredients, except reserved raspberries, reserved raspberry sauce, and confectioners' sugar, and allow to rest. Spread paste in 10 3-inch circles on silicon. Bake in 350°F oven until edges are brown. Curve the circles on a rolling pin while still hot to make a decorative tuile garnish.
7. Cool and dust with confectioners' sugar.

8. Pool sauce on plate. Unmold cake on sauce. Place cookie on top of cake and garnish with raspberries.

NUTRITIONAL ANALYSIS:

Kcalories	Protein (gm)	Fat (gm)	Carbo (gm)	Sodium (mg)	Chol (mg)
235	9	2	45	106	3

Recipe 8-29 *Raspberry Creamed Ice*

Category: Dessert **Yield: 1 quart or 4 1-cup servings**

INGREDIENTS

1 pound soft tofu
1 pint raspberry purée (or substitute any other fruit)
6 ounces egg whites
1 ounce turbinado sugar

STEPS

1. Cream tofu in processor until smooth.
2. Add raspberry purée slowly to achieve creamy texture.
3. Whip egg whites until they form soft peaks. Add sugar until the meringue forms peaks.
4. Churn in an ice-cream machine until the mixture reaches the desired consistency.
5. Freeze in an airtight container.

NUTRITIONAL ANALYSIS:

Kcalories	Protein (gm)	Fat (gm)	Carbo (gm)	Sodium (mg)	Chol (mg)
148	12	5	17	79	0

Recipe 8-30 *Angel Food Savarin*

Category: Dessert **Yield: 8 4-inch savarin rings**

INGREDIENTS

2 ounces cake flour
6 ounces 10X sugar
6 ounces egg whites
1 teaspoon cream of tartar
1/2 teaspoon lemon rind

STEPS

1. Sift flour with 3 ounces of sugar. Repeat, and set aside.
2. Whip egg whites and cream of tartar to soft peaks, then gradually add remaining 3 ounces of sugar. Whip until stiff and glossy.
3. Gently fold in sifted ingredients and lemon rind.
4. Pipe into savarin molds that have been sprayed with cold water.
5. Bake in 350°F oven until light golden brown on top.
6. Cool completely, then remove from mold.

NUTRITIONAL ANALYSIS:

Kcalories	Protein (gm)	Fat (gm)	Carbo (gm)	Sodium (mg)	Chol (mg)
119	3	0	27	35	0

Recipe 8-31 *Fresh Berry Phyllo Cones*

Category: Dessert **Yield: 8 portions**

INGREDIENTS

Phyllo

Vegetable oil cooking spray

8 phyllo sheets

8 teaspoons melted butter

Bavarian Mix

7 ounces maple syrup

9 ounces skim-milk ricotta cheese

13 ounces nonfat plain yogurt

1 teaspoon vanilla extract

2 teaspoons gelatin

4 teaspoons water

Fruit and Garnish

4 cups fresh berries (strawberries, blueberries, raspberries)

Powdered sugar, for dusting

8 each mint leaves

STEPS

1. Make 2-1/2-inch circle cones out of aluminum foil. (Wrap foil around a small juice or soup can.)
2. Spray foil lightly with vegetable-oil spray.
3. Cut phyllo into 2-1/2-inch widths. Paint with melted butter between layers. Use 1 sheet per cone.
4. Wrap phyllo around foil and take the cone off the can.
5. Place cone on baking sheet. Repeat process 8 times. These can be made one day in advance.
6. Bake phyllo cones at 375°F for about 10 minutes. Let cool. Take off foil and reserve it for future use.
7. Mix maple syrup, ricotta cheese, yogurt, and vanilla in blender. Whip until smooth.
8. Soften gelatin in water. Warm to dissolve.
9. Add a little of the dessert base to the dissolved gelatin to temper the mixture. Mix the rest of the dessert base into tempered mixture. Fold in 2 cups of fresh berries. Reserve other fruit for garnish.
10. Place a cone on a dessert dish. Scoop 3 ounces of the Bavarian mixture into the cone.
11. Top with fresh berries. Dust with powdered sugar. Garnish with mint leaf.

NUTRITIONAL ANALYSIS:

Kcalories	Protein (gm)	Fat (gm)	Carbo (gm)	Sodium (mg)	Chol (mg)
262	9	8	39	212	22

Recipe 8-32 *Heart-Healthy Oatmeal Raisin Cookies*

Category: Dessert Yield: 40 1-ounce cookies

INGREDIENTS

1 pound brown sugar
14 ounces applesauce
1/2 ounce canola oil
1 ounce egg whites
12 ounces all-purpose flour
7 ounces oatmeal
1 tablespoon baking soda
2-1/2 cups raisins
1 teaspoon cinnamon

STEPS

1. Mix together the sugar, applesauce, oil, and egg whites until smooth.
2. Combine the flour, oatmeal, baking soda, raisins, and cinnamon and add to the applesauce mixture. Mix until well blended.
3. Spoon onto a parchment-lined sheet pan and bake at 325°F until golden brown.

NUTRITIONAL ANALYSIS:

Kcalories	Protein (gm)	Fat (gm)	Carbo (gm)	Sodium (mg)	Chol (mg)
130	2	1	30	142	0

Breakfasts

Recipe 8-33 *Crunchy Cinnamon Granola*

Category: Breakfast Yield: About 6 cups, or 12 1/2-cup servings

INGREDIENTS

2 cups rolled oats
1 cup wheat flakes
1 cup wheat germ
1/4 cup unsalted sunflower seeds
1 tablespoon cinnamon
3/4 cup apple juice
1/2 cup prune juice
1 cup assorted dried fruit (such as raisins, apricots, dates, figs, pears)
1 tablespoon honey
1/2 cup skim milk

STEPS

1. Mix all dry ingredients together in a large mixing bowl.
2. In a saucepan, place juices over a medium heat and reduce by one-third.
3. Add dried fruits to hot liquid and cook slowly for one minute.
4. Pour fruits and juices over dry mixture. Add honey and skim milk, and toss to moisten oats.

5. Pour onto a vegetable-oil-sprayed cookie sheet with sides. Place in 325°F oven to toast, and stir to keep oats from burning and to brown the granola evenly. This takes about 30 minutes.
6. Let cool and store in airtight container. You can grind in the food processor after cooling to get a fine-textured cereal. Serve at breakfast with sliced fresh fruit and nonfat yogurt.

NUTRITIONAL ANALYSIS:

Kcalories	Protein (gm)	Fat (gm)	Carbo (gm)	Sodium (mg)	Chol (mg)
169	6	3	32	31	0

Recipe 8-34 *Stuffed French Toast Layered with Light Cream Cheese and Bananas*

Category: Breakfast **Yield: 4 servings**

INGREDIENTS

8 slices whole-wheat bread
2 ounces light cream cheese
4 small bananas
Batter
1-1/2 cups egg substitute
3/4 cup skim milk
1 teaspoon vanilla
1/2 teaspoon cinnamon
Syrup
1 cup fresh strawberries, cleaned and cut
1/4 cup strawberry all-fruit jam
1 tablespoon lemon juice

STEPS

1. Lay out bread and spread all 8 slices with cream cheese evenly.
2. Slice bananas paper-thin and layer on 4 slices of bread, overlapping slightly, and top each slice with other side of bread; press down lightly.
3. Whip together batter ingredients to a smooth consistency.
4. Heat a nonstick pan and spray with vegetable-oil spray.
5. Dip the banana sandwiches carefully in batter on both sides to absorb batter. Place in nonstick pan and brown nicely on both sides, about 1-1/2 to 2 minutes per side. Do not let the pan get too hot, or the toast will brown without cooking in the middle. Transfer to an oven-safe dish.
6. Warm slightly in the oven to crisp before serving.
7. While the French toast is in the oven, blenderize strawberries, jam, and lemon juice until smooth. Serve as syrup on the side.

NUTRITIONAL ANALYSIS:

Kcalories	Protein (gm)	Fat (gm)	Carbo (gm)	Sodium (mg)	Chol (mg)
526	23	11	86	552	10

Recipe 8-35 *Glazed Grapefruit and Orange Slices with Maple Vanilla Sauce*

Category: Breakfast Yield: 4 portions

INGREDIENTS

2 grapefruit, peeled and sliced
3 oranges, peeled and sliced
1 ounce maple syrup
2 ounces apple juice
1 ounce water
1 tablespoon granola
4 sprigs mint

STEPS

1. Alternate orange slices and grapefruit slices on individual plates like a wheel, until you have 4 portions.
2. Heat maple syrup, apple juice, and water. Reduce glaze by one half. Pour lightly over fruits. Top with granola and very thin strips of mint.

NUTRITIONAL ANALYSIS:

Kcalories	Protein (gm)	Fat (gm)	Carbo (gm)	Sodium (mg)	Chol (mg)
124	2	1	30	2	0

Chapter 9

Marketing Healthy Menu Options

Marketing nutrition has become increasingly common as interest has increased. According to a National Restaurant Association report, more than half of consumers 35 and older look for lower-fat menu options when eating out. Also, two out of every five 18- to 34-year-olds are doing the same. Restaurateurs report that their customers are increasingly requesting meatless meals. Add to this discussion the number of customers, particularly seniors over 65 years of age, who make requests such as no salt or no butter.

Restaurants offer much more to nutrition-conscious customers than the old-fashioned diet plate consisting of cottage cheese with fruit on top of a lettuce leaf. Menus now carry items ranging from low-fat, low-kcalorie tostadas to full-course meals that do it all for under 600 kcalories. Restaurants market their nutritionally modified dishes on menus and table tents, by radio advertising, and in other ways. In 1999, 91 percent of surveyed noncommercial foodservices promoted nutritious items on menus, menu boards, and newsletters and by other methods.

Marketing—The process of finding out what your customers need and want, and then developing, promoting, and selling the products and services they desire.

Marketing means finding out what your customers need and want, and then developing, promoting, and selling the products and services they desire. Keeping in touch with your customers is crucial, and can be done without much fuss, by talking with customers, finding out from waitstaff what customers have been requesting, and doing a survey.

This chapter will help you to:

- Describe two methods a foodservice operator can use to gauge customers' needs and wants
- Give three examples of how to draw attention to healthy menu options
- Discuss effective ways to communicate and promote a nutrition menu program
- Explain the importance and extent of staff training needed to successfully implement a healthy menu program
- Describe two methods used to evaluate healthy menu options
- Discuss how nutrition labeling laws regulate nutrient content or health claims on restaurant menus

Gauging Customers' Needs and Wants

Most foodservice operators who have successfully implemented healthy menu options have done so through reviewing eating trends, examining what other operators are doing, and keeping abreast of their customers' requests for healthy foods. To determine customer wants, foodservice operators could interview the waitstaff about customer requests, for example, for light foods such as broiled meat, poultry, or fish; dishes prepared without salt; sauces and gravies removed or served on the side; butter substitutes; reduced-calorie salad dressing; or skim or low-fat milk.

Another way to gauge customer needs is to do a survey, as shown in Figure 9-1. At the same time, answers to the following questions need to be considered.

Figure 9-1
Customer survey

1. How often do you visit this restaurant?
   - First visit
   - Once or twice a year
   - Once every three months
   - Once every two months
   - Once a month
   - Two or three times a month
   - Once a week
   - More than once a week
2. Today I came for:
   - Breakfast
   - Lunch
   - Dinner
   - Snack
3. Are you here during your workday?
   Yes No
4. Are you here for social reasons?
   Yes No
5. Have you ever been to a restaurant that offers light and nutritious menu choices?
   Yes No
6. Would you order light and nutritious foods if they were offered here?
   - Yes, frequently
   - Yes, sometimes
   - No
   - Not sure
7. How likely would you be to try the following nutritious menu choices?

Menu Choice	Very Likely	Likely	Unlikely
A. Broiled fish without butter			
B. Reduced-calorie salad dressing			
C. Vegetables with no added salt			

1. What are the majority of requests made during a particular meal?
2. Which items are most frequently requested?
3. How much time does your cooking staff and waitstaff have available to meet these special requests?
4. Which requests are easy to meet? Which are very time-consuming?

Answers to these questions can help you decide which types of healthy menu items to offer.

If market research demonstrates a sizable need for healthy entrées and the like, and there is enough time and staff to commit to this project, then now may be the time to do more than meet customers' special requests.

■ MINI-SUMMARY

Customer interest in nutritious menu items can be gauged through waitstaff feedback and customer surveys/feedback.

Developing and Implementing Healthy Menu Options

Various personnel are normally involved in the development and implementation phase: foodservice operators, directors, and managers; chefs and cooking staff; and nutrition experts such as registered dietitians. Chefs and cooking staff are valuable resources in modifying recipes or creating new ones, and may be given much of this responsibility. Nutrition experts are needed to provide accurate nutrient analysis data as well as suggestions for modifying dishes. In larger companies, personnel responsible for training, advertising and publicity, marketing, menu planning, and recipe development may also be involved.

Chapter 8 covered the basics of developing healthy menu items and modifying recipes. Once you know what you want to offer, you need to think about how to inform your clientele of these options. Here are some suggestions.

1. Simply give a good description of your menu item so that nutrition-conscious customers can see that the item is healthy.
2. Use the waitstaff to offer and describe nutritious menu options. In some instances, healthier preparation methods can be suggested for regular menu items.
3. Highlight nutritious menu selections with symbols or words such as "light." For example, put a picture of wheat next to nutrition selections that meet specific nutrition goals, usually described at the bottom of the menu (Figure 9-2).
4. Include a special, separate section on the regular menu. With this format, customers are certain to see the nutritious options, and see them as being integrated into the foodservice concept. A heading for this section might be "Fit Fare."
5. Add a clip-on to the regular menu and/or a blackboard or lightboard. This method requires no alterations to the menu and is particularly useful in that it is flexible and inexpensive. Healthy selections can be changed without involving much time or money. In some operations, treating the nutrition selections like daily specials has increased their selling power.

No matter which method you use to include healthy selections on the menu, you must consider how thorough a description is appropriate. In general, customers do not want kcalorie counts, fat, cholesterol, or sodium content on the menu, but prefer simply a good description of the ingredients, portion size, and preparation method. Menu items are more effectively promoted by giving customers this information and by emphasizing quality and variety rather than nutrition.

Figure 9-2
St. Andrew's Café menu. Courtesy the Culinary Institute of America, Hyde Park, NY

St. Andrew's Dinner

Soups

Louisiana Chicken and Shrimp Soup 4.

Roasted Butternut Squash Soup with Spiced Crème Fraîche 4.

Lentil Soup with Sausage and Spinach 4.

St. Andrew's Soup Sampler 4.

****

Salads

St. Andrew's Garden Salad 5.
Goat Cheese, Hazelnut Crisp and Citrus Vinaigrette

Arugula Salad with Toasted Walnuts and Pear Vinaigrette 6.
Champagne and Pear Vinaigrette

****

Starters

Wood Fired Sea Bass 6.
Tomato-Fennel Ragut

Steamed Wontons 6.
Shrimp, Brown Rice and Soy Dipping Sauce

Risotto Special of the Day 6.

Seared Tuna Loin 6.
Wakame Seaweed Salad and Miso Sauce

Indicates Vegetarian Selections

St. Andrew's Dinner

Main Courses

Grilled Tuna with Soba Noodles 18.
Asian Scented Vegetables

Tea Cured Pork Loin 16.
Sweet Onion Jam and Dried Morello Cherry Sauce

Pan-seared Sea Scallops 18.
Edamame Soy Beans, Shiitake Mushrooms and Orange-Soy Glaze

Udon Noodles with Tofu Soy-Ginger Broth 16.
with Cashews

Grilled Beef Tenderloin, Steamed Wild-Pecan Rice 21.
Wild Mushroom and Burgundy Wine Glaze

Marinated-Grilled Chicken Breast 16.
Spicy Apple Chutney, Whole Wheat Spaetzle and Mustard Jus

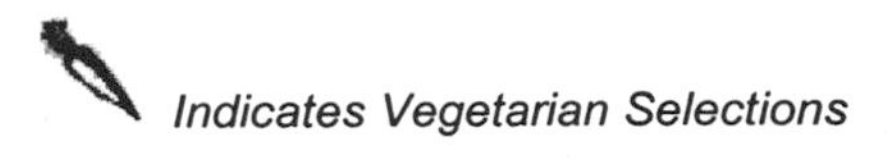
Indicates Vegetarian Selections

Michael Garnero, Chef-Instructor
Carleen von Eikh, Maître d'Hôtel Instructor

The St. Andrew's Cafe is a non-smoking restaurant. Thank you for your cooperation.

VISIT OUR OTHER RESTAURANTS:
the American Bounty Restaurant, the Caterina de Medici Restaurant & the Escoffier Restaurant

A 12% service charge has been added to your check. It is used to fund student scholarships, purchase graduation jackets and support student activities. Tips in excess of 12 % will go directly to the students serving you. The service charge amount reflected on your check is not mandatory. If you are not satisfied with your dining experience, the charge may be adjusted. If you have any questions, please see the Maître d'Hôtel Instructor. Thank you.

Figure 9-2 *(continued)*

Figure 9-2 *(continued)*

St. Andrew's Dinner

Desserts

Chocolate Mousse Cake with Raspberry Coulis and Whipped Cream 5.00
with a glass of Fonseca Tawny Porto 7.50

Sorbet of the Day with Oatmeal Crisp 5.00
with a glass of Nivole Moscato d'Asti 8.00

Panna Cotta with Strawberries and Aged Balsamic Vinaigrette 5.00
with a glass of "Electra" Orange Muscat 7.50

Strawberry-Rhubarb Crisp with Ricotta Glaze
and Caramel Sauce 5.00
with a glass of "Elysium" Black Muscat 8.50

Dessert Sampler 7.00
with a glass of Nivole Moscato d'Asti 10.00

****

Dessert Beverages

	glass
Moscato d'Asti "Nivole"	3.50
"Electra" Orange Muscat	3.00
"Elysium" Black Muscat	4.00
Harvey's Bristol Cream Sherry	4.00
Fonseca Tawny Porto	3.00

K K K

Millstone Coffee	1.75
Decaffeinated Coffee	1.75
Serendipitea Tea and Herbal Infusions	2.75
~ from our cart ~	
Illy Cafe Espresso	2.50
Illy Cafe Cappuccino	3.00
Cappuccino of the Day made with Torani Syrups	3.50

Marketing healthy menu items can be done in a positive manner. When you tout foods as "heart-healthy," you may be approaching some customers in a negative manner. When you market freshly squeezed fruit juices, you are approaching customers in a positive manner.

Promotion

Advertising—Any paid form (such as over radio) of calling public attention to goods, services, or ideas of a company or sponsor.

Three methods of promoting a nutrition program are advertising, sales promotion, and publicity. **Advertising** can be done through magazines and newspapers, radio and television, outdoor displays (posters and signs), indoor table tents and posters, direct mail, and novelties (such as matchboxes). Direct mail works well when targeted to current customers.

Advertising messages should say something desirable, beneficial, distinctive, and believable about the nutritious dining program. For example, the new menu selections could be advertised as healthy and using only the freshest, most exotic ingredients. Because foodservice operators need to get the best advertising for the money, hiring a reputable advertising company may be the best option.

Sales promotion—Marketing activities other than advertising and public relations that offer an extra incentive.

Publicity—Obtaining free space or time in various media to get public notice of a program, book, and so on.

Sales promotions can include coupons, point-of-purchase displays (such as a blackboard at the dining room entrance listing the nutrition selections), or contests (such as having customers guess the number of kcalories in a nutritious dining entrée to win a free meal).

Publicity involves obtaining free editorial space or time in various media. Many foodservice operators do their own publicity. However, if you wish to obtain the advice of outside publicity consultants, O'Dwyer's Directory includes most public-relations firms. Here are some ideas for publicizing your nutrition program.

1. Send a **press release** about your healthy dining options to the appropriate contact person by name, not title, as indicated in the following list:

   Television and radio news: Assignment editor or specialty reporters appropriate to your story, such as health and food editors

   Television and radio talk shows: Producer

   Newspapers: Section editors (food, health and science, lifestyle), or city desk editor for special events

   Magazines and trade publications: Managing editor, articles editor, or specialty editors appropriate to your story

   Local publications and newsletters: Corporate employee or customer newsletters, or supermarket, utility company, bank, school, or church publications

Follow up each press release with a phone call. Editors are always looking for article ideas and just may pick up on your story.

2. Offer to write a column on nutritious meal preparation for a local newspaper.
3. Offer cooking demonstrations or on-site classes, or volunteer to conduct classes for health associations, retail stores, or supermarkets.
4. Invite local media and community leaders for the opening day of your new program and let them taste some nutritious menu selections.
5. Contact the foodservice director of a medical center or the public relations director of a health maintenance organization and offer to cosponsor a health or nutrition event such as a bike race or health fair. Check for local health and sporting events in which you can participate.
6. Contact your local American Heart Association and ask if it has a dining-out guide in which you may feature your restaurant.
7. Develop a newsletter for your operation and use it to publicize the new program (include some of your nutritious recipes). Newsletters help to build loyal customers.

There are many sources for promotional materials, such as table tents, posters, buttons, menu clip-ons, point-of-sale materials, and artwork (Figure 9-3). Food manufacturers, foodservice distributors, and food marketing boards and associations are excellent sources of promotional materials.

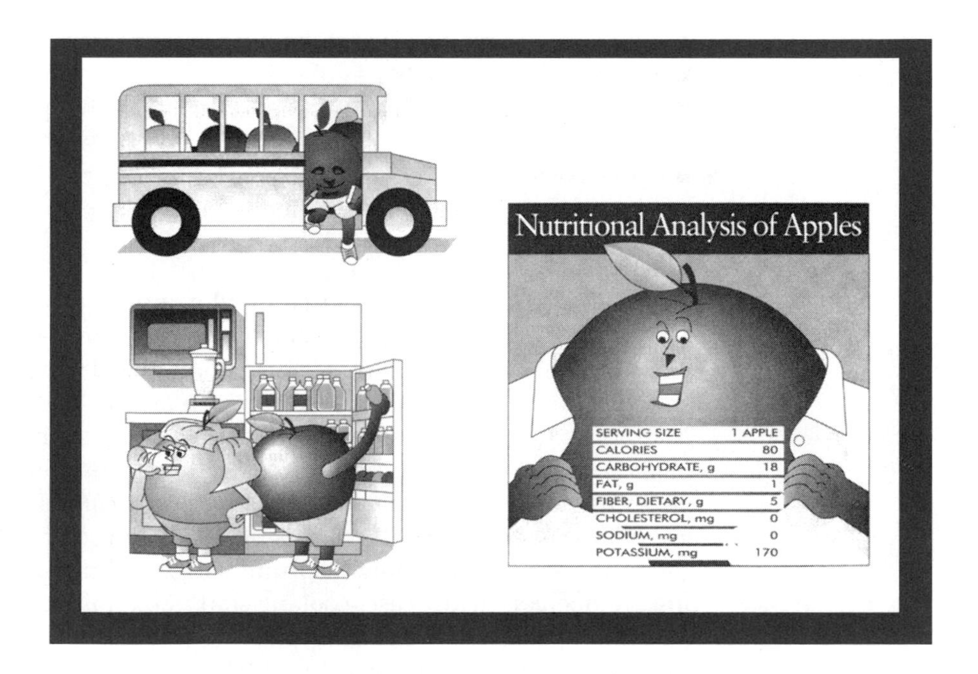

Figure 9-3 Artwork from International Apple Institute

Staff Training

Staff training centers on the waitstaff and the cooking staff. Before training begins, involve the waitstaff as much as possible in the development of your nutrition program so that they feel a part of it and take some ownership. They can be a valued resource in designing the program, because they make daily contact with the customers both in selling and serving as well as in listening to requests, compliments, and complaints. During training, the waitstaff needs to understand

- The scope and rationale for the nutrition program
- Grand-opening details
- The ingredients, preparation, and service for each menu item
- Some basic food and nutrition concepts so they can help guests with special dietary concerns, such as food allergies
- How to handle special customer requests, such as orders for half portions
- Merchandising and promotional details

Table 9-1 gives specific learning objectives for the service staff.

A poorly trained waitstaff will confuse the customer and, quite frankly, doom the program instead of knowledgeably promoting it. Conversely, a properly trained waitstaff can function as excellent sales agents and solicit feedback, including customer recommendations.

The cooking staff also needs training. Their training needs center on

- The scope and rationale for the nutrition program
- Grand-opening details
- The ingredients, preparation, portion size, and plating of each new menu item
- Some basic food and nutrition concepts so they can help guests with special dietary concerns, such as food allergies
- How to respond to special dietary requests

Table 9-2 outlines seven ways that employees learn best.

The cooking staff need to understand the prime importance of using only the freshest ingredients, using standardized recipes, measuring and weighing accurately, and attractive presentation.

Training the cooking staff to prepare healthy dishes correctly can be challenging. As managers have found during nutrient-content analysis, cooks do not always prepare recipes exactly as called for. Perhaps a key ingredient was unavailable, time was tight, or the cook forgot a step. Healthy menu items often are more labor-intensive, and more training and coaching are needed. In any case, cooking staff need training not only in making new menu items but also in understanding the importance of following recipes and serving the correct portion size.

TABLE 9-1 Learning Objectives for Service Staff

1. Servers must be able to respond to consumer health concerns by providing menu suggestions that meet their dietary needs. They should be able to make menu suggestions for the following dietary restrictions:
   Low calorie
   Low sodium
   Low cholesterol and low fat
   Low sodium, low cholesterol, and low fat
   High fiber
2. The waitstaff should be able to describe healthful dining options in straightforward, appropriate language to patrons. Servers must be able to provide information on ingredients, methods of preparation, portion sizes, and how the menu items are served.
   Ingredients: The waitstaff should be knowledgeable about details regarding ingredient usage: addition of fat or the type of fat used in cooking, the use of salt or high-sodium seasonings, cuts of meats used, the type of liquids used to prepare menu items, fats and thickening agents used in sauces, and the use of sugar or sugar substitutes.
   Cooking methods: Patrons commonly need to know not only what the composition of a menu item is but how it was prepared. Was the food fried in vegetable oil or animal shortening? Was the food prepared by pan frying, broiling, baking, poaching, or sautéing? Can the item be broiled without added butter? Is fat removed from meat juices or stocks before using them for sauces or soups?
   Presentation and portions: Patrons frequently want to know how the item will be served when making their menu selections. The waitstaff should be prepared to answer the following questions: What is the portion size of the item? What accompaniments are served with the item? Are special food items available to accompany the item? For example, is light syrup available for the light pancakes? Is a fruit spread available instead of jam for the whole-wheat breads? Can toast be served dry instead of buttered? Can salad dressing be served on the side?
3. The waitstaff should be able to explain the nutritional basis for menu items designated as light or healthful in terms of caloric, fat, cholesterol, and/or sodium content. They may need to answer questions about the program rationale. They should know the nutritional guidelines (U.S. Dietary Guidelines, American Heart Association guidelines, and/or National Cancer Institute recommendations) that provide the basis for the program.
4. Servers should be able to respond to patron inquiries about the availability of special foods or beverages. Does the restaurant serve brewed decaffeinated coffee? Are diet salad dressings available for the light salad entrées? Is margarine available instead of butter? Are herb seasonings available instead of salt for adjusting seasonings at the table? Does the restaurant serve skim milk?
5. The waitstaff should be able to respond to questions concerning substitutions of meal accompaniments. Can an entrée be served with two vegetables instead of a vegetable and a starch? Can a salad be substituted for the starch or vegetable served with the entrée? Can a fruit appetizer be served for dessert?
6. The waitstaff need to know what special requests the foodservice operation can accommodate. For example, can margarine or vegetable oil be used instead of butter in preparing foods? Can entrées be broiled instead of fried? Are smaller portion sizes available? Can sauces and salad dressing be served on the side?
7. The waitstaff should be able to recommend other foods and beverages that complement menu choices, including appetizers, soups, salads, desserts, and beverages that are light or meet the dietary restrictions of the patron.
8. The waitstaff should be knowledgeable about what is served with light menu items and what the correct portion sizes are for these items. Servers will act as the final quality-control agents prior to the serving of the foods. If light menu items are similar to traditional offerings, the waitstaff should be able to distinguish between the two items.

TABLE 9-1 *(continued)*

9. Staff members should respond politely and accurately to guest questions about healthful dining options. It should be emphasized that the proper response to a patron's inquiry is "I can find out for you," not "I don't know." When uncertain of the answer, the staff member should ask the kitchen manager, manager, or chef.

Source: Ganem, Beth Carlson. 1990. *Nutritional Menu Concepts for the Hospitality Industry*. New York: Van Nostrand Reinhold. Reprinted with permission.

Studies of healthy menu items have shown that, although the healthy items had less fat, fewer kcalories, and more vegetables and fruit than other menu items, sometimes an entrée winds up containing a lot more fat than in the recipe. For example, a fajita from the menu of a Mexican restaurant was supposed to have only 17 grams of fat but actually had 30 grams of fat because the appropriate lean meat was not used. Problems such as this point to the importance of training and retraining cooking staff.

■ MINI-SUMMARY

There are a variety of ways of using the menu to inform your clientele of healthy menu items, including simply using good descriptive language of how the item is prepared, what ingredients are used, and the portion size. Healthy menu items should be promoted in a positive manner. Three methods of promoting a nutrition program are advertising, sales promotion, and publicity (using press releases and other methods). There are many sources for promotional materials from food manufacturers, foodservice distributors, and food marketing boards and associations. Staff training is crucial and centers on the waitstaff and cooking staff. Training the cooking staff to prepare healthy dishes correctly can be challenging and is very important for success.

Program Evaluation

The healthy menu program should be evaluated much like any other program. Key questions for evaluation include:

1. How did the program do operationally? Did the cooks prepare and plate foods correctly? Did the waitstaff promote the program and answer questions well?
2. Did the food look good and taste good?
3. How well did each of the healthy menu options sell? How much did each item contribute to profits? How did the overall program affect profitability?

TABLE 9-2 How Employees Learn Best

1. When employees participate in their own training, they tend to identify with and retain the concepts being taught. To get employees involved, choose appropriate training methods.
2. Employees learn best when training material is practical, relevant, useful, and geared to an appropriate level. Learning is facilitated, too, when the material is well organized and presented in small, easy-to-grasp steps. Adult learners are selective about what they will spend time learning, and learning must be especially pertinent and rewarding for them. Adults also need to be able to master new skills at their own pace.
3. Employees learn best in an informal, quiet, and comfortable setting. Your effort in selecting and maintaining an appropriate training environment shows employees that you think their training is important. When employees are stuffed into a crowded office or a noisy part of the kitchen, or when the trainer is interrupted by phone calls, they may rightly feel that their training isn't really important. Employees like to feel special; so, when possible, find a quiet setting. Of course, much training, such as on-the-job training, necessarily takes place in the work environment.
4. Employees learn best when they are being paid for time spent in training.
5. Employees learn best with a good trainer. Although you may not ever find a person with all these qualities, you can use this list to evaluate potential trainers:

   A Successful Trainer
   - Is knowledgeable
   - Displays enthusiasm
   - Has a sense of humor
   - Communicates clearly, concisely, straightforwardly
   - Is sincere, caring, respectful, responsive to employees
   - Encourages employee performance; is patient
   - Sets an appropriate role model
   - Is well organized
   - Maintains control, frequent eye contact with employees
   - Listens well
   - Is friendly and outgoing
   - Keeps calm; is easygoing
   - Tries to involve all employees
   - Facilitates the learning process
   - Positively reinforces employees

6. Employees learn best when they receive awards or incentives. For example, when an employee has completed training for the position of cook, you can send a letter of recognition to the employee, which can also be put into the personnel file. The largest franchisee of Arby's awards employees a progression of bronze, silver, and gold name tags, as well as pay increases, as they learn each area of the restaurant. When the employee has learned all areas, he or she is promoted to the position of crew leader.
7. Employees learn best when they are coached on their performance on the job.

Source: Drummond, Karen. 1992. *Retaining Your Foodservice Employees.* © 1992 by John Wiley & Sons, Inc. Adapted by permission of John Wiley & Sons, Inc.

4. Did the program increase customer satisfaction? What was the overall feedback from customers? Did the program create repeat customers?

Proper program evaluation requires much time observing and talking with staff and customers, as well as going over written records, such as sales records.

Once a program has been evaluated, certain changes to fine-tune the program might be necessary. Here are some suggestions:

- Develop ongoing promotions to maintain customer interest.
- Add, modify, or delete certain menu items.
- Change pricing.
- Improve the appearance of healthy items.
- Listen to customers more to get future menu and merchandising ideas.

■ MINI-SUMMARY

Evaluation is needed to determine program effectiveness from customer, employee, and management viewpoints.

Restaurants and Nutrition-Labeling Laws

Foods prepared and served in restaurants or other foodservice operations are exempt from mandatory nutrition labeling found on packaged foods. However, restaurants are not exempt from Food and Drug Administration (FDA) rules concerning nutrient claims and health claims (discussed in Chapter 2) when used on menus, table tents, posters, or signs. In addition, restaurants must have nutrition information available upon request for any menu item using nutrient or health claims.

Nutrient content claims such as "good source of calcium" or "fat-free" can appear only if they follow legal definitions (see Table 2-10 in Chapter 2). For example, a food that is a good source of calcium must provide 10 to 19 percent of the Daily Value for calcium in one serving. Nutrient content claims are based on what the FDA has defined as a standardized serving size, called a Reference Amount. Standardized serving sizes or Reference Amounts are frequently measured in grams, milliliters, or cups in order to be very accurate. For example, the Reference Amount for cookies is 30 grams.

In addition to nutrient content claims based on a single serving, there are nutrient content claims for main dishes, such as lasagna, and meals. A main dish must weigh at least six ounces, be represented on the menu as a main dish, and contain no less than 40 grams each of at least three different foods from at least two food groups. Meals are defined as weighing at least 10 ounces and containing no less than 40 grams each of at least three different foods from at least two food groups. In general, for main dishes and meal products the nutrient content claim is based on the nutrient amount per 100 grams of the food. For example, a "low-fat" food must contain 3 grams of fat or less. Main dishes and meals must therefore contain 3 grams of fat or less per 100 grams, and not more than 30 percent of kcalories from fat.

Claims that promote a health benefit must meet certain criteria, as described in Chapter 2 and Table 2-10. In addition, any food being used in a health claim may not contain more than 20 percent of the Daily Value for fat, saturated fat, cholesterol, or sodium. Table 9-3 shows the maximum amount of these nutrients that are permitted in a single food serving, main dish, or meal for foods with health claims.

When providing nutrition information for a nutrient or health claim, restaurants do not have to provide the standard nutrition information profile and more exacting nutrient content values required in the Nutrition Facts panel of packaged foods. Instead, restaurants can present the information in any format desired, and they have to provide only information about the nutrient or nutrients that the claim is referring to. Restaurants also are not required to do chemical analyses to determine the nutrient values of their foods. They can use nutrient analysis software, books with nutrient composition information, or cookbooks with reliable nutrient analysis data.

Restaurants may use symbols on the menu to highlight the nutritional content of specific menu items. When doing so, they are required to explain the criteria used for the symbols. Restaurants may also highlight dishes that meet criteria set down by recognized organizations, such as the American Heart Association or a medical center. In these cases, the menu must explain that the items meet the dietary guidelines of that organization. Lastly, menus can use references or symbols to show that a food or meal is based on the Dietary Guidelines for Americans.

TABLE 9-3 Maximum Allowable Amount of Fat, Saturated Fat, Cholesterol, and Sodium for Foods with Health Claims

	Total Fat	Saturated Fat	Cholesterol	Sodium
Single Serving*	13 grams	4 grams	60 milligrams	480 milligrams
Main Dish	19.5	6	90	720
Meal	26	8	120	960

* Per Reference Amount and per 50 g when Reference Amount is 30 grams or less *or* 2 tablespoons or less.

Check-Out Quiz

1. Marketing includes:
   a. finding out what consumers want and need
   b. developing a product consumers want and need
   c. promoting the product
   d. all of the above
2. When hearing descriptions of healthy menu entrées, most customers want:
   a. complete nutrient information
   b. fat, saturated fat, and cholesterol information
   c. good descriptions of the ingredients, portion size, and method of preparation
   d. none of the above.
3. When you tout foods as "heart-healthy," you are approaching customers in a negative manner.
   a. True b. False
4. An example of publicity is:
   a. radio advertising
   b. a press release
   c. recipes from a food association
   d. a point-of-purchase display
5. When you evaluate the success or failure of healthy menu options, you need to get feedback from:
   a. customers
   b. staff
   c. managers
   d. all of the above

Activities and Applications

1. Restaurant Menu Check

Study the menus from five different foodservices, including a quick-service business, and identify any menu items that are healthy. How do menu items appear on the menu? Would customers know they are healthy? What information is included?

2. Restaurant Promotion

Check restaurant advertisements in the newspaper, on radio, or in other media and watch for any advertising of healthy and nutritious foods. What do the advertisements state? To which market segments are these advertisements targeted?

3. Restaurant Visit

Visit a local restaurant/foodservice that offers healthy menu options. Find out how much it sells of its healthy menu options, and the profile of the typical customer buying these items. Also find out how these products are marketed and evaluated.

Nutrition Web Explorer

A restaurant chain such as TGI Friday's or Chili's www.Fridays.com or www.chilis.com

Go to the website of a restaurant chain, such as TGI Fridays or Chili's, and print out their menu items, which come with descriptions. Read the descriptions. Circle the menu items that appear directed to nutritionally conscious customers. For two of the recipes given in Chapter 8, write menu descriptions that make the foods sound appealing and also describe their nutrient contribution.

Food marketing boards and associations

American Egg Board www.aeb.org

Mann Packing www.broccoli.com

Grains Nutrition Information Center www.wheatfoods.org

Milk www.whymilk.com

National Pork Producers Council www.nppc.org

National Turkey Federation www.turkeyfed.org

Produce Marketing Association www.pma.com

Pick one of these food marketing board/association websites to visit. Write a brief report about the website, including the name of the board/association, website address, and list of items available that a foodservice operator could use (such as recipes), and attach sample material.

Chapter 10

Light Beverages and Foods for the Beverage Operation

Heavy foods and strong alcoholic beverages were a way of life in most beverage operations until about 1970. At that time, the sale of spirits began to decline, whereas sales of "lighter" drinks, such as beer and wine, began to increase. Next, beer went light—a well-received change—and wine by the glass was promoted. Mocktails—nonalcoholic beverages—began to appear as well, as part of the movement against alcohol abuse. In addition, the drinking age was raised to 21 in many states, warning labels appeared on liquor bottles, and successful lawsuits were waged against beverage operations by victims of drunk drivers—all part of a movement aimed not at the use but at the abuse of alcohol.

There is a big difference between alcohol abuse and moderate use. In fact, moderate drinking is associated with a lower risk for coronary heart disease in some individuals (mainly among men over age 45 and women over age 55). Moderate consumption provides little, if any, health benefit for younger people. Risk of alcohol abuse increases when drinking starts at an early age.

Drinking in moderation is defined as no more than one drink per day for women and no more than two drinks per day for men. This limit is based on differences between the sexes in both weight and metabolism. One drink is 12 ounces of regular beer, 5 ounces of wine, or 1.5 ounces of 80-proof distilled spirits.

In an effort to curb drunk-driving accidents, some beverage operations have changed their emphasis from all-you-can-drink specials (and the like) to foods, entertainment, music, contests, games, and theme parties, in order to overcome an association with heavy drinking. Themes might be international, such as Mexican, with tacos and burritos and staff dressed in Mexican outfits. A Workout Night theme might involve asking customers to come dressed in their workout clothes.

This chapter first looks at lower-kcalorie and lower-alcohol beverages and nutritious food choices, then goes on to discuss marketing these selections. This chapter will help you to:

- List and describe lower-kcalorie and lower-alcohol drink options
- State the number of kcalories in one gram of alcohol
- Define various types of bottled water
- List healthy snacks/appetizers
- Outline the steps a foodservice operator should follow to effectively market light beverages and foods

Lower-Kcalorie and Lower-Alcohol Drink Options

Bar customers who want a lower-kcalorie beverage or a drink with little or no alcohol have a number of choices: fruit juices and fruit-juice-based beverages, creamy drinks, **alcohol-free wines, light beers** and **nonalcoholic**

Alcohol-free wines—Wines whose alcohol is removed after fermentation; they must be 99.5 percent alcohol-free.

Light beer—Beer with one-third to one-half less alcohol and kcalories than regular beers.

Nonalcoholic malt beverage—A beerlike product with only 0.5 percent or less alcohol.

Low-alcohol refreshers—A category of drinks made with fruit juices, carbonated water, and sugar, with an alcohol base of wine, distilled spirits, or malt; the group includes wine coolers.

Mocktails—Drinks made to resemble mixed alcoholic drinks but containing no alcohol.

Fruit smoothies—Frozen blends of fruit, milk, and/or fresh or frozen yogurt.

malt beverages, low-alcohol refreshers, mocktails, and other beverages. As Table 10-1 indicates, when the alcohol content of a drink increases, so do kcalories. Alcohol contains 7 kcalories per gram, compared to 4 kcalories per gram of carbohydrate or protein, and 9 kcalories per gram of fat.

Fruit Juices and Fruit-Juice-Based Beverages

Fruit juices serve as a base for many different nonalcoholic drinks, such as fruit juice served with seltzer (water with bubbles or carbonation). Fruit may serve as the basis for blended drinks, such as strawberry coolers made with fresh strawberries, sugar, sparkling water, and ice cubes.

Nonalcoholic sparkling cider, a soft, fruity, and light drink, is made in France and in the United States. Most sparkling ciders contain no alcohol, but it is available with 5 to 12 percent alcohol. Serve it in stemmed glasses with a fresh fruit garnish to merchandise its fresh, fruity taste.

Creamy Drinks

Creamy drinks are made with milk, yogurt, cream, ice cream, or other dairy products. A good example is a category of creamy drinks called **fruit smoothies,** frozen blends of fruit, milk, and/or fresh or frozen yogurt. When regular dairy products such as cream and ice cream are used, these drinks are not low in kcalories; however, low-fat or nonfat yogurt, ice milk, and skim or 1 percent milk can be used to make drinks with fewer kcalories and less fat.

Alcohol-Free Wines

Alcohol-Free wine has had its alcohol removed after fermentation. It contains up to 0.5 percent alcohol and is available in red, pink, and white varieties, either still or sparkling. Varietal wines with one predominant grape variety, such as gamay or Riesling, are available as well.

Light Beers and Nonalcoholic Malt Beverages

Several varieties of beer with varying alcohol and calorie contents are available, as seen in Table 10-1. Different brands of light beer can vary tremendously in kcalorie content, but most have one-third to one-half the alcohol and kcalories of regular beers. Nonalcoholic brews, produced by removing the alcohol after brewing or by stopping the fermentation process before alcohol forms, contain one-half of 1 percent or less of alcohol. In either case, these products have about half the kcalories of light beer and one-third the kcalories of regular beer.

TABLE 10-1 Kcalories in Alcoholic and Nonalcoholic Beverages

Beverage	Amount (fluid ounces)	Kcalories	Alcohol
Mixed Drinks			
Bloody Mary	5 fl. oz.	116	11.7%
Bourbon and soda	4 fl. oz.	105	16.1%
Daiquiri	2 fl. oz.	111	28.3%
Gin and tonic	7.5 fl. oz.	171	8.8%
Manhattan	2 fl. oz.	128	36.9%
Martini	2.5 fl. oz.	156	38.4%
Piña colada	4.5 fl. oz.	262	12.3%
Screwdriver	7 fl. oz.	174	8.2%
Tequila sunrise	5.5 fl. oz.	189	13.5%
Tom Collins	7.5 fl. oz.	121	9.0%
Whiskey sour	3 fl. oz.	123	20.6%
Distilled Liquors			
Gin, 90 proof	1.5 fl. oz.	110	45%
Rum, 80 proof	1.5 fl. oz.	97	40%
Vodka, 80 proof	1.5 fl. oz.	97	40%
Whiskey, 86 proof	1.5 fl. oz.	105	43%
Wine			
Table wine, red	3.5 fl. oz.	74	11.5%
Table wine, rosé	3.5 fl. oz.	73	11.5%
Table wine, white	3.5 fl. oz.	70	11.5%
Beer			
Beer	12 fl. oz.	146	4.5%
Light beer	12 fl. oz.	100	2.2–4.4%
Nonalcoholic malt beverage	12 fl. oz.	50	Less than 0.5%
Nonalcoholic Beverages			
Cola	12 fl. oz.	151	
Ginger ale	12 fl. oz.	124	
Lemon-lime soda	12 fl. oz.	149	
Orange soda	12 fl. oz.	177	
Root beer	12 fl. oz.	152	
Tonic water	12 fl. oz.	125	
Club soda, seltzer, mineral, or sparkling waters	12 fl. oz.	0–10	
Cranberry-apple juice drink	6 fl. oz.	123	
Cranberry juice cocktail	6 fl. oz.	108	
Grape juice drink	6 fl. oz.	94	
Lemonade	8 fl. oz.	100	
Tomato vegetable juice cocktail	8 fl. oz.	44	

Source: United States Department of Agriculture Handbook #8–14.

Low-Alcohol Refreshers

Wine coolers—Mixes of wine, fruit flavor, and carbonated water or plain water.

Wine spritzers—Drinks made with wine and club soda.

This group started out as wine coolers but has grown to include other drinks. **Wine coolers** are a mix of wine, fruit flavor, and carbonated or plain water. They usually contain fructose or high-fructose corn syrup for sweetening, and a few also have preservatives added. They generally have half the alcohol content of wine. Many flavors are on the market, and some brands are available in kegs. Wine coolers can also be made on site and served on the rocks or over shaved ice. **Wine spritzers** are made with wine and club soda and are mixed at the bar. Other low-alcohol refreshers are made with fruit juices, carbonated water, sugar, and distilled spirits or malt instead of wine.

Mocktails

Mocktails are much more than just a Virgin Mary, a bloody Mary without the vodka, or simply tomato juice! Mocktails attempt to imitate the real thing. For instance, a mockarita, an imitation margarita, is blenderized with lemonade, lime juice, and ice cubes and served in a margarita glass with a salted rim. Hot mocktails, such as mock Irish coffee, can be made as well. Some bartenders substitute lower-proof products in cocktails, such as Chablis for tequila to make a white wine margarita; such drinks are not true mocktails, but this substitution does lower both alcohol and kcalories. Nonalcoholic "liqueurs," such as creme de menthe and creme de cacao, are also available for use in mocktails or lower-alcohol cocktails.

Other Choices

Other lower-kcalorie and/or no-alcohol alternatives include bottled waters (see the next section in this chapter), vegetable juices, diet soft drinks, iced tea flavored with lemon or other fruit juice, and some specialty coffees such as espresso or cafe latte (see Table 10-2). Other flavorings for lower-alcohol drinks include vanilla and rum extracts or nonalcoholic fruit syrups such as cassis.

■ MINI-SUMMARY

Drinks with fewer kcalories and/or less alcohol include fruit juices and fruit-juice-based beverages, creamy drinks, alcohol-free wines, light beers and nonalcoholic malt beverages, low-alcohol refreshers, mocktails, bottled waters, vegetable juices, diet soft drinks, iced tea, and specialty coffees.

TABLE 10-2 Espresso Drinks

Name	Description
Espresso	The word **espresso** refers to the unique process used to brew this coffee. Hot water is forced under high pressure through finely ground coffee that has been compacted. The coffee is made from beans that have been roasted until very dark. The serving size is about 1-1/2 fluid ounces (45 ml).
Cappuccino	1-1/2 fluid ounces (45 ml) of espresso, topped by frothed milk that is separated into hot milk and foam, in an approximate ratio of one part espresso, one part milk, and one part foam. Usually served in a glass mug.
Latte	A single or double serving of espresso, topped by frothed milk, in a ratio of one part espresso to three parts milk. Usually served in a wide-mouthed glass.
Mocha	A single serving of espresso mixed with about 5 fluid ounces of steamed milk and chocolate syrup to taste. Often topped with whipped cream.

Bottled Waters

People buy **bottled water** for what it does not have—calories, sugar, caffeine, additives, preservatives, and, in most cases, not too much sodium. The Food and Drug Administration (FDA) has established standards of quality for bottled drinking water. Bottled water is defined as water that is intended for human consumption sealed in bottles or other containers with no added ingredients except that it may optionally contain safe and suitable antimicrobial agents. Also, fluoride may be optionally added within limits. The FDA has established maximum allowable levels for physical, chemical, microbiological, and radiological contaminants in the bottled water quality standard regulations.

Bottled water—Water intended for human consumption sealed in bottles or other containers with no added ingredients except that it may optionally contain safe and suitable antimicrobial agents and fluoride.

Artesian well water—Water from a well that taps an aquifer—layers of porous rock, sand, and earth that contain water—which is under pressure from surrounding upper layers of rock or clay.

Bottled water is different from tap water in that it has more consistent quality and taste. While bottled water originates from protected sources (75 percent from underground springs and aquifers), tap water comes mostly from lakes and rivers. Also, whereas bottled water is regulated by the FDA as a food product, tap water is regulated by the U.S. Environmental Protection Agency (EPA) and is regarded as a utility.

The FDA has published standard definitions for different types of bottled water to promote honesty and fair dealing in the marketplace.

- **Artesian well water** is water from a well that taps an aquifer—layers of porous rock, sand, and earth that contain water—which is under pressure from surrounding upper layers of rock or clay. When tapped, the pressure in the aquifer, commonly called artesian pressure, pushes the water above the level of the aquifer, sometimes to the surface. Other means may be used to help bring the water to the surface. According to the EPA, water from artesian aquifers often is more pure because the confining layers of rock and clay impede the movement of contamination. However, despite the claims of some bottlers, there is no guarantee that artesian waters are any cleaner than ground water from an unconfined aquifer.

- **Mineral water** is water from an underground source that contains at least 250 parts per million total dissolved solids. Minerals and trace elements must come from the source of the underground water. They cannot be added later.
- **Spring water** is derived from an underground formation from which water flows naturally to the earth's surface. Spring water must be collected only at the spring or through a borehole tapping the underground formation feeding the spring. If some external force is used to collect the water, the water must have the same composition and quality as the water that naturally flows to the surface.
- **Well water** is water from a hole drilled into the ground, which taps into an aquifer.

Mineral water—Water from an underground source that contains at least 250 parts per million in total dissolved solids. Minerals and trace elements must come from the source of the underground water.

Spring water—Water derived from an underground formation from which water flows naturally to the earth's surface.

Well water—Water from a hole drilled into the ground, which taps into an aquifer.

Purified water—Water produced by reverse osmosis, ozonation, or other suitable processes and that meets the definition set by the *U.S. Pharmacopoeia.*

Sparkling water—Any carbonated water.

Seltzer—Filtered, artificially carbonated tap water that generally has no added mineral salts.

Some bottled water also comes from municipal sources, in other words, the tap. Municipal water is usually treated before it is bottled. Bottled water that has been treated by one of the following methods and that meets the definition of purified water in the *U.S. Pharmacopeia* can be labeled as **purified water.** Examples of water treatments include:

- Reverse osmosis—Water is forced through membranes to remove minerals in the water.
- Absolute 1 micron filtration—Water flows through filters that remove particles larger than 1 micron in size, such as Cryptosporidium, a parasitic protozoan.
- Ozonation—Bottlers of all types of waters typically use ozone gas, an antimicrobial agent, to disinfect the water instead of chlorine, since chlorine can leave residual taste and odor to the water.

Like all other foods regulated by the FDA, bottled water must be processed, packaged, shipped, and stored in a safe and sanitary manner and be truthfully and accurately labeled.

Beverages labeled as containing sparkling water, seltzer water, soda water, tonic water, or club soda are not included as bottled water under the FDA's regulations, because these beverages have historically been considered soft drinks.

- **Sparkling water** is any carbonated water.
- **Seltzer** is filtered, artificially carbonated tap water that generally has no added mineral salts. It is available with assorted flavor essences, such as

Club soda—Filtered, artificially carbonated tap water to which mineral salts are added.

Tonic water—A carbonated water containing lemon, lime, sweeteners, and quinine.

black cherry and orange. If seltzer contains sweeteners (and therefore calories), it must be called a flavored soda.

- **Club soda,** sometimes called soda water or plain soda, is filtered, artificially carbonated tap water to which mineral salts are added to give it a unique taste. Most average 30 to 70 milligrams of sodium per 8 ounces.
- **Tonic water** is not really water or low in calories. It contains 84 calories per 8 ounces. Diet tonic water uses sugar substitutes.

Different mineral and carbonation levels of waters make them appeal to different customers and eating situations. For instance, a heavily carbonated sparkling water such as Perrier is excellent as an aperitif, yet some customers may prefer the lighter sparkle of San Pellegrino. Still waters such as Evian are generally more popular and appropriate to have on the table during the meal.

■ MINI-SUMMARY

Bottled water cannot contain sweeteners and must be kcalorie-free and sugar-free. Bottled waters include artesian well, mineral, spring, well, and purified waters. Bottled water is different from tap water in that it has more consistent quality and taste. While bottled water originates from protected sources (75 percent from underground springs and aquifers), tap water comes mostly from lakes and rivers. The FDA regulates bottled water. Beverages labeled as sparkling water, seltzer water, soda water, tonic water, or club soda are not included as bottled water under the FDA's regulations, because these beverages have historically been considered soft drinks.

Healthy Snacks and Appetizers

Nutritious food choices for beverage operations need not be dull, monotonous, or unappetizing. Many snack foods and appetizers can easily fit in, such as a variety of dips served with crudités (raw, cut vegetables), fresh fruit, and/or baked tortilla chips. Other choices may include some from this list.

Afternoon Snacks

Fresh Fruit Kabobs with Yogurt Dipping Sauce
Vegetable Tortilla Rolls with Dill Dressing
Whole Fresh Fruits with Sliced Banana Breads and Reduced-Fat Cream Cheese
Air-Popped Popcorn Sprinkled with Parmesan Cheese

Macédoine of Fresh Fruits Marinated with Mint and Sliced Angel-Food Cake
Baked Assorted Tortilla Chips with Tofu Dips
Cinnamon Oatmeal Cookies, Date Granola Bars
Turkey and Stone-Ground Wheat Bread Sandwiches with Dijon Mustard Sauce

Canapés

Baked Pita Chips with Caponata
Stuffed Celery with Herbed Cream Cheese
Tomato, Spinach, and Turkey Rolls with White-Bean Spread
Oat-Bran Mini Muffins with Whole-Grain Honey Mustard Spread and Fresh Turkey
Oven-Dried Vegetable Chips with Hummus

Hot Hors d'Oeuvres

Pan-Seared Vegetable Spring Rolls
Spinach and Feta Wrapped in Phyllo
Teriyaki Steak Skewers
Hummus Bruschetta with Roasted Peppers
Grilled Chicken Sausages with Whole-Grain Breads
Jalapeño Pancakes with Corn Relish
Grilled Pork Quesadillas with Onions, Peppers, and Mushrooms
Mini Meatballs (Asian or Swedish)

Late-Night Snacks

Meatless Chili with Baked Corn Tortillas
Carrot and Celery Sticks with Garlic Yogurt Sauce
Cornflake-Crusted Chicken Tenders with Pineapple Chili Dipping Sauce
Assorted Oat-Bran Fruit Bars with Fruit Leather
Baked Stuffed Potato Skins with Broccoli, Cheddar Cheese, and Tomato Vinaigrette

■ MINI-SUMMARY

Nutritious food choices for beverage operations need not be dull, monotonous, or unappetizing. Many snack foods and appetizers can easily fit in.

Marketing Light Beverages and Foods

The first step in marketing light beverages and foods involves analyzing the surrounding market and clientele needs and wants; projecting sales volume, pricing, and profit structures; and seeing how such options would fit into the operation in terms of ordering, inventory, and menus. A beverage operator might start by choosing five items, perhaps a wine cooler, a dealcoholized wine and beer, a mineral water, and a juice-based drink. In a more developed market, the choices might be enlarged. Using cookbooks or other resources, you can select, implement, and evaluate lower-kcalorie foods that fit well into your operation. The production staff must be involved in these decisions. For both drinks and food, variety and quality are of the utmost importance.

Presentation of lower-kcalorie and/or lower-alcohol drinks is just as important as it is for regular drinks. For instance, serve sparkling cider in stemmed glasses or champagne flutes, and garnish with an apple ring and a sprig of mint.

Bartenders and servers need to be trained about the new offerings as well as tasting them. A well-trained staff can suggest options appropriately, but they must learn to present lower-kcalorie and/or lower-alcohol drinks as positively as regular drinks. If servers lose tips by selling these new items, a bonus or incentive may be needed in order for the program to be successful.

The proposed program must be merchandised with flair during good selling times to enhance its success. During these times, menu clip-ons, blackboards, lightboards, table tents, posters, or other techniques let customers know what is available. A special dinner menu may be used, and promotions such as water tastings and "nonalcoholic beer of the month" offerings can be effective.

Once the program is launched, getting customer feedback is very important, as well as analyzing sales and profits. After enough time, the success of the program needs to be evaluated, just like any other program.

■ MINI-SUMMARY

As with other programs, light beverages and foods must first be selected and developed to meet clientele needs and wants and, hopefully, to make money. Staff must be trained, and merchandising and promotion techniques selected, before the program is launched. When the program has been in place long enough, sales and profits can be evaluated.

Check-Out Quiz

1. Which is lower in kcalories?
   a. 12 fluid ounces of light beer
   b. 7 fluid ounces of gin and tonic
   c. 6 fluid ounces of white wine
   d. 12 fluid ounces of tonic water
2. Drinks that taste like mixed drinks but contain no alcohol are called:
   a. coolers
   b. de-alcoholized spirits
   c. mocktails
   d. smoothies
3. The emphasis in beverage operations has changed to:
   a. food
   b. theme events
   c. entertainment
   d. all of the above
4. Which of the following foods could not be served as a nutritious snack?
   a. raw vegetables with yogurt dip
   b. hot, soft pretzels
   c. vegetarian pizza on pita bread
   d. deep-fried mozzarella sticks
5. Ordinary water that has been filtered and artificially carbonated is:
   a. club soda
   b. mineral water
   c. seltzer
   d. spring water

Activities and Applications

1. Supermarket Sleuth

Go to a supermarket and examine the selection of bottled waters and de-alcoholized beers and wines. Read the labels carefully.

2. Taste Testing: De-Alcoholized Beer and Wine

Have a taste test of de-alcoholized beers and wines. Examine the de-alcoholized beers in terms of how close they taste to real beer, kcalories per bottle, appearance, foam, and flavor. Examine the de-alcoholized wines in terms of how close they taste to real wine, calories per serving, sweetness or dryness, and appearance.

3. Develop Light Beverage and Food Options

You are in charge of developing five food and five beverage selections to be featured in a special Friday-night promotion using a workout theme. Customers will be asked to come in their workout clothes, and more healthful food and beverage selections will be featured.

4. Taste Testing: Bottled Waters

Have a taste-testing of bottled water, including at least two to four products from each of these categories: spring water, purified water, and carbonated water. Use the following form to collect pertinent information. You will compare your information with others in your class.

Type of Water
(circle one): Spring Water   Purified Water   Carbonated Water
Brand Name: ______________________

1. Read about the quality of this water at: www.bottledwaterweb.com/bott/index/html
Water quality notes:
____

2. Taste the water and rate its flavor on a scale of 1 to 5 (1 being poor, 5 being excellent).
1   2   3   4   5

Spring water and purified water should taste clean and fresh with no off-tastes or plastic taste.

Carbonated water should be bubbly and a little sour (due to the gas making the bubbles). There will be a touch of bitterness and slight saltiness.

3. Calculate the cost per 8 fluid ounces of the water. ___$___.___

Nutrition Web Explorer

International Bottled Water www.bottledwater.org
When you visit this website, click on “Bottled Water Facts,” then click on “FAQ.” Find out how to store bottled water.

Part Three

Nutrition's Relationship to Health and Life Span

Chapter 11
Nutrition and Health

The two leading causes of death in the United States are cardiovascular disease (including coronary heart disease, strokes, and high blood pressure) and cancer. Since 1900, cardiovascular disease has been the number-one killer in the United States every year except one (1918). In 1999, over 725,000 people died due to heart disease. In addition to cardiovascular disease and cancer, more and more people are being diagnosed with diabetes.

These diseases all have two things in common: they have great financial and emotional costs, and their prevention and treatment have a dietary component. This chapter will look at coronary heart disease, stroke, high blood pressure, cancer, and diabetes, as well as the healthfulness of the vegetarian diet. This chapter will help you to:

- Define risk factor and list the three major risk factors for cardiovascular disease
- List and describe three common forms of cardiovascular disease and their causes
- Explain how diet can play a role in the prevention and treatment of cardiovascular disease and cancer
- List five lifestyle modifications for hypertension control
- List five menu-planning guidelines to lower cardiovascular risk
- Define cancer
- Outline six menu-planning guidelines to lower cancer risk
- Distinguish between type 1 and type 2 diabetes mellitus and the principles of planning meals for people with diabetes
- Name the major types of vegetarian eating styles and state the health benefits of the vegetarian diet
- List six menu-planning guidelines for vegetarians

Nutrition and Cardiovascular Disease

Cardiovascular disease (CVD) is a general term for diseases of the heart and blood vessels, as seen in the following:

- Coronary artery disease
- Stroke
- High blood pressure
- Rheumatic heart disease
- Congenital heart defects

This section will discuss the first three diseases.

Smoking, high blood pressure, and high blood cholesterol are three major **risk factors** for cardiovascular disease. A risk factor is a habit, trait, or condition associated with an increased chance of developing a disease. Preventing or controlling risk factors generally reduces the probability of illness. These three risk factors are modifiable to some extent. Other risk fac-

Cardiovascular disease—Diseases of the heart and blood vessels such as coronary artery disease, stroke, and high blood pressure.

Risk factor—A habit, trait, or condition associated with an increased chance of developing a disease.

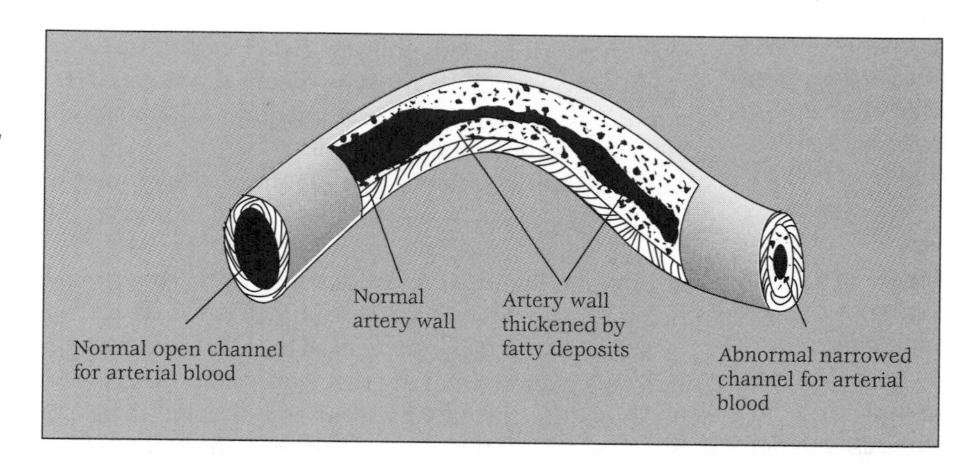

Figure 11-1 Cross-sectional representation of a coronary artery partially closed with plaque.

tors are diabetes, age (45 years or older for men, 55 years or older for women), obesity, low HDL, a family history of premature CVD (heart attack in a father before age 55, or before 65 in a mother), and physical inactivity.

The two medical conditions that lead to most cardiovascular disease are atherosclerosis and high blood pressure. Let's take a look at atherosclerosis first. **Atherosclerosis,** a condition characterized by plaque buildup along the artery walls, is the most common form of artery disease. (*Arteriosclerosis* is a general medical term that includes all diseases of the arteries involving hardening and blocking of the blood vessels.) Atherosclerosis affects primarily the larger arteries of the body. In this condition, arterial linings become thickened and irregular with deposits called **plaque.** Plaque contains cholesterol, fat, fibrous scar tissue, calcium, and other biological debris. Why plaque deposits are formed and what role fat and cholesterol play in its formation are questions with only partial answers.

Atherosclerosis—A condition characterized by plaque buildup along the artery walls; it is the most common form of arteriosclerosis.

Plaque—Deposits on arterial walls that contain cholesterol, fat, fibrous scar tissue, and other biological debris.

Atherosclerosis develops by a process that is totally silent. At birth the blood vessels are clear and smooth. As time goes on, plaque builds up, resulting in narrower passages and less elasticity in the vessel wall, both of which contribute to high blood pressure (Figure 11-1). What's even more dangerous is when the plaque closes off blood flow completely, or the rough plaque surface provides a place for a blood clot to form and block blood flow in an already partially closed artery. If the artery takes blood to the brain, then a stroke occurs. If the blocked artery is near the heart, then a heart attack occurs, the next topic.

Coronary Heart Disease

The heart is like a pump, squeezing and forcing blood throughout the body. Like all muscles in the body, the heart must have oxygen and nutrients in order to do its work. The heart cannot use oxygen and nutrients directly from the blood it pumps within its chambers. Instead, nutrients and

Coronary heart disease (CHD)—Damage to or malfunction of the heart caused by narrowing or blockage of the coronary arteries.

Angina—Symptoms of pressing, intense pain in the heart area caused by insufficient blood flow to the heart muscle.

Myocardial ischemia—A temporary injury to heart cells caused by a lack of blood flow and oxygen.

oxygen are furnished by three main blood vessels on the heart, referred to as coronary arteries. **Coronary heart disease (CHD)** is a broad term used to describe damage to or malfunction of the heart caused by narrowing or blockage of the coronary arteries.

More than two-thirds of a coronary artery may be filled with fatty deposits without causing symptoms. Symptoms may manifest themselves as chest pain, as in **angina,** or as a heart attack. Angina refers to the symptoms of pressing, intense pain in the area of the heart when the heart muscle is not getting enough blood. Sometimes stress or exertion can cause angina.

Most heart attacks are caused by a clot in a coronary artery at the site of narrowing and hardening that stops the flow of blood. In healthy individuals, blood clots normally form and dissolve in response to injuries in the blood vessels. But in atherosclerosis, blood clots appear to form in response to plaque when they are not needed. If an area of the heart is supplied by more than one vessel, the heart muscle may live for a period of time even if one vessel becomes blocked. The extent of heart muscle damage after a heart attack depends on which vessel is blocked, whether it is big or small, and on the remaining blood supply to that area. When the heart muscle does not get adequate oxygen and nutrients, it may die. This is a heart attack. A further area of heart muscle may be deprived of blood flow and oxygen to a lesser degree, causing a temporary injury called **myocardial ischemia.** This dead or injured heart muscle causes the heart to lose some of its effectiveness as a pump, because reduced muscle contraction means reduced blood flow.

Coronary heart disease is the number-one killer of women. Every year, as many women as men die from coronary heart disease. Whereas most men's heart attacks are experienced at 40 years of age and older, women do not usually experience heart attacks until after menopause.

■ MINI-SUMMARY

Cardiovascular disease includes diseases of the heart such as coronary artery disease, stroke, and high blood pressure. Major risk factors include smoking, high blood pressure, and high blood cholesterol. Atherosclerosis is characterized by plaque buildup on arterial walls, which can eventually close off blood circulation.

Nutrition and Coronary Heart Disease

Much research over the past 40 years has shown that a diet high in saturated fat, trans fatty acids, and cholesterol contributes to high blood cholesterol. CHD risk can be assessed by measuring total blood cholesterol as well as the proportions of the various types of lipoproteins. Total cholesterol refers to the overall level of cholesterol in the blood. High-density lipoprotein (HDL) is often referred to as "good" cholesterol, because high levels of HDL are associated with lowered CHD risk. High levels of low-density lipoprotein (LDL)—often referred to as "bad" cholesterol—increase CHD risk.

The National Cholesterol Education Program (NCEP) recommends that all adults (age 20 and older) have a lipoprotein profile once every five years. A lipoprotein profile measures levels of total cholesterol, LDL, HDL, and triglycerides. A total blood cholesterol level of less than 200 mg/dL is considered desirable; from 200 to 239 is borderline high; and 240 or more is high. An HDL level of less than 40 mg/dL is defined as low, and is considered a CHD risk factor. Optimal LDL is less than 100 mg/dL, and near/above optimum is 100–129 mg/dL. Borderline high LDL is 130–159 mg/dL, and high is 160–189 mg/dL. Triglycerides can also raise heart disease risk. Levels that are borderline high (150–199 mg/dL) or higher may need treatment in some people.

The main goal of cholesterol-lowering treatment is to lower the LDL level enough to reduce the risk of developing heart disease or having a heart attack. The higher the risk, the lower the LDL goal will be. There are two main ways to lower cholesterol.

TLC Diet—A low saturated fat, low cholesterol eating plan designed to fight cardiovascular disease and lower LDL; the diet calls for less than 7 percent of kcalories from saturated fat and less than 200 milligrams of cholesterol daily and also recommends only enough kcalories to maintain a desirable weight.

- Therapeutic lifestyle changes (TLC)—includes a cholesterol-lowering diet (called the **TLC diet**), physical activity, and weight management. TLC is for anyone whose LDL is above optimum levels.
- Drug treatment—if cholesterol-lowering drugs are needed, they are used together with TLC treatment to help lower LDL.

To reduce risk for heart disease or keep it low, it is very important to control any other risk factors such as high blood pressure and smoking.

TLC is a set of things you can do to help lower LDL cholesterol. The main parts of TLC are:

- **The TLC diet.** This is a low-saturated-fat, low-cholesterol eating plan that calls for less than 7 percent of kcalories from saturated fat and less than 200 milligrams of dietary cholesterol per day. The TLC diet recommends only enough kcalories to maintain a desirable weight and avoid weight gain. If LDL is not lowered enough by reducing saturated fat and cholesterol intakes, the amount of soluble fiber in the diet can be increased. Certain food products that contain plant stanols or plant sterols (for example, cholesterol-lowering margarines and salad dressings) can also be added to the TLC diet to boost its LDL-lowering power.
- **Weight management.** Losing weight if you are overweight can help lower LDL and is especially important for those with a cluster of risk factors that includes high triglyceride and/or low HDL levels and being

overweight with a large waist measurement (more than 40 inches for men and more than 35 inches for women).

- **Physical activity.** Regular physical activity (60 minutes on most, if not all, days) is recommended for everyone. It can help raise HDL and lower LDL and is especially important for those with high triglyceride and/or low HDL levels who are overweight with a large waist measurement.

Even with drug treatment to lower cholesterol, lifestyle changes are important. This will keep the dose of medicine as low as possible, and lower risk in other ways as well. There are several types of drugs available for cholesterol lowering, including statins, bile acid sequestrants, nicotinic acid, and fibric acids.

Once the LDL goal has been reached, a physician may prescribe treatment for high triglycerides and/or a low HDL level, if present. The treatment includes losing weight if needed, increasing physical activity, quitting smoking, and possibly taking a drug.

■ MINI-SUMMARY

The National Cholesterol Education Program (NCEP) recommends that all adults have a lipoprotein profile once every five years. A lipoprotein profile measures levels of total cholesterol, LDL, HDL, and triglycerides. The main goal of cholesterol-lowering treatment is to lower LDL enough to reduce the risk of developing heart disease or having a heart attack. There are two main ways to lower cholesterol: therapeutic lifestyle changes (TLC—includes a cholesterol-lowering diet, physical activity, and weight management) and drugs.

Stroke

Stroke—Damage to brain cells resulting from an interruption of blood flow to the brain.

A **stroke** is damage to brain cells resulting from an interruption of the blood flow to the brain. The brain must have a continual supply of blood rich in oxygen and nutrients for energy. Although the brain constitutes only 2 percent of the body's weight, it uses about 25 percent of the oxygen and almost 75 percent of the glucose circulating in the blood. Unlike other organs, the brain cannot store energy. If deprived of blood for more than a few minutes, brain cells die from energy loss and from certain chemical interactions that are set in motion. The functions these cells control—speech, muscle movement, comprehension—die with them. Dead brain cells cannot be revived.

The majority of strokes are caused by blockages in the arteries that supply blood to the brain. These blockages may be caused by a clot that forms on the inner lining of a brain or neck artery already partly clogged by plaque. The most serious kinds of stroke occur not from blockage but from hemorrhage, when a spot in a brain artery weakened by disease—usually atherosclerosis or high blood pressure—ruptures or begins to leak blood. If an artery inside the brain ruptures, it is called a **cerebral hemorrhage.** Hemorrhagic strokes account for less than 20 percent of all types of strokes but are far more lethal, with a death rate of over 50 percent. Strokes caused by clots or hemorrhage usually strike suddenly, with little or no warning, and do all their damage in a matter of seconds or minutes.

Cerebral hemorrhage—A stroke due to a ruptured brain artery.

Because blood clots play a major role in causing strokes, drugs that inhibit blood coagulation may prevent clot formation. Physicians have several drugs at their disposal—including aspirin—to treat those at risk. Aspirin works by preventing blood platelets from sticking together. Controlling blood pressure is also important.

Most people who have had mild strokes, and about half of those who have had moderate or severe paralysis on one side, recover enough to walk out of the hospital under their own steam or with some mechanical aid and resume their lives, though with certain limitations. Others are not so lucky.

Five treatable risk factors associated with stroke include high blood pressure, cigarette smoking, heart disease, history of stroke, and diabetes. High blood pressure is by far the most important risk factor.

■ **MINI-SUMMARY**

The majority of strokes are caused by blockages in the arteries that supply blood to the brain. Another type of stroke is called a hemorrhagic stroke, or cerebral hemorrhage.

High Blood Pressure

As many as 50 million Americans have high blood pressure (also called **hypertension**) or are taking antihypertensive medications. Because high blood pressure usually doesn't give early warning signs, it is known as the "silent killer." High blood pressure is one of the major risk factors for coronary heart disease and stroke. All stages of hypertension are associated with increased risk of nonfatal and fatal cerebrovascular disease and renal disease.

Hypertension—High blood pressure.

Arterial blood pressure—The pressure of blood within arteries as it is pumped through the body by the heart.

Arterial blood pressure is the pressure of blood within arteries as it's pumped through the body by the heart. Whether your blood pressure is high, low, or normal depends mainly on several factors: the output from your heart, the resistance to blood flow by your blood vessels, the volume of your blood, and blood distribution to the various organs.

Everyone experiences hourly and even moment-by-moment blood pressure changes. For example, your blood pressure will temporarily rise with strong emotions such as anger and frustration, with water retention caused by too much salty food that day, and with heavy exertion, which pushes more blood into your arteries. These transient elevations in blood pressure usually don't indicate disease or abnormality.

Systolic pressure—The pressure of blood within arteries when the heart is pumping—the top blood pressure number.

Diastolic pressure—The pressure in the arteries when the heart is resting between beats—the bottom number in blood pressure.

Blood pressure is represented as a fraction, as in 120/80. The top number, 120, is called the **systolic pressure**—the pressure of blood within arteries when the heart is pumping. The bottom number, 80, is called the **diastolic pressure**—the pressure in the arteries when the heart is resting between beats. Both blood pressure numbers are measured in millimeters of mercury, abbreviated mm Hg.

Normal blood pressure varies from person to person. High blood pressure occurs when the blood pressure stays too high, and is defined as systolic pressure greater than 140 mm Hg and/or a diastolic pressure greater than or equal to 90 mm Hg. Table 11-1 classifies blood pressure readings for adults. Even though hypertension starts with a systolic reading of 140 or higher, and/or a diastolic reading of 90 or higher, optimal blood pressure is about 120/80. People with high normal pressure account for more than one-third of preventable deaths related to blood pressure.

Systolic blood pressure is the key determinant for assessing the presence and severity of high blood pressure for middle-aged and older adults. In the past, many physicians relied on diastolic blood pressure to diagnose hypertension. But research has found that diastolic blood pressure rises until about age 55 and then declines, while systolic blood pressure increases steadily with age. Controlling systolic hypertension significantly reduces heart attack, heart failure, and stroke.

TABLE 11-1 Classification of Blood Pressure for Adults Age 18 and Older*

Category	Systolic mm Hg	Diastolic mm Hg
Normal	<130	<85
High normal	130–139	85–89
Hypertension**		
Stage 1 (Mild)	140–159	90–99
Stage 2 (Moderate)	160–179	100–109
Stage 3 (Severe)	180–209	110–119

* Not taking antihypertensive drugs and not acutely ill. When systolic and diastolic pressure fall into different categories, the higher category should be selected to classify the individual's blood pressure status.

** Based on the average of two or more readings taken at each of two or more visits following an initial screening.

Source: National Institutes of Health and National Heart, Lung, and Blood Institute.

Secondary hypertension—Persistently elevated blood pressure caused by a medical problem.

Primary (essential) hypertension—A form of hypertension whose cause is unknown.

When persistently elevated blood pressure is due to a medical problem, such as hormonal abnormality or an inherited narrowing of the aorta (the largest artery leading from the heart), it is called **secondary hypertension.** That is, the high blood pressure arises secondary to another condition. Only 5 percent of individuals with hypertension have secondary hypertension. The remaining 95 percent have what is called **primary (essential) hypertension.** The cause of essential hypertension is not well understood.

The prevalence of high blood pressure increases with age, is greater for blacks than for whites, and in both races is greater in less-educated than in more-educated people. It is especially prevalent and devastating in lower socioeconomic groups. In young adulthood and early middle age, high blood pressure prevalence is greater for men than for women. Thereafter, more women have high blood pressure than men.

The following lifestyle modifications offer some hope for prevention of hypertension and are effective in lowering the blood pressure of many people who follow them (Table 11-2):

- Weight reduction
- Increased physical activity
- Adequate dietary intake of potassium, calcium, magnesium, and vitamin C (found in fruits, vegetables, and dairy products)
- Moderation of alcohol intake
- Moderation of dietary sodium

These lifestyle modifications can also reduce other risk factors for premature cardiovascular disease. Because of their ability to improve the cardiovascular risk profile, lifestyle modifications offer many benefits at little cost and with minimal risk.

Excess body weight is correlated closely with increased blood pressure. Weight reduction reduces blood pressure in a large proportion of hypertensive individuals who are more than 10 percent above ideal weight.

TABLE 11-2 Lifestyle Modifications for Hypertension Control and/or Overall Cardiovascular Risk

- Lose weight if overweight.
- Limit alcohol intake to no more than 1 ounce of ethanol per day (24 ounces of beer, 8 ounces of wine, or 2 ounces of 100-proof whiskey).
- Exercise (aerobic) regularly.
- Reduce sodium intake to less than 2300 milligrams per day.
- Maintain adequate dietary potassium, calcium, magnesium, and vitamin C intake.
- Stop smoking, and reduce dietary saturated fat and cholesterol intake for overall cardiovascular health. Reducing fat intake also helps reduce caloric intake—important for control of weight and type 2 diabetes.

Source: National Institutes of Health and National Heart, Lung, and Blood Institute.

Excessive alcohol intake can raise blood pressure and cause resistance to antihypertensive therapy. Hypertensive patients who drink alcohol-containing beverages should be counseled to limit their daily intake to 1 ounce of ethanol (2 ounces of 100-proof whiskey), 8 ounces of wine, or 24 ounces of beer.

Regular aerobic physical activity, adequate to achieve at least a moderate level of physical fitness, may be beneficial for both prevention and treatment of hypertension. Sedentary and unfit individuals with normal

TABLE 11-3 Dietary Approaches to Stop Hypertension (DASH) Diet

This meal plan is based on 2000 calories a day. Depending on your calorie needs, your number of daily servings may vary from those listed. Consult your doctor or a dietitian to determine your calorie needs.

Food Group	Daily Servings	Serving Size
Grains and grain products	7 to 8	1 slice bread 2 to 1-1/4 cup dry cereal 2 cup cooked rice, pasta, or cereal
Vegetables	4 to 5	1 cup raw leafy vegetables 2 cup cooked vegetable 6 oz vegetable juice
Fruits	4 to 5	6 oz fruit juice 1 medium fruit 1/4 cup dried fruit 2 cup fresh, frozen, or canned fruit
Low-fat or nonfat dairy foods	2 to 3	8 oz milk 1 cup yogurt 1-1/2 oz cheese
Meats, poultry, fish	2 or fewer	3 oz cooked lean meat, poultry (skinless white meat), or fish
Nuts, seeds, and dry beans	4 to 5 per week	1/3 cup nuts 2 Tbsp seeds 1/2 cup legumes
Fats and oils	2 to 3	1 tsp soft margarine or butter 1 tsp regular mayonnaise *or* 1 Tbsp low-fat mayonnaise 1 Tbsp salad dressing *or* 2 Tbsp light salad dressing 1 tsp oil (olive, corn, canola, safflower, or other)
Sweets	5 per week	1 Tbsp maple syrup, sugar or jelly 1/2 cup sherbet 3 pieces of hard candy

Source: National Heart, Lung, and Blood Institute.

blood pressure have a 20 to 50 percent increased risk of developing hypertension during follow-up when compared with their more active and fit peers.

Salt affects blood pressure. The lower the amount of sodium in the diet, the lower the blood pressure, for both those with and without hypertension.

Calcium, potassium, magnesium, and vitamin C also need to be in the diet in adequate supply to normalize blood pressure. In fact, the DASH diet (Table 11-3, Dietary Approaches to Stop Hypertension) recommends 8 to 10 servings of fruits and vegetables (great sources of potassium, vitamin C, and magnesium) and 2 to 3 servings of dairy products daily.

When lifestyle modifications do not succeed in lowering blood pressure enough, drugs are the next step. Reducing blood pressure with drugs clearly decreases the incidence of cardiovascular death and disease.

■ MINI-SUMMARY

High blood pressure occurs when the blood pressure stays too high and is defined as a systolic pressure greater than 140 mm Hg and/or a diastolic pressure greater than or equal to 90 mm Hg. Table 11-2 lists the components of lifestyle modifications that can lower blood pressure. Table 11-3 shows the DASH diet, which has been shown to reduce blood pressure. If they do not bring blood pressure down enough, drug treatment is the next choice.

Menu Planning for Cardiovascular Diseases

Menu planning for cardiovascular diseases revolves around offering dishes rich in complex carbohydrates and fiber and using small amounts of fat, saturated fat, cholesterol, and sodium.

General Recommendations

- Decrease or replace salt in recipes by using vegetables, herbs, spices, and flavorings.
- Offer salt-free seasoning blends and lemon wedges.
- Read the Nutrition Facts label—especially the percent Daily Value of fat, saturated fat, and sodium.

Breakfast

- Offer fresh and canned fruits and juices.
- Almost all cold and hot cereals are great choices. Granola cereals tend to be high in fat unless labeled as reduced-fat.

- Most breads are low in fat except for croissants, brioche, cheese breads, and many biscuits. Bagels, low-fat muffins, and baguettes are good choices.
- Have reduced-fat margarine and light cream cheese available to spread on bagels or toast.
- Serve turkey sausage, low-fat and low-sodium ham slices, or fish as leaner sources of protein than the traditional bacon and pork sausage.
- Offer egg substitutes for scrambled eggs and other egg-based items. Egg substitutes taste better when herbs, flavorings, and/or vegetables are cooked with them. Instead of egg substitutes, you can offer to make scrambled eggs and omelets by mixing one whole egg to two egg whites.
- Serve an omelet with blanched vegetables such as chopped broccoli or spinach and low-fat cheese instead of regular cheese.
- As spreads or toppings on pancakes and waffles, offer sauces combining low-fat or nonfat yogurt with a fruit puree.
- Serve a breakfast buffet with loads of fruits, low-fat dairy products, and cereals.

Appetizers and Soups

- Offer juices and fresh sliced fruits. Fresh sliced fruits can be served with a yogurt dressing flavored with fruit juices.
- Offer raw vegetables with dips using low-fat yogurt, low-fat cottage cheese, or ricotta cheese as the base, rather than dips using sour cream, cheeses, cream, or cream cheese. Try hummus, a chickpea-based dip, or salsa made from tomatoes, onions, hot peppers, garlic, and herbs.
- Offer grilled chicken, broiled Buffalo-style chicken wings, or steamed seafood such as shrimp.
- Use baked (rather than fried) potato skins and baked corn tortillas for tortilla chips. Sprinkle with grated cheese, or garlic, onion, or chili powder.
- Feature soups that use stock as the base and vegetables and grains as the ingredients. Dried beans, peas, and lentils make great soups when cooked and puréed, without using cream or high-fat thickeners such as roux.

Salads

- Offer salads with lots of vegetables and fruits.
- Use only small amounts, if any, of bacon, meat, cheese, eggs, or croutons. Choose cooked beans and peas or low-fat cheeses.
- Offer reduced-calorie or nonfat salad dressings. Place on the side when desired.
- Make tuna fish salad and similar salads with low-fat mayonnaise.
- Use cooked salad dressing that contains little fat for Waldorf and other salads. It has a tarter flavor than mayonnaise.

Breads

- Most breads are low in fat. Breads with more fat include biscuits, cheese breads, croissants, popovers, brioche, corn bread, and many commercial crackers (although low-fat varieties are available or can be made with less fat).

Entrées

- Serve combination dishes with small amounts of meat, poultry, or seafood with whole grains such as rice, legumes, vegetables, and/or fruits.
- Offer moderate portions of broiled, baked, stir-fried, or poached seafood, white-meat poultry without skin, and lean cuts of meat (see Chapter 4).
- Offer fresh meat, poultry, or seafood instead of canned, cured, smoked, and otherwise salty items (such as ham, corned beef, smoked turkey, dried cod, and most luncheon meats). Cheeses are high in sodium, so use only small amounts in sandwiches.
- Feature freshly made entrées instead of processed or prepared foods.
- Offer a hamburger, meat loaf, or other ground-beef dishes made with low-fat ground beef.
- Feature one or more meatless entrées, such as vegetarian burgers. Vegetarian burgers either try to imitate beef burgers (usually through the use of soy products) or are real veggie burgers (made of vegetables, especially mushrooms).
- For sauced entrées (or side dishes), feature sauces thickened with flour, cornstarch, or vegetable purées. Salsas, chutneys, relishes, and coulis also work well.
- Offer sandwiches made with roasted turkey, chicken, water-packed tuna fish salad, lean roast beef made from the round, or a spicy bean or lentil spread.
- For sandwich spreads, use reduced-fat mayonnaise, French or Russian-style salad dressing, mustard, ketchup, barbecue sauce, or salsa.
- Feature lots of different vegetables in sandwiches.
- Instead of high-sodium accompaniments to sandwiches such as pickles, olives, and potato chips, serve fresh vegetables, coleslaw made with reduced-calorie mayonnaise, or another healthful salad.

Side Dishes

- Most side dishes of vegetables, grains, and pasta are good choices as long as little fat is added during preparation.
- Serve grilled potato halves instead of french fries, as well as other grilled vegetables.

Desserts

- Offer fruit-based desserts such as apple cobbler.
- Spotlight sorbets, sherbets, frozen yogurt, and ice milk. All contain less fat than ice cream does.
- Feature desserts made from fat-free egg whites, such as angel food cake and meringues. Serve with a fruit sauce.
- Offer puddings made with skim milk.
- Serve low-fat cookies such as ladyfingers, biscotti, gingerbread, and fruit bars.

Beverages

- Offer 1-percent or skim milk.

■ MINI-SUMMARY

Menu planning for cardiovascular diseases revolves around offering dishes rich in complex carbohydrates and fiber and using small amounts of fat, saturated fat, cholesterol, and sodium.

Nutrition and Cancer

Cancer—A group of diseases characterized by unrestrained cell division and growth that can disrupt the normal functioning of an organ and also spread beyond the tissue in which it started.

Carcinogen—Cancer-causing substance.

Promoters—Substances such as fat that advance the development of mutated cells into a tumor.

Metastasis—The condition when a cancer spreads beyond the tissue in which it started.

Following heart disease, **cancer** is the second leading cause of death in the United States. The good news about cancer is that people being treated for cancer are living longer. The bad news is that more of it is being diagnosed; however, there is no evidence of a cancer epidemic. The most prevalent form of cancer is skin cancer, which is quite curable when treated early. More than 90 percent of skin cancers are completely cured. The most frequent malignant cancer found in women is breast cancer.

Cancer is a group of diseases characterized by unrestrained cell division and growth that can disrupt the normal functioning of an organ and spread beyond the tissue in which it started. Figure 11-2 is a diagram of this process. Cancer is basically a two-step process. First, a **carcinogen,** such as an X-ray, initiates the sequence by altering the genetic material of a cell, the deoxyribonucleic acid (DNA), and causing a mutation. Such cells are generally repaired or replaced. When repair or replacement does not occur, however, **promoters** such as alcohol can advance the development of the mutated cell into a tumor. Promoters do not initiate cancer but enhance its development once initiation has occurred. The tumor may disrupt normal body functions and leave the tissue for other sites, a process called **metastasis.**

Cancer develops as a result of interactions between environmental factors (such as diet, smoking, alcohol, and radiation) and genetic factors. Research suggests that diet plays a role in the cause of certain cancers.

Figure 11-2
Process of cancer.

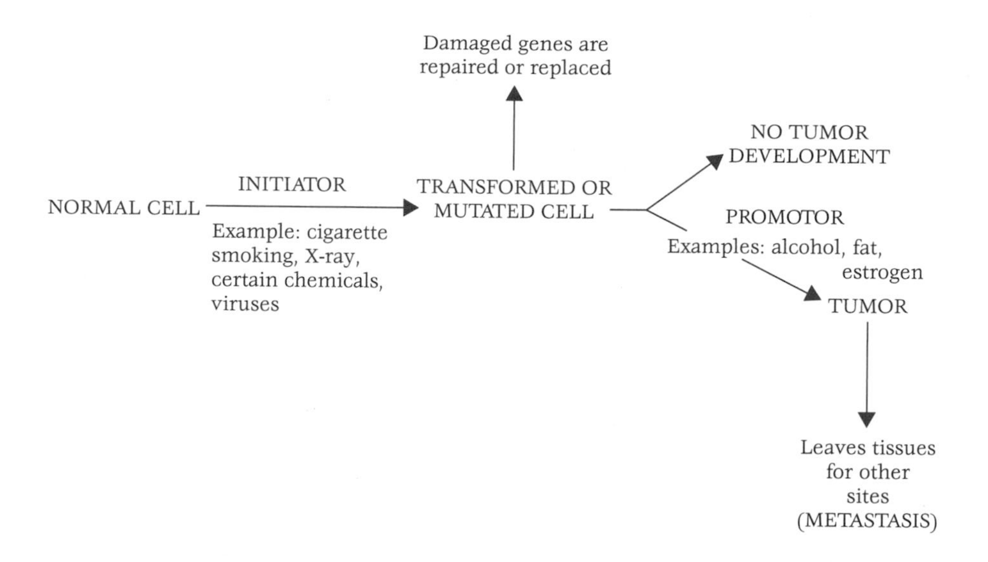

Perhaps up to 33 percent of all cancer deaths are associated with the typical American diet, which is high in fat, red meat, and kcalories, and low in fiber, fruits, and vegetables. The most common cancers are cancer of the lung, breast, colon and rectum, prostate, bladder, and skin. Of these cancers, all but one (skin cancer) is associated with diet.

The American Cancer Society has four guidelines to reduce cancer risk.

1. Eat a variety of healthful foods, with an emphasis on plant sources.
   - Eat five or more servings of a variety of vegetables and fruits each day.
   - Choose whole grains in preference to processed (refined) grains and sugars.
   - Limit consumption of red meats, especially those high in fat and processed.
   - Choose foods that help maintain a healthful weight.
2. Adopt a physically active lifestyle.
   - Adults: engage in at least moderate activity for 30 minutes or more on 5 or more days of the week; 45 minutes or more of moderate to vigorous activity on 5 or more days per week may further enhance reductions in the risk of breast and colon cancer.
   - Children and adolescents: engage in at least 60 minutes per day of moderate to vigorous physical activity at least 5 days per week.
3. Maintain a healthful weight throughout life.
   - Balance caloric intake with physical activity.
   - Lose weight if currently overweight or obese.
4. If you drink alcoholic beverages, limit consumption.

Phytochemicals—Minute substances in plants that may reduce risk of cancer and heart disease when eaten often.

Fruits and vegetables, as well as other plant foods, contain **phytochemicals,** minute plant compounds that fight cancer formation. For instance, broccoli contains the chemical sulforaphane, which seems to initiate increased production of cancer-fighting enzymes in the cells. Isoflavonoids, found mostly in soy foods, are known as plant estrogens or phytoestrogens because they are similar to estrogen and interfere with its actions (estrogen seems to promote breast tumors). Members of the cabbage family (cabbage, broccoli, cauliflower, mustard greens, kale), also called **cruciferous vegetables,** contain phytochemicals such as indoles and dithiolthiones. They activate enzymes that destroy carcinogens.

Cruciferous vegetables—Members of the cabbage family containing phytochemicals that might help prevent cancer.

Some consumers are concerned about eating more fruits and vegetables that may contain carcinogenic pesticides. The National Academy of Sciences, along with other organizations, feels that the health benefits of eating fresh fruits and vegetables far outweigh any risk associated with pesticide residues. The federal government strictly regulates the kinds and amounts of pesticides used on field crops. The tiny amounts of pesticide residues found on produce are set hundreds of times lower than the amounts that would actually pose any health threat.

■ **MINI-SUMMARY**

Cancer begins as depicted in Figure 11-2. The components of American diet that raise our cancer risk include too much fat, too much red meat, too little fiber, and too few fruits and vegetables. Fiber, fruits, and vegetables lower cancer risk. Fruits and vegetables also contain phytochemicals thought to fight the formation of cancer.

Menu Planning to Lower Cancer Risk

In the long run, prevention of cancer, and therefore eating a healthy diet, will play the major role in its control. Use the following guidelines to plan menus to lower cancer risk.

1. Offer lower-fat menu items. See Chapter 4 for tips on lowering fat and saturated fat. Also offer more plant-based menu items.
2. Avoid salt-cured, smoked, and nitrite-cured foods. These foods, which are also high in fat, include anchovies, bacon, corned beef, dried chipped beef, herring, pastrami, processed lunch meats such as bologna and hot dogs, sausage such as salami and pepperoni, and smoked meats and cheeses. Conventionally smoked meats and fish contain tars that are thought to be carcinogenic due to the smoking process. Nitrites are known carcinogens.
3. Offer high-fiber foods. For example:
   - Use beans and peas as the basis for entrées, and add them to soups, stews, casseroles, and salads. Nuts and seeds are high in fiber but also contain a significant amount of fat and kcalories, so use them sparingly.

- Serve whole-grain breads, rolls, crackers, cereals, and muffins. Bran or wheat germ can be added to some baked goods to increase the fiber content.
- High-fiber grains such as brown rice and bulgur (cracked wheat) can be used as side dishes instead of white rice.
- Leave skins on potatoes, fruits, and vegetables as much as possible.
- Offer salads using lots of fresh fruits and vegetables. Omit chopped eggs and bacon bits, both of which contribute fat.

4. Include lots of vegetables, especially cruciferous vegetables, and fruits. Cruciferous vegetables contain substances that are natural anti-carcinogens. These vegetables include broccoli, brussels sprouts, cabbage, cauliflower, bok choy, kale, collards, kohlrabi, mustard, rutabagas, spinach, and watercress. Wash fruits and vegetables well.
5. Offer foods that are good sources of beta-carotene and vitamins C and E, all antioxidants. Excellent sources of beta-carotene include dark green, yellow, and orange vegetables and fruits such as broccoli, cantaloupe, carrots, spinach, squash, and sweet potatoes. Good sources include apricots, beet greens, brussels sprouts, cabbage, nectarines, peaches, tomatoes, and watermelon. Sources of vitamin C include citrus fruits and juices, any other juices with vitamin C added, berries, tomatoes, broccoli, brussels sprouts, cabbage, melons, cauliflower, and potatoes. Vitamin E is found in vegetable oils and margarines, whole-grain cereals, wheat germ, soybeans, leafy greens, and spinach.
6. Offer alternatives to alcoholic drinks. Heavy drinkers are more likely to develop cancer in the gastrointestinal tract, such as cancer of the esophagus and stomach. Chapter 10 presents nonalcoholic alternatives.

■ MINI-SUMMARY

The keys to menu planning are lower-fat foods; foods that have not been salt-cured, nitrite-cured, or smoked; high-fiber foods; lots of fruits and vegetables, including cruciferous vegetables; and nonalcoholic drinks for heavy drinkers.

Nutrition and Diabetes Mellitus

Diabetes mellitus—A disorder of carbohydrate metabolism characterized by high blood sugar levels and inadequate or ineffective insulin.

Diabetes mellitus gets its name from the ancient Greek word for siphon (tube), because early physicians noted that diabetics tend to be unusually thirsty and to urinate a lot, as if a tube quickly drained out everything they drank. *Mellitus* is from the Latin version of the ancient Greek word for honey, used because doctors in centuries past diagnosed the disease by the sweet taste of the patient's urine.

The number of people diagnosed with diabetes has increased more than sixfold, from 1.6 million in 1958 to 10 million in 1997. Today, some 16 million people have the disease, yet 5 million don't know they have it. And nearly 800,000 new cases of diabetes are diagnosed each year.

Diabetes is a disease in which there is insufficient or ineffective insulin, a hormone that helps regulate blood sugar level. When the blood sugar rises, such as after eating a meal, the pancreas releases insulin. The insulin facilitates the entry of glucose into body cells to be used for energy. If there is no insulin or if the insulin is not working, sugar cannot enter the cells. Thus, high blood sugar levels (called **hyperglycemia**) result and sugar spills into the urine.

Hyperglycemia—High levels of blood sugar.

The life expectancy for a diabetic is only two-thirds that of the nondiabetic, and diabetes is the fourth leading cause of death. Risk factors for the most common type of diabetes include advanced age, family history, obesity, high waist-to-hip ratio, and a high-fat/low-carbohydrate diet.

People with diabetes are more vulnerable to many kinds of infections and to deterioration of the kidneys, heart, blood vessels, nerves, and vision. The National Institutes of Health estimates that more than 250,000 Americans per year die from the complications of this illness, largely because it doubles their chances of having heart attacks and strokes. In addition, diabetes is the nation's leading cause of kidney failure and adult blindness. Because of the damage diabetes can do to the blood vessels and nerves of the lower limbs, only accidents necessitate more amputations of the toes, feet, and legs.

There are two types of diabetes.

1. **Type 1 diabetes (insulin-dependent).** This form of diabetes is seen mostly in children and adolescents. These patients produce no insulin at all and therefore require frequent insulin injections to maintain a normal level of blood glucose. Fewer than 10 percent of Americans who have diabetes have type 1.
2. **Type 2 diabetes (non-insulin-dependent).** This form of diabetes is a separate disease from type 1. Patients are usually older and obesity is common; in fact, 80 percent of patients weigh 20 percent or more than they should. Obese children may also be at risk of developing Type 2 diabetes. In type 2, the person's beta cells do, in fact, produce insulin and may make too much. The problem here is that the patient's tissues aren't sensitive enough to the hormone and so use it inefficiently. Some of these patients require insulin, but many do not. Treatment is with diet, weight reduction when appropriate, exercise, and if necessary oral hypoglycemic agents (medications taken by mouth that can stimulate the release of insulin and improve the body's sensitivity to insulin). Most diabetics fall into this category.

Type 1 diabetes—A form of diabetes seen mostly in children and adolescents. These patients make no insulin and therefore require frequent injections of insulin to maintain a normal level of blood glucose.

Type 2 diabetes—A form of diabetes seen most often in overweight adults. These patients make insulin but their tissues aren't sensitive enough to the hormone and so use it inefficiently.

Type 2 diabetes accounts for more than 90 percent of cases in the United States. Type 2 diabetes is nearing epidemic proportions in the United States, due to an increased number of older Americans and a greater prevalence of obesity and sedentary lifestyles. Characteristics of type 2 diabetes include the following.

- Age of onset over 40 years old in most cases, although alarming numbers of obese adolescents are being diagnosed, too
- Most frequently occurs in overweight individuals; occasionally occurs in people of normal weight
- Usually slow onset with thirst, frequent urination, and weight loss symptoms developing over weeks to months, or even years
- Can be a "silent" disease
- Usually runs in families
- Most often treated first with diet and exercise, progressing to pills and later to insulin if needed
- Easier to control without fluctuating blood sugar range
- May respond to weight loss and/or change in diet and exercise
- Can be caused by combination of heredity, insulin resistance, and deficiency of the insulin-producing beta cells of the pancreas

Although both types of diabetes are popularly called "sugar diabetes," they are not caused by eating too many sweets. High sugar levels in the blood and urine are a result of these illnesses; their exact causes are unknown. What is clear is that type 2 runs in families far more often than type 1 does, and—unlike type 1, which cannot be prevented— it can frequently be avoided by staying in shape.

Symptoms of type 1 diabetes typically appear abruptly and include excessive, frequent urination, insatiable hunger, and unquenchable thirst. Unexplained weight loss is also common, as are blurred vision (or other vision changes), nausea and vomiting, weakness, drowsiness, and extreme fatigue.

The most immediate and life-threatening aspect of type 1 diabetes is the formation of poisonous acids called *ketone bodies.* They occur as an end product of burning fat for energy, because glucose does not get into the cells to be burned for energy. Like glucose, ketone bodies accumulate in the blood and spill into the urine. Ketone bodies are acidic and make the blood acidic, resulting in a dangerous condition called ketoacidosis.

Symptoms of type 2 diabetes may include any or all of those of type 1 except that the problem with ketone bodies is rare. Symptoms are often overlooked because they tend to come on gradually and be less pronounced. Other symptoms are tingling or numbness in the lower legs, feet, or hands; skin or genital itching; and gum, skin, or bladder infections that recur and are slow to clear up. Again, many people fail to connect these symptoms with possible diabetes.

Measuring glucose levels in samples of the patient's blood is key to diagnosing both types of diabetes. This is first done early in the morning on an empty stomach. For more information, the blood test is repeated, usually on another day, before the patient has drunk a liquid containing a known amount of glucose and at intervals thereafter.

Treatment for either type seeks to do what the human body normally does naturally: maintain a proper balance between glucose and insulin. The guiding principle is that food makes the blood glucose level rise, whereas insulin and exercise make it fall. The trick is to juggle the three factors to avoid both hyperglycemia (a blood glucose level that is too high) and hypoglycemia (one that is too low). Either condition causes problems for the patient.

The cornerstone of treatment is diet, exercise, and medication. The exact nature of the diet has changed over the years from starvation diets to more liberal diets. The current diet is based on the following principles:

- There is no one diet suitable for every person with diabetes. The diet needs to be individualized based on each person's type of diabetes, food preferences, culture, age, lifestyle, medication, other health concerns, education, nutrition status, medical treatment goals, and other factors.
- The goals for meal planning are to maintain the best glucose control possible, keep blood levels of fat and cholesterol in normal ranges, maintain or get body weight within a desirable range, and meet all nutrient needs.
- Sugars can be incorporated into the meal pattern in moderation as part of the total carbohydrate allowance. The total amount of carbohydrate consumed, rather than the source of carbohydrate, should be the priority. Because sugars appear in foods that usually contain a lot of kcalories and fat, sweets can be eaten occasionally.
- Instead of setting rigid percentages of protein, fat, and carbohydrates, guidelines recommend that protein make up 10 to 20 percent of kcalories consumed. Saturated fat and polyunsaturated fat should be maintained at less than 10 percent each. The remaining 60 to 70 percent of kcalories should come from carbohydrates and monounsaturated fats. The meal pattern should provide sufficient fiber. Kcaloric distribution depends on the individual's nutritional assessment and treatment goals.

The *Exchange Lists for Meal Planning* have been developed by the American Diabetes Association and the American Dietetic Association for use primarily by people with diabetes, who need to regulate what and how much they eat. There are seven exchange lists of like foods. Each food on a list has approximately the same amount of kcalories, carbohydrate, fat, and protein as another in the portions listed, so that any food on a list can be exchanged, or traded, for any other food on the same list (see Table 11-4). The seven exchange lists are starch, fruit, milk, other carbohydrates, vegetables, meat and meat substitutes, and fat. People with diabetes can exchange starch, fruit, or milk choices within their meal plans because they all have about the same amount of carbohydrates per serving.

TABLE 11-4 Nutrient Content of Exchange Lists

Groups/Lists	Typical Item	Carbohydrate (grams)	Protein (grams)	Fat (grams)	Kcalories
Carbohydrate Group					
Starch	1 slice bread	15	3	1 or less	80
Fruit	1 small apple	15	—	—	60
Milk					
Skim	1 cup	12	8	0–3	90
Low-fat	1 cup	12	8	5	120
Whole	1 cup	12	8	8	150
Other carbohydrates	2 small cookies	15	varies	varies	varies
Vegetables	1/2 cup cooked carrots	5	2	—	25
Meat and Meat Substitute Group					
Very lean	1 oz. chicken (no skin, white)	—	7	0–1	35
Lean	1 oz. lean beef	—	7	3	55
Medium-fat	1 oz. ground beef	—	7	5	75
High-fat	1 oz. pork sausage	—	7	8	100
Fat Group		—	—	5	45

Adapted from: *Exchange Lists for Meal Planning,* 1995, American Diabetes Association and American Diabetic Association.

Each exchange list has a typical item with an easy-to-remember portion size:

Starch—1 slice bread, 80 kcalories
Meat—1 ounce lean meat, 55 kcalories
Vegetable—1/2 cup cooked vegetable, 25 kcalories
Fruit—1 small apple, 60 kcalories
Milk—1 cup skim milk, 90 kcalories
Other carbohydrates—2 small cookies, kcalories vary
Fat—1 teaspoon margarine, 45 kcalories

The meat exchange is broken down into very lean, lean, medium-fat, and high-fat meat and meat alternatives. Very lean and lean meats are encouraged. The milk exchange contains skim, low-fat, and whole-milk exchanges. Fats are divided into three groups, based on the main type of fat they contain: monounsaturated, polyunsaturated, or saturated. There is also a listing of free foods that contain negligible kcalories.

■ MINI-SUMMARY

There are two classifications of diabetes: type 1 (insulin-dependent), seen mostly in children, and type 2 (non-insulin-dependent), seen mostly in overweight older adults. The life expectancy for a person with diabetes is only two-thirds that of the non-diabetic, and diabetes is the fourth leading cause of death. People with diabetes are more vulnerable to many kinds of infections and to deterioration of the kidneys, heart, blood vessels, nerves, and vision. The cornerstone of treatment is diet, exercise, and medication. Diets are individualized for each person. Sugar need not be avoided but can be incorporated into the diabetic diet as part of the total carbohydrate allowance.

Vegetarian Eating

The number of vegetarians in the United States has been increasing. Instead of eating the meat entrées that have traditionally been the major source of protein in the American diet, they dine on main dishes emphasizing legumes (dried beans and peas), grains, and vegetables. Vegetarian entrées, such as red beans and rice, can supply adequate protein with less fat and cholesterol, and more fiber than their meat counterparts (see Table 11-5).

Whereas vegetarians do not eat meat, poultry, or fish, the largest group of vegetarians, referred to as **lacto-ovo vegetarians,** do consume animal products in the form of eggs (*ovo-*) and milk and milk products (*lacto-*).

Another group of vegetarians, **lacto vegetarians,** consume milk and milk products but forgo eggs. Most vegetarians are either lacto-ovo vegetarians or lacto vegetarians. **Vegans,** a third group of vegetarians, do not eat eggs or dairy products and therefore rely exclusively on plant foods to meet protein and other nutrient needs. Vegans are a small group, and it is estimated that only 4 percent of vegetarians are vegans.

In addition, some vegetarians **(pesco vegetarians)** eat seafood. Also, some vegetarian diets restrict certain foods and beverages, such as highly processed foods containing certain additives and preservatives, foods that contain pesticides and/or have not been grown organically, or caffeinated or alcoholic beverages.

The number-one reason people give for being vegetarian is health. Being a vegetarian has health benefits. Vegetarians tend to be leaner and to keep their body weight and blood lipid levels closer to desirable levels than non-vegetarians. Vegetarians tend to have a lower incidence of the following diseases:

1. Hypertension
2. Coronary artery disease
3. Colon and lung cancer
4. Type 2 diabetes
5. Diverticular disease of the colon

Lacto-ovo vegetarians— Vegetarians who do not eat meat, poultry, or fish but do consume animal products in the form of eggs, milk, and milk products.

Lacto vegetarians— Vegetarians who do not eat meat, poultry, or fish but do consume animal products in the form of milk and milk products.

Vegans—Individuals eating a type of vegetarian diet in which no eggs or dairy products are eaten; their diet relies exclusively on plant foods.

Pesco vegetarians— Vegetarians who eat fish.

TABLE 11-5 Fat, Saturated Fat, Protein, and Cholesterol in Animal and Plant Foods

Animal Foods	Fat (grams)	Saturated Fat (grams)	Protein (grams)	Cholesterol (milligrams)
Beef, ground, broiled, 3 oz.	16	6	21	74
Chicken breast, roasted, 3 oz.	3	1	27	73
Cod, baked, 3 oz.	1	0	19	47
Milk, 2%, 8 fl. oz.	5	3	8	18
Cheese, American, 1 oz.	9	6	6	27
Egg, 1	6	2	6	274
Plant Foods	**Fat (grams)**	**Saturated Fat (grams)**	**Protein (grams)**	**Cholesterol (milligrams)**
Lentils, cooked, 1/2 cup	0	0	9	0
Brown rice, cooked, 1/2 cup	1	0	2	0
Spaghetti, whole-wheat, 1 cup	1	0	7	0
Whole-wheat bread, 2 slices	2	0	5	0
Broccoli, chopped 1 cup	0	0	6	0
Apple, 1 medium	0	0	0	0

Sources: United States Department of Agriculture Handbooks. Numbers 8-1, 8-5, 8-13, 8-15, 8-20.

A comprehensive study of rural Chinese suggested that eating much less animal protein and fat (and more complex carbohydrates) results in reductions of blood cholesterol levels and the chronic degenerative diseases (such as heart disease and cancer) associated with high blood cholesterol levels.

Being vegetarian does not mean that you automatically get these benefits. There are vegetarians who eat well-balanced and varied diets, and then there are vegetarians who eat 3 eggs for breakfast, 1/4 cup of peanut butter for lunch, and 3 slices of pizza for supper. In other words, it is possible to be a vegetarian and still eat too much fat, saturated fat, and cholesterol. It's probably the exception, rather than the rule, but it is still possible. Being vegetarian does not guarantee that your diet will meet current dietary recommendations. Some other reasons for becoming vegetarian include the following:

1. **Ecology.** For ecological reasons, vegetarians choose plant protein because livestock and poultry require much land, energy, water, and plant food (such as soybeans), which they consider wasteful. According to the North American Vegetarian Society, grains and soybeans that are fed to U.S. livestock could feed 1.3 billion people. Livestock also waste loads of water—it takes 2,500 gallons of water to produce one pound of meat, but only 25 gallons to produce one pound of wheat.
2. **Economics.** A vegetarian diet is more economical—in other words, less expensive. This can be easily demonstrated by the fact that in a typical

foodservice operation, the largest component of food purchases is for meats, poultry, and fish.

3. **Ethics.** Vegetarians do not eat meat for ethical reasons; they believe that animals should not suffer or be killed unnecessarily. They feel that animals suffer real pain in crowded feed lots and cages, and that both their transportation to market and their slaughter are traumatic.
4. **Religious beliefs.** Some vegetarians, such as the Seventh-Day Adventists, practice vegetarianism as a part of their religion, which also encourages exercise and forbids smoking and drinking alcohol.

Nutritional Adequacy of Vegetarian Diets

Vegetarian diets can be nutritionally adequate when varied and adequate in kcalories (except for vegan diets, which need supplementation with vitamin B_{12}). Most vegetarians get enough protein, and their diets are typically lower in fat, saturated fat, and cholesterol.

As discussed in Chapter 5, most plant proteins are considered incomplete, but this doesn't mean they are low in quality. When plant proteins are eaten with other foods, the food combinations usually result in complete protein. For example, when peanut butter and whole-wheat bread are eaten over the course of a day, the limiting amino acid in each of these foods is supplied by the other food. Such combinations are called complementary proteins. Eating complementary proteins at different meals during the day generally ensures a balance of dietary amino acids. Some vegetable proteins, such as those found in amaranth, quinoa, and soybeans, are complete proteins.

Let's take a look at some of the nutrients that need some special attention.

1. **Vitamin B_{12}.** Vitamin B_{12} is found only in animal foods. Lacto-ovo vegetarians usually get enough of this vitamin unless they limit their intake of dairy products and eggs. Vegans definitely need either a supplement or vitamin B_{12}–fortified foods, such as most ready-to-eat cereals, most meat analogs, some soy beverages, and some brands of nutritional yeasts.
2. **Vitamin D.** Milk is fortified with vitamin D, and vitamin D can be made in the skin with sunlight. Generally, only vegans without enough exposure to sunlight need a supplementary source of vitamin D. Some ready-to-eat breakfast cereals and some soy beverages are fortified with vitamin D.
3. **Calcium.** Lacto vegetarians and lacto-ovo vegetarians generally don't have a problem here, but vegans sometimes do if they don't eat enough other calcium-rich foods. Good choices include calcium-fortified soy milk or orange juice and tofu made with calcium sulfate. Some green leafy vegetables (such as spinach, beet greens, Swiss chard, sorrel, and parsley) are rich in calcium, but they also contain a binder (called oxalic

acid) that prevents some of the calcium from being absorbed. Dried beans and peas are moderate sources of calcium. Without calcium-fortified drinks or calcium supplements, it can be difficult to consume enough calcium.

4. **Iron.** Interestingly enough, vegetarians do not experience any more problems with iron-deficiency anemia than their meat-eating counterparts—don't forget, meat is rich in iron. Iron is widely distributed in plant foods, and its absorption is greatly enhanced by vitamin-C-containing fruits and vegetables. Vegetarians get iron from eating dried beans and peas, green leafy vegetables, dried fruits, many nuts and seeds, and enriched and whole-grain products.
5. **Zinc.** Zinc is found in many plant foods, such as whole grains, legumes, and nuts and seeds (especially peanut butter). Its absorption into the body is reduced by certain plant substances, such as phytate. Children may need zinc supplements.

Infants, children, and adolescents can follow vegetarian diets, even vegan diets. For growing youngsters, however, vegetarian diets need to be well planned, varied, and adequate in kcalories. In the case of a vegan diet, special attention should be focused on getting enough kcalories, vitamin B_{12}, vitamin D, calcium, iron, zinc, and linolenic acid. A reliable source of vitamin B_{12} is particularly essential. Soy milk that has been fortified with calcium and vitamin D will help meet the needs for those vitamins. Meat analogs, soy products, legumes, and nut butters are valuable sources of protein, and some provide iron, zinc, and/or linolenic acid. If inadequate amounts of these nutrients are taken in from food, supplements are always available.

■ **MINI-SUMMARY**

Reasons for becoming vegetarian may be related to health benefits, economics, ecology, ethics, or religious beliefs. Vegetarian diets can be nutritionally adequate when varied and adequate in kcalories, except for vegan diets, which need supplementation with vitamin B_{12}. Nutrients of special interest to vegetarians are vitamin B_{12}, vitamin D, calcium, iron, zinc, and linolenic acid.

Vegetarian Food Pyramid

The Food Guide Pyramid in Chapter 2, when modified as described here, works well for vegetarians.

1. **Breads, Cereals, Rice, and Pasta** (6–11 servings per day). This group stays the same. Whole-grain products are recommended. Vitamin B_{12}–fortified breakfast cereals are important for vegans. Some vegans with large kcalorie needs may eat more than 11 servings.

2. **Fruits** (2 or more servings). This group also stays the same, except that no limit is placed on consumption.
3. **Vegetables** (3 or more servings). As with fruits, no limit is placed on the number of servings.
4. **Legumes, Nuts, Seeds, Eggs, and Other Meat Substitutes** (2–3 servings). This group obviously omits any meats, poultry, or fish and instead concentrates on substitutes such as cooked dry beans, peas, or lentils; tofu and other soybean products; nuts and seeds, and butters made from them; meat analogs; and eggs for lacto-ovo vegetarians. One-half of a serving is 1/2 cup of cooked beans, peas, or lentils; 4 ounces of tofu; 1/4 cup of shelled nuts; 1/8 cup of seeds; 2 tablespoons of peanut butter; or 1 egg. One serving of legumes should be consumed daily.
5. **Milk, Cheese, and Yogurt** (2 to 3 servings for adults; 3 servings for pregnant and lactating women, teenagers, and young adults up to age 24). This group is enlarged to include soy milk fortified with calcium and vitamin D (and vitamin B_{12} for vegans), and soy cheese fortified with calcium and vitamin D. One serving is 1 cup of fortified soy milk or 1-1/2 ounces of soy cheese. One cup of cooked broccoli or greens can be substituted for 1 cup of milk. If a vegan does not drink soy milk or eat soy cheese, a carefully planned diet with sources of the missed nutrients, or supplements, is necessary.

Menu-Planning Guidelines for Vegetarians

Variety is a key word to remember when planning vegetarian menu items.

1. **Use a variety of plant protein sources at each meal: legumes, grain products (preferably whole-grain), nuts and seeds, and/or vegetables.** Vegetarian entrées commonly use cereal grains such as rice or bulgur (precooked and dried whole wheat) in combination with legumes and/or vegetables. Use small amounts of nuts and seeds in dishes.
2. **Use a wide variety of vegetables.** Steaming, stir-frying, or microwaving vegetables retains flavor, nutrients, and color.
3. **Offer entrées that are acceptable to lacto vegetarians and lacto-ovo vegetarians, and an entrée for vegans.** Although lactovegetarians and lacto-ovo vegetarians will eat vegan entrees, vegans won't eat entrées with any dairy products or eggs.
4. **Choose low-fat and nonfat varieties of milk and milk products and limit the use of eggs.** This is important to prevent a high intake of saturated fat, found in whole milk, low-fat milk, regular cheeses, eggs, and other foods.
5. **Offer dishes made with soybean-based products, such as tofu and tempeh.** Soybeans are unique in that they contain the only plant protein that is nutritionally equivalent to animal protein.

TABLE 11-6 Good Sources of Vitamin B_{12}, Vitamin D, Calcium, Iron, Zinc, and Linolenic Acid

Vitamin B_{12}	Dairy products, eggs, fortified cereals, and meat analogs
Vitamin D	Fortified milk, eggs, fortified cereals, and soy milk
Calcium	Milk and milk products, canned salmon and sardines (with bones), oysters, calcium-fortified juice or soy milk, broccoli, collards, kale, greens
Iron	Liver, meats, breads and cereals, green leafy vegetables, legumes, dried fruits
Zinc	Whole grains, legumes, nuts and seeds, peanut butter
Linolenic acid	Walnuts, walnut oil, canola oil, soybean oil, soybeans

6. **Provide foods that contain nutrients of special importance to vegetarians.** Table 11-6 lists good sources of vitamin B_{12}, vitamin D, calcium, iron, zinc, and linolenic acid.

For menu ideas, don't forget to look at the cuisine of other countries.

- Mexican—vegetable quesadillas, enchiladas
- Carbibbean—spinach and potato croquettes, black beans and rice
- Asian—stir-fries, noodle bowls
- Italian—vegetable lasagne, risotto, pasta and bean soup
- Indian—vegetable curries, dal (lentil stew)
- Middle Eastern—falafel, tabbouleh, hummus

■ MINI-SUMMARY

Legumes, grains, nuts, and seeds are important foods for vegetarians. Vegetarians can still follow the Food Pyramid concept by concentrating on meat substitutes in the Meat/Meat Substitute group and adding fortified soy milk and soy cheeses to the dairy group. Menu-planning guidelines are given.

Check-Out Quiz

1. The two medical conditions that lead to most cardiovascular diseases are atherosclerosis and high blood pressure.
   a. True b. False
2. Atherosclerosis develops by a process that is totally silent.
   a. True b. False

3. The most effective dietary method to lower the level of your blood cholesterol is to eat less cholesterol.
   a. True b. False
4. The TLL diet is used to treat diabetes mellitus.
   a. True b. False
5. Blood pressure increases with age.
   a. True b. False
6. Both heart attacks and strokes are usually due to clots caught in arteries.
   a. True b. False
7. Dietary protein probably promotes cancer.
   a. True b. False
8. Phytochemicals may protect against cancer.
   a. True b. False
9. Symptoms of diabetes include increased hunger and thirst.
   a. True b. False
10. Vegetarians enjoy certain health benefits from their diets.
    a. True b. False

Activities and Applications

1. A Diet for Disease Prevention

You have read about several diseases and the dietary means for preventing and treating each one. Using this information, write what you would consider to be an ideal diet for disease prevention (other than vegetarian). Focus on different groups of foods and what role each would play (and why) in disease prevention.

2. Vegetarian Meal Planning

Using vegetarian cookbooks and magazines or trade publications, suggest two vegetarian entrées and desserts for use at each of the following establishments: a casual-themed restaurant for younger people, a college cafeteria, and a dining room for an investment bank.

3. Vegetarian Meal

With your classmates, plan and carry out a vegetarian meal that uses foods you may not be familiar with. Plan the meal so there are items for both lacto-ovo vegetarians and vegans.

4. Taste Testing: Veggie Burgers

A variety of nonmeat burgers are on the market. Ingredients vary and may include vegetables, soy, and other ingredients. Select a number of veggie burgers and prepare according to the package directions. Serve on a bun with lettuce, tomato, and condiments. Find out which one is the class favorite.

Nutrition Web Explorer

National Heart, Lung, and Blood Institute (NHLBI)
http://hp2010.nhlbihin.net/emails/apt3_6.htm
At this website, you can use an interactive tool to assess your 10-year risk for heart disease.

American Cancer Society www.cancer.org
On the home page for the American Cancer Society, click on "Prevention and Early Detection." Then click on "Nutrition for Risk Reduction," and read the dietary guidelines for cancer prevention.

Center for Science in the Public Interest www.cspinet.org
To prevent disease, you need to eat a healthy diet. Click on "Nutrition Quizzes," then click on "Rate Your Diet." How does your diet rate?

American Diabetes Association www.diabetes.org
On the home page of the American Diabetes Association, click on "Healthy Fast Foods" and read about good fast-food choices.

National Restaurant Association www.restaurant.org
Read the article "Vegetarian Cuisine" that appeared in the National Restaurant Association's magazine, *Restaurants USA,* in January 1999 on this website.

Food Facts *Caffeine*

Check out how much you know, or don't know, about caffeine.

True or false?

1. Tea has more caffeine than coffee.
2. Brewed coffee has more caffeine than instant coffee.
3. Some nonprescription drugs contain caffeine.
4. Caffeine is a nervous-system stimulant.
5. Withdrawing from regular caffeine use causes physical symptoms.
6. Caffeine is a diuretic.

Check your answers as you read on.

Caffeine, a stimulant, is the most widely used psychoactive substance in the world. Eighty percent of Americans drink at least one caffeine-containing beverage each day, and average caffeine consumption is about 280 milligrams. Caffeine is present in over 60 plant species in various parts of the world, such as the coffee bean in Arabia, the tea leaf in China, the kola nut in West Africa, and the cocoa bean in Mexico.

Coffee, tea, cola, and cocoa are the most common sources of caffeine in the American diet (Table 11-7), with coffee being the chief source. The caffeine content of coffee or tea depends on the variety of coffee bean or tea leaf, the particle size, the brewing method, and the length of brewing or steeping time. Brewed coffee always has more caffeine than instant coffee, and espresso always has more caffeine than brewed coffee. Espresso is made by forcing hot pressurized water through finely ground, dark-roast beans. Because it is brewed with less water, it contains more caffeine than regular coffee.

In soft drinks, caffeine is both a natural and an added ingredient. The Food and Drug Administration requires caffeine as an ingredient in colas and pepper-flavored beverages and allows it to be added to other soft drinks as well. About 5 percent of the caffeine in colas and pepper-flavored soft drinks is obtained naturally from cola nuts; the remaining 95 percent is added. Caffeine-free soft drinks contain virtually no caffeine and make up a small part of the soft-drink market.

Numerous prescription and nonprescription drugs also contain caffeine. It is often used in alertness or stay-awake tablets, headache and pain-relief remedies, cold products, and diuretics. When caffeine is an ingredient, it must be listed on the product label.

Caffeine is rapidly absorbed into the bloodstream. For most people, caffeine raises blood pressure and heart rate, increases attentiveness and performance, and gives relief from fatigue. In high doses it can produce insomnia, nervousness, a racing heart, and other troublesome symptoms. Many people build a tolerance to caffeine's effects that may then lead to increased usage.

It is easy to become dependent on caffeine. When caffeine is withdrawn, symptoms include headache, fatigue, irritability depression, and poor concentration. The symptoms peak on day one or two and progressively decrease over the course of a week. It has been shown that even moderate consumption—about one or two 10-ounce cups of coffee daily (or the equivalent of caffeine from other sources)—often causes people to experience these debilitating symptoms of caffeine withdrawal. To minimize withdrawal symptoms, experts recommend reducing caffeine intake by

TABLE 11-7 Caffeine Content of Beverages, Foods, and Drugs

Item	Milligrams Caffeine Average	Range
Coffee (5-ounce cup)		
Brewed, drip method	115	110–150
Instant	65	30–120
Decaffeinated, brewed	3	2–5
Decaffeinated, instant	2	1–5
Tea (5-ounce cup)		
Brewed, major U.S. brands	40	20–90
Brewed, imported brands	60	25–110
Instant	30	25–50
Iced (12-ounce glass)	70	67–76
Soft drinks (12-ounce can)		
Cola, pepper		30–46
Decaffeinated cola, pepper		0–2
Cherry cola		36–46
Lemon-lime		0
Other citrus		0–64
Root beer		0
Ginger ale		0
Tonic water		0
Other regular soda		0–44
Juice added		0
Diet cola, pepper		0.6
Decaffeinated diet cola, pepper		0–0.2
Diet cherry cola		0–46
Diet lemon-lime, diet root beer		0
Other diets		0–70
Club soda, seltzer, sparkling water		0
Cocoa		
Cocoa beverage (5-ounce cup)	4	2–20
Chocolate milk beverage (8 ounces)	5	2–7
Milk chocolate (1 ounce)	6	1–15
Dark chocolate, semisweet (1 ounce)	20	5–35
Baker's chocolate (1 ounce)	26	26
Chocolate-flavored syrup (1 ounce)	4	4

Source: Lecos, Chris W. 1987–1988. "Caffeine jitters: Some safety questions remain." *FDA Consumer* 21(10):22–27.

about 20 percent a week over four to five weeks.

Although caffeine use has stirred fears in the past (it reportedly caused pancreatic cancer and birth defects), moderate use probably confers the benefits of caffeine with few of the risks. There are, however, two groups of individuals who should abstain from caffeine. In some susceptible people with heart disease, caffeine can cause irregular heartbeat. Caffeine is also not recommended for individuals with peptic ulcers, because caffeine increases the production of stomach acid. Pregnant and lactating women should consume caffeine in moderation—less than two cups of coffee a day. Studies on caffeine and pregnancy have shown conflicting results, so moderation is recommended.

High intakes of caffeine may be linked to heart attacks and bone loss in women. Once an individual drinks more than two 10-ounce cups of coffee daily, there is added cardiac risk among both smokers and nonsmokers, and especially in people with high blood pressure. Women who consume caffeine and drink little or no milk lose more calcium in their urine and have less dense bones than do women who don't consume any caffeine. For optimum bone density, women should moderate their caffeine intake and get at least two to three servings from the milk group daily.

Caffeine had a reputation until recently of being a diuretic, a substance that increases the flow of urine. We now know that caffeine does not flush water out of the body, so it does not have a dehydrating effect.

Hot Topic Biotechnology

BACKGROUND

Potatoes with built-in insecticide. Rice with extra vitamin A. Decaf coffee beans fresh off the tree. What do these foods have in common? They have all been created using biotechnology and genetic engineering. **Biotechnology** is a collection of scientific techniques, including genetic engineering, that are used to create, improve, or modify plants, animals, and microbe (such as bacteria). **Genetic engineering** is a process in which genes are transferred to a plant, animal, or microbe to have a certain effect, such as produce a soybean with built-in insecticide.

Farmers and scientists have been genetically modifying plants for hundreds of years, most often using a process called crossbreeding. Crossbreeding involves cross-fertilizing two related plants, and all 100,000 or so genes of each, to produce offspring that expresses the best of both. Since farmers usually only want a few genes transferred, crossbreeding is quite hit-or-miss. On the other hand, genetic engineering involves introducing one gene or a group of genes that are known quantities into a plant, animal, or microbe.

In order to understand genetic engineering, it is necessary to have an understanding of chromosomes, genes, and DNA. Each cell in your body contains a complete copy of your genetic plan or blueprint. The genetic material is packaged into long strands (called *chromosomes*) made of the chemical substance called *DNA* (deoxyribonucleic acid). Each chromosome contains *genes.* Genes contain the information for making proteins, which perform all the critical functions of the cell, and the blueprint for traits such as height, color, and disease resistance. Some traits are produced from the code contained in one gene, more complex traits depend on several genes. However, not all genes are switched on in every cell. The genes active in a liver cell are different from the genes active in a brain cell because the cells have different functions.

The language of DNA is common to all organisms. Humans share 7,000 genes with a worm named *C. elegans!* The main difference between organisms lies in their total number of genes, how the genes are arranged, and which ones are turned on or off in different cells. A gene from Arctic flounder that keeps the fish from freezing was introduced into strawberries to extend their growing season in northern climates. This did not make the strawberries fishy because it is a cold tolerance gene (not a flounder gene) that was introduced into the strawberries.

Did you know that 3.3 million people with diabetes use insulin produced by genetically modified (GM) bacteria? When given a copy of the gene for human insulin, a certain type of bacteria can make insulin that is purified and used to treat diabetes. In 1982, human insulin became the first GM product made commercially and approved for use. Since then, genetic engineering has developed additional medicines and vaccines.

PLANT APPLICATIONS

Genetic engineering has been used with plants to make:

1. Fruits or vegetables that ripen differently.
2. Plants that are resistant to disease, pests, selected herbicides, or environmental conditions (such as drought).
3. Plant foods with desirable nutritional characteristics.

One of the first GM foods to appear in the supermarket was the fresh tomato, called Flav Savr. If picked when ripe, tomatoes rot quickly so they are usually picked when green. The Flav Savr tomato was engineered to remain on the vine longer to ripen to full flavor before harvest. Once harvested, it did not soften as quickly as other tomatoes and was still firm and flavorful when it arrived at the supermarket. Eventually, the Flav Savr tomato was withdrawn from the market due to shipping problem and flavor problems because of the tomato varieties used.

Plants can also be made resistant to disease, pests, selected herbicides, or environmental conditions. For example, GM squash and potatoes are protected from viruses that normally affect these plants.

For years, organic farmers used a spray containing the microbe *Bacillus thuringiensis* as an effective pesticide. The microbe, which naturally occurs in soil, makes a protein (known as Bt) that poisons certain insect pests but is harmless to humans, animals, and most other insects. Scientists transferred the Bt gene that makes the insect-killing protein to corn, cotton, and soybeans. In this manner, farmers useless pesticide because the plant that once was a food source for the insect now kills it. Most corn and soybean crops in the United States are fed to animals.

About two-thirds of the American soybean crop planted in 2002 carried a gene that made it resistant to a popular herbicide used to control weeds. When the farmer sprays the fields to kill weeds, the plant is not harmed. This gene has also been introduced into corn, canola, and cotton seeds as well. Farmers can till (or plow) the soil less often (a common technique for reducing weeds), which results in less soil erosion.

Another use of genetic engineering has been to enhance the nutritional profile of foods. For example, by inserting two genes from a daffodil and one from a bacterium into a rice plant, scientists have created a rice with beta carotene, which the body makes into vitamin A. Called golden rice, it is still being tested and could eventually be used in developing countries where vitamin A deficiency is common. Other examples include canola oil that is lower in saturated fat, and peanuts and soybeans that don't cause allergic reactions.

ANIMAL APPLICATIONS

Using the same principle of gene transfer that gives plants more desirable traits, scientists have started working with GM animals, also called transgenic animals. Animal biotechnology research has largely focused on producing transgenic animals for the study of human disease or to produce drugs. There are many more barriers to using genetic engineering in the breeding and production of animal foods than with plant foods, such as difficulties working with living tissues and life cycles.

Currently seeking Food and Drugs Administration approval is a variety of Atlantic salmon that grows to market weight in about 18 months, compared to the 24 to 30 months that it normally takes for a fish to reach that size. New genes cause a continuous supply of salmon growth hormones that accelerates the fish's development. For fish farmers, raising these salmon is cheaper and faster because it takes less feed and about half the time to produce a crop they can send to market.

One future area of research is to produce transgenic cows with differing milk nutritional quality. For example, cow's milk could be made to have reduced lactose or a high whey:casein ratio so the milk is more similar to breast milk and could be used to make infant formulas.

REGULATIONS

Bioengineered foods are regulated by three federal agencies: the Food and Drug Administration (FDA), the U.S. Department of Agriculture (USDA), and the Environmental Protection Agency (EPA). The FDA is the lead agency for assessing the safety of GM plants and animals. Although the FDA only has to formally approve GM animals. Companies that produce GM crops are asked to voluntarily submit test results, which is normally done. The USDA ensures that new plant varieties pose no threat to agriculture or to the environment during cultivation. The USDA requires extensive field trials that are controlled and contained. The EPA regulates any pesticide that may be present in food and sets tolerance levels for them. For new plants that protect themselves against insects or disease, the EPA is responsible for assessing the safety of the protein or trait for human and animal consumption as well as for the environment and non-target organisms.

The FDA requires labeling of biotech foods significantly different from its conventional counterpart. The FDA considers whether a GM orange, for example, is "substantially equivalent" to a traditional orange. If the orange has a higher or lower level of vitamin C, for example, then the FDA requires that the label state the food is genetically engineered.

ISSUES: PROS AND CONS

The whole idea of using biotechnology brings up many issues and pros and cons. Here is a partial listing for you to consider.

- **Environmental.** Outcrossing is when a domestic crop breeds with a related "wild" species. Wild weeds can incorporate bioengineered genes, potentially making the weeds stronger and more resistant to pests and/or herbicides. Luckily, corn, soybeans, and cotton have no wild relatives with which to interbreed, but other crops, such as squash and canola, do have wild relatives in the United States. Likewise, if transgenic salmon escaped and crossbred with their wild cousins, this could adversely affect the wild populations as the transgenic fish might out-produce (they are bigger and might get more food and mates) and push aside wild salmon. There is also the possibility of contaminating non-GM crops, including organic crops, with GM crops (due to pollen carried by the wind, birds, etc.), or GM crops adversely affecting wildlife such as butterflies and worms.
- **Health.** GM foods pose the same risk to human health as do other foods: allergens, toxins (produced by the gene or the protein made by the gene), and compounds known as antinutrients that inhibit the absorption of nutrients. Allergic reactions are an important consideration for bioengineered foods because there is some possibility that a new protein in a food could be an allergen. To date, all new proteins in bioengineered foods have been shown to lack the characteristics of food allergens. If fewer pesticides are used, there will be less pesticide residue on plants. The accidental mixing of StarLink(tm) corn, a GM product approved as animal feed only, with corn intended for human consumption demonstrates another concern.
- **Agricultural.** Using biotechnology, it is possible to have increased crop yield, greater flexibility in growing environments, less erosion due to less tillage, and decreased use of pesticides and herbicides.
- **Nutrition.** Genetic engineering can certainly enhance the nutrient profile of foods, but there is also potentially the possibility of decreased nutrients in a product such as a banana that takes longer to ripen but loses nutrients.
- **Ethical and Moral.** Some people have religious and philosophical concerns regarding the transfer of genes into plants or animals and may accuse the food industry of playing God.

Chapter 12

Weight Management and Exercise

Overweight and obesity are among the most pressing new health challenges in the United States. Just take a look at these statistics:

- In 1999, an estimated 61 percent of American adults were overweight, along with 13 percent of children and adolescents.
- Obesity among adults has doubled since 1980, while overweight among adolescents has tripled.
- Only 3 percent of all Americans meet at least four of the five Food Guide Pyramid recommendations for the intake of grains, fruits, vegetables, dairy products, and meats.
- Less than one-third of Americans meet the federal recommendations to engage in at lest 30 minutes of moderate physical activity at least five days a week, while 40 percent of adults engage in no leisure-time physical activity at all.
- Over 50 million Americans are currently on a diet, and only 5 percent will get the weight off over the long term.

As the prevalence of overweight and obesity has increased in the United States, so have related health-care costs—both direct and indirect. Direct health-care costs refer to preventive, diagnostic, and treatment services such as physician visits, medications, and hospital care. Indirect costs are the value of wages lost by people unable to work because of illness or disability, as well as the value of future earnings lost by premature death. The cost of overweight and obesity is $99.2 billion a year. The direct cost is $51.6, and the indirect cost is $47.6 billion.

Obesity is a disease, not a moral failing. It is a very complex disease that has social, behavioral, cultural, physiological, metabolic, and genetic factors. It occurs when energy intake exceeds the amount of energy expended over time. In a small number of cases, obesity is caused by illnesses such as hypothyroidism or is the result of taking medications that can cause weight gain.

Because so many factors affect how much or how little food a person eats and how food is metabolized by the body, losing weight is not simple. This chapter discusses obesity's impact on health, theories about what causes obesity, and treatment to lose weight and maintain weight loss. Exercise, which is important for both obese and nonobese individuals, is also examined, along with nutrition for athletes. This chapter will help you to:

- Define obesity and overweight
- List one advantage and one disadvantage of each of the three methods of measuring obesity
- List the health implications of obesity
- Explain possible causes of obesity
- List the six components of a comprehensive treatment program for obesity
- Describe seven basic concepts of nutrition education to consider when planning weight-reducing diets

- Explain the relationship between exercise and weight loss
- Outline behavior and attitude modification and discuss how they can be used to help someone lose weight
- Discuss strategies that appear to support weight maintenance
- Identify nutrient needs for athletes and plan menus for athletes

"How Much Should I Weigh?"

Obesity can be measured using various methods. **Height-weight tables** are easy to use and understand. For every height, there is a recommended weight range. One concern with height-weight tables is that the weights say nothing about body composition. For instance, a 250-pound linebacker at 6 feet 2 inches would be considered overweight according to the tables, yet his excess pounds are not fat but muscle. Conversely, a person whose weight is within the appropriate range may indeed have excess fat stores.

Height-weight tables—Tables that show an appropriate weight for a given height.

Body Mass Index—A method of measuring degree of obesity that is a more sensitive indicator than height-weight tables.

Overweight—A state of having a Body Mass Index of 25 or greater.

Obesity—A state of having a Body Mass Index of 30 or greater.

A second method of measuring degree of obesity is **body mass index** (BMI), the measurement of choice. BMI is a direct calculation based on height and weight, and it applies to both men and women. BMI does not directly measure percentage of body fat, but it provides a more accurate measure of overweight and obesity than relying on weight alone. BMI is found by dividing a person's weight in kilograms by height in meters squared. To determine your BMI refer to Table 12-1. Find your height along the first column (on the left), then look across the row until you find the number that is closest to your weight. The number at the top of that column is your BMI.

The National Institutes of Health defines **overweight** as a BMI of 25–29.9. A BMI of 30 or greater is considered **obese.** These cutoff points were chosen because studies show that as BMI rises above 25, blood pressure and blood cholesterol rise, HDL levels fall, and maintaining normal blood sugar levels becomes more difficult. Thus, individuals with a BMI of 25 or higher run a greater risk of coronary artery disease, high blood pressure, heart attacks, stroke, type 2 diabetes, and certain cancers, such as breast cancer.

Using BMI is simple, quick, and inexpensive—but it does have limitations. One problem with using BMI as a measurement tool is that very muscular people may fall into the "overweight" category when they are actually healthy and fit. Another problem with using BMI is that people who have lost muscle mass, such as the elderly, may be in the healthy weight category when they actually have reduced nutritional reserves. Further evaluation of a person's weight and health status are necessary.

Because BMI doesn't tell you how much of your excess weight is fat and where that fat is located, the National Institutes of Health has asked physicians to measure patients' waistlines. Studies show that excessive abdominal

TABLE 12-1 Body Mass Index*

BMI	19	20	21	22	23	24	25	26	27	28	29	30	31	32	33	34	35	36	37	38	39	40
Height (inches)	Body Weight (pounds)																					
58	91	96	100	105	110	115	119	124	129	134	138	143	148	153	158	162	167	172	177	181	186	191
59	94	99	104	109	114	119	124	128	133	138	143	148	153	158	163	168	173	178	183	188	193	198
60	97	102	107	112	118	123	128	133	138	143	148	153	158	163	168	174	179	184	189	194	199	204
61	100	106	111	116	122	127	132	137	143	148	153	158	164	169	174	180	185	190	195	201	206	211
62	104	109	115	120	126	131	136	142	147	153	158	164	169	175	180	186	191	196	202	207	213	218
63	107	113	118	124	130	135	141	146	152	158	163	169	175	180	186	191	197	203	208	214	220	225
64	110	116	122	128	134	140	145	151	157	163	169	174	180	186	192	197	204	209	215	221	227	232
65	114	120	126	132	138	144	150	156	162	168	174	180	186	192	198	204	210	216	222	228	234	240
66	118	124	130	136	142	148	155	161	167	173	179	186	192	198	204	210	216	223	229	235	241	247
67	121	127	134	140	146	153	159	166	172	178	185	191	198	204	211	217	223	230	236	242	249	255
68	125	131	138	144	151	158	164	171	177	184	190	197	203	210	216	223	230	236	243	249	256	262
69	128	135	142	149	155	162	169	176	182	189	196	203	209	216	223	230	236	243	250	257	263	270
70	132	139	146	153	160	167	174	181	188	195	202	209	216	222	229	236	243	250	257	264	271	278
71	136	143	150	157	165	172	179	186	193	200	208	215	222	229	236	243	250	257	265	272	279	286
72	140	147	154	162	169	177	184	191	199	206	213	221	228	235	242	250	258	265	272	279	287	294
73	144	151	159	166	174	182	189	197	204	212	219	227	235	242	250	257	265	272	280	288	295	302
74	148	155	163	171	179	186	194	202	210	218	225	233	241	249	256	264	272	280	287	295	303	311
75	152	160	168	176	184	192	200	208	216	224	232	240	248	256	264	272	279	287	295	303	311	319
76	156	164	172	180	189	197	205	213	221	230	238	246	254	263	271	279	287	295	304	312	320	328

*Underweight: <18.5

Healthy Weight: 18.5–24.9

Overweight: 25–29.9

Obese: >30

Source: The Practical Guide: Identification, Evaluation, and Treatment of Overweight and Obesity in Adults. (NIH Pub. No. 00-4084) National Institutes of Health, National Heart, Lung, & Blood Institute. 2000.

fat is more health-threatening than fat in the hips or thighs. A woman whose waist measures more than 35 inches and a man whose waist measures more than 40 inches may be at particular risk for developing health problems. Studies indicate that increased abdominal or upper-body fat is related to the risk of developing heart disease, diabetes, high blood pressure, gallbladder disease, stroke, and certain cancers, and is associated with overall likelihood of death, especially premature death. Body fat concentrated in the lower body (around the hips for example) may be less harmful.

Another way to measure obesity is to examine the percentage of your body that is fat. For men, a desirable percentage of body fat is 13 to 25 percent, for women about 17 to 29 percent. When a man's body fat goes over 25 percent or a woman's over 29 percent, health risks increase. Body fat is most often measured by using special calipers to measure the skinfold thickness of the triceps and other parts of the body. Because half of all your fat is under the skin, this method is quite accurate when performed by an experienced professional. Other methods of estimating body fatness include underwater weighing and bioelectrical impedance.

■ MINI-SUMMARY

Obesity can be measured using various methods. The BMI is a more sensitive indicator than height-weight tables. A BMI of 25 to 29.9 is considered overweight and over 30 is considered obese. Health risks increase as BMI rises. Men are at increased risk for disease if they have a waist circumference greater than 40 inches; women are at increased risk if they have a waist circumference greater than 35 inches.

Health Implications of Obesity

An estimated 300,000 deaths per year may be attributable to obesity. Individuals who are obese have a 50 to 100 percent increased risk of premature death from all causes, compared to individuals with a healthy weight.

An obese individual is at increased risk for:

- Hypertension
- High blood cholesterol levels
- Type 2 diabetes
- Coronary heart disease
- Stroke
- Gallbladder disease
- Osteoarthritis

- Sleep apnea (interrupted breathing during sleeping) and respiratory problems
- Some types of cancer

Conditions aggravated by obesity include arthritis, varicose veins, gallbladder disease, and pregnancy. In addition, surgery is riskier for obese individuals.

Obesity creates a psychological burden that, in terms of suffering, may be its greatest adverse effect. In American and other Westernized societies there are powerful messages that people, especially women, should be thin and that to be fat is a sign of poor self-control. Negative attitudes about the obese have been reported in children and adults, in health-care professionals, and in the overweight themselves. People's negative attitudes toward the obese often translate into discrimination in employment opportunities, job earnings, and other areas.

Losing weight often decreases blood pressure and blood cholesterol levels, and brings diabetes under better control. Although obesity does not cause these medical conditions, losing weight can help to reduce some of their negative aspects.

■ MINI-SUMMARY

An obese individual is at increased risk for hypertension, high blood cholesterol levels, Type 2 diabetes, coronary heart disease, stroke, gallbladder disease, osteoarthritis, sleep apnea and respiratory problems, and some types of cancer (endometrial, breast, prostate, and colon). Obesity (BMI over 30) creates a psychological burden and discrimination. Losing weight, and keeping it off, improves health.

Theories of Obesity

As researchers try to figure out why some people get fat and others don't, it is becoming increasingly apparent that obesity is caused by an interaction of genetic (inherited), environmental (social and cultural), metabolic (physical and chemical), and behavioral (psychological and emotional) factors—and, therefore, no single cure is available.

The body has an almost limitless capacity to store fat. Not only can each fat cell balloon to more than 10 times its original size, but should the available cells get filled to the brim, new ones will propagate. As the body stores more fat, weight and girth increase.

Studies have shown that genetics often influences the development of overweight and obesity. A large number of twin, adoption, and family studies have explored the extent to which obesity is inherited. Recent studies of individuals with a wide range of BMIs, together with information obtained on their parents, siblings, and spouses, suggest that about 25 to 40 percent of the individual differences in body mass or body fat may depend on genetic factors. However, studies with identical twins reared apart suggest that the genetic contribution to BMI may be higher, about 70 percent.

The environment is also a major determinant of overweight and obesity. Environmental influences on overweight and obesity are primarily related to food intake and physical activity behaviors. In countries such as the United States, there is an overall abundance of tasty, kcalorie-dense food. In addition, aggressive and sophisticated food marketing promote high kcalorie consumption. Many of our sociocultural traditions promote overeating and the preferential consumption of high kcalorie foods. For many people, even when caloric intake is not above the recommended level, the number of kcalories expended in physical activity is insufficient to offset consumption. Many people are stuck in daily routines that are completely sedentary.

■ MINI-SUMMARY

Obesity is caused by an interaction of genetic (inherited), environmental (social and cultural), metabolic (physical and chemical), and behavioral (psychological and emotional) factors—and, therefore, no single cure is available.

Treatment of Obesity

Before discussing the treatment of obesity, it is a good idea to discuss treatment goals. The goals of most weight-loss programs have focused on short-term weight loss. Critics of this type of goal feel that short-term weight loss is not a valid measure of success, because it is not associated with health benefits (as is long-term weight loss). Also, weight-loss goals tend to reinforce the American preoccupation with being slender, especially for women. Lastly, critics point out that weight-loss goals are often set too high, since even a small weight loss can reduce the risks of developing chronic diseases. Overweight individuals who lose even relatively small amounts of weight are likely to:

- Lower their blood pressure, and thereby the risks of high blood pressure
- Reduce abnormally high levels of blood glucose associated with diabetes
- Bring down blood levels of cholesterol and triglycerides associated with cardiovascular disease
- Reduce sleep apnea or irregular breathing during sleep
- Decrease the risk of osteoarthritis of the weight-bearing joints
- Decrease depression
- Increase self-esteem

Current recommendations from the Food and Nutrition Board favor shifting from a weight-loss perspective to one of weight management, in which success is judged more by the program's effect on an individual's health status than by its effect on weight.

Treatment of obesity generally consists of one or more of the following: diet, exercise, behavior modification, and drug therapy. Treatment programs often concentrate on two aspects of losing weight: diets and exercise. Individuals in treatment programs that just offer diets tend to regain any lost weight within two or three years.

The next section will discuss each of the following components of treatment: diet and nutrition education, exercise, behavior and attitude modification, social support, maintenance support, drugs, and surgery. But first, let's take a look at a newer approach to treating obesity that doesn't use diets.

More health professionals are adopting a nondieting approach to treating obesity that includes eating less fat and exercising more. This new health-centered approach steers clear of dieting and emphasizes helping obese people adopt a healthier lifestyle. In many cases, diets simply don't work. By restricting food intake, diets often cause dieters to become obsessed with food, which may then lead to binge eating. Eating fewer fat kcalories and exercising can help many obese people to lose some weight and keep it off.

Diet and Nutrition Education

Basic nutrition education is crucial for dieters. They need education about fat, carbohydrate, and protein in foods and about balancing them in a lower-kcalorie and lower-fat diet. They need to understand variety, moderation, and nutrient density, particularly because they have to pack the same amount of nutrients into fewer kcalories.

Dieters need to understand seven basic concepts of nutrition education before planning their actual diets:

1. Kcalories should not be overly restricted during dieting because this practice decreases the likelihood of success. A dieter who normally eats 2,500 kcalories daily and goes on a 1,200-kcalorie diet is eating less than half of what he or she normally eats. Reducing kcalories by 500 each day amounts to about one pound lost in a week. In any case, kcalories should not be restricted below 1,200, because getting adequate nutrients is impossible below this level. A progressive weight loss of 1 to 2 pounds a week is considered safe.
2. Fat should be restricted to 30 percent or less of total kcalories, protein to 15 percent, and carbohydrate to 55 percent or more. During weight loss, attention should be given to maintaining an adequate intake of vitamins and minerals.

3. No foods should be forbidden, as that only makes them more attractive.
4. Eating three meals and one or two snacks each day is crucial to minimizing the possibility of getting hungry. People tend to overeat when hungry.
5. Portion control is vital. Measuring and weighing foods is important, because "eyeballing" is not always accurate.
6. Variety, balance, and moderation are crucial to satisfying all nutrient needs.
7. Weighing is important but should not be done every day, because 1- to 2-pound weight gains and losses can occur on a daily basis due to fluid shifts. Weekly weigh-ins are more accurate and less likely to cause disappointment.

Exercise

Exercise is a vital component of any weight-control program. Research consistently shows that time spent exercising is a major predictor of long-term weight loss. Exercise not only facilitates weight loss through direct energy expenditure, but burns fat both during and after exercise. Regular exercise also helps control or suppress appetite and builds and tones muscles, which in turn raises your basal metabolic rate. But regular exercise has many benefits beyond simply losing pounds and keeping them off. Indeed, the Centers for Disease Control and Prevention and the American College of Sports Medicine recommend that every American adult accumulate 30 minutes or more of moderate-intensity physical activity on most, preferably all, days of the week. The Food and Nutrition Board, which develops the DRIs, recommends one hour of physical activity daily, including normal activities such as climbing stairs as well as moderate to vigorous exercise such as walking. Additional advantages of regular physical activity include:

1. Improved functioning of the cardiovascular system
2. Reduced levels of blood lipids associated with cardiovascular disease
3. Increased ability to cope with stress, anxiety, and depression
4. Increased stamina
5. Increased resistance to fatigue
6. Improved self-image

A consistent pattern of exercise is vital to achieving these beneficial results.

The key to a successful exercise program is choosing an enjoyable activity. Some questions people need to answer to develop a good exercise program include:

1. Do you like to exercise alone or with others?
2. Do you prefer to exercise outdoors or indoors?
3. Do you prefer to exercise at home?

4. What activities are particularly enjoyable?
5. How much money are you willing to spend for sports equipment or facilities if needed?
6. When can you best fit the activity into your schedule?

An obese person may resist starting an exercise program for several reasons. Many have had bad experiences in school physical education classes and want to avoid such activity. Some tend to be self-conscious and may not want to be seen exercising. Activity is also harder for obese people and requires more effort.

Aerobic activities such as walking, jogging, cycling, and swimming are ideal as the major component of an exercise program. Aerobic activities must be brisk enough to raise heart and breathing rates, and they must be sustained, meaning that they must be done at least 15 to 30 minutes without interruption. Activities such as baseball, bowling, and golf are not vigorous or sustained. They still have certain benefits—they can be enjoyable, help improve coordination and muscle tone, and help relieve tension—but they are not aerobic.

Any exercise program for sedentary and overweight people must be started slowly, with enjoyment and commitment as the major goals. A buildup in intensity and duration should be gradual and progressive, depending to a large extent on how overweight the individual is. Aerobic capacity improves when exercise increases the heart rate to a target zone of 65 to 80 percent of the maximum heart rate, which is the fastest the heart can beat. Maximum heart rate can be calculated by subtracting your age from 220. One goal of the exercise program should be to build up to this intensity of exercise. Exercise that increases the heart rate to between 65 and 80 percent of its maximum conditions the heart and lungs, besides burning kcalories (Table 12-2). To determine whether your heart rate is in the target zone, take your pulse as follows:

1. When you stop exercising, quickly place the tip of your third finger lightly over one of the blood vessels on your neck located to the left or right of your Adam's apple. (Another convenient pulse spot is the inside of your wrist just below the base of your thumb.)
2. Count your pulse for 30 seconds and multiply by 2.
3. If your pulse is below your target zone, exercise a little harder the next time. If you are above your target zone, exercise a little easier. If it falls within the target zone, you are doing fine.
4. Once you are exercising within your target zone, you should check your pulse at least once each week.

Exercise should take place most days and include a warm-up period, the exercise itself, and then a cooling-down period. To warm up before exercising, do stretching exercises slowly and in a steady, rhythmic way. Start at a medium pace and gradually increase. Next, begin jumping rope or jogging

TABLE 12-2 Kcalories Burned per Hour by a 150-Pound Person (Reduce the Calories 1/3 for a 100-Pound Person and Multiply by 1-1/3 for a 200-Pound Person)

Activity	Kcalories Burned per Hour
Bicycling at 6 mph	240
Bicycling at 12 mph	410
Cross-country skiing	700
Jogging at 5.5 mph	740
Jogging at 7 mph	920
Running in place	650
Jumping rope	750
Swimming 25 yards/minute	275
Swimming 50 yards/minute	500
Tennis, singles	400
Walking at 2 mph	240
Walking at 3 mph	320
Walking at 4.5 mph	440

Source: U.S. Department of Health and Human Services. 1983. *Exercise and Your Heart.* Washington, D.C.: U.S. Government Printing Office.

in place slowly before starting any vigorous activities to ease the cardiovascular system into the aerobic exercise. The exercise part of the session should burn at least 300 kcalories, which can be achieved with 15 to 30 minutes of aerobic activity in the target zone or 40 to 60 minutes of lower-intensity activity, such as leisurely walking. After exercising, slowing down the exercise or changing to a less vigorous activity for 5 to 10 minutes is important to allow the body to relax gradually.

Beyond the exercise program, obese people should be encouraged to schedule more activity into their daily routines. For instance, they should use stairs, both up and down, instead of elevators or escalators. They can also park or get off public transportation farther away from their destination to allow more walking.

Behavior and Attitude Modification

Behavior modification deals with identifying and changing behaviors that affect weight gain, such as raiding the refrigerator at midnight. Elements of behavior and attitude modification can be grouped into several categories: self-monitoring, stimulus or cue control, eating behaviors, reinforcement or self-reward, self-control, and attitude modification.

Self-monitoring involves keeping a food diary or daily record of types and amounts of foods and beverages consumed, as well as time and place of eating, mood at the time, and degree of hunger felt. Its purpose is, of course, to increase awareness of what is actually being eaten and whether it is in response to hunger or other stimuli. Once harmful patterns that encourage overeating are identified, negative behaviors can be changed to more positive ones. Figure 12-1 contains a sample food diary page.

Through self-monitoring, cues or stimuli to overeating can be identified. For example, passing a bakery may be a cue for someone to stop and buy a dozen cookies. Examples of behavioral modification techniques for cue or stimulus control follow.

Food Purchasing, Storage, and Cooking

1. Plan meals a week or more ahead.
2. Make a shopping list.
3. Do food shopping after eating, on a full stomach.
4. Do not shop for food with someone who will pressure you to buy foods you do not need.
5. If you feel you must buy high-kcalorie foods for someone else in the family who can afford the kcalories, buy something you do not like or let them buy, store, and serve the particular food.
6. Store food out of sight and limit storage to the kitchen.
7. Keep low-kcalorie snacks on hand and ready to eat.
8. When cooking, keep a small spoon such as a half-teaspoon measure on hand to use if you must taste while cooking.

Mealtime

1. Do not serve food at the table or leave serving dishes on the table.
2. Leave the table immediately after eating.

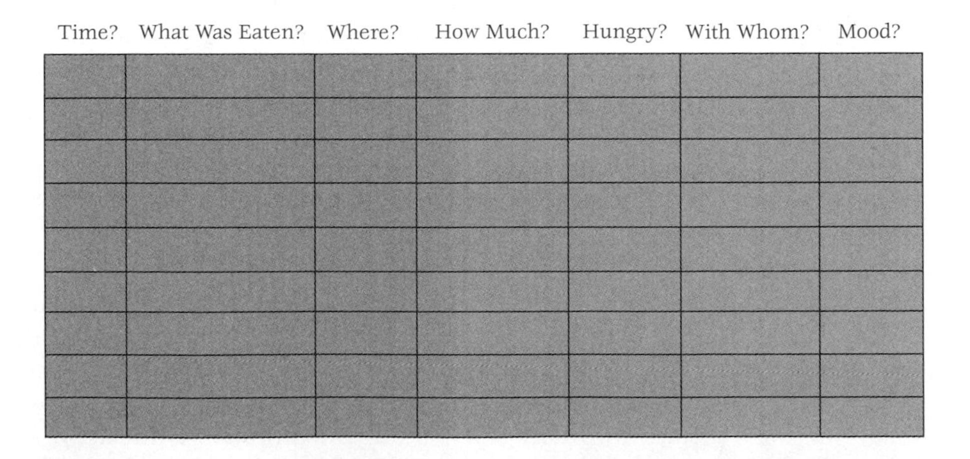

Time?	What Was Eaten?	Where?	How Much?	Hungry?	With Whom?	Mood?

Figure 12-1
Food diary form.

Holidays and Parties

1. Eat and drink something before you go to a party.
2. Drink fewer alcoholic beverages.
3. Bring a low-kcalorie food to the party.
4. Decide what you will eat before the meal.
5. Stay away from the food as much as possible.
6. Concentrate on socializing.
7. Be polite but firm and persistent when refusing another portion or drink.

Eating behaviors need to be modified to discourage overeating. First, one or two eating areas, such as the kitchen and dining-room tables, need to be set up so that eating occurs in only these designated locations. Eat only while sitting at the table, and make the environment as attractive as possible. Do not read, watch television, or do anything else while you eat, because you can easily form associations between certain activities and food, such as television and snacking. Do not eat while standing at cabinets or refrigerators.

Second, plan three meals and at least two snacks daily, preferably at certain times of the day. Third, eat slowly by putting your fork down between bites, eating your favorite foods first, talking to others at the table, eating a high-fiber food that requires time to chew, and drinking a no-kcalorie beverage to help you fill up. In addition, take smaller bites, savor each bite, use a smaller plate to make the food look bigger, and leave a bite or two on your plate. If you clean your plate, you are responding to the sight of food, not to real hunger. When you want a snack, postpone it for 10 minutes.

Reward yourself for positive steps taken to lose weight, but do not use food as a reward. Other rewards could include reading a good book or magazine, doing a hobby such as biking, fishing, woodworking; taking a long bubble bath, or calling a friend.

Overeating sometimes occurs in reaction to stressful situations, emotions, or cravings. The food diary is very useful in identifying these situations. Then you can handle these situations in new ways. You can express your feelings verbally if you are overeating in response to frustration or similar stressful emotions. You can exercise or use relaxation techniques to relieve stress, or you can switch to a new activity, such as taking a walk, knitting, reading, or engaging in a hobby to help you take your mind off food. If you allow yourself five minutes before getting something to eat, you often will go on to something else and forget about the food. Positive self-talk is important for good self-control. Instead of repeating a negative statement such as "I cannot resist that cookie," say, "I will resist that cookie."

The most common attitude problem obese people have is thinking of themselves as either on or off a diet. Being on a diet implies that at some point the diet will be over, resulting in weight gain if old habits are resumed. Dieting should not be so restrictive or have such unrealistic goals that the person cannot wait to get off the diet. When combined with exercise, behavior and attitude modification, social support, and a maintenance plan, dieting is really a plan of sensible eating that allows for periodic indulgences.

Another attitude that needs modification revolves around using words such as *always, never,* or *every.* The following are examples of unrealistic statements using these terms.

- I will always control my desire for chocolates.
- I will never eat more than 1500 kcalories each day.
- I will exercise every day.

Goals stated in this manner decrease the likelihood that you will ever accomplish them, and thus result in discouragement and thoughts of failure.

Setting realistic goals, followed by monitoring and self-reward when appropriate, is crucial to the success of any weight-loss program. Through goal-setting, complex behavior changes can be broken down into a series of small, successive steps. Goals need to be reasonable and stated in a positive, behavioral manner. For example, if a problem behavior is buying a chocolate bar every afternoon at work, a goal may be to bring an appropriate afternoon snack from home. If this goal is not truly attainable, perhaps one chocolate bar per week should be allowed and worked into the diet.

Even with reasonable goals, occasional lapses in behavior occur. This is when having a constructive attitude is critical. After eating and drinking too much at a party one night, for example, feelings of guilt and failure are common. However, they do nothing to help people get back on their feet. Instead, the dieter must stay calm, realize that what is done is done, and understand that no one is perfect.

Two other attitudes that often need correcting concern hunger and foods that are "bad for you." Hunger is a physiological need for food, whereas appetite is a psychological need. Eating should be in response to hunger, not to appetite. Although people frequently regard certain foods as "good" or "bad," they must realize that no food is inherently good or bad. Some foods do contain more nutrients per kcalorie, and some are mostly empty kcalories with few nutrients. However, no food is so bad that it can never be eaten.

Social Support

In general, obese people are more likely to lose weight when their families and friends are supportive and involved in their weight-loss plans. As social support increases, so do a person's chances of maintaining weight loss. When possible, obese people need to enlist the help of someone who is easy to talk to, understands and empathizes with the problems of losing weight, and is genuinely interested in helping. Supporters can model good eating habits and give praise and encouragement. The obese person needs to tell others exactly how to be supportive by, for example, not offering high-kcalorie snacks. Requests need to be specific and positive.

Maintenance Support

Not enough is known about factors associated with weight-maintenance success or what support is needed during the first few months of weight maintenance, when a majority of dieters begin to relapse. Being at a normal, or more normal, weight can bring about stress as adjustments are made. Food is no longer a focal point, and old friends and activities may not fit into the new lifestyle. Support and encouragement from significant others will probably diminish. A formal maintenance program can help deal with those issues as well as others.

Strategies that appear to support weight maintenance include:

1. Determining how many kcalories are needed for weight maintenance and working out a livable diet
2. Learning skills for dealing with high-risk situations when a lapse in eating behavior may occur, and knowing what to do when a relapse occurs
3. Continued self-monitoring
4. Continued exercise
5. Continued social support
6. Continued use of other strategies that were useful during weight loss
7. Dealing with unrealistic expectations about being thin

These strategies can be used during formal maintenance programs and after treatment is terminated.

Drugs

Weight-loss drugs approved by the Food and Drug Administration for long-term use may be used for people with a BMI of 30 or more, or people with a BMI of 27 or more who also have diseases such as diabetes or other risk factors. Drugs are used only as part of a program that includes diet, physical activity, and behavior therapy. Because of the tendency to regain weight after weight loss, the use of long-term medication to aid in the treatment of obesity may be indicated for some people. On average, individuals who use prescribed weight-loss drugs lose about 5 to 10 percent of their original weight.

All the prescription weight-loss drugs (except Xenical) work by suppressing the appetite. Xenical (the brand name for orlistat) was approved by the FDA in 1999. It is the first in a new class of anti-obesity drugs known as lipase inhibitors. Lipase is the enzyme that breaks down dietary fat. Xenical interferes with lipase function, decreasing dietary fat absorption by 30 percent. Since undigested fats are not absorbed, there is less caloric intake, which may help in controlling weight. The main side effects of Xenical are cramping, diarrhea, flatulence, intestinal discomfort, and leakage of oily stool. Also, this drug causes decreased absorption of fat-soluble vitamins.

Meridia (brand name for sibutramine) was approved in 1997, and it increases the levels of certain brain chemicals that help reduce appetite. Because it may increase blood pressure and heart rate, Meridia should not be used by people with a history of heart disease, stroke, or uncontrolled high blood pressure. Other common side effects of Meridia include headache, dry mouth, constipation, and insomnia.

Both Meridia and Xenical are approved for long-term use. Other antiobesity prescription drugs include products such as Bontril (phendimetrazine tartrate); these products are only to be taken for a few weeks.

Over-the-counter weight-control drugs have for years contained a drug called phenylpropanolamine. The FDA has asked drug manufacturers to discontinue marketing diet pills (and nasal decongestants, too) containing phenylpropanolamine, based on evidence linking it to an increased risk of hemorrhagic stroke (bleeding in the brain). The FDA is taking steps to remove this drug from all pharmaceutical products.

Herbal preparations are not recommended as part of a weight-loss program. These preparations have unpredictable amounts of active ingredients and unpredictable—and potentially harmful—effects.

Surgery

Weight-loss surgery is an option for weight reduction in people with clinically severe obesity, which is defined as a BMI of 40 or higher, or a BMI of 35 or higher with other risk factors. This treatment is only used for people in whom other methods of treatment have failed. Weight-loss surgery provides medically significant sustained weight loss for more than 5 years in most people.

Two types of operations have proven to be effective: those that restrict stomach volume (banded gastroplasty) and those that, in addition to limiting food intake, alter digestion (gastric bypass). An integrated program that provides guidance on diet, physical activity, and psychosocial concerns before and after surgery is necessary. Most patients fare well, with control of hypertension, marked improvement in mobility, and other improvements.

■ MINI-SUMMARY

A comprehensive treatment plan for obesity needs to include diet and nutrition education, exercise, behavior and attitude modification, social support, and maintenance support. A newer, nondieting approach deemphasizes dieting because dieting works in so few cases. This approach emphasizes exercise and fewer fat kcalories. Drugs and surgery may also be used when necessary and can have excellent results.

Menu Planning for Weight Loss and Maintenance

The basics of menu planning for weight loss and maintenance are covered in Chapter 8. Figure 12-2 lists some of the lower-fat foods you can use in place of foods higher in fat. Figures 12-3 to 12-7 give examples of reduced-kcalorie one-day menus.

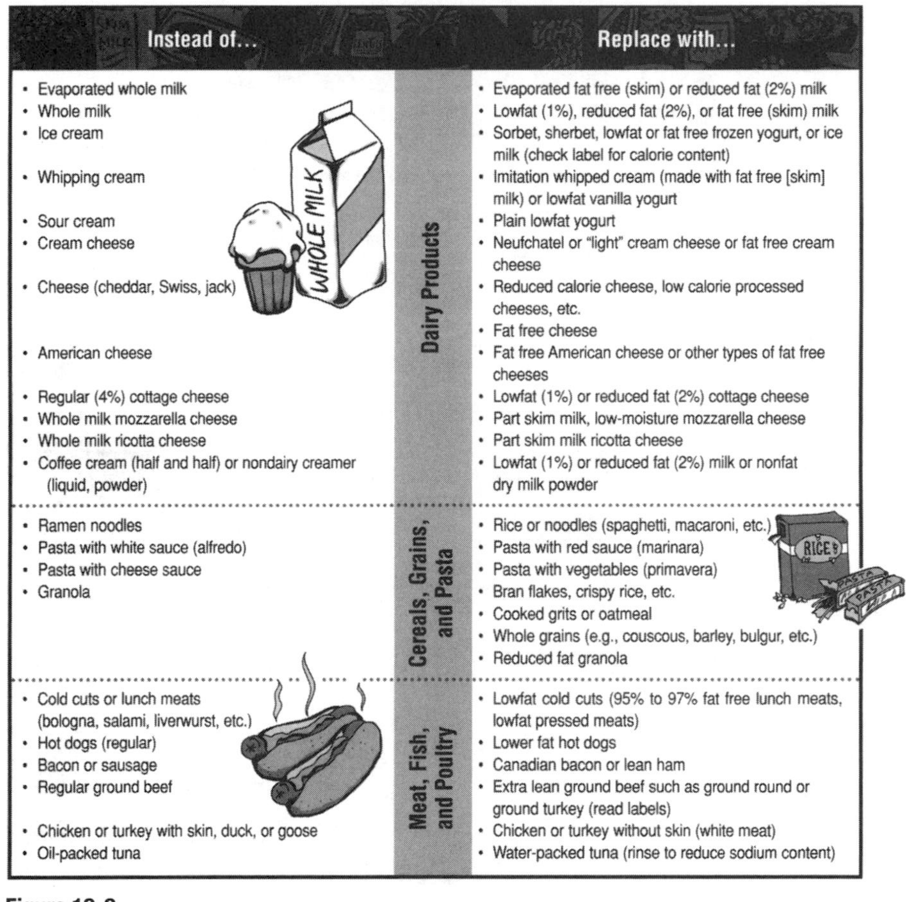

Instead of...		Replace with...
• Evaporated whole milk	Dairy Products	• Evaporated fat free (skim) or reduced fat (2%) milk
• Whole milk		• Lowfat (1%), reduced fat (2%), or fat free (skim) milk
• Ice cream		• Sorbet, sherbet, lowfat or fat free frozen yogurt, or ice milk (check label for calorie content)
• Whipping cream		• Imitation whipped cream (made with fat free [skim] milk) or lowfat vanilla yogurt
• Sour cream		• Plain lowfat yogurt
• Cream cheese		• Neufchatel or "light" cream cheese or fat free cream cheese
• Cheese (cheddar, Swiss, jack)		• Reduced calorie cheese, low calorie processed cheeses, etc.
		• Fat free cheese
• American cheese		• Fat free American cheese or other types of fat free cheeses
• Regular (4%) cottage cheese		• Lowfat (1%) or reduced fat (2%) cottage cheese
• Whole milk mozzarella cheese		• Part skim milk, low-moisture mozzarella cheese
• Whole milk ricotta cheese		• Part skim milk ricotta cheese
• Coffee cream (half and half) or nondairy creamer (liquid, powder)		• Lowfat (1%) or reduced fat (2%) milk or nonfat dry milk powder
• Ramen noodles	Cereals, Grains, and Pasta	• Rice or noodles (spaghetti, macaroni, etc.)
• Pasta with white sauce (alfredo)		• Pasta with red sauce (marinara)
• Pasta with cheese sauce		• Pasta with vegetables (primavera)
• Granola		• Bran flakes, crispy rice, etc.
		• Cooked grits or oatmeal
		• Whole grains (e.g., couscous, barley, bulgur, etc.)
		• Reduced fat granola
• Cold cuts or lunch meats (bologna, salami, liverwurst, etc.)	Meat, Fish, and Poultry	• Lowfat cold cuts (95% to 97% fat free lunch meats, lowfat pressed meats)
• Hot dogs (regular)		• Lower fat hot dogs
• Bacon or sausage		• Canadian bacon or lean ham
• Regular ground beef		• Extra lean ground beef such as ground round or ground turkey (read labels)
• Chicken or turkey with skin, duck, or goose		• Chicken or turkey without skin (white meat)
• Oil-packed tuna		• Water-packed tuna (rinse to reduce sodium content)

Figure 12-2
Lower-Calorie, Lower-Fat Alternatives.

Source: The Practical Guide: Identification, Evaluation, and Treatment of Overweight and Obesity in Adults. NIH Pub. No. 00-4084. National Institutes of Health, National Heart, Lung, & Blood Institute. 2000.

Instead of...		Replace with...
• Beef (chuck, rib, brisket) • Pork (spareribs, untrimmed loin) • Frozen breaded fish or fried fish (homemade or commercial) • Whole eggs • Frozen TV dinners (containing more than 13 grams of fat per serving) • Chorizo sausage	Meat, Fish, and Poultry (continued)	• Beef (round, loin) (trimmed of external fat) (choose select grades) • Pork tenderloin or trimmed, lean smoked ham • Fish or shellfish, unbreaded (fresh, frozen, canned in water) • Egg whites or egg substitutes • Frozen TV dinners (containing less than 13 grams of fat per serving and lower in sodium) • Turkey sausage, drained well (read label) • Vegetarian sausage (made with tofu)
• Croissants, brioches, etc. • Donuts, sweet rolls, muffins, scones, or pastries • Party crackers • Cake (pound, chocolate, yellow) • Cookies	Baked Goods	• Hard french rolls or soft "brown 'n serve" rolls • English muffins, bagels, reduced fat or fat free muffins or scones • Lowfat crackers (choose lower in sodium) • Saltine or soda crackers (choose lower in sodium) • Cake (angel food, white, gingerbread) • Reduced fat or fat free cookies (graham crackers, ginger snaps, fig bars) (compare calorie level)
• Nuts • Ice cream, e.g., cones or bars • Custards or puddings (made with whole milk)	Snacks and Sweets	• Popcorn (air-popped or light microwave), fruits, vegetables • Frozen yogurt, frozen fruit, or chocolate pudding bars • Puddings (made with skim milk)
• Regular margarine or butter • Regular mayonnaise • Regular salad dressings • Butter or margarine on toast or bread • Oils, shortening, or lard	Fats, Oils, and Salad Dressings	• Light-spread margarines, diet margarine, or whipped butter, tub or squeeze bottle • Light or diet mayonnaise or mustard • Reduced calorie or fat free salad dressings, lemon juice, or plain, herb-flavored, or wine vinegar • Jelly, jam, or honey on bread or toast • Nonstick cooking spray for stir-frying or sautéing • As a substitute for oil or butter, use applesauce or prune puree in baked goods
• Canned cream soups • Canned beans and franks • Gravy (homemade with fat and/or milk) • Fudge sauce • Avocado on sandwiches • Guacamole dip or refried beans with lard	Miscellaneous	• Canned broth-based soups • Canned baked beans in tomato sauce • Gravy mixes made with water or homemade with the fat skimmed off and fat free milk included • Chocolate syrup • Cucumber slices or lettuce leaves • Salsa

Figure 12-2 *(continued)*

	Calories	Fat (grams)	% Fat	Exchange for:
Breakfast				
• Whole wheat bread, 1 medium slice	70	1.2	15.4	(1 bread/starch)
• Jelly, regular, 2 tsp	30	0	0	(1/2 fruit)
• Cereal, shredded wheat, 1 cup	207	2	8	(2 bread/starch)
• Milk, 1%, 1 cup	102	3	23	(1 milk)
• Orange juice, 3/4 cup	78	0	0	(1 1/2 fruit)
• Coffee, regular, 1 cup	5	0	0	(free)
• Milk, 1%, 1 oz	10	0.3	27	(1/8 milk)
Breakfast total	**502**	**6.5**	**10**	
Lunch				
• Roast beef sandwich:				
Whole wheat bread, 2 medium slices	139	2.4	15	(2 bread/starch)
Lean roast beef, unseasoned, 2 oz	60	1.5	23	(2 lean protein)
American cheese, lowfat and low sodium, 1 slice, 3/4 oz	46	1.8	36	(1 lean protein)
Lettuce, 1 leaf	1	0	0	
Tomato, 3 medium slices	10	0	0	(1 vegetable)
Mayonnaise, low calorie, 2 tsp	30	3.3	99	(2/3 fat)
• Apple, 1 medium	80	0	0	(1 fruit)
• Water, 1 cup	0	0	0	(free)
Lunch total	**366**	**9**	**22**	
Dinner				
• Salmon, 3 ounces edible	155	7	40	(3 lean protein)
• Vegetable oil, 1 1/2 tsp	60	7	100	(1 1/2 fat)
• Baked potato, 3/4 medium	100	0	0	(1 bread/starch)
• Margarine, 1 tsp	34	4	100	(1 fat)
• Green beans, seasoned, with margarine, 1/2 cup	52	2	4	(1 vegetable) (1/2 fat)
• Carrots, seasoned, with margarine, 1/2 cup	52	2	4	(1 vegetable) (1/2 fat)
• White dinner roll, 1 medium	80	3	33	(1 bread/starch)
• Ice milk, 1/2 cup	92	3	28	(1 bread/starch) (1/2 fat)
• Iced tea, unsweetened, 1 cup	0	0	0	(free)
• Water, 2 cups	0	0	0	(free)
Dinner total	**625**	**28**	**38**	
Snack				
• Popcorn, 2 1/2 cups	69	0	0	(1 bread/starch)
• Margarine, 1/2 tsp	58	6.5	100	(1 1/2 fat)
Total	**1,613**	**50**	**28**	

Calories 1,613
Total carbohydrate, % kcals 55
Total fat, % kcals 29
*Sodium, mg 1,341

Saturated fat, % kcals 8
Cholesterol, mg 142
Protein, % kcals 19

Note: Calories have been rounded.
1,600: 100% RDA met for all nutrients except vitamin E 99%, iron 73%, and zinc 91%.
* No salt added in recipe preparation or as seasoning. Consume at least 32 ounces of water.

Figure 12-3

Reduced-Calorie Menu—Traditional American Cuisine—1600 Calories.

Source: The Practical Guide: Identification, Evaluation, and Treatment of Overweight and Obesity in Adults. NIH Pub. No. 00-4084. National Institutes of Health, National Heart, Lung, & Blood Institute. 2000.

	1,600 Calories	1,200 Calories
Breakfast		
• Banana	1 small	1 small
• Whole wheat bread	2 slices	1 slice
• Margarine	1 tsp	1 tsp
• Orange juice	3/4 cup	3/4 cup
• Milk 1%, lowfat	3/4 cup	3/4 cup
Lunch		
• Beef noodle soup, canned, low sodium	1/2 cup	1/2 cup
• Chinese noodle and beef salad:		
Roast beef	3 oz	2 oz
Peanut oil	1 1/2 tsp	1 tsp
Soy sauce, low sodium	1 tsp	1 tsp
Carrots	1/2 cup	1/2 cup
Zucchini	1/2 cup	1/2 cup
Onion	1/4 cup	1/4 cup
Chinese noodles, soft type	1/4 cup	1/4 cup
• Apple	1 medium	1 medium
• Tea, unsweetened	1 cup	1 cup
Dinner		
• Pork stir-fry with vegetables:		
Pork cutlet	2 oz	2 oz
Peanut oil	1 tsp	1 tsp
Soy sauce, low sodium	1 tsp	1 tsp
Broccoli	1/2 cup	1/2 cup
Carrots	1 cup	1/2 cup
Mushrooms	1/4 cup	1/2 cup
• Steamed white rice	1 cup	1/2 cup
• Tea, unsweetened	1 cup	1 cup
Snack		
• Almond cookies	2 cookies	—
• Milk 1%, lowfat	3/4 cup	3/4 cup

	1,600 Calories	1,200 Calories
Calories	1,609	1,220
Total carbohydrate, % kcals	56	55
Total fat, % kcals	27	27
*Sodium, mg	1,296	1,043
Saturated fat, % kcals	8	8
Cholesterol, mg	148	117
Protein, % kcals	20	21

1,600: 100% RDA met for all nutrients except zinc 95%, iron 87%, and calcium 93%.
1,200: 100% RDA met for all nutrients except vitamin E 75%, calcium 84%, magnesium 98%, iron 66%, and zinc 77%.
* No salt added in recipe preparation or as seasoning. Consume at least 32 ounces of water.

Figure 12-4

Asian-American Cuisine—Reduced-Calorie Menu.

Source: The Practical Guide: Identification, Evaluation, and Treatment of Overweight and Obesity in Adults. NIH Pub. No. 00-4084. National Institutes of Health, National Heart, Lung, & Blood Institute. 2000.

Breakfast	1,600 Calories	1,200 Calories
• Oatmeal, prepared with 1% milk, lowfat	1/2 cup	1/2 cup
• Milk, 1%, lowfat	1/2 cup	1/2 cup
• English muffin	1 medium	—
• Cream cheese, light, 18% fat	1 T	—
• Orange juice	3/4 cup	1/2 cup
• Coffee	1 cup	1 cup
• Milk, 1%, lowfat	1 oz	1 oz
Lunch		
• Baked chicken, without skin	2 oz	2 oz
• Vegetable oil	1 tsp	1/2 tsp
• Salad:		
Lettuce	1/2 cup	1/2 cup
Tomato	1/2 cup	1/2 cup
Cucumber	1/2 cup	1/2 cup
• Oil and vinegar dressing	2 tsp	1 tsp
• White rice	1/2 cup	1/4 cup
• Margarine, diet	1/2 tsp	1/2 tsp
• Baking powder biscuit, prepared with vegetable oil	1 small	1/2 small
• Margarine	1 tsp	1 tsp
• Water	1 cup	1 cup
Dinner		
• Lean roast beef	3 oz	2 oz
• Onion	1/4 cup	1/4 cup
• Beef gravy, water-based	1 T	1 T
• Turnip greens	1/2 cup	1/2 cup
• Margarine, diet	1/2 tsp	1/2 tsp
• Sweet potato, baked	1 small	1 small
• Margarine, diet	1/2 tsp	1/4 tsp
• Ground cinnamon	1 tsp	1 tsp
• Brown sugar	1 tsp	1 tsp
• Corn bread prepared with margarine, diet	1/2 medium slice	1/2 medium slice
• Honeydew melon	1/4 medium	1/8 medium
• Iced tea, sweetened with sugar	1 cup	1 cup
Snack		
• Saltine crackers, unsalted tops	4 crackers	4 crackers
• Mozzarella cheese, part skim, low sodium	1 oz	1 oz

1,600: 100% RDA met for all nutrients except vitamin E 97%, magnesium 98%, iron 78%, and zinc 90%.
1,200: 100% RDA met for all nutrients except vitamin E 82%, vitamin B_1 & B_2 95%, vitamin B_3 99%, vitamin B_6 88%, magnesium 83%, iron 56%, and zinc 70%.
* No salt added in recipe preparation or as seasoning. Consume at least 32 ounces of water.

	1,600	1,200
Calories	1,653	1,225
Total carbohydrate, % kcals	53	50
Total fat, % kcals	28	31
*Sodium, mg	1,231	867
Saturated fat, % kcals	8	9
Cholesterol, mg	172	142
Protein, % kcals	20	21

Figure 12-5

Southern Cuisine—Reduced-Calorie Menu.

Source: The Practical Guide: Identification, Evaluation, and Treatment of Overweight and Obesity in Adults. NIH Pub. No. 00-4084. National Institutes of Health, National Heart, Lung, & Blood Institute. 2000.

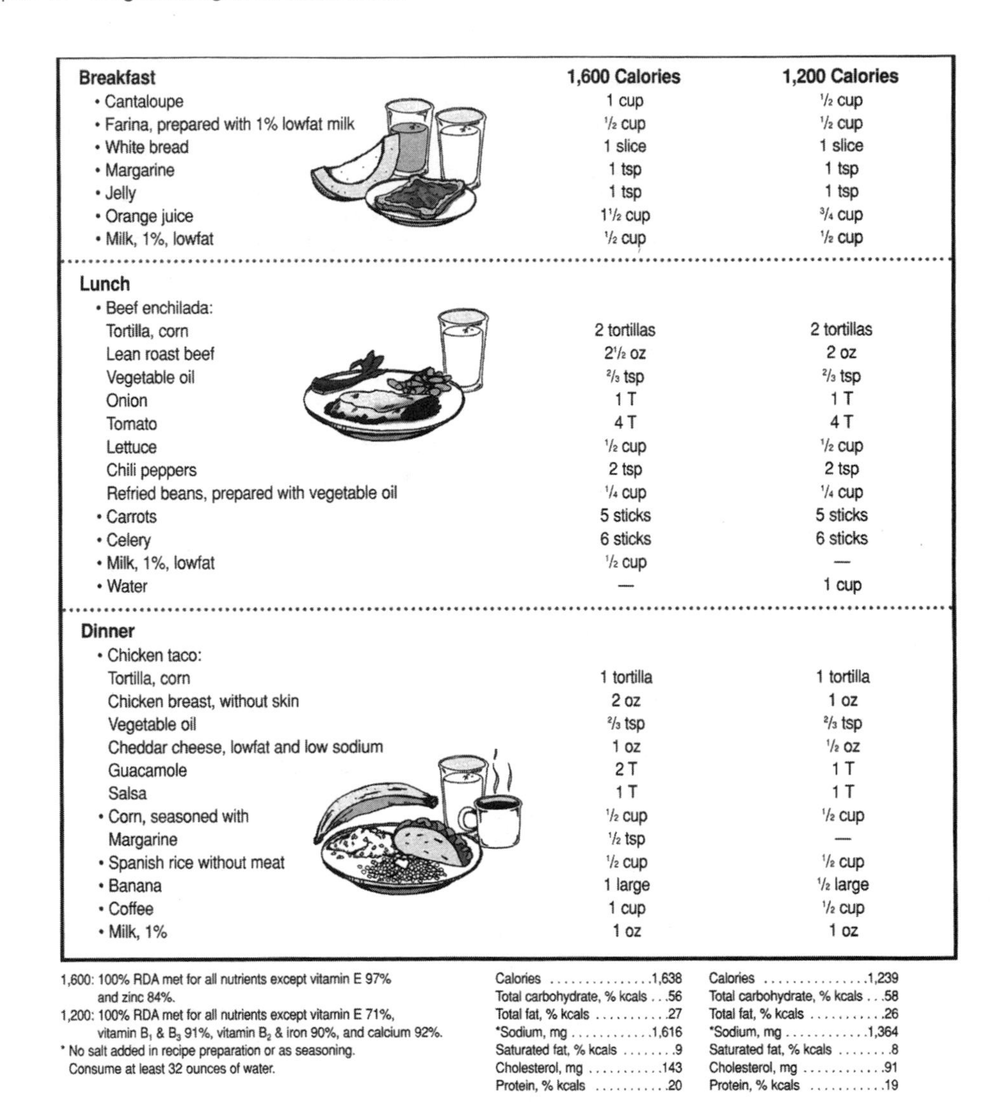

Breakfast	1,600 Calories	1,200 Calories
• Cantaloupe	1 cup	1/2 cup
• Farina, prepared with 1% lowfat milk	1/2 cup	1/2 cup
• White bread	1 slice	1 slice
• Margarine	1 tsp	1 tsp
• Jelly	1 tsp	1 tsp
• Orange juice	1 1/2 cup	3/4 cup
• Milk, 1%, lowfat	1/2 cup	1/2 cup
Lunch		
• Beef enchilada:		
Tortilla, corn	2 tortillas	2 tortillas
Lean roast beef	2 1/2 oz	2 oz
Vegetable oil	2/3 tsp	2/3 tsp
Onion	1 T	1 T
Tomato	4 T	4 T
Lettuce	1/2 cup	1/2 cup
Chili peppers	2 tsp	2 tsp
Refried beans, prepared with vegetable oil	1/4 cup	1/4 cup
• Carrots	5 sticks	5 sticks
• Celery	6 sticks	6 sticks
• Milk, 1%, lowfat	1/2 cup	—
• Water	—	1 cup
Dinner		
• Chicken taco:		
Tortilla, corn	1 tortilla	1 tortilla
Chicken breast, without skin	2 oz	1 oz
Vegetable oil	2/3 tsp	2/3 tsp
Cheddar cheese, lowfat and low sodium	1 oz	1/2 oz
Guacamole	2 T	1 T
Salsa	1 T	1 T
• Corn, seasoned with	1/2 cup	1/2 cup
Margarine	1/2 tsp	—
• Spanish rice without meat	1/2 cup	1/2 cup
• Banana	1 large	1/2 large
• Coffee	1 cup	1/2 cup
• Milk, 1%	1 oz	1 oz

1,600: 100% RDA met for all nutrients except vitamin E 97% and zinc 84%.
1,200: 100% RDA met for all nutrients except vitamin E 71%, vitamin B_1 & B_3 91%, vitamin B_2 & iron 90%, and calcium 92%.
* No salt added in recipe preparation or as seasoning. Consume at least 32 ounces of water.

	1,600	1,200
Calories	1,638	1,239
Total carbohydrate, % kcals	56	58
Total fat, % kcals	27	26
*Sodium, mg	1,616	1,364
Saturated fat, % kcals	9	8
Cholesterol, mg	143	91
Protein, % kcals	20	19

Figure 12-6
Mexican-American Cuisine—Reduced-Calorie Menu.

Source: The Practical Guide: Identification, Evaluation, and Treatment of Overweight and Obesity in Adults. NIH Pub. No. 00-4084. National Institutes of Health, National Heart, Lung, & Blood Institute. 2000.

Breakfast	**1,600 Calories**	**1,200 Calories**
• Orange	1 medium	1 medium
• Pancakes, made with 1% lowfat milk and eggs whites	3 4" circles	2 4" circles
• Pancake syrup	2 T	1 T
• Margarine, diet	1 1/2 tsp	1 1/2 tsp
• Milk, 1%, lowfat	1 cup	1/2 cup
• Coffee	1 cup	1 cup
• Milk, 1%, lowfat	1 oz	1 oz
Lunch		
• Vegetable soup, canned, low sodium	1 cup	1/2 cup
• Bagel	1 medium	1/2 medium
• Processed American cheese, lowfat	3/4 oz	–
• Spinach salad:		
Spinach	1 cup	1 cup
Mushrooms	1/2 cup	1/2 cup
• Salad dressing, regular calorie	2 tsp	2 tsp
• Apple	1 medium	1 medium
• Iced tea, unsweetened	1 cup	1 cup
Dinner		
• Omelette:		
Egg whites	4 large eggs	4 large eggs
Green pepper	2 T	2 T
Onion	2 T	2 T
Mozzarella cheese, made from part skim milk, low sodium	1 oz	1/2 oz
Vegetable oil	1 T	1/2 T
• Brown rice, seasoned with	1/2 cup	1/2 cup
Margarine, diet	1/2 tsp	1/2 tsp
• Carrots, seasoned with	1/2 cup	1/2 cup
Margarine, diet	1/2 tsp	1/2 tsp
• Whole wheat bread	1 slice	1 slice
• Margarine, diet	1 tsp	1 tsp
• Fig bar cookie	1 bar	1 bar
• Tea	1 cup	1 cup
• Honey	1 tsp	1 tsp
• Milk, 1%, lowfat	3/4 cup	3/4 cup

1,600: 100% RDA met for all nutrients except vitamin E 92%, vitamin B_3 97%, vitamin B_6 67%, iron 73%, and zinc 68%.
1,200: 100% RDA met for all nutrients except vitamin E 75%, vitamin B_1 92%, vitamin B_3 69%, vitamin B_6 59%, iron 54%, and zinc 46%.
* No salt added in recipe preparation or as seasoning.
Consume at least 32 ounces of water.

1,600 Calories		1,200 Calories	
Calories	1,650	Calories	1,205
Total carbohydrate, % kcals	56	Total carbohydrate, % kcals	60
Total fat, % kcals	27	Total fat, % kcals	25
*Sodium, mg	1,829	*Sodium, mg	1,335
Saturated fat, % kcals	8	Saturated fat, % kcals	7
Cholesterol, mg	82	Cholesterol, mg	44
Protein, % kcals	19	Protein, % kcals	18

Figure 12-7

Lacto-Ovo Vegetarian Cuisine—Reduced-Calorie Menu.

Source: The Practical Guide: Identification, Evaluation, and Treatment of Overweight and Obesity in Adults. NIH Pub. No. 00-4084. National Institutes of Health, National Heart, Lung, & Blood Institute. 2000.

The Problem of Underweight

A person is considered *underweight* if he or she has a BMI below 18.5. Although anyone who has seriously dieted may think the underweight person is problem-free, this is hardly the case. Underweight persons who have trouble gaining weight have very real concerns. Just as some people cannot seem to lose weight, so some people have trouble putting on a few extra pounds. The cause could be genetics, metabolism, or environment. However, some thin people, if they were to gain weight, would feel uncomfortable. Anyone who is underweight due to wasting diseases such as cancer or eating disorders also has a problem: malnutrition.

The following list contains tips on gaining weight.

1. If you find you cannot eat large meals, do not get discouraged. You can increase your intake by eating smaller meals frequently.
2. Avoid drinking low-kcalorie beverages such as coffee, tea, or water, especially with meals. Try fruit juices, milk, and milkshakes for more calories.
3. Add additional calories to your meals by using margarine, mayonnaise, oil, salad dressing, or other fats that do not contain much saturated fat. For example, spread margarine on bread or use oil-packed tuna fish.
4. Add skim milk powder to soups, sauces, gravies, casseroles, scrambled eggs, and hot cereals. It adds both kcalories and protein. It can also be blended with milk at the rate of 2 to 4 tablespoons of powder to 1 cup of milk.
5. Add cheese to favorite sandwiches. Use grated cheese on top of casseroles, salads, soups, sauces, and baked potatoes.
6. Try breaded foods.
7. Eat regular yogurt, peanut butter or cheese with crackers, nuts, milkshakes, and whole-grain cookies and muffins as snacks.
8. Add regular cottage cheese to casseroles or egg dishes such as quiche, scrambled eggs, and soufflés. Add it to spaghetti or noodles.
9. Make every mouthful count!

Nutrition for the Athlete

Many athletes require between 3000 and 6000 kcalories daily. The amount of energy required by the athlete depends on the type of activity and its duration, frequency, and intensity. In addition, the athlete's basal metabolic rate, body composition, age, and environment must be taken into account.

Carbohydrate (from glycogen and blood glucose) and fat are the primary fuel sources for exercise. Protein plays a minor role. The availability of carbohydrate, more specifically the amount of glycogen stores, heavily influences athletic performance. Glucose is the main source of energy for intense exercise, while fat is the main source of energy during low to moderate exercise. An appropriate diet for many athletes consists of 60 to 65 percent of kcalories as carbohydrates, 30 percent or less as fat, and enough protein to provide 1.2 to 1.6 grams per kilogram of body weight for endurance athletes, and 1.6 to 1.7 grams for power (strength or speed) athletes.

Although many athletes take vitamin and mineral supplements, these will not enhance performance unless there is a deficiency. Most athletes get plenty of vitamins and minerals in their regular diets, although young athletes and women need to pay special attention to iron and calcium.

Water is the most crucial nutrient for athletes. They need about 1 liter of water for every 1000 kcalories consumed. For moderate exercise without extreme temperatures or duration, cold water is the choice for replacing fluids. Cold water both cools the body and empties more quickly from the stomach. Athletes need 2 cups of water about 15 to 20 minutes before endurance exercise and at regular intervals during exercise. A good way to determine how much fluid to replace after exercising is to weigh in before and after exercise and also the next morning. For every pound that is lost, the athlete needs to drink 2 cups, or 16 ounces, of water.

For endurance events, some carbohydrates, such as table sugar or glucose, in fluids taken before and during competition may be helpful in maintaining normal blood-sugar levels.

Carbohydrate or glycogen loading—A regimen involving both decreased exercise and increased consumption of carbohydrates before an athletic event to increase the amount of glycogen stores.

Carbohydrate or glycogen loading is a regimen involving three or more days of decreasing amounts of exercise and increased consumption of carbohydrates before an event to increase glycogen stores. The theory is that increasing glycogen stores by 50 to 80 percent will enhance performance by providing more energy during lengthy competition. It is most appropriate for endurance athletes.

Here are some menu-planning guidelines for athletes.

1. Offer a variety of foods from the Food Guide Pyramid.
2. Good sources of complex carbohydrates to emphasize on menus include pasta, rice, other grain products such as breads and cereals, legumes, and fruits and vegetables. On the eve of the New York City Marathon each year, marathon officials typically host a pasta dinner for runners featuring spaghetti with marinara sauce and cold pasta primavera. Complex carbohydrates such as pasta also provide needed B vitamins, minerals, and fiber. Whole-grain products such as whole-wheat bread contain more nutrients than refined products such as white bread. If using refined products, be sure they are enriched (the thiamin, riboflavin, niacin, and iron have been replaced). Here are some ways to include complex carbohydrates in your menu:
   - At breakfast, offer a variety of pancakes, waffles, cold and hot cereals, breads, and rolls.

- At lunch, make sandwiches with different types of bread, such as pita pockets, raisin bread, onion rolls, and brown bread. Also, have a variety of breads and rolls available for non-sandwich items.
- Serve pasta and rice as a side or main dish with, for example, chicken and vegetables. Cold pasta and rice salads are great, too.
- Potatoes, whether baked, mashed, or boiled, are an excellent source of carbohydrates.
- Always have available as many types of fresh fruits and salads as possible.
- Don't forget to use beans and peas in soups, salads, entrées, and side dishes.
- Nutritious desserts emphasizing carbohydrates are frozen yogurt with fruit toppings, oatmeal cookies, and fresh fruit.

3. Don't offer too much protein and fat in the belief that athletes need the extra kcalories. They do, but much of those extra calories should come from complex carbohydrates. The days of steak-and-egg dinners are over for athletes. The protein and fat present in these meals do nothing to improve performance. Here are some ways to moderate the amount of fat and protein:
   - Use lean, well-trimmed cuts of beef.
   - Offer chicken, turkey, and fish—all lower in fat than beef. Broiling, roasting, and grilling are preferred cooking methods, with frying being acceptable occasionally.
   - Offer larger serving sizes of meat, poultry, and fish, perhaps 1 to 2 ounces more, but don't overdo it!
   - Offer fried food in moderation.
   - Offer low-fat and skim milk.
   - Offer high-fat desserts, such as ice cream and many types of sweets, in moderation. Frozen yogurt and ice milk generally contain less fat than ice cream and can be topped with fruit or crushed oatmeal cookies. Fruit ice and sorbet contain no fat.
4. Offer a variety of fluids, not just soft drinks and other sugared drinks. Good beverage choices include fruit juices, iced tea and iced coffee (preferably freshly brewed decaffeinated), plain and flavored mineral and seltzer water, spritzers (fruit juice and mineral water), and milkshakes made with yogurt or ice milk and fruit. Soft drinks and juice drinks—both loaded with sugar—should be offered in moderation.
5. Make sure iodized salt is on the table.
6. Be sure to include sources of iron, calcium, and zinc at each meal. Good iron sources include liver, red meats, legumes, and iron-fortified breakfast cereal. Moderate iron sources include raisins, dried fruit, bananas, nuts, and whole-grain and fortified grain products. Be sure to include good vitamin C sources at each meal, as vitamin C assists in iron absorption.

Vitamin C sources include citrus fruits and juices, cantaloupe, strawberries, broccoli, potatoes, and brussels sprouts. Calcium (found in milk and dairy products) and zinc (found in shellfish, meat and poultry, legumes, dairy foods, whole grains, and fortified cereals) are also important.

7. The most important meal is the one closest to the competition, commonly called the **precompetition meal.** The functions of this meal include getting the athlete fueled up, both physically and psychologically, helping to settle the stomach, and preventing hunger. The meal should consist of mostly complex carbohydrates (they are digested easier and faster and help maintain blood sugar levels), and should be low in fat. High-fat foods take longer to digest and can cause sluggishness. Substantial precompetition meals are usually served three to four hours prior to competition, in order to allow enough time for stomach emptying (to avoid cramping and discomfort during the competition). Menus might include cereals with low-fat or skim milk topped with fresh fruit, low-fat yogurt with muffins and juice, or one or two eggs with toast and jelly and juice. The meal should include 2 to 3 cups of fluid for hydration and typically provide 300 to 1,000 kcalories. Smaller precompetition meals may be served two to three hours before competition. Many athletes have specific "comfort" foods that they enjoy before competition.

Precompetition meal— The meal closest to the time of a competition or event.

8. After competition and workouts, again emphasize complex carbohydrates to ensure glycogen restoration. The sooner an athlete fuels up after exercising, the more glycogen will be stored in the muscle. Food is also important to restore the minerals lost in sweating.

■ MINI-SUMMARY

Athletes have increased needs for calories. Carbohydrate (from glycogen and blood glucose) and fat are the primary fuel sources for exercise. Protein plays a minor role. Glucose is the main source of energy for intense exercise, while fat is the main source of energy during low to moderate exercise. An appropriate diet for many athletes consists of 60 to 65 percent of kcalories as carbohydrates, 30 percent or less as fat, and enough protein to provide 1.2 to 1.7 grams per kilogram of body weight depending on type of athletic activity. Although many athletes take vitamin and mineral supplements, these will not enhance performance unless there is a deficiency. Water is the most crucial nutrient for athletes. They need about 1 liter of water for every 1000 kcalories consumed. Athletes need 2 cups of water about 15 to 20 minutes before endurance exercise and at regular intervals during exercise. Carbohydrate or glycogen loading is a regimen involving three or more days of decreasing amounts of exercise and increased consumption of carbohydrates before an event to increase glycogen stores. Menu-planning guidelines are given.

Check-Out Quiz

1. Obesity is due simply to overeating.
   a. True b. False
2. An obese person is at increased risk for coronary heart disease.
   a. True b. False
3. Obesity is defined as having a BMI greater than 25.
   a. True b. False
4. When trying to lose weight, omitting junk foods that you like from your diet is a good idea.
   a. True b. False
5. Carbohydrate loading is a regimen involving three days of increased exercise and increased intake of carbohydrates.
   a. True b. False
6. Serious athletes need about 8 grams of protein per kilogram of body weight.
   a. True b. False
7. Body mass index tells you about your weight more accurately than height/weight tables.
   a. True b. False
8. An example of using stimulus control to lose weight is to leave the table immediately after eating.
   a. True b. False
9. A person is considered underweight if he or she weighs 10 percent below his or her desirable weight.
   a. True b. False
10. Fat that collects around the abdomen presents an increased risk of developing serious health problems.
    a. True b. False

Activities and Applications

1. Your Desirable Weight

Using Table 12-1, determine your BMI. Are you overweight (a BMI of 25 or more)? If so, find out if you have a family history of any of the medical conditions discussed in the section on health and obesity.

2. Low-Kcalorie Menu Planning

A local steak and seafood restaurant has asked you to design lower-kcalorie menu items as follows: two appetizers, three entrées, and one dessert.

Their emphasis is freshly made traditional American cooking. Provide recipes and kcalorie information if possible.

3. Using a Food Diary

Using the food diary form, complete a three-day food diary. Then examine it to increase your awareness of how much and how often you are eating and whether you are eating in response to moods, people, and/or activities. Write down two insights you gained about your eating habits.

4. Box Lunches

Design three complete box lunches for kcalorie- and fat-conscious customers of a gourmet take-out deli with complete kitchen facilities located in San Francisco. The meal must be well balanced and provide a main dish, side dish, dessert, and beverage.

5. Precompetition Meals

Devise a menu for a noontime precompetition meal for long-distance runners on a university track team who will compete at 4 P.M. Have two selections for each category you choose.

Nutrition Web Explorer

National Institute of Diabetes and Digestive and Kidney Diseases www.niddk.nih.gov

On the home page for NIDDK, click on "Weight Loss and Control." Under "Weight Loss and Control Topics," click on "Weight Loss and Nutrition Myths." Write down five myths presented.

Sports Science News www.sportsci.org

On the home page, click on the index on the left for "Sports Nutrition." Read any one of the articles listed. Summarize in one paragraph what you read.

American College of Sports Medicine www.acsm.org

Read the ACSM position stand on "Losing Weight: Keeping It Off." What do they suggest to maintain weight loss?

Food Facts *Sports Drinks*

A topic of much interest to athletes is whether sports drinks, such as Gatorade or Exceed, are needed during an event or workout. Sports drinks contain a dilute mixture of carbohydrate and electrolytes. Most contain about 50 kcalories per cup (or about 12 grams of carbohydrate) and small amounts of sodium and potassium. Sports drinks are purposely made to be weak solutions so they can empty faster from the stomach, and the nutrients they contain are therefore available to the body more quickly. They are primarily designed to be used during exercise, although there are some specially formulated sports drinks with slightly more sugar that can be used just prior to exercising.

During exercise lasting 90 minutes or more, sports drinks can help replace water and electrolytes and provide some carbohydrates for energy. During an endurance event or workout, you increasingly rely on blood sugar for energy as your muscle glycogen stores diminish. Carbohydrates taken during exercise can help you maintain a normal blood-sugar level and enhance (as well as lengthen) performance. Athletes often consume 1/2 to 1 cup of sports fluids every 15 to 20 minutes during exercise.

Some sports drinks claim that they contain glucose polymers, which are chains of glucose. It was thought that sports drinks with glucose polymers emptied from the stomach faster than solutions with sucrose or glucose, but it turned out that each of these solutions empties at approximately the same rate and provides the same positive effects on performance as long as the carbohydrate concentration is between 5 and 8 percent (which it usually is).

Although sports drinks clearly can help the athlete in lengthy events or workouts, are there other products that can do much the same? Long before sports drinks were available, athletes had their own homemade sports drinks: diluted juices with a pinch of salt, tea with honey, and dilute lemonade. Which works best? Whichever satisfies the athlete both physically and psychologically.

Hot Topic Fad Diets

It is estimated that over 50 million Americans are currently on a diet. Some succeed in taking weight off, but far fewer—maybe just 5 percent—manage to keep the weight off over the long term. With half of the adult population in the United States considered to be overweight, it's little wonder that consumers are constantly searching for the "magic bullet" to help them lose weight quickly and effortlessly.

There is little scientific research to corroborate the theories expounded in the majority of diet books currently on the market. Many promise weight-loss programs that are easy, allow favorite foods or foods traditionally limited in weight loss diets without limitations, and do not require a major shift in exercise habits. Authors may simplify or expand upon biochemistry and physiology in order to help support their theories and provide a plethora of scientific jargon that people do not understand but that seems to make sense. Few, if any, offer solid scientific support for their claims in the form of published research studies. Instead, most evidence is based on anecdotal findings, theories, and testimonials of short-term results.

Some of the most popular diets to hit the news wires these days are those that promote low-carbohydrate and high-protein intakes and promise significant weight loss. Variations of low-carbohydrate diets have been around since the 1960s. These diets are nothing more than low-kcalorie diets in disguise, but with some potentially serious consequences. Following a low-carbohydrate, high-protein diet will encourage the body to burn its own fat. Without carbohydrates, however, fat is not burned completely and substances called ketones are formed and released into the bloodstream. Abnormally high ketone levels in the body, or ketosis, may indeed make dieting easier, since they typically decrease appetite and cause nausea. Although these diets may not be harmful when used by healthy people for a short period of time, they restrict healthful foods that provide essential nutrients. Individuals who follow these diets for more than just a short period of time run the risk of compromised vitamin and mineral intake, as well as potential heart, kidney, bone, and liver abnormalities. There are no long-term studies that have found these diets to be effective and safe.

Here's a rundown on some of the more popular high-protein, low-carbohydrate diets.

1. **Sugar Busters!** by H. Leighton Steward, Sam S. Andrews, MD, Morrison C. Bethea, MD, and Luis A. Balart, MD

Premise/Theory: Sugar and certain carbohydrates are toxic to the body, causing blood-sugar levels to rise and increasing the levels of insulin production, thereby prompting fat storage and weight gain. Supposedly, decreasing sugar intake can help people lose weight and decrease body fat, no matter what other foods are eaten.

Dietary Recommendations:

- Eliminates refined and processed carbohydrates—especially sugar and white flour and foods made with these ingredients. Also eliminates foods such as potatoes, corn, white rice, carrots, and soft drinks.
- Encourages consumption of whole grains, high-fiber fruits and vegetables, and lean meats, with no restrictions on protein foods.
- Authors claim that washing food down with liquid does not allow for proper chewing. Claims excess fluid with meals also dilutes digestive juices and can result in partially digested food.
- Average intake of calories is 1200 kcalories per day distributed as 30 percent carbohydrates, 32 percent protein, and 28 percent fat.

Concerns:

- There is no scientific basis or published data for this diet's theory. The explanation of insulin's role in weight gain that is provided is simplistic. The body does produce insulin in response to a rise in blood-sugar levels, but it does not promote storage of fat unless excess calories are consumed.
- There is no scientific evidence supporting the claim that the consumption of fluids during meals negatively affects digestion.
- The diet is low in some vitamins and minerals, including calcium and vitamin A.

2. **Dr. Atkins' New Diet Revolution** by Robert Atkins, MD
   Premise/Theory: Excess carbohydrate intake prevents the body from burning fat efficiently. Eating too many carbohydrates causes production of excessive amounts of insulin, leading to obesity and a variety of other health problems. Drastically decreasing dietary intake of carbohydrates forces the body to burn reserves of stored fat for energy, causing a buildup of ketones that lead to decreased hunger.

Dietary Recommendations:

- Limits carbohydrates to 20 grams per day for the start of the diet and 0–60 grams per day in the ongoing weight loss phase. Carbohydrate intake ranges from 25 to 90 grams per day in the maintenance diet.
- Unlimited quantities of protein foods and fat—steak, bacon, eggs, chicken, fish, butter, and vegetable oil—are allowed. Avoid or limit carbohydrates, specifically breads, pasta, most fruits and vegetables, milk, and yogurt.

Concerns:

- No published scientific studies support the diet claims.
- Offers extremely limited food choices. Diet is nutritionally unbalanced and excessively high in protein, fat, saturated fat, and cholesterol.
- Promotes ketosis as a means of weight loss.
- Suggests that a high-saturated-fat, low-carbohydrate diet does not have an effect on blood fat levels.
- Dehydration is possible if large amounts of water are not consumed.
- Diet is low in calcium, magnesium, potassium, vitamin C, and folate (dietary supplements are recommended).

3. **Enter the Zone** by Barry Sears, PhD

   Premise/Theory: The "zone" is a metabolic state in which the mind is relaxed and focused and the body is strong and works at peak efficiency. A person in the "zone" will allegedly experience permanent body-fat loss, optimal health, greater athletic performance, and improved mental productivity. Insulin is released as a result of eating carbohydrates and leads to weight gain. Because food has a potent, druglike effect on the hormonal systems that regulate the body's physiological processes, eating the right combination of foods leads to a metabolic state in which the body works at peak performance and experiences decreased hunger, weight loss, and increased energy.

Dietary Recommendations:

- To get into the "zone," rigid quantities of food, apportioned in blocks and at prescribed times, are recommended in a distribution of 40 percent carbohydrate, 30 percent protein, and 30 percent fat. Meals should provide no more than 500 kcalories and snacks less than 100 kcalories.
- Food should be treated like a medical prescription or drug.
- Menus suggest lots of egg whites, nuts, olives, peanut butter, and monounsaturated fats, and large amounts of allowable fruits and vegetables. Alcohol is okay in moderation, but "zone" followers are advised to avoid or limit carbohydrates, especially pasta, bread, and fruits and vegetables such as carrots and bananas, which cause blood sugar to rise more than others. Saturated fat should also be avoided.
- Diet averages 1300 kcalories per day, although some menus may run as low as 850 kcalories.

Concerns:

- Oversimplifies complicated body physiology.
- The metabolic pathways explained in the book are not found in standard biochemistry or nutrition books. The premise that any type of diet completely controls insulin secretion is not supported by current nutrition knowledge, nor is the theory that insulin and glucagon

control the production of eicosanoids (a biologically active class of compounds that are involved in a wide range of regulatory processes in the body).

- Relies upon unproven claims based on case histories, testimonials, and uncontrolled studies that are not published in peer-review journals.

Although these diets may promote short-term weight loss, their long-term effectiveness is a different story. In the first major review of popular diets by the federal government in 2001, they found that only traditional moderate-fat, high-carbohydrate regiments seem to keep dieters slim. They recommend consuming no more than 30 percent of kcalories as fat, limiting protein to about 20 percent of the diet, and consuming more fruits, vegetables, and complex carbohydrates to help satisfy hunger with fewer calories. Along with exercise, this type of diet can work in the long term as well.

Source: Adapted from "Fad Diets: Look Before You Leap," *Food Insight: Current Topics in Food Safety and Nutrition* (March-April 2000), published by the International Food Information Council.

Chapter 13

Nutrition Over the Life Cycle

Although much nutrition advice we hear on television or read about in magazines is for adults—and indeed, the first part of this book is mostly for adults—other groups have their own special nutrition needs and concerns. What do pregnant women, babies, children, and even teenagers have in common? They are all growing, and growth demands more nutrients. Did you know that compared with an adult, a one-month-old baby needs twice the amount, proportionally, of many vitamins and minerals? At the other end of the age spectrum, the fastest-growing age group in the United States is the over-85 group. As people age, many new factors affect their nutrition status: the aging process, onset of chronic diseases such as heart disease, living alone, dentures, and inability to get out to shop for food.

This chapter takes you from pregnancy through infancy, childhood, and adolescence and on to the golden years. Along the way, we will explore the nutritional needs and factors affecting nutrition status for each group, along with menu-planning guides. This chapter will help you to:

- Explain the benefits of good nutrition to mother and baby during pregnancy
- Identify nutrients that must be increased during pregnancy
- Plan menus for women during pregnancy and lactation
- Describe what an infant should be fed during the first year, including the progression of solid foods
- Plan menus for preschool and school-age children
- Describe influences on children's and adolescents' eating habits
- Plan menus for adolescents
- Distinguish among anorexia nervosa, bulimia nervosa, binge eating disorder, and female athlete triad
- Describe factors that influence the nutrition status of adults and older adults
- Plan menus for healthy older adults

Pregnancy

Fetus—The infant in the mother's uterus from eight weeks after conception until birth.

Amniotic sac—The protective bag, or sac, that cushions and protects the fetus during pregnancy.

From a modest one-cell beginning, an actual living and breathing baby is born after 40 weeks. From eight weeks after conception until birth, the infant in the mother's uterus is called a **fetus.** At eight weeks the fetus is about 1 inch long and has a beating heart and gastrointestinal and nervous systems. To cushion and protect the fetus, it floats in a protected bag, or sac, called the **amniotic sac.** From the second to eighth week after conception, the infant is called an **embryo.**

During the first month of pregnancy, an organ called the **placenta** develops to provide an exchange of nutrients and wastes between fetus and mother and to secrete the hormones necessary to maintain pregnancy. If a

Embryo—The name of the fertilized egg from conception to the eighth week.

Placenta—The organ that develops during the first month of pregnancy, which provides for exchange of nutrients and wastes between fetus and mother and secretes the hormones necessary to maintain pregnancy.

mother is not sufficiently nourished during early pregnancy (when she probably doesn't even know she's pregnant), the placenta will not perform properly and the fetus will not get optimal nourishment.

Pregnancy lasts for nine months. The first three month period is called the first trimester. Likewise, the next three months is the second trimester and the final three months is the third trimester.

Nutrition During Pregnancy

The nutritional status of women before and during pregnancy influences both the mother's and the baby's health. Factors that place a woman at nutritional risk during pregnancy include an inadequate diet, smoking, and other influences described in Table 13-1.

Table 13-2 shows optimum weight gain during pregnancy. Underweight women (10 percent or more below their desirable weight) must either gain weight before or gain more weight during pregnancy. Overweight women need to lose weight before pregnancy or gain less weight during pregnancy, otherwise they are at greater risk of developing gestational diabetes and hypertension, both of which can cause complications. Table 13-2 also shows that, of the weight gained, about 8 pounds is actually baby, with the rest serving to support the baby's growth.

Low-birth-weight baby—A newborn who weighs less than 5-1/2 pounds; these infants are at higher risk for disease.

Both prepregnancy weight and weight gain during pregnancy directly influence infant birth weight. The newborn's weight is the number-one indicator of his or her future health status. A newborn who weighs less than 5-1/2 pounds is referred to as a **low-birth-weight baby.** Low-birth-weight babies, as well as babies born before 37 weeks of gestation, are at higher risk for complications and experience more difficulties surviving the first year. Often the mother of a low-birth-weight baby has a history of poor

TABLE 13-1 Nutrition Risk Factors for Pregnant Women

- Pregnancy during adolescence
- Inadequate diets
- Multiple birth (twins, triplets, and so on)
- Use of cigarettes, alcohol, or illicit drugs
- Lactose intolerance
- Underweight or overweight at time of conception
- Gaining too few or too many pounds during pregnancy
- Health conditions such as diabetes mellitus, hypertension, and HIV infection

TABLE 13-2 Optimum Weight Gain in Pregnancy

Weight at Conception	Optimum Weight Gain
Normal weight	25–35 pounds
Underweight	28–40 pounds
Overweight	15–25 pounds
Components of Weight Gain	**Pounds**
Fetus	8 pounds
Placenta	1.5 pounds
Amniotic fluid	2.0 pounds
Increase in size of uterus, breast, fluid, and blood volume	12 pounds
Fat	2 to 8 pounds

nutrition status before and/or during pregnancy. Other factors associated with low birth weight are smoking, alcohol use, drug use, and certain disease conditions.

For healthy babies, pregnant women need to eat more kcalories, but not a whole lot more. If a pregnant woman "eats for two" during pregnancy, she is likely to put on too much weight. Pregnancy does increase kcalorie needs, but only by an additional 340 kcalories during the second trimester and an additional 450 kcalories during the third trimester. Within that measly 340 to 450 kcalories, however, the pregnant woman must pack more protein and more of most vitamins and minerals! See Appendix B for a comparison of the Dietary Reference Intakes for nonpregnant and pregnant women.

During the first 13 weeks of pregnancy, the total weight gain is between 2 and 4 pounds. Thereafter, about 1 pound per week is normal. Corresponding with the timing of weight gain, it makes sense that the greatest need for kcalories begins around the tenth week of pregnancy and continues until birth. The two major factors influencing kcalorie requirements during pregnancy are the woman's activity level and basal metabolic rate (BMR). The BMR increases to support the growth of the fetus. Pregnancy is no time to diet, or especially to follow a fad diet, which could have dangerous implications for the fetus.

Pregnancy-induced hypertension— Hypertension during pregnancy that can cause serious complications.

Protein needs increase 25 grams for pregnant women. The requirements for protein are generous during pregnancy and probably high for some women. Meeting protein needs is rarely a problem.

During pregnancy, calcium, phosphorus, and magnesium are necessary for the proper development of the skeleton and teeth. In adequate amounts, calcium may help reduce the incidence of **pregnancy-induced**

Edema—Swelling due to an abnormal accumulation of fluid in the intercellular spaces.

hypertension, a sometimes deadly disorder marked by high blood pressure (called hypertension) and **edema** (swelling due to an abnormal accumulation of fluid in the intercellular spaces). On the positive side, much more calcium (about double) is absorbed through the intestine. On the negative side, many women do not eat enough calcium-rich foods during pregnancy or lactation. The need for calcium (1000 milligrams) can be met by having at least three servings from the dairy group or calcium-fortified foods each day. Magnesium is found in green leafy vegetables, nuts, seeds, legumes, and whole grains.

Neural tube—The embryonic tissue that develops into the brain and spinal cord.

Spina bifida—A birth defect in which parts of the spinal cord are not fused together properly, so gaps are present where the spinal cord has little or no protection.

The need for folate increases 50 percent during pregnancy. This makes perfect sense when you realize that folate is needed to sustain the growth of new cells and the increased blood volume that occur during pregnancy. Folate is critical in the first four to six weeks of pregnancy (when most women don't even know they are pregnant) because this is when the **neural tube,** the tissue that develops into the brain and spinal cord, forms. Without enough folate, birth defects of the brain and spinal cord, such as **spina bifida,** can occur. In spina bifida, parts of the spinal cord are not properly fused, so gaps are present (Figure 13-1). Not every woman who has insufficient folate during early pregnancy will have a child with such a birth defect. However, if all women of childbearing age consumed enough folate, 50 to 70 percent of birth defects of the brain and spinal cord could be prevented, according to the U.S. Centers for Disease Control and Prevention.

Figure 13-1
Spina bifida aperta.

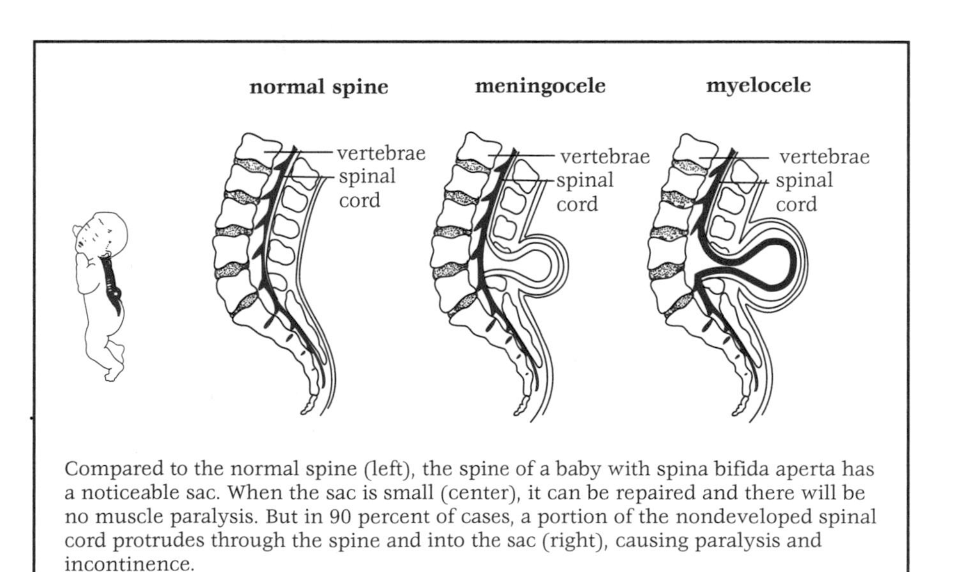

Compared to the normal spine (left), the spine of a baby with spina bifida aperta has a noticeable sac. When the sac is small (center), it can be repaired and there will be no muscle paralysis. But in 90 percent of cases, a portion of the nondeveloped spinal cord protrudes through the spine and into the sac (right), causing paralysis and incontinence.

Folate is also critical during the entire pregnancy. Women consuming less than 240 micrograms per day of folate have about a two- to threefold greater risk of preterm delivery and low-birth-weight babies. Folate is found in leafy green vegetables, beans, and enriched breads and cereals.

Since vitamin B_{12} works with folate to make new cells, increased amounts of this vitamin are also needed. As long as animal products, such as meat and milk, are being consumed, vitamin B_{12} deficiency is not a concern.

Iron also helps in the formation of blood—it is necessary for hemoglobin in both maternal and fetal red blood cells. After 34 weeks of pregnancy, a woman's blood volume has increased 50 percent from the time of conception. Although iron absorption does increase during pregnancy, whether the diet can supply enough iron is questionable. The National Academy of Sciences recommends iron supplements during the second and third trimesters to prevent anemia (a decreased number of red blood cells in the blood).

In the past it was thought that sodium restriction was necessary for women with edema, or tissue swelling. It is now known that moderate swelling is normal during pregnancy and that sodium restriction is unnecessary and could actually be harmful for healthy pregnant women.

Table 13-3 lists the current supplementation recommendations for pregnant women. The only nutrient suggested for all pregnant women is iron, although a prenatal vitamin and mineral supplement is prescribed routinely by most physicians.

Menu Planning During Pregnancy

To plan menus properly, certain diet-related concerns that occur during pregnancy need to be discussed. These include morning sickness, changes in taste and smell, constipation, heartburn, and the intake of alcohol, saccharin, aspartame, mercury, and polychlorinated biphenyls.

Nausea and vomiting, commonly referred to as morning sickness, can occur at any time of the day. Fatigue and aversion to certain odors also accompany the nausea and vomiting. Constipation may also be a concern. Each woman experiences it a little differently. Morning sickness lasts for an average of 17 weeks, but for some women it lasts until delivery. About 50 percent of women experience morning sickness. Health concerns with morning sickness are that it can cause dehydration, which in turn causes nausea, and then make it difficult for a woman to consume an adequate diet.

No one is exactly sure why morning sickness occurs. During pregnancy a woman's immune system is naturally suppressed to keep her from rejecting the fetus. Due to the suppressed immune system, she is more sus-

TABLE 13-3 Supplementation Recommendations: National Academy of Sciences (1990)

Nutrient	Candidates for Supplementation	Level of Nutrient Supplementation
Iron	All pregnant women (2nd and 3rd trimesters)	30 mg ferrous iron daily
Folate	Pregnant women with suspected dietary inadequacy of folate	300 μg/day
Vitamin D	Complete vegetarians and others with low intake of vitamin D-fortified milk	10 μg/day
Calcium	Women under age 25 whose daily dietary calcium intake is less than 600 mg	600 mg/day
Vitamin B_{12}	Complete vegetarians	2 μg/day
Zinc/copper	Women under treatment with iron for iron-deficiency anemia	15 mg zinc/day 0.2 mg copper/day
Multivitamin-mineral supplements	Pregnant women with poor diets and for those who are considered high risk: multiple gestation, heavy smokers, alcohol/drug abusers, other	Preparation containing iron—30 mg zinc—15 mg copper—2 mg calcium—250 mg vitamin B_6—2 mg folate—300 μg vitamin C—50 mg vitamin D— 5 μg

Source: Institute of Medicine. 1990. *Nutrition During Pregnancy.* Washington, DC: National Academy Press.

ceptible to infectious and toxic organisms in foods such as meat, poultry, fish, and eggs, which are often the foods a pregnant women with morning sickness does not want to eat.

Dietary advice in the past concentrated on small, carbohydrate-rich meals, and tea and crackers. For many women, this dietary advice doesn't work. More-recent advice centers on eating whatever foods you can keep down, even foods that aren't terribly nutritious, such as potato chips. The logic behind this recommendation is that tastes change when you are sick and you often crave something when you feel ready to eat. It's better to eat that food and keep it down than to eat something that is not appealing and throw it up. During pregnancy, women often develop a fine-tuned sense of smell, which often adds to their nausea. The smell of foods and cooking can make them sick.

Pregnant women commonly report changes in taste and smell. They may prefer saltier foods and crave sweets and dairy products such as ice cream. Certain foods they may have aversions to include alcohol, caffeinated drinks, and meats. Their cravings and aversions do not necessarily reflect actual physiological needs.

Constipation is not uncommon, due to the relaxation of gastrointestinal muscles. It can be counteracted by eating more high-fiber foods, drinking more fluids, and getting additional exercise.

Heartburn—A painful burning sensation in the esophagus caused by acidic stomach contents flowing back into the lower esophagus.

Heartburn is a common complaint toward the end of pregnancy, when the growing uterus crowds the stomach. This condition has nothing to do with the heart but is actually a painful burning sensation in the esophagus. It occurs when stomach contents, which are acidic, flow back into the lower esophagus. Possible solutions include eating small and frequent meals, eating slowly and in a relaxed atmosphere, avoiding caffeine, wearing comfortable clothes, and not lying down after eating.

Alcohol consumption and environmental contaminants such as mercury and polychlorinated biphenyls (PCBs) found in foods may affect the course and outcome of pregnancy. The following appears on the label on any bottle of beer, wine, or liquor:

> Government Warning: According to the Surgeon General, women should not drink alcoholic beverages during pregnancy because of the risk of birth defects.

Alcohol and pregnancy don't mix. During pregnancy, alcohol crosses the placenta, and high alcohol levels can build up in the fetus. Alcohol can limit the amount of vital oxygen delivered to the fetus, as well as slow the growth of cells. It can also produce abnormal cells.

Fetal alcohol syndrome (FAS)—A set of symptoms occurring in newborn babies that are due to alcohol use of the mother during pregnancy; symptoms may include mental retardation, brain damage, etc.

The heavy consumption of alcohol during pregnancy may cause a variety of symptoms called **fetal alcohol syndrome (FAS).** FAS children may show signs of mental retardation, growth retardation, brain damage, and facial deformities. Newborns with FAS are generally small in size and irritable because of alcohol withdrawal. The most serious concern with FAS infants is their impaired physical and mental development. They have problems gaining weight and are frequently mentally impaired.

You don't have to be a chronic alcoholic to have problems with FAS. Even moderate drinkers can have babies that show subtle signs of FAS. These women also have a higher rate of miscarriages and low-birth-weight babies. The American Academy of Pediatrics recommends that women stop drinking alcohol as soon as they plan to become pregnant, because harm can be done during the first six to eight weeks, when a woman may not know for sure whether she is pregnant.

Seafood is also a concern due to possible high levels of methyl mercury and PCBs (polychlorinated biphenyls). Pregnant women (and also nursing mothers) should not eat shark, swordfish, king mackerel, and tilefish.

These long-lived larger fish contain the highest levels of methyl mercury, which may harm an unborn baby's developing nervous system. Pregnant women should select a variety of other kinds of fish—shellfish, canned fish, smaller ocean fish, or farm-raised fish. They can safely eat 12 ounces of cooked fish per week. Pregnant women should also avoid eating fish that are likely to be contaminated with PCBs (freshwater carp, wild catfish, lake trout, whitefish, bluefish, mackerel, and striped bass) and raw fish (to reduce the risk of viral and bacterial illness). Although banned, PCBs are still present as pollutants in some freshwater lakes and rivers.

TABLE 13-4 Daily Food Guide for Pregnancy and Lactation

Food Group	Servings	Serving Size
Meat/Meat Alternative	3 (pregnancy, to include 1 serving legumes) 3 (lactation)	2 ounces cooked lean meat, poultry, or fish 2 eggs 2 ounces cheese 1/2 cup cottage cheese 1 cup dried beans or peas 4 tablespoons peanut butter
Milk and Dairy	3 (pregnancy and lactation) 4 (pregnant/lactating teenagers)	1 cup milk, yogurt, pudding or custard 1-1/2 ounces cheese 1-1/2 to 2 cups cottage cheese
Vegetables*	4 (5 during pregnancy and lactation)	1/2 cup cooked or juice 1 cup raw
Fruits*	3 (4 during pregnancy and lactation)	Portion commonly served, such as a medium apple or banana
Grain	7 (11 during pregnancy and lactation)	1 slice whole-grain or enriched bread 1 cup ready-to-eat cereal 1/2 cup cooked cereal or pasta 1/2 bagel or hamburger roll 6 crackers 1 small roll 1/2 cup rice or grits
Fats and sweets**		Includes butter, margarine, salad dressings, mayonnaise, oils, candy, sugar, jams, jellies, syrups, soft drinks, and any other fats and sweets

* A source rich in vitamin C (citrus, strawberries, melons, tomatoes) is needed daily; a source rich in vitamin A (dark green and deep yellow vegetables) is needed every other day.

** In general, the amount of these foods to use depends on the number of calories you require. Get your essential nutrients in the other food groups first before choosing foods from this group.

Aspartame is not yet known to cause problems during pregnancy. However, its use should be moderated in pregnancy, as should the use of caffeine. The fetus can't detoxify caffeine, and large amounts of caffeine (5 cups of coffee per day or more) may increase the chance of miscarriage in early pregnancy and increase the possibility of low-birth-weight babies. During pregnancy, caffeine-containing beverages should be limited to 2 cups a day.

Table 13-4 is a daily food guide for pregnancy. Problems can arise when an individual omits entire or substantial parts of certain groups. For instance, vegetarians who do not eat any food of animal origin need varied, adequate diets and supplements to obtain adequate vitamin B_{12}, calcium, zinc, and, unless getting adequate sunshine, vitamin D. Individuals who avoid the dairy group may need calcium and vitamin D supplements unless they eat foods fortified with these nutrients, such as calcium-fortified orange juice or calcium- and vitamin-D-fortified soy milk.

The following are some menu-planning guidelines for pregnant (and lactating) women.

1. Offer a varied and balanced selection of nutrient-dense foods. Because energy needs increase less than nutrient needs, empty kcalories are rarely an acceptable choice.
2. In addition to traditional meat entrées, choose entrées based on legumes and/or grains and dairy products. Beans, peas, rice, pasta, and cheese can be used in many entrées. Chapter 11 covers vegetarianism and has much information on meatless entrées.
3. Be sure to offer dairy products made with skim or low-fat milk.

TABLE 13-5 Sources of Problem Nutrients During Pregnancy

Nutrient	Food Sources
Fiber	Bran, dried beans and peas, whole-grain breads and cereals, fruits and vegetables, nuts, seeds
Folate	Organ meats, legumes, dark green leafy vegetables, orange juice, whole-wheat breads and cereals
Vitamin D	Vitamin D fortified milk
Iron	Red meat, liver, shellfish, poultry, dried fruit, beans, nuts, whole-grain and enriched breads and cereals
Calcium	Milk, dairy products, calcium-fortified orange juice
Magnesium	Green leafy vegetables, nuts, seeds, whole grains, and legumes
Zinc	Red meat, liver, poultry, fish, legumes

4. Use a variety of whole-grain and enriched breads, rolls, cereals, rice, pasta, and other grains.
5. Use assorted fruits and vegetables in all areas of the menu, including appetizers, salads, entrées, side dishes, and desserts.
6. Be sure to have good sources of problem nutrients: fiber, folate, vitamin D, iron, calcium, magnesium, and zinc (see Table 13-5).
7. Be sure to use iodized salt.

■ MINI-SUMMARY

Women's nutritional status before and during pregnancy influences both the mother and baby's health. Both prepregnancy weight and weight gain during pregnancy directly influence infant birth weight, the most important indicator of the baby's future health status. Although a pregnant woman should consume only 340 or 450 additional kcalories per day (during the second and third trimesters, respectively), she must take in more nutrients such as iron, folate, zinc, protein, and vitamin B_{12}. Iron supplements are likely to be prescribed because the diet just doesn't provide enough. When planning menus, keep in mind that pregnant women need nutrient-dense foods, an extra serving from the dairy and meat/meat-alternative groups, and plenty of fruits and vegetables. Advice for morning sickness is to let women eat whatever foods they feel they can keep down, without worrying excessively about how nutritious they are. Moderate use of aspartame is advised during pregnancy, but it is important to stay away from alcoholic beverages.

Nutrition and Menu Planning During Lactation

Table 13-4 shows the daily food guide for breast-feeding mothers. During lactation, the period of milk production, 330 additional kcalories are necessary for the first six months which increases to 400 additional kcalories from the 7th to 12th month. Actually, more than 330 extra kcalories are needed during the first 6 months, but some are supplied by extra fat stored during pregnancy. Lactating mothers, who normally produce about 25 ounces of milk a day, also need at least 2 to 3 quarts of water each day to prevent dehydration. Extra cups of coffee contain extra fluid but also contribute excess caffeine, which can cross to the baby and cause irritability.

A balanced, varied, and adequate diet (at least 1800 kcalories per day) is critical to successful breast-feeding and infant health. If the mother is not eating properly, any nutritional deficiencies are more likely to affect the quantity of milk she makes, rather than the quality. Menu-planning guidelines for lactating women are the same as for pregnant women, with emphasis on fluids, dairy products, fruits, and vegetables. Occasional consumption of small amounts of alcohol will probably have no consequences. The National Research Council suggests iron supplementation for the mother to replenish stores depleted during pregnancy. Lactating vegetarian mothers who eat no food of animal origin need to pay special attention to getting enough calories, iron, zinc, calcium, vitamin D, and vitamin B_{12}.

■ MINI-SUMMARY

During the first six months of lactation, mothers need 330 extra kcalories a day. During the 7th to 12th month, 400 additional kcalories a day are needed. Lactating mothers also need plenty of fluids, more grains, fruits, vegetables, and milk. If the mother is not eating a nutritionally adequate diet, the quantity of milk will be adversely affected. Small amounts of alcohol or caffeine are permissible.

Infancy: The First Year of Life

The nutrient needs of infants are about double those of an adult when viewed in proportion to their weights. Little wonder, considering that infants generally double their birth weight in the first four to five months and then triple their birth weight by the first birthday. An infant will also grow 50 percent in length by the first birthday. (In other words, a baby who was 20 inches at birth grows to 30 inches in one year.)

Nutrition During Infancy

Newborns need a plentiful supply of all nutrients, especially those necessary for growth, such as protein, vitamins C and D, folate, vitamin B_{12}, calcium, and iron. The DRI is set for infants from 0 to 6 months, and then from 7 to 12 months. By 6 months, growth occurs at a slower rate.

For the first 4 to 6 months of life, the source of all nutrients is breast milk or formula. Breast milk is recommended for all infants in the United States under ordinary circumstances from birth to 12 months. A baby needs breast milk for the first year of life, and as long as desired after that.

The number of women who are choosing to breast-feed is increasing. Current estimates are that more than 50 percent of American mothers breast-feed their babies in the hospital, but only 19 percent are still breastfeeding six months later. The reasons behind this increase include research findings that show definite health benefits of breast milk, as well as support and information groups that communicate breast-feeding guidelines and advantages. However, too few mothers breast-feed. Unfortunately, women who are young, unemployed, and on low incomes are the least likely to breast-feed. Their babies could greatly benefit from breast-feeding because they typically face the highest risk of health problems.

The following list shows the advantages of breast-feeding, compared to formula-feeding.

1. Breast milk is nutritionally superior to any formula or other type of feeding. It provides exactly the right proportion and form of calories and nutrients needed for optimum growth, brain development, and digestion. Cow's milk contains a different type of protein than breast milk, and infants can have difficulty digesting it. It also provides more of the essential fatty acids compared to formula. The composition of breast milk changes to meet the needs of the growing infant.
2. Newborns are less apt to be allergic to breast milk than to any other food.
3. Suckling promotes the development of the infant's jaw and teeth. It's harder work to get milk out of a breast than from a bottle; the exercise strengthens the jaw and encourages the growth of straight, healthy teeth. The baby at the breast can control the flow of milk by sucking and stopping. With a bottle, the baby must constantly suck or react to the pressure of the nipple in the mouth.
4. Breast-feeding promotes a close relationship—a bonding between mother and child. At birth, infants see only 12 to 15 inches, the distance between a nursing baby and its mother's face.
5. Breast milk is less likely to be mishandled. Some formulas require accurate dilutions, and all are much more apt to be mishandled, which can result in food-borne illness.
6. Breast milk helps the infant build up immunities to infectious disease, because it contains the mother's antibodies to disease. Breast-fed infants are much less likely to develop serious respiratory and gastrointestinal illnesses.
7. Breast-feeding may reduce the risk of breast cancer for the mother.
8. Breast-feeding is less expensive.
9. Breast-fed infants have lower rates of hospital admissions, ear infections, diarrhea, and other medical problems than do bottle-fed babies.

Breast-feeding is not recommended if the mother uses addictive drugs, drinks more than a minimal amount of alcohol, is on certain medications, or is HIV-positive (has the virus that causes AIDS).

If the infant or mother is not exposed regularly to sunlight or if the mother's intake of vitamin D is low, the breast-fed infant may need vitamin D. Vitamin D supplements are generally recommended for breast-fed infants.

Formula-feeding is an acceptable substitute for breast-feeding and has some advantages. Some women find formula-feeding more convenient (others find breast-feeding more convenient). Other family members can take part in formula feeding. For some women who are uncomfortable with breast-feeding, even after education, formula feeding is the method of choice.

All formulas must meet nutrient standards set by the American Academy of Pediatrics. Recently the Food and Drug Administration approved the addition of two fatty acids to infant formula. The fatty acids are docosahexaenoic acid (DHA) and arachidonic acid (AA), which are made in the body from the essential fatty acids. DHA and AA are present in breast milk and are thought to enhance the mental and visual development of infants. Major formula manufacturers now have one brand that includes DHA and AA. These formulas are more expensive than their traditional counterparts.

The three forms of formulas on the market are ready-to-feed formula, liquid concentrate that needs to be mixed with equal amounts of water, and powdered formulas, which also need to be mixed with water. All formulas must be handled in a sanitary manner to prevent contamination and possible food poisoning.

Cow-milk-based formulas are normally used unless the baby is allergic to the protein or sugar in milk. In that case, a soy-based formula is used. For the baby who is allergic to both milk-based and soy-based formulas, predigested formulas are available. Symptoms of allergies usually include diarrhea and/or vomiting. Cow's milk has too much protein and minerals and too little essential fatty acids, vitamin C, and iron. Therefore, it is not recommended until 12 months of age, when the baby is less likely to be allergic to it. Babies are normally switched slowly from formula to cow's milk.

Whether infants are breast-fed or formula fed, their iron stores are relatively depleted by 4 to 6 months, at which time they typically start to eat iron-fortified cereals. Fluoride supplements may also be prescribed for infants after six months, unless the infant receives formula made with fluoridated water.

Feeding the Infant

Successful infant feeding requires cooperative functioning between the mother and her baby. Feeding time should be a pleasurable period for both parent and child, so be sure to be comfortable and relaxed to better enjoy the experience. Ideally, the feeding schedule should be based on reasonable self-regulation by the baby. By the end of the first week of life, most infants want six to eight feedings a day. Formula-fed babies are fed about every four hours and breast-fed babies about every two or three hours.

Colostrum—A yellowish fluid that is the first secretion to come from the mother's breast a day or so after delivery of a baby; it is rich in proteins, antibodies, and other factors that protect against infectious disease.

Transitional milk—The type of breast milk produced from about the third to the tenth day after childbirth, when mature milk appears.

Milk letdown—The process by which milk comes out of the mother's breast to feed the baby; sucking causes the release of a hormone that allows milk letdown.

The mother must breast-feed the child as soon as possible after delivery to enhance success. **Colostrum,** a yellowish fluid, is the first secretion to come from the breast a day or so after delivery. It is rich in proteins, antibodies, and other factors that protect against infectious disease. Colostrum changes to **transitional milk** between the third and sixth days, and by the tenth day the major changes are finished.

The breast-feeding process begins with the infant using a sucking action that stimulates hormones to move milk into the ducts of the breast. This process is referred to as **milk letdown** and is hindered if the mother is tired or anxious. A baby will often empty a breast in about 10 minutes of nearly continuous sucking. The baby should empty at least one breast per feeding in order to stimulate the breast to produce more milk. To ensure that the newborn is getting enough milk, he or she needs to be nursed frequently. In order to nurse the child successfully, the mother needs adequate rest, nutrition, and fluids, as well as education and support to decrease anxiety. Table 13-6 lists tips for breast-feeding success.

Babies are ready to eat semisolid foods such as cereal when they can sit up and open their mouths. This usually occurs between five and seven months of age. Other signs that babies are ready for spoon-feeding are when they:

- Have doubled their birth weight
- Drink more than a quart of formula per day
- Seem hungry often
- Open their mouths in response to seeing food coming

Although some parents think that feeding of solids will help the baby to sleep through the night, this is not often so. Feeding of solid food before a baby is ready can create problems, because the baby's digestive system is not ready for it. Feeding solids early also increases the risk of allergies and the chance of choking, and may encourage overfeeding.

Most babies can digest starchy foods at around four months of age. Once a baby starts on solid foods, it is important to make sure that the baby gets sufficient fluids. Up to this point, breast milk or formula met the baby's need for fluids. Now, however, drinking water or other fluid is needed to prevent dehydration. Proportionally, babies have more water in their bodies than adults. They can become dehydrated very quickly due to hot weather, diarrhea, or vomiting, so extra fluids need to be offered at these times.

Although eating solid food is certainly simple for an adult, it involves a number of difficult steps for the baby. First the infant must have enough muscle control to close his or her mouth over the spoon, scrape the food from the spoon with the lips, and then move the food from the front to the back of the tongue. By about 16 weeks, a baby generally has these skills, but

TABLE 13-6 Tips for Breast-Feeding Success

- ***Get an early start:*** Nursing should begin within an hour after delivery if possible, when an infant is awake and the suckling instinct is strong. Even though the mother won't be producing milk yet, her breasts contain colostrum, a thin fluid that contains antibodies to disease.
- ***Proper positioning:*** The baby's mouth should be wide open, with the nipple as far back into his or her mouth as possible. This minimizes soreness for the mother. A nurse, midwife, or other knowledgeable person can help her find a comfortable nursing position.
- ***Nurse on demand:*** Newborns need to nurse frequently, at least every two hours, and not on any strict schedule. This will stimulate the mother's breasts to produce plenty of milk. Later, the baby can settle into a more predictable routine. But because breast milk is more easily digested than formula, breast-fed babies often eat more frequently than bottle-fed babies.
- ***Delay artificial nipples:*** It's best to wait a week or two before introducing a pacifier, so that the baby doesn't get confused. Artificial nipples require a different sucking action than real ones. Sucking at a bottle could also confuse some babies in the early days.
- ***Air dry:*** In the early postpartum period or until her nipples toughen, the mother should air-dry them after each nursing to prevent them from cracking, which can lead to infection. If her nipples do crack, the mother can coat them with breast milk or other natural moisturizers to help them heal. Vitamin E oil and lanolin are commonly used, although some babies may have allergic reactions to them. Proper positioning at the breast can help prevent sore nipples. If the mother is very sore, the baby may not have the nipple far enough back in his or her mouth.
- ***Watch for infection:*** Symptoms of breast infection include fever and painful lumps and redness in the breast. These require immediate medical attention.
- ***Expect engorgement:*** A new mother usually produces lots of milk, making her breasts big, hard, and painful for a few days. To relieve this engorgement, she should feed the baby frequently and on demand until her body adjusts and produces only what the baby needs. In the meantime, the mother can take over-the-counter pain relievers, apply warm, wet compresses to her breasts, and take warm baths to relieve the pain.
- ***Eat right, get rest:*** To produce plenty of good milk, the nursing mother needs a balanced diet that includes extra calories and at least eight glasses of fluid. She should also rest as much as possible to prevent breast infections, which are aggravated by fatigue.

Source: Adapted from "Breast-Feeding Best Bet for Babies." FDA Consumer. 1995.

probably no teeth! The baby's first teeth will cut through the gums between six and ten months of age. If a baby can't swallow well enough to get the food from the back of the tongue into the pharynx, the baby will gag. The **gag reflex** prevents choking and sometimes results in vomiting.

Gag reflex—The ability to cough or vomit up food (or anything) that can't be swallowed properly.

Foods are generally introduced as follows. Keep in mind that the order of introducing different types and textures of foods is tied to the baby's developmental stages.

4 to 6 months: Iron-fortified baby cereals
5 to 7 months: Strained or puréed vegetables and fruits

7 to 9 months: Strained or soft protein foods (meat, chicken, fish, cheese, yogurt, beans, egg yolk)
Finger foods such as crackers
Fruit juice
9 to 12 months: Soft, chopped foods (finely chopped at first)
Breads and grain products
12 months: Cut-up table foods
Whole milk
Whole eggs

The first solid food is iron-fortified baby cereal mixed with breast milk or formula. Usually, rice cereal is offered first because it is the least likely to cause an allergic reaction. Barley and oatmeal cereals follow. The iron found in these cereals is very important to meet the infant's high iron needs. Avoid putting cereals or any other solids into the infant's bottle.

Once the baby is used to various cereals, puréed or mashed vegetables and fruits can be tried at about five to six months. It is a good idea to start with vegetables so that the baby does not become accustomed to the sweet taste of fruits (babies like sweets) and then reject the vegetables. When adding new foods to the infant's diet, always do so one at a time (and in small quantities) so that if there is an allergic reaction (such as hives or diarrhea), you will know which food caused it. Introduce new vegetables and fruits about one week apart. Babies adjust differently to new tastes and new textures. If the baby does not like a certain food, offer it a few days later. If you offer new foods when the baby is hungry, as at the beginning of a meal, he or she is more likely to eat them.

Fruit juice that is fortified with vitamin C can be started about the fifth to seventh month. Although some babies get two or more bottles a day of apple juice (or other type of juice), it is a good idea to limit juice to a half cup, or 4 fluid ounces, daily. More than a half cup of juice daily can result in growth failure if substituted for breast milk or formula. Another problem with fruit juice can occur when you let your baby go to sleep with a bottle in his or her mouth. The natural sugars in the juice can cause serious tooth decay, called **baby bottle tooth decay.** Letting a baby go to bed with a bottle of formula, cow's milk, or breast milk will also cause baby bottle tooth decay.

Before a baby can move on to finger foods, he or she has to be able to grab them. At about eight months, a baby discovers and starts to use the thumb and forefinger together to pick things up (called the **pincer grasp**). From about six months, the baby has been using the palm (called the **palmar grasp**) to do this. Suitable finger foods include chopped ripe bananas, dry cereal, and pieces of cheese. About this time infants can also start eating protein foods. Poultry and fish must be very tender, and meat will have to be chopped or cut very fine and possibly moistened.

Baby bottle tooth decay—Serious tooth decay in babies caused by letting a baby go to bed with a bottle of juice, formula, cow's milk, or breast milk.

Pincer grasp—The ability of a baby at about eight months of age to use the thumb and forefinger together to pick things up.

Palmar grasp—The ability of a baby from about six months of age to grab objects with the palm of the hand.

Between 10 and 12 months of age, babies may have four to six sharp teeth, and many are eating soft, chopped foods with the family. At this time it is appropriate to let your child begin drinking from a cup. It takes time, but sooner or later your child will get the idea. By one year of age, infants can enjoy cut-up table foods as well as whole milk. By 12 months, a baby should be almost entirely self-feeding. Children should not be switched to low-fat milk until they are at least two years old, because they need the fat in whole milk for proper growth and development. Table 13-7 is a food guide for infants from birth to 12 months.

Because honey and liquid corn syrup may be contaminated with botulism, these foods may cause food poisoning or food-borne illness in children younger than one year.

Certain foods are also more apt than others to cause choking and block the child's airway. These foods are unsafe for infants and toddlers and include nuts, seeds, raisins, hot dogs, popcorn, whole grapes, hard candies (including jelly beans), peanut butter, cherry tomatoes, raw carrots and celery, cherries with pits and large chunks of any food including meat, potatoes, and raw vegetables and fruit. Grapes and cherry tomatoes can be cut in quarters. Hot dogs can be cut into tiny pieces.

Foods more apt to cause allergies include milk, eggs, wheat, nuts, chocolate, and shellfish. Whole milk and eggs are usually introduced at about 12 months.

■ MINI-SUMMARY

The growth rate during the first year will never be duplicated again. The nutrient needs of infants are about double those of an adult when viewed in proportion to their weight. For the first four to six months, the only food an infant should get is either breast milk or formula. Breast milk is recommended for many reasons. It is nutritionally superior to formula, and newborns are less likely to be allergic to it. Breast milk contains antibodies, to help babies build up immunities, and breast-feeding promotes bonding between mother and infant. Breast-fed babies are generally given vitamin D supplements. For formula-fed babies, the formula is generally cow-milk-based. If the baby is allergic to it, then a soy-based formula is used. A baby is generally ready to eat semisolid foods between five and seven months. The progression of foods starts with iron-fortified baby cereals, strained or puréed vegetables and fruits, strained or soft protein foods, finger foods, and fruit juice, to soft chopped foods, and finally to cut-up table foods by age one. Whole milk and whole eggs are not recommended until 12 months because of possible allergic reactions. Infants start getting baby teeth after six months and many have four to six teeth by their first birthday. Foods that cause choking, such as nuts and raisins, should be avoided.

TABLE 13-7 Food Guide for Infants

Age	Food*	Amount
0–4 months	Breast milk or formula**	21–29 ounces, formula, 5–8 feedings daily. 6–8 nursings.
4–6 months	Breast milk or formula	27–39 ounces formula, 4–6 feedings daily. 4–5 nursings.
	Iron-fortified infant cereal (usually starts at 5 months)	Give 1 tablespoon with mother's milk/formula to start. Start with rice cereal. Give once to twice daily. Can work up to 1-1/2 tablespoons twice daily.
5–7 months	Strained vegetables and fruits	Give 1–2 teaspoons once to twice daily. First fruits can be applesauce, pears, peaches, and bananas. First vegetables can be carrots, squash, and sweet potatoes. Slowly increase to 2 tablespoons twice daily.
7–9 months	Breast milk or formula	30–32 ounces formula, 3–5 feedings daily. 3–5 nursings.
	Iron-fortified infant cereal	3 tablespoons plus mother's milk/formula twice daily.
	Strained fruits and vegetables	3 tablespoons twice daily.
	Strained plain meats	1 to 2 tablespoons twice daily.
	Crackers, plain toast, or teething biscuit	When baby has teeth, offer these foods after other foods are eaten.
	Fruit juice (vitamin C fortified, non-acid; usually starts at 5 months)	Start with 2 ounces watered-down juice, usually apple juice. Limit fruit juice to 1/2 cup daily.
9–12 months	Breast milk or formula (Your physician may suggest switching to whole milk at 10 months or after)	24–32 ounces formula, 3–4 times daiy. 3–4 nursings.
	Fruit juice (vitamin C fortified)	1/2 cup daily.
	Iron-fortified infant cereal	3–4 tablespoons plus mother's milk/formula twice daily.
	Vegetables, cut up	3–4 tablespoons twice daily.
	Fruits, cut up	3–4 tablespoons twice daily.
	Meats, cut up	2–3 tablespoons twice daily.
	Egg (usually at 12 months)	1 egg = 1 serving of meat.
	Bread and grain products	1/2 slice four times daily.

* Avoid the following foods in the first year because of possible allergic reactions: chocolate, nuts, berries, tomatoes, shellfish.

** Physician may request iron-fortified formula by third or fourth month.

Childhood

Around age one, the baby's growth rate decreases markedly. Yearly weight gain now approximates 4 to 6 pounds per year. Children can expect to grow about 3 inches per year between ages one and seven, and then 2 inches per year until the pubertal growth spurt. Until adolescence, growth will come in spurts, during which the child will grow more and eat more.

Appendix D shows growth charts for boys and girls that compare the child's height and weight to national percentiles. If a girl is in the 90 percentile height for age, it means that, out of 100 girls her age, she is taller than 89 and shorter than 10. These are commonly used by physicians as an indicator of the child's overall health and adequacy of calorie intake for children from 2 to 18 years of age. Growth charts are available for ages birth to 36 months and then 2 to 18 years of age.

After age 1, children start to lose baby fat and become leaner, with muscle accounting for a larger percentage of body weight. The legs become longer, and the baby now starts to look like a child and to walk, run, and jump like a child. By age 2 brain growth is 75 percent complete. A child's head size in relation to body size starts to decrease and look more normal, and by age 6 to 10, the brain becomes adult size.

By about age 1, the child has six to eight teeth, and by age 2 the baby teeth are almost all in. Between ages 6 and 12, these are gradually replaced with permanent teeth. After the first birthday, as children's physical capabilities and desire for independence increase, they are more capable of feeding themselves. By age 18 months, many children can successfully use a spoon without too much spilling, and by 24 months many children can drink properly from a cup.

Nutrition During Childhood

Table 13-8 shows the EER for kcalories and RDA for protein for children. A 2 year-old needs about 1000 kcalories a day. By age 6, a child needs closer to 1700 kcalories daily. Energy needs of children of similar age, sex, and size can vary due to differing BMRs (basal metabolic rates), growth rates, and activity levels. Energy and protein needs decline gradually per pound of body weight.

Growth spurt—Periods of rapid growth.

During growth spurts, the requirements for kcalories and nutrients are greatly increased. Appetite fluctuates tremendously, with a good appetite during **growth spurts** (periods of rapid growth) and a seemingly terrible appetite during periods of slow growth. Parents may worry and force a child to eat more than needed at such times, when the child appears to be "living on air." A decreased appetite in childhood is perfectly normal. As long as the child is choosing nutrient-dense kcalories, nutritional problems

TABLE 13-8 Energy and Macronutrients for Children

Estimated Energy Requirement (EER) and Recommended Dietary Allowance (RDA) for Protein				
Gender and Age	Height	Weight	EER*	Protein
Male 1–3 years	34 inches	27 pounds	1046 kcal	1.1 grams/kilogram body weight
Female 1–3 years	34	27	992	1.1
Male 4–8 years	45	44	1742	0.95
Female 4–8 years	45	44	1642	0.95

*The EER is for healthy moderate active Americans and Canadians.

Acceptable Macronutrient Distribution Ranges			
Age	Carbohydrate	Fat	Protein
1–3 years	45–65%	30–40%	5–20%
4–18 years	45–65%	25–35%	10–30%
Over 18 years	45–65%	20–35%	10–35%

Source: Adapted with permission from the *Dietary References Intakes for Energy, Carbohydrates, Fiber, Fat, Protein, and Amino Acids (Macronutrients).* © 2002 by the National Academy of Sciences. Courtesy of the National Academy Press, Washington, D.C.

are unlikely. In preparation for the adolescent growth spurt, children accumulate stores of nutrients, such as calcium, that will be drawn upon later, as intake cannot meet all the demands of this intensive growth spurt.

Although kcalorie and protein intakes are rarely inadequate in American children, there are concerns about iron intake. Lack of iron can cause decreased energy and affect behavior, mood, and attention span. A balanced diet with adequate consumption of iron-rich foods such as lean meat (ground meat is easier for younger children to chew), enriched breads and cereals, and legumes is important to get enough iron. A source of vitamin C, such as citrus fruits, increases the amount of iron absorbed.

To prevent coronary heart disease early in life, medical authorities generally agree that by age five all healthy children should comply with the following guidelines for adults:

1. Reduce total fat intake to 30 percent of kcalories.
2. Reduce saturated fat intake to less than 10 percent of kcalories.
3. Consume no more than 300 milligrams of cholesterol daily.

Food jag—A habit of young children in which they have favorite foods they want to eat frequently.

Some parents have overzealously interpreted these guidelines and restricted the fat content of their children's' diets to the point where the children received inadequate calories, which then interferes with normal growth. After the child reaches age two parents may want to limit fatty meats and cheeses and use reduced fat milk.

Preschoolers exhibit some food-related behaviors that drive their parents crazy, such as **food jags** (eating mostly one food for a period of time). Food jags usually don't last long enough to cause any harm. Preschoolers often pick at foods or refuse to eat vegetables or drink milk. Lack of variety, erratic appetites, and food jags are typical of this age group. Toddlers (ages one to three) tend to be pickier eaters than older preschoolers (ages four to five). Toddlers are just starting to assert their independence and love to say no to parental requests. They may wage a control war, and parents need to set limits without being too controlling or rigid. Here are several tactics for dealing with preschoolers' (and school-age children's) food habits.

1. Make mealtime as relaxing and enjoyable as possible.
2. Don't nag, bribe, force, or even cajole a child to eat. Stay calm. Pushing or prodding children almost always backfires. Children learn to hate the foods they are encouraged to eat and to desire the foods used as rewards, such as cake and ice cream. Once children know that you won't allow eating to be made into an issue of control, they will eat when they're hungry and stop when they're full. Your child is the best judge of when he or she is full.
3. Allow children to choose what they will eat from two or more healthy choices. You are responsible for choosing which foods are offered, and the child is responsible for deciding how much he or she wants of those foods.
4. Let children participate in food selection and preparation. Table 13-9 lists cooking activities for children of various ages.
5. Respect your child's preferences when planning meals, but don't make your child a quick peanut butter sandwich, for instance, if he or she rejects your dinner.
6. Make sure your child has appropriately sized utensils and can reach the table comfortably.
7. Preschoolers love rituals, so start them early with the habit of eating three meals plus snacks each day at fairly regular times. Also, eat with your preschooler and model good eating habits.
8. Expect preschoolers to reject new foods at least once, if not many times. Simply continue presenting the new food, perhaps prepared differently, and one day they will try it (usually after 12 to 15 exposures).
9. Let the child serve himself or herself small portions.
10. Do not use desserts as a reward for eating meals. Make dessert a normal part of the meal, and make it nutritious.
11. Ask children to try new foods (just a little bite!) and praise them when they try something different. Encourage them by telling them about someone who really likes the food or relating the food to something they think is fun. Realize, though, that some children are less likely to try new things, including new foods.

TABLE 13-9 Age-Appropriate Cooking Activities

2-1/2–3-Year-Olds
- Wash fruits and vegetables
- Peel bananas
- Stir batters
- Slice soft foods with table knife (cooked potatoes, bananas)
- Pour
- Fetch cans from low cabinets
- Spread with a knife (soft onto firm)
- Use rotary egg beater (for a short time)
- Measure (e.g., chocolate chips into 1-cup measure)

4–5-Year-Olds
- Grease pans
- Open packages
- Peel carrots
- Set table (with instruction)
- Shape dough for cookies/hamburger patties*
- Snip fresh herbs for salads or cooking
- Wash and tear lettuce for salad, separate broccoli, cauliflower
- Place toppings on pizza or snacks

6–8-Year-Olds
- Take part in planning part of or entire meal
- Set table (with less supervision)
- Make a salad
- Find ingredients in cabinet or spice rack
- Shred cheese or vegetables
- Garnish food
- Use microwave, blender, or toaster oven (with previous instruction)
- Measure ingredients
- Present prepared food to family at table
- Roll and shape cookies

9–12-Year-Olds
- Depending on previous experience, plan and prepare an entire meal

* Children should not put their hands in their mouths while handling raw hamburger meat or dough with eggs. It can carry harmful bacteria. They should wash hands after shaping patties.

12. If all else fails, keep in mind that children under six have more taste buds (which may explain why youngsters are such picky eaters) and that this, too, will pass.

Luckily, school-age children are much better eaters. Although they generally have better appetites and will eat a wider range of foods, they often dislike vegetables and casserole dishes.

Both preschoolers and school-age children learn about eating by watching others: their parents, their siblings, their friends, and their teachers. Parents, siblings, and friends provide role models for children and influence children's developing food patterns. Parents' interactions with their children will also influence what foods they or will not accept.

When children go to school, their peers influence their eating behaviors as well as what they eat for lunch. Lunch for school-age children often consists of the school lunch or a packed lunch from home.

Having breakfast makes a difference in how children perform at school. Breakfast also makes a significant contribution to the child's intake of kcalories and nutrients for the day. Children who skip breakfast usually don't make up for the calories at other meals. If a child gets both breakfast and lunch at school, these meals together typically contribute about 50 percent of the day's kcalories.

Preschoolers and school-age children also learn about food by watching television. Research shows that children who watch a lot of television are more apt to be overweight. It not only takes them away from more robust activities but exposes them to commercials that are often for sugared cereals, candy, and other empty-calorie foods. Both obesity and inactivity are currently on the rise in school-age children, especially among adolescents.

So what can parents do to make sure their children eat nutritious diets and get exercise? Be a good role model by eating a well-balanced and varied diet. Have nutritious food choices readily available at home and serve a regular, nutritious breakfast. Maintain regular family meals as much as possible. Family meals are an appropriate time to model healthy eating habits and try out new foods. Also, limit television watching and encourage physical activity. Eating behaviors are learned during childhood (and adolescence) and are maintained into adulthood.

Menu Planning for Children

By the time children are four years old, they can eat amounts that count as regular Food Guide Pyramid servings eaten by older family members—that is, 1/2 cup fruit or vegetable, 3/4 cup of juice, 1 slice of bread, and 2 to 3 ounces of cooked lean meat, poultry, or fish. Children two to three years of age need the same variety of foods as four- to six-year-olds but may need smaller portions, about two-thirds of what counts a regular Food Guide Pyramid serving (except for milk). Two- to six-year-old children need a total of two full servings from the milk group each day. Following are additional menu-planning guidelines.

Preschoolers

1. Offer simply prepared foods and avoid casseroles or any foods that are mixed together, as children need to identify what they are eating.
2. Offer at least one colorful food, such as carrot sticks.
3. Preschoolers like nutritious foods in all food groups but are often reluctant to eat vegetables. Part of this problem may be due to the difficulty involved in getting them onto a spoon or fork. Vegetables are more likely to be accepted if served raw and cut up as finger foods. However, when serving celery, be sure to take off the strings. Serve cooked vegetables somewhat undercooked, so they are a little crunchy. Brightly colored, mild-flavored vegetables such as peas and corn are more popular with children.

4. Provide at least one soft or moist food that is easy to chew at each meal. A crisp or chewy food is important, too, to develop chewing skills.
5. Avoid strong-flavored and highly salted foods. Children have more taste buds than adults, so these foods taste too strong to them.
6. Preschoolers love carbohydrate foods, including cereals, breads, and crackers, as they are easy to hold and chew.
7. Smooth-textured foods such as pea soup or mashed potatoes should not have any lumps—children find this unusual.
8. Before age four, when food-cutting skills start to develop, the child needs food served in bite-size pieces that are either eaten as finger foods or with utensils. For example, cut meat into strips or use ground meat, cut fruit into wedges or slices, and serve pieces of raw vegetables instead of a mixed salad. Other good finger foods include cheese sticks, wedges of hard-boiled eggs, dry ready-to-eat cereal, fish sticks, arrowroot biscuits, and graham crackers.
9. Serve foods warm, not hot; a child's mouth is more sensitive to hot and cold than an adult's. Also, little children need little plates, utensils, and cups, as well as seats that allow them to reach the table comfortably.
10. Cut-up fruit and vegetables make good snacks. Let preschoolers spread peanut butter on crackers or use a spoon to eat yogurt. Snacks are important to preschoolers because they need to eat more often than adults.
11. To minimize choking hazards for children under four:
    - Avoid large chunks of any food.
    - Slice hot dogs in quarters lengthwise.
    - Shred hard raw vegetables and fruits.
    - Remove pits from apples, cherries, plums, peaches, and other fruits.
    - Cut grapes and cherry tomatoes in quarters.
    - Spread peanut butter thin.
    - Chop nuts and seeds fine.
    - Check to make sure fish is boneless.
    - Avoid popcorn and hard candies.
12. Children learn to like new foods by being presented with them repeatedly.

School-Age Children

1. Serve a wide variety of foods, including children's favorites: tuna fish, pizza (use vegetable toppings), macaroni and cheese, hamburgers (use lean beef combined with ground turkey breast), hot dogs (use low-fat varieties), and peanut butter.
2. Good snack choices are important, as children do not always have the desire or the time to sit down and eat. Snacks can include fresh fruits

and vegetables, dried fruits, fruit juices, breads, cold cereals, popcorn (without excessive fat), pretzels, tortillas, muffins, milk, yogurt, cheese, pudding, sliced lean meats and poultry, and peanut butter.
3. Balance menu items that are higher in fat with those containing less fat.
4. Pay attention to serving sizes.
5. Children's most common nutritional problem is iron-deficiency anemia. Offer iron-rich foods such as meat in hamburgers or roast beef sandwiches, peanut butter, baked beans, chili, dried fruits, and fortified dry cereals.
6. As children grow, they need to eat more high-fiber foods, such as fruits, vegetables, beans and peas, and whole-grain foods. See Table 3-6 for the AI for fiber for childen.

All children up to age 10 need to eat every four to six hours to keep their blood glucose at a desirable level; therefore, snacking is necessary between meals. Nutritious snack choices for both preschoolers and school-age children are noted above.

Snacks as well as meals should provide good sources of calcium, such as dairy foods, and iron and zinc, such as meats, legumes, and fortified cereals. If a child drinks little or no milk, try adding flavorings to milk such as chocolate or strawberry, or make cocoa, milkshakes, puddings, and custards. Milk can be fortified with powdered milk by blending 2 cups of fluid milk with 1/3 cup powdered milk. One cup of fortified milk is equal to 1-1/2 cups of regular milk. Powdered milk can also be added in baking and to casseroles, soups, sauces, gravies, ground meats, mashed potatoes, and scrambled eggs. Cheese and yogurt, of course, are also good sources of calcium.

■ MINI-SUMMARY

By a child's first birthday, the growth rate decreases markedly, and yearly weight gain until puberty is about 4 to 6 pounds. By the second birthday, the baby teeth are almost all in. During growth spurts, children's appetites are good; otherwise, their appetites may seem poor. Preschoolers can be fussy eaters, often have food jags, and can take the pleasure out of mealtime. Guidelines for eating with preschoolers and menu planning for them are detailed. Children's eating habits are influenced by family, friends, teachers, availability of school breakfast and lunch programs, and television. Menu-planning guidelines for school-age children are detailed. The most common nutritional problem of children is iron-deficiency anemia.

Adolescence

Puberty, the process of physically developing from a child to an adult, starts at about age 10 or 11 for girls and 12 or 13 for boys. In girls, it peaks at age 12 and is completed by age 15. In males, it peaks at age 14 and is completed by age 19. The timing of puberty and rates of growth show much individual variation. During the five to seven years of pubertal development, adolescents gain about 20 percent of adult height and 50 percent of adult weight. Most of the body organs double in size, and almost half of total bone growth occurs.

Whereas before puberty the proportion of fat and muscle was similar in males and females, males now put on twice as much muscle as females, and females gain proportionately more fat. In adolescent girls, an increasing amount of fat is being stored under the skin, particularly in the abdominal area. The male also experiences a greater increase in bone mass than does the female.

Nutrition During Adolescence

Table 13-10 compares the EER for kcalories and RDA for protein for adolescent males and females to adult needs. The highest levels of nutrients are for individuals growing at the fastest rate.

TABLE 13-10 Energy and Macronutrients for Adolescents

Estimated Energy Requirement (EER) and Recommended Dietary Allowance (RDA) for Protein

Gender and Age	Height	Weight	EER*	Protein
Male 9–13 years	57 inches	79 pounds	2279 kcal	0.95 grams/kilogram body weight
Female 9–13 years	57	81	2071	0.95
Male 14–18 years	68	134	3152	0.85
Female 14–18 years	64	119	2368	0.85
Male 19–30 years	70	154	3067	0.80
Female 19–30 years	64	126	2403	0.80

*The EER is for healthy moderate active Americans and Canadians.

Acceptable Macronutrient Distribution Ranges

Age	Carbohydrate	Fat	Protein
4–18 years	45–65%	25–35%	10–30%
Over 18 years	45–65%	20–35%	10–35%

Source: Adapted with permission from the *Dietary References Intakes for Energy, Carbohydrates, Fiber, Fat, Protein, and Amino Acids (Macronutrients).* © 2002 by the National Academy of Sciences. Courtesy of the National Academy Press, Washington, D.C.

Males now need more kcalories, protein, calcium, iron, and zinc for muscle and bone development than females; however, females need increased iron due to the onset of menstruation. Owing to their big appetites and calorie needs, teenage boys are more likely than girls to get sufficient nutrients. Females have to pack more nutrients into fewer kcalories, which can become difficult if they decrease their food intake to lose weight.

With their increased independence, adolescents assume responsibility for their own eating habits. Teenagers are not fed; they make most of their own food choices. They eat more meals away from home, such as at fast-food restaurants, and skip more meals than previously. Irregular meals and snacking are common due to busy social lives and after-school activities and jobs. Teenagers will tell you that they lack the time or discipline to eat right, although many are pretty well informed about good nutrition practices. Ready-to-eat foods such as cookies, chips, and soft drinks are readily available, and teenagers pick them up as snack foods. Studies show that snacks contribute one-quarter to one-third of total daily kcalories and make substantial nutrient contributions except for iron.

Adolescents often have a variety of lunch options when they are at school. These choices may include leaving the school to buy lunch, eating a lunch from the National School Lunch Program, buying à la carte foods in the school cafeteria that do not qualify as a lunch in the National School Lunch Program, or buying food from a school store or vending machines. Although federal regulations prohibit the sale of carbonated beverages, chewing gum, water ices, and most hard candies in the foodservice area or cafeteria during mealtimes, vending machines with these foods are often found just outside of the cafeteria. State, local, or school rules may close vending machines during mealtime and other times during the school day. Two professional associations, the American Dietetic Association and the American School Food Service Association, have concerns that the foods sold in vending machines, school stores, and à la carte cafeterias discourage students from eating meals provided by the National School Lunch and Breakfast programs.

The media have a powerful influence on adolescents' eating patterns and behaviors. Advertising for not-so-nutritious foods and fast foods permeates television, radio, and billboards. In addition, questionable eating habits are portrayed on television shows. In a study of 12- to 17-year-olds, the prevalence of obesity increased 2 percent for each additional hour of television watched.

A typical meal at a fast-food restaurant—a 4-ounce hamburger, french fries, and a regular soft drink—is high in calories, fat, and sodium. However, more nutritious choices are available at fast-food restaurants. Smaller hamburgers, milk, salads, and grilled chicken sandwiches are examples of more nutritious options.

Parents can positively influence adolescents' eating habits by being good role models and by having dinner and nutritious breakfast and snack foods available at home. Adolescents can become involved in food purchasing and preparation. Parents can also influence their children's fitness level by limiting sedentary activities and encouraging exercise.

Both adolescent boys and girls are influenced by their body image. Adolescent boys may take nutrition supplements and fill up on protein in hopes of becoming more muscular. Adolescent girls who feel they are overweight may skip meals and modify their food choices in hopes of losing weight. Teens who need to lose weight should limit the amount of high-fat food and/or substitute lower-fat choices, such as skim milk for whole milk or nonfat frozen yogurt instead of ice cream. High-fat foods such as french fries and candy bars that have no low-fat substitutes should be eaten only once in a while or in very small amounts. Whether overweight or not, teens need regular exercise.

Menu Planning for Adolescents

1. Emphasize complex carbohydrates such as assorted breads, rolls, cereals, fruits, vegetables, potatoes, pasta, rice, and dried beans and peas. These foods supply kcalories along with needed nutrients. Whole-grain products are preferred.
2. Offer well-trimmed lean beef, poultry, and fish. Don't think that just because adolescents need more calories that fatty meats are in order. Their fat calories should come from less saturated food choices.
3. Low-fat and skim milk need to be offered at all meals. Girls are more likely to need to select skim milk than males. Other forms of calcium also need to be available, such as pizza, macaroni and cheese, and other entrées using cheese; yogurt, frozen yogurt, ice milk, puddings, and custards made with skim milk.
4. Offer margarine; many adolescents are probably used to eating it at home.
5. Have nutritious choices available for hungry on-the-run adolescents looking for a snack. Nutritious snack choices include fresh fruit, muffins and other quick breads, crackers or rolls with low-fat cheese or peanut butter, vegetable-stuffed pita pockets, yogurt or cottage cheese with fruit, or fig bars.
6. Emphasize quick and nutritious breakfasts, such as whole-grain pancakes or waffles with fruit, juices, whole-grain toast or muffins with low-fat cheese, cereal topped with fresh fruits, or bagels with peanut butter.
7. The nutrients most often lacking in adolescent diets are iron, folate, and calcium. Significant iron sources include meats, poultry, fish, eggs, legumes, and dried fruits. Vitamin C aids the absorption of iron from

legumes and dried fruits. Folate is found in leafy green vegetables, orange juice, and beans and peas. Calcium may be lacking for those who have an inadequate intake of milk and other dairy products. If teenagers frequently drink soft drinks instead of milk, they may not have enough calcium in their diets to support bone growth. Teenagers need three daily servings from the milk group in the Food Guide Pyramid.

■ **MINI-SUMMARY**

The pubertal growth spurt starts at about age 10 for girls and 12 for boys. During the five to seven years of pubertal development, the adolescent gains about 20 percent of adult height and 50 percent of adult weight. Boys gain twice as much muscle and more bone mass than girls, who gain proportionately more fat. Males now need more kcalories, protein, calcium, iron, and zinc for muscle and bone development than females, who need increased amounts of iron due to menstruation. Teenagers make most of their own food choices, which are influenced by their body image, peers, and the media. Menu-planning guidelines for adolescents are detailed.

Eating Disorders

Each year millions of Americans develop serious and sometimes life-threatening eating disorders. The vast majority—more than 90 percent—of those afflicted with eating disorders are adolescent and young adult women. One reason that women in this age group are particularly vulnerable to eating disorders is their tendency to go on strict diets to achieve an "ideal" figure. Researchers have found that such stringent dieting can play a key role in triggering eating disorders. The actual cause of eating disorders is not entirely understood, but many risk factors have been identified. Risk factors may include a high degree of perfectionism, low self-esteem, genetics, or family preoccupation with dieting and weight.

Eating-disorder patients deal with two sets of issues: those surrounding their eating behaviors and those surrounding their interactions with others and themselves. Eating disorders are considered a mental disorder, and both psychotherapy and medical nutrition therapy are cornerstones of treatment.

Approximately 1 percent of adolescent girls develop **anorexia nervosa,** a dangerous condition in which they can literally starve themselves to death. Another 2 to 3 percent of young women develop **bulimia nervosa,** a destructive pattern of excessive overeating followed by vomiting or other "purging" behaviors to control their weight. A more recently recognized eat-

Anorexia nervosa—An eating disorder most prevalent in adolescent females who starve themselves.

Bulimia nervosa—An eating disorder characterized by a destructive pattern of excessive overeating followed by vomiting or other "purging" behaviors to control weight.

Binge eating disorder—An eating disorder characterized by episodes of uncontrolled eating or binging.

ing disorder, **binge eating disorder,** could turn out to be the most common. With this disorder, binges are not followed by purges, so these individuals often become overweight. Eating disorders also occur in men and older women, but much less frequently.

The consequences of eating disorders can be severe, with 1 in 10 cases leading to death from starvation, cardiac arrest, or suicide over the course of 10 years. The outlook is better for bulimia than for anorexia; anorexia patients tend to relapse more. Many patients with anorexia or bulimia also suffer from other psychiatric illnesses such as clinical depression, anxiety, obsessive-compulsive disorder, or substance abuse. Fortunately, increasing awareness of the dangers of eating disorders—sparked by medical studies and extensive media coverage of the illness—has led many people to seek help. Nevertheless, some people with eating disorders refuse to admit that they have a problem and do not get treatment. Family members and friends can help recognize the problem and encourage the person to seek treatment. The earlier treatment is started, the better the chance of a full recovery.

Anorexia Nervosa

People who intentionally starve themselves suffer from anorexia nervosa. This disorder, which usually begins in young people around the time of puberty, involves extreme weight loss—at least 15 percent below the individual's normal body weight. Many people with the disorder look emaciated but are convinced they are overweight. Sometimes they must be hospitalized to prevent starvation. Let's look at a typical case.

> Deborah developed anorexia nervosa when she was 16. A rather shy, studious teenager, she tried hard to please everyone. She had an attractive appearance but was slightly overweight. Like many teenage girls, she was interested in boys but concerned that she wasn't pretty enough to get their attention. When her father jokingly remarked that she would never get a date if she didn't take off some weight, she took him seriously and began to diet relentlessly—never believing she was thin enough, even when she became extremely underweight.
>
> Soon after the pounds started dropping off, Deborah's menstrual periods stopped. As anorexia tightened its grip, she became obsessed with dieting and food, and developed strange eating rituals. Every day she weighed all the food she would eat on a kitchen scale, cutting solids into minuscule pieces and precisely measuring liquids. She would then put her daily ration in small containers, lining them up in neat rows. She also exercised compulsively, even after she weakened and became faint.
>
> No one could convince Deborah that she was in danger. Finally, her doctor insisted that she be hospitalized and carefully monitored for treatment of her illness. While in the hospital, she secretly continued her

exercise regimen in the bathroom. It took several hospitalizations and a good deal of individual and family outpatient therapy for Deborah to face and solve her problems.

One of the most frightening aspects of the disorder is that people with anorexia continue to think they are overweight, even when they are bone-thin. Food and weight become obsessions. For some, the compulsiveness shows up in strange eating rituals or the refusal to eat in front of others. It is not uncommon for anorexics to collect recipes and prepare gourmet feasts for family and friends but not partake in the meals themselves.

In patients with anorexia, starvation can damage vital organs such as the heart and brain. To protect itself, the body shifts into "slow gear": menstrual periods stop, and breathing, pulse, and blood pressure rates drop. Nails and hair become brittle. The skin dries, yellows, and becomes covered with soft hair called **lanugo.** Reduced body fat leads to lowered body temperature and the inability to withstand cold.

Lanugo—Downy hair on the skin.

Mild anemia, swollen joints, reduced muscle mass, and light-headedness are also common. If the disorder becomes severe, patients may lose calcium from the bones, making them brittle and prone to breakage. They may also experience irregular heart rhythms and heart failure.

Bulimia Nervosa

People with bulimia nervosa consume large amounts of food and then rid their bodies of excess calories by vomiting, abusing laxatives or diuretics, taking enemas, or exercising obsessively. Some use a combination of all these forms of purging. Because many individuals with bulimia "binge and purge" in secret and maintain normal or above-normal body weight, they can often successfully hide their problem for years. Let's take a look at Lisa.

Lisa developed bulimia at age 18. As with Deborah, her strange eating behavior began when she started to diet. She too dieted and exercised to lose weight, but unlike Deborah, she regularly ate huge amounts of food and maintained her normal weight by forcing herself to vomit. Lisa often felt like an emotional powder keg—angry, frightened, and depressed.

Unable to understand her own behavior, she thought no one else would either, so she felt isolated and lonely. Typically, when things were not going well, she would be overcome with an uncontrollable desire for sweets. She would eat pounds of candy and cake at a time and often not stop until she was exhausted or in severe pain. Then, overwhelmed with guilt and disgust, she would make herself vomit.

While recuperating in a hospital from a suicide attempt, she was referred to an eating disorders clinic, where she got into group therapy. She also received medications to treat the illness and the understanding and help she so desperately needed from others who had the same problem.

Individuals with this disorder may binge and purge once or twice a week or as much as several times a day. Dieting stringently between episodes of binging and purging is also common.

As with anorexia, bulimia typically begins during adolescence. The condition occurs most often in women but is also found in men. Many individuals with bulimia, ashamed of their strange habits, do not seek help until they reach their thirties or forties. By this time, their eating behavior is deeply ingrained and more difficult to change.

Bulimic patients—even those of normal weight—can severely damage their bodies by frequent binge eating and purging. Vomiting causes serious problems: The acid in vomit wears down the outer layer of the teeth and can cause scarring on the backs of hands when fingers are pushed down the throat to induce vomiting. Further, the esophagus becomes inflamed and the glands near the cheeks become swollen.

Binge Eating Disorder

Binge eating disorder resembles bulimia in that it is characterized by episodes of uncontrolled eating, or binging. However, binge eating disorder differs from bulimia in that its sufferers do not purge their bodies of excess food. Binge eating was recognized as a mental disorder in 1994. This is not to say that binge eating is a recent development—it's been around for a long time, sometimes called compulsive overeating—but only lately has it been categorized as a mental disorder.

Binge eaters feel that they lose control of themselves when eating. They eat large quantities of food and do not stop until they are uncomfortably full. They binge at least two times a week. Usually, they have more difficulty losing weight and keeping it off than do people with other serious weight problems. Most people with this disorder are obese and have a history of weight fluctuations. Binge eating disorder is found in about 2 to 5 percent of the general population-more often in women than men. Binge eating disorder occurs in about 30 percent of people participating in medically supervised weight control programs.

Female Athlete Triad

Female athlete triad— An eating disorder found among female college athletes in which they have disordered eating, osteoporosis, and no menstruation.

A disorder called **female athlete triad** was recently named by the American College of Sports Medicine to describe female college athletes who have:

- Disordered eating (eating too little) and osteoporosis (weak bones)
- No menstruation (due to food restriction and stress)

Women with this disorder may be swimmers, gymnasts, or other types of athletes. Treatment consists of increasing meals and snacks to establish regular menstrual periods and increase bone density.

Treatment

The sooner a disorder is diagnosed, the better the chances that treatment can work. The longer abnormal eating behaviors persist, the more difficult it is to overcome the disorder and its effects on the body. In some cases, long-term treatment is required.

Once an eating disorder is diagnosed, the clinician must determine whether the patient is in immediate medical danger and requires hospitalization. Although most patients can be treated as outpatients, some need hospital care, as in the case of severe purging or risk of suicide.

Eating-disorder patients commonly work with a treatment team that includes an internist, a nutritionist, an individual psychotherapist, and someone who is knowledgeable about psychoactive medications used in treating these disorders. Treatment usually includes individual psychotherapy, family therapy, cognitive-behavioral therapy, medical nutrition therapy, and possibly medications such as antidepressant drugs.

Eating disorders, unfortunately, have a very high death rate; one out of every ten patients will die. With that in mind, prevention of these diseases needs to be seriously examined. Research has identified the community groups most important to reach: junior high school students, coaches, and parents.

Table 13-11 lists questions to help individuals determine whether they have an eating disorder.

TABLE 13-11 Do You Have an Eating Disorder?

A positive answer to one or more of these questions may indicate an eating disorder.

1. Do you eat large amounts of food in a very short period while feeling out of control and by yourself?
2. Do you frequently eat a lot of food when you are not hungry and usually when you are alone?
3. Do you feel guilty after overeating?
4. Do you make yourself vomit or use laxatives or diuretics to purge yourself?
5. Do you carefully make sure you eat only a small number of calories each day, such as 500 kcalories or less, and exercise a lot?
6. Do you avoid going out to maintain your eating and exercise schedule?
7. Do you feel that food controls your life?

■ MINI-SUMMARY

Table 13-12 gives some characteristics of anorexia nervosa, bulimia, and binge eating disorder. Recently another eating disorder, female athlete triad, has been identified. Female athele triad is characterized by disordered eating, stopped menstruation, and osteoporosis. Most people afflicted with these problems are adolescent girls and young women. The sooner the disorder is diagnosed, the better the chances for successful treatment. Treatment usually includes individual psychotherapy, family therapy, cognitive-behavior therapy, medical nutrition therapy, and possibly medications. Of all types of mental illness, eating disorders have one of the highest death rates.

TABLE 13-12 Common Symptoms of Eating Disorders

Symptoms	Anorexia Nervosa*	Bulimia Nervosa*	Binge Eating Disorder
Excessive weight loss in relatively short period of time	√		
Continuation of dieting although bone-thin	√		
Dissatisfaction with appearance; belief that body is fat, even though severely underweight	√		
Loss of monthly menstrual periods	√	√	
Unusual interest in food and development of strange eating rituals	√	√	
Eating in secret	√	√	√
Obsession with exercise	√	√	
Serious depression	√	√	√
Bingeing—consumption of large amounts of food		√	√
Vomiting or use of drugs to stimulate vomiting, bowel movements, and urination		√	
Bingeing but no noticeable weight gain		√	
Disappearance into bathroom for long periods of time to induce vomiting		√	
Abuse of drugs or alcohol		√	√

* Some individuals suffer from anorexia and bulimia and have symptoms of both disorders.

Older Adults

The older population—persons 65 years or older—numbered 35.0 million in 2000. They represented 12.4 percent of the U.S. population, about one in every eight Americans. The number of older Americans increased by 3.7 million or 12.0 percent since 1990, compared to an increase of 13.3 percent for the under-65 population. However, the number of Americans age 45–64—who will reach 65 over the next two decades—increased by 34 percent during this period. In 2000, persons reaching age 65 had an average life expectancy of an additional 17.9 years (19.2 years for women and 16.3 years for men).

The older population will continue to grow significantly in the future (Figure 13-2). This growth slowed somewhat during the 1990s because of the relatively small number of babies born during the Great Depression of the 1930s. But the older population will grow dramatically between the years 2010 and 2030, when the baby boom generation reaches age 65. Minority populations are projected to represent 25.4 percent of the elderly population in 2030, up from 16.4 percent in 2000.

Before looking at nutrition during aging, let's take a look at what happens when we age. Studies suggest that the maximum efficiency of many organ systems occurs between ages 20 and 35. After age 35, the functional capability of almost every organ system declines. Similar changes occur in adults as they age, but the rate of decline shows great individual variation. Both genetics and environmental factors such as nutrition affect the rate of aging. Conversely, changes brought about by the aging process affect nutrition status. Of particular importance are changes that affect digestion, absorption, and metabolism of nutrients.

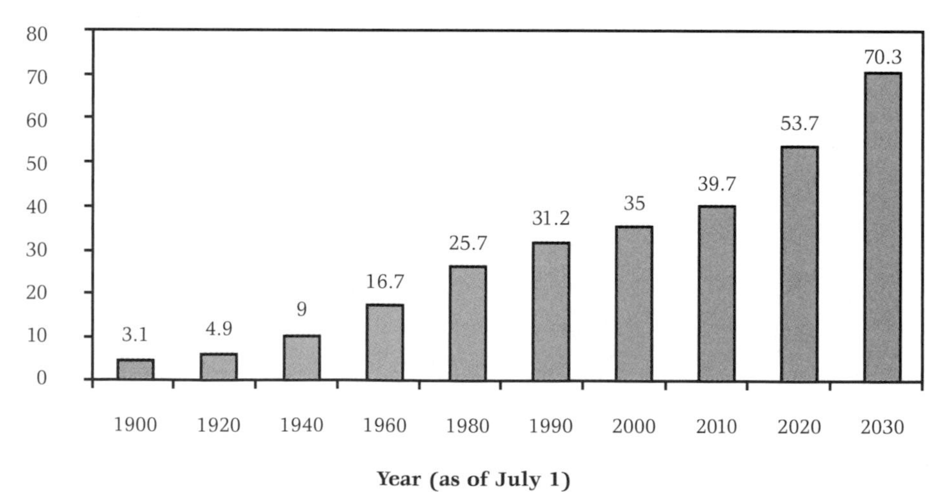

Figure 13-2
Number of persons age 65 and over, 1900–2030 (numbers in millions). Note: Increments in years are uneven. Based on data from the U.S. Bureau of the Census.

The basal metabolic rate declines between 8 and 12 percent from age 30 to 70 and is accompanied by a 25 to 30 percent loss in muscle mass. Combined with a general decrease in activity level, these factors clearly indicate a need for decreased kcalorie intake, which generally does take place during aging. But the elderly need not lose all that muscle mass. Studies have shown that when the elderly do regular weight-training exercises, they increase their muscular strength, increase basal metabolism, improve appetite, and improve blood flow to the brain.

Overall, the functioning of the cardiovascular system declines with age. The workload of the heart increases due to atherosclerotic deposits and less elasticity in the arteries. The heart does not pump as hard as before, and cardiac output is reduced in elderly people who do not remain physically active. Blood pressure increases normally with age. Pulmonary capacity decreases by about 40 percent throughout life. This decrease does not restrict the normal activity of healthy older persons but may limit vigorous exercise. Kidney function deteriorates over time, and the aging kidney is less able to excrete waste. Adequate fluid intake is important, as is avoiding megadoses of water-soluble vitamins because they will put a strain on the kidneys to excrete them. Last, loss of bone occurs normally during aging, and osteoporosis is common (see Chapter 7).

Factors Affecting Nutrition Status

The nutrition status of an elderly person is greatly influenced by many variables, including physiological, psychosocial, and socioeconomic factors.

Physiological Factors

- **Disease.** The presence of disease, both acute and chronic, and use of modified diets can affect nutrition status. Most older persons have at least one chronic condition, and many have multiple conditions, such as arthritis, hypertension, heart disease, and diabetes. Their chronic conditions often require modified diets. Certain chronic diseases such as gastrointestinal disease, congestive heart failure, renal disease, and cancer, are associated with **anorexia** (lack of appetite). Other diseases, such as stroke, are not associated with anorexia but can cause the individual to take in little food.

Anorexia—Lack of appetite.

- **Less muscle mass.** With aging, there is less muscle mass, so the basal metabolic rate decreases. As the BMR slows down, the number of kcalories needed by the elderly decreases.
- **Activity level.** Because active individuals tend to eat more calories than their sedentary counterparts do, they are more likely to ingest more nutrients.

- **Dentition.** Approximately 30 percent of Americans have lost their teeth by age 65. Despite widespread use of dentures, chewing still presents problems for many of the elderly.
- **Functional disabilities.** Functional disabilities interfere with the ability of the elderly to perform daily tasks, such as the purchasing and preparation of food and eating. These disabilities may be due to arthritis or rheumatism, stroke, visual impairment, heart trouble, or dementia. One study reported that 39 percent of the elderly subjects needed help when food shopping and 26 percent needed help making meals.
- **Taste and smell.** Sensitivity to taste and smell decline slowly with age. The taste buds become less sensitive, and the nasal nerves that register aromas need extra stimulation to detect smells. That's why seniors may find ordinarily seasoned foods too bland. Medications also may alter an individual's ability to taste.
- **Changes in the gastrointestinal tract.** The movement of food through the gastrointestinal tract slows down over time, causing problems such as constipation, a frequent complaint of older people. Constipation may also be related to low fiber and fluid intake, medications, or lack of exercise. Other frequent complaints include nausea, indigestion, and heartburn. (Heartburn, a burning sensation in the area of the throat, has nothing to do with the heart. It occurs when acidic stomach contents are pushed into the lower part of the esophagus or throat.)
- **Medications.** More than half of all seniors take at least one medication daily, and many take six or more a day. Medications may alter appetite or the digestion, absorption, and metabolism of nutrients (Table 13-13).
- **Thirst.** Many of the elderly suffer a diminished perception of thirst-especially problematic when they are not feeling well. Because the aging kidney is less able to concentrate urine, more fluid is lost, setting the stage for dehydration.

TABLE 13-13 Nutrients Depleted by Selected Drugs

Drug Group	Drug	Nutrients Depleted
Analgesics	Uncoated aspirin	Iron
Antacids	Aluminum or magnesium hydroxide Sodium bicarbonate	Phosphate, calcium, and folate calcium, folate
Antiulcer drugs	Cimetidine	B_{12}
Chemotherapeutic agents	Methotrexate	Folate
Cholesterol-lowering agents	Cholestyramine	Fat, vitamins A and K
Diuretics	Lasix	Potassium

Psychosocial Factors

- **Cognitive functioning.** Poor cognitive functioning may affect nutrition, or perhaps poor nutritional status contributes to poor cognitive functioning.
- **Social support.** An individual's nutritional health results in part from a series of social acts. The purchasing, preparing, and eating of foods are social events for most people. For example, elderly people may rely on one another for rides to the supermarket, cooking, and sharing meals. The benefits of social networks or support are largely due to the companionship and emotional support they provide. It is anticipated that this has a positive effect on appetite and dietary intake.

Socioeconomic Factors

- **Education.** Higher levels of education are positively associated with increased nutrient intakes.
- **Income.** Almost one in five elderly have incomes at or below the federal poverty level. Lower-income elderly tend to eat fewer calories and fewer servings from the Food Guide Pyramid food groups than higher-income elderly.
- **Living arrangements.** The elderly, particularly women, are more likely to be widowed. The trend has been for widows and widowers in the United States to live alone after the spouse dies. Research focusing on the impact of living arrangements on dietary quality showed that living alone is a risk factor for dietary inadequacy for older men, especially those over age 75 years of age, and for women in the youngest age group (55 to 64).
- **Availability of federally funded meals.** The availability of nutritious meals through federal programs such as Meals on Wheels, in which meals are delivered to the home, is crucial to the nutritional health of many elderly. Another popular elderly feeding program is the Congregate Meals Program, in which the elderly go to a senior center to eat.

Nutrition for Older Adults

A survey completed for the Nutrition Screening Initiative—targeted at improving the nutritional health status of the elderly—shows that although 85 percent of seniors surveyed believe that nutrition is important for their health and well-being, few act on their beliefs. Because the elderly consume fewer kcalories, this means there is less room in the diet for empty-calorie foods such as sweets, alcohol, and fats. At a time when good nutrition is so important to good health, there are many obstacles to fitting more nutrients into fewer kcalories, such as medical conditions, dentures, and medications.

Nutrients of concern to the elderly include the following:

- **Water.** Due to decreased thirst sensation and other factors, fluid intake is more important for older adults than for younger people. It is also important to prevent constipation (as is fiber).
- **Vitamin B_6.** Deficiency of vitamin B_6, which may occur in some women and older adults, can cause symptoms such as fatigue, depression, and irritability. Symptoms can become much more serious if the deficiency continues.
- **Vitamin B_{12} and folate.** The elderly often have a problem with vitamin B_{12}, even if they take in enough. The stomach of an elderly person secretes less gastric acid and pepsin, both of which are necessary to break vitamin B_{12} from its polypeptide linkages in food. The result is that less vitamin B_{12} is absorbed. Vitamin B_{12} is necessary to convert folate into its active form so that folate can do its job of making new cells, such as new red blood cells. When vitamin B_{12} is not properly absorbed, pernicious anemia develops. *Pernicious* means ruinous or harmful, and this type of anemia is marked by a megaloblastic anemia (as with folate deficiency), also called macrocytic anemia, in which there are too many large, immature red blood cells. Symptoms include extreme weakness and fatigue. Nervous system problems also erupt. The cover surrounding the nerves in the body becomes damaged, making it difficult for impulses to travel along them. This causes a poor sense of balance, numbness and tingling sensations in the arms and legs, and mental confusion. Pernicious anemia can result in paralysis and death if not treated. Because deficiencies in folate or vitamin B_{12} cause macrocytic anemia, a physician may mistakenly administer folate when the problem is really a vitamin B_{12} deficiency. The folate would treat the anemia, but not the deterioriation of the nervous system due to a lack of vitamin B_{12}. If untreated, this damage can be significant and sometimes irreversible, although it takes many years to occur. When vitamin B_{12} is deficient due to an absorption problem, injections of the vitamin must be given.
- **Vitamin D.** Several factors adversely affect the vitamin D status of the elderly. First, the elderly tend to be outside less, so they make less vitamin D from exposure to the sun. Also, they have less of the vitamin D precursor in the skin necessary to make vitamin D, and older women absorb less vitamin D from food. Vitamin D-fortified milk and cereals are important to getting enough vitamin D. If milk intake is low, supplements may be recommended.
- **Calcium.** Current intakes for calcium are below the recommendations for individuals over 65 years of age (1200 milligrams). To meet this recommendation, an elderly person would need to eat three servings of

dairy products or other calcium-rich foods daily. Because this can be difficult, supplements may be recommended.

- **Zinc.** Because the elderly are at risk for taking in less than the RDA for zinc, and due to the importance of zinc in cell production, wound healing, the immune system, and taste, attention needs to be placed on getting enough of this mineral.

Here are some ways in which the elderly can improve the nutrient content of what they eat.

1. Eat with other people. This usually makes mealtime more enjoyable and stimulates appetite. Taking a walk before eating also stimulates appetite.
2. Prepare larger amounts of food and freeze some for heating up at a later time. This saves cooking time and is helpful for someone who is reluctant to cook.
3. If big meals are too much, eat small amounts more frequently during the day. Eat regular meals.
4. If getting to the supermarket is a bother, go at a time when it is not busy or engage a delivery service (this is more expensive though).
5. Use unit pricing and sales to cut back on the amount of money spent on food.
6. Take advantage of community meal programs for the elderly, such as Meals on Wheels and meals at senior centers.
7. To perk up a sluggish appetite, increase use of herbs, spices, lemon juice, vinegar, and garlic.

Menu Planning for Older Adults

Figure 13-3 is a modified Food Guide Pyramid for people over 70 developed by researchers at the USDA Human Nutrition Research Center on Aging at Tufts University. This pyramid is designed for healthy individuals over 70 years of age. You will notice the following differences.

1. The Pyramid is narrower than the traditional Pyramid to illustrate that older adults have decreased kcalorie needs and therefore fewer food selections. Their selections must be more nutrient-dense than for other groups of people.
2. A small supplement flag at the top of the Pyramid suggests supplements for those nutrients (calcium, vitamin D, vitamin B_{12}) that may be insufficient in the diet due to smaller and fewer servings of food and medical conditions such as lactose intolerance.
3. At the base of the Pyramid you will find a suggestion to drink at least eight servings of water (fluid) daily.
4. A symbol for fiber within the Pyramid emphasizes the importance of high-fiber foods, which are especially important to avoid constipation.

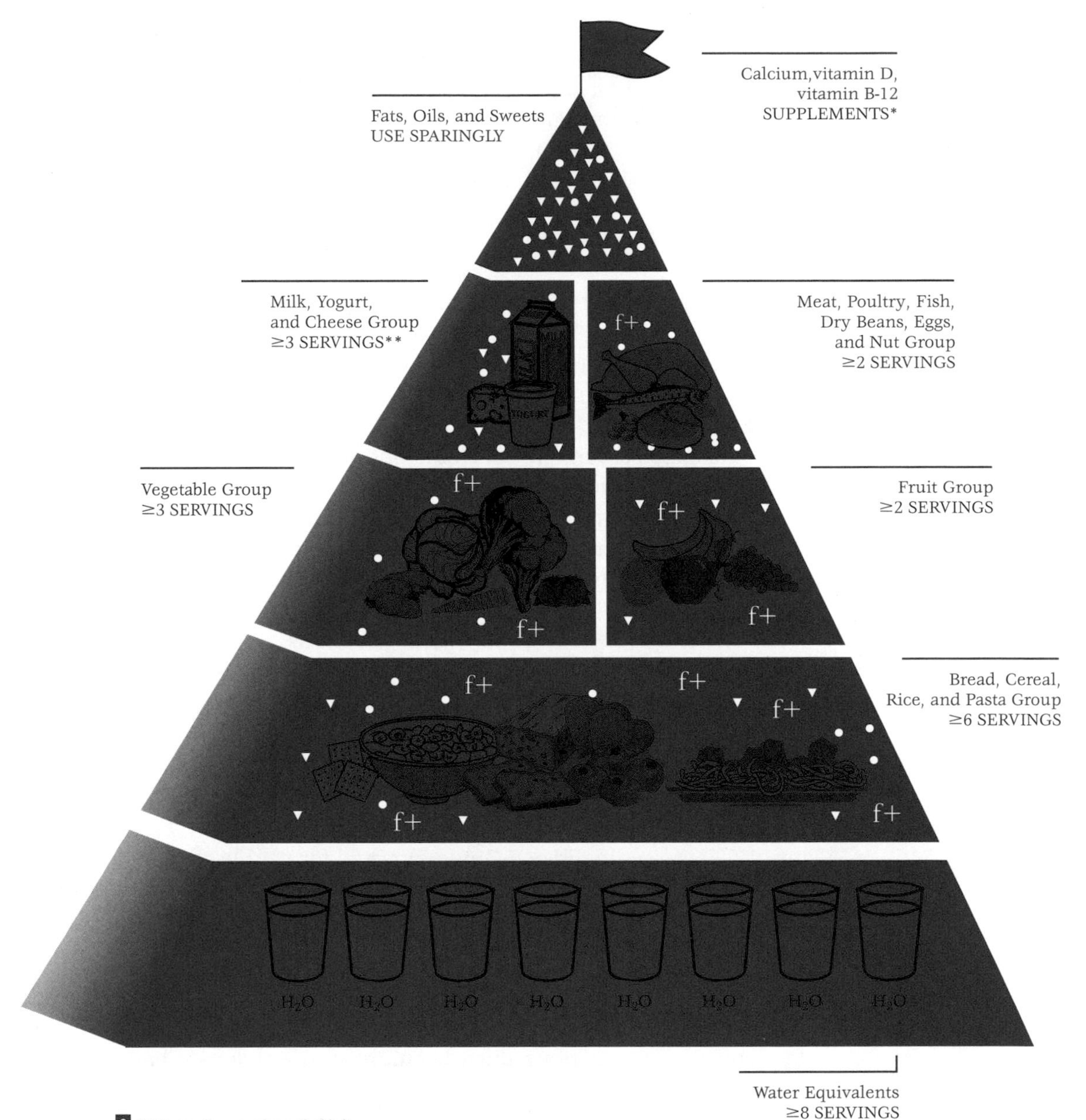

■ **Fat** (naturally occurring and added)
▼ **Sugars** (added)
f+Fiber (should be present)
These symbols show fat, added sugars, and fiber in foods
*Not all individuals need supplements, consult your healthcare provider
**≥ Greater than or equal to

Figure 13-3
Modified food pyramid for adults age 70 and over. © 1999 Tufts University. Reprinted with permission.

When planning meals for older adults, use these guidelines.

1. Offer moderately sized meals and half-portions. Older adults frequently complain when given too much food because they hate to see waste. Restaurants might reduce the size of their entrées by 15 to 25 percent.
2. Emphasize complex carbohydrates and high-fiber foods such as fruits, vegetables, grains, and beans. Older people requiring softer diets may have problems chewing some high-fiber foods. High-fiber foods that are soft in texture include cooked beans and peas, bran cereals soaked in milk, canned prunes and pears, and cooked vegetables such as potatoes, corn, green peas, and winter squash.
3. Moderate the use of fat. Many seniors don't like to see the entrée swimming in a pool of butter. Use lean meats, poultry, or fish and sauces prepared with vegetable or fruit purées. Have low-fat dairy products available, such as skim or low-fat milk.
4. Dairy products, such as milk and yogurt, are important sources of calcium, vitamin D, protein, potassium, vitamin B_{12}, and riboflavin.
5. Offer adequate protein but not too much. Use a variety of both animal and vegetable sources. Providing protein on a budget, as in a nursing home, need not be a problem. Lower-cost protein sources include beans and peas, cottage cheese, macaroni and cheese, eggs, liver, dried skim milk, chicken, and ground beef.
6. Moderate the use of salt. Many seniors are on low-sodium diets and recognize a salty soup when they taste it. Avoid highly salted soups, sauces, and other dishes. It is better to let them season food to taste.
7. Use herbs and spices to make foods flavorful. Seniors are looking for tasty foods just like anyone else, and they may need them more than ever!
8. Offer a variety of foods, including traditional menu items and cooking from other countries and regions of the United States.
9. Fluid intake is critical, so offer a variety of beverages. Diminished sensitivity to dehydration may cause older adults to drink less fluid than needed. Special attention must be paid to fluids, particularly for those who need assistance with eating and drinking. Beverages such as water, milk, juice, coffee, or tea, and foods such as soup contribute to fluid intake.
10. Intake of the following vitamins and minerals may be inadequate in older adults and needs to be considered in menu planning: vitamin B_6, vitamin B_{12}, folate, vitamin D, calcium, and zinc. See Chapters 6 and 7 for food sources.
11. If chewing is a problem, softer foods can be chosen to provide a well-balanced diet. Following are some guidelines for soft diets.
    - Use tender meats, and if necessary, chop or grind them. Ground meats can be used in soups, stews, and casseroles. Cooked beans and peas, soft cheeses, and eggs are additional softer protein sources.

- Cook vegetables thoroughly and dice or chop them by hand if necessary after cooking.
- Serve mashed potatoes or rice, with gravy if desired.
- Serve chopped salads.
- Soft fruits such as fresh or canned bananas, berries, peaches, pears, or melon, as well as applesauce, are some good choices.
- Soft breads and rolls can be made even softer by dipping them briefly in milk.
- Puddings and custard are good dessert choices.
- Many foods that are not soft can be easily chopped by hand or puréed in a blender or food processor to provide additional variety.

■ MINI-SUMMARY

During aging, the functional capability of almost every organ declines, and muscle tissue and mass decreases. Along with a declining basal metabolic rate, the need for kcalories decreases. The nutrition status of an elderly person is greatly influenced by many variables: presence of disease, activity level, quality of dentition, functional disabilities, decline in taste and smell acuity, changes in the gastrointestinal tract, use of medications, diminished sense of thirst, level of cognitive functioning, available social support, level of education and income, living arrangements, and availability of federally funded meals. The modified Food Guide Pyramid for people over 70, as well as guidelines given, will help you plan menus for older adults.

Check-Out Quiz

1. Low-birth-weight babies are at increased health risk.
   a. True b. False
2. Morning sickness occurs only in the morning.
   a. True b. False
3. Nutritional deficiencies during lactation are more likely to affect the quantity of milk the mother makes, rather than the quality.
   a. True b. False
4. For the first year, the newborn's only source of nutrients is either breast milk or formula.
   a. True b. False
5. A deficiency of iron during the first weeks of pregnancy can cause birth defects.
   a. True b. False
6. Breast-feeding is considered to be more nutritious than formula feeding.
   a. True b. False

7. Moderate drinking during pregnancy may cause fetal alcohol syndrome.
   a. True b. False
8. Baby's first food is strained vegetables.
   a. True b. False
9. School-age children tend to be better eaters than preschoolers.
   a. True b. False
10. After one year of age, children start to lose baby fat and become leaner.
    a. True b. False
11. A good way to get a child to finish dinner is to promise dessert once all foods are eaten.
    a. True b. False
12. A child will ask for and need more food during growth spurts than during slower periods of growth.
    a. True b. False
13. Breast-fed babies need supplementation with vitamin A.
    a. True b. False
14. Children's most common nutritional problem is iron-deficiency anemia.
    a. True b. False
15. During adolescence, females put on proportionately more muscle than fat.
    a. True b. False
16. The elderly are major users of modified diets.
    a. True b. False
17. Certain drugs can impair absorption and metabolism of certain nutrients.
    a. True b. False
18. After age 35, functional capability of almost every organ system declines.
    a. True b. False
19. As you get older, energy needs decrease.
    a. True b. False
20. Two vitamins of concern to the elderly are thiamin and riboflavin.
    a. True b. False

Activities and Applications

1. Media Watch

On a Saturday morning, watch children's television for one hour and record the name of each featured product. How many of the total number of advertisers were selling food? Were the majority of the advertised foods healthy foods or junk foods?

2. Childhood Eating Habits

Think back to when you were a child and teenager. What influenced what you put in your mouth? Consider influences such as home, school, friends, and relatives. Which positively influenced your eating style? Which negatively? Discuss this with someone else in your class who is close in age.

3. Preschoolers' Eating Habits
Visit a preschool or day-care center when a meal is being served. Observe the children while they eat. Ask the caregivers about the children's food preferences and eating habits. Ask how well the children accept new foods and how the caregivers introduce new foods.

4. School Lunch Menu
Write a five-day lunch menu for elementary, middle school, or high school students. The menu must provide one-third of the DRI/RDA for all nutrients. Only 30 percent of total calories are allowed from fat. Saturated fat can provide no more than 10 percent of total calories. Use the resource "A Menu Planner for Healthy School Meals" available at this web address: http://schoolmeals.nal.usda.gov/Recipes/menuplan/menuplan.html

5. Eldercare Menu
Visit or phone the foodservice director of a local continuing-care retirement community, congregate meals feeding center, or nursing home. Ask about the type of menu being used (restaurant-style or cycle menu) and meals and foods being offered. What are the major meal-planning considerations used in planning meals for the elderly? What special circumstances come up that are unique to them?

Nutrition Web Explorer

Medline Plus www.nlm.nih.gov/medlineplus/pregnancy.html
Click on one of the articles listed under "Nutrition." Read the article and write a one-paragraph summary of what you read.

Food and Nutrition Services, USDA www.fns.usda.gov/fns
On this home page, click on "Team Nutrition." Find out what Team Nutrition is, and what they do.

National Eating Disorders Association www.nationaleatingdisorders.org
At the top of this home page, click on "Eating Disorders Info" and read one of the articles. Write a one-paragraph summary of what you read.

Federal Interagency Forum on Aging-Related Statistics www.agingstats.gov
On the home page, click on "Older Americans 2000: Key Indicators of Well-Being." This website lists 31 indicators of well-being for older adults. Write down any indicators that you think are nutrition-related.

FirstGov for Seniors www.seniors.gov
This portal site of FirstGov includes many topics of interest to seniors. Click on USDA Seniors Farmers' Market and find out what this program does.

Food Facts *Creative Puréed Foods*

In the past, puréed foods had the reputation of looking pretty miserable when served in most hospitals and nursing facilities. Puréed meat and vegetables were often scooped into small bowls (which for some reason are called "monkey" dishes), covered with gravy, and sent away to some unfortunate patient as supper.

Puréed diets are often necessary for individuals with chewing or swallowing disorders. But that doesn't mean they need runny, liquid foods. On the contrary, the hardest foods for these individuals to swallow are runny foods. The easiest texture to swallow resembles that of mashed potatoes.

Luckily, times have changed. Many cooks are using thickeners to help shape puréed foods so they look like the original foods (Figure 13-4). Thickeners are often powdered and can be mixed directly with liquids and puréed foods. Although there are several commercial thickeners available, such as Thick & Easy, some cooks use thickeners such as cornstarch or instant mashed-potato flakes.

Preparing puréed foods is a challenge not only because the foods must look good and be the right consistency for swallowing, but also because the volume of puréed foods differs from that of regular foods. Fruits and vegetables tend to decrease in volume when puréed, so half a cup of puréed peaches, for example, would contain more calories and nutrients than half a cup of regular peaches. On the other hand, meats almost double in volume because of the liquid required to purée them. Using standardized recipes ensures that puréed foods

- Are nutritionally adequate
- Are the right consistency
- Look and taste appropriate
- Are not too expensive (recipes cut down on waste)

Figure 13-4
Creative and attractive puréed foods using thickeners. Puréed peaches, puréed ham on slurred pumpernickel with puréed lettuce and tomato wedges, garnished with puréed cantaloupe thickened with Menu Magic's Thicken Right. Courtesy Menu Magic, Indianapolis, Indiana.

Hot Topic Food Allergies

Do you start itching whenever you eat peanuts? Does seafood cause your stomach to churn? Symptoms like those cause millions of Americans to suspect they have a **food allergy,** when indeed most probably have a **food intolerance.** True food allergies affect a relatively small percentage of people. Experts estimate that only 2 percent of adults, and from 4 to 8 percent of children are truly allergic to certain foods. So what's the difference between a food intolerance and a food allergy? A food allergy involves an abnormal immune-system response. If the response doesn't involve the immune system, it is called a food intolerance. Symptoms of food intolerance may include gas, bloating, constipation, dizziness, or difficulty sleeping.

Food allergy symptoms are quite specific. **Food allergens,** the food components that cause allergic reactions, are usually proteins. These food protein fragments are not broken down by cooking or by stomach acids or enzymes that digest food. When the allergen passes from the mouth into the stomach, the body recognizes it as a foreign substance and produces antibodies to halt the invasion. As the body fights off the invasion, symptoms begin to appear throughout the body. The most common sites (Figure 13-5) are the mouth (swelling of the lips or tongue,

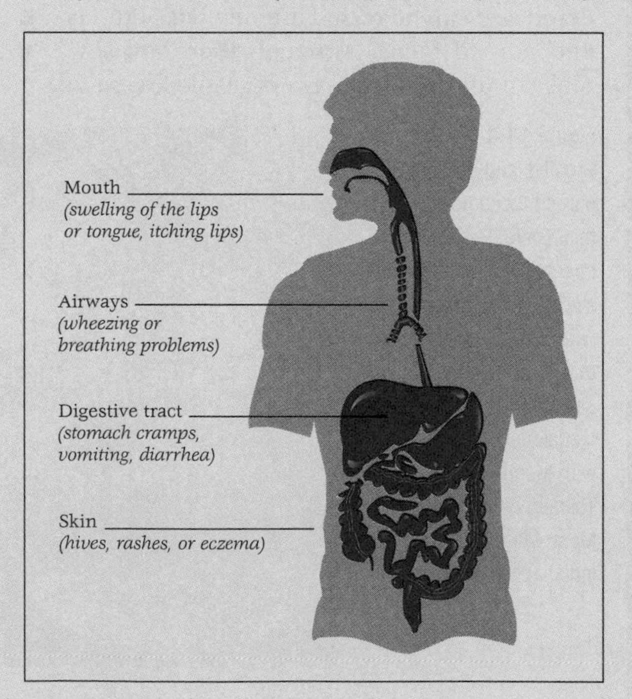

Figure 13-5
Common sites for allergic reactions

itching lips), digestive tract (stomach cramps, vomiting, diarrhea), the skin (hives, rashes, or eczema), and the airways (wheezing or breathing problems). Allergic reactions to foods usually begin within minutes to a few hours after eating.

Food intolerance may produce symptoms similar to those of food allergies, such as abdominal cramping. But whereas people with true food allergies must avoid offending foods altogether, people with food intolerance can often eat small amounts of the offending food without experiencing symptoms.

Food allergies are much more common in infants and young children, who may outgrow them later. Cow's milk, peanuts, eggs, wheat, and soy are the most common food allergies in children. Children typically outgrow their allergies to milk, egg, soy, and wheat, while allergies to peanuts, tree nuts, fish, and shrimp usually are not outgrown. In general, the more severe the first allergic reaction, the longer it takes to outgrow. The most common foods to cause allergies in adults are shrimp, lobster, crab, and other shellfish; peanuts (one of the chief foods responsible for severe anaphylaxis—described in a moment); walnuts and other tree nuts; fish; and eggs. Adults usually do not lose their allergies.

Most cases of allergic reactions to foods are mild, but some are violent and life-threatening. The greatest danger in food allergy comes from **anaphylaxis,** also known as **anaphylactic shock,** a rare allergic reaction involving a number of body parts simultaneously. Like less serious allergic reactions, anaphylaxis usually occurs after a person is exposed to an allergen to which he or she was sensitized by previous exposure. That is, it does not usually occur the first time a person eats a particular food. Anaphylaxis can produce severe symptoms in as little as five to fifteen minutes. Signs of such a reaction include difficulty breathing, swelling of the mouth and throat, drop in blood pressure, and loss of consciousness. The sooner anaphylaxis is treated, the greater the person's chance of surviving.

Although any food can trigger anaphylaxis, peanuts, nuts, shellfish, milk, eggs, and fish are the most common culprits. Peanuts are the leading cause of death from food allergies. As little as 1/5 to 1/5000 of a teaspoon of the offending food has caused death.

There is no specific test to predict the likelihood of anaphylaxis, although allergy testing may help determine which foods a person may be allergic to and provide some guidance as to the severity of the allergy. Experts advise people who are prone to anaphylaxis to carry medication—usually injectable epinephrine—with them at all times and to check the medicine's expiration date regularly.

Diagnosing a food allergy begins with a thorough medical history to identify the suspected food, the amount that must be eaten to cause a reaction, the amount of time between food consumption and development of symptoms, how often the reaction occurs, and other detailed information. A complete physical examination and selected laboratory tests are conducted to rule out underlying medical conditions not related to food allergy. Several tests, such as skin-testing and blood tests, are available to determine whether a person's immune system is sensitized to a specific food. In skin-prick testing, a diluted extract of the suspected food is placed on the skin, which is then scratched or punctured. If no reaction at the site occurs, then the skin test is negative and allergy to the food is unlikely. If a bump surrounded by redness (similar to a mosquito bite) forms within 15 minutes, then the skin test is positive and the person may be allergic to the tested food.

Once diagnosed, most food allergies are treated by avoidance of the food allergen. Avoiding allergens is relatively easy to do at home where you can read food labels. Manufacturers are asked voluntarily to list each allergen by its recognizable name, such as "milk" instead of "casein." However, this task becomes harder when you go out to eat. Even in the home, the most diligent label readers and ingredient checkers likely will be inadvertently exposed to proteins that elicit an allergic response at some point. For example, commercially produced foods that do not contain peanuts or egg may contain traces of these foods through improper cleaning of utensils.

As a foodservice professional, you can do the following to help customers with food allergies.

1. Have recipe/ingredient information available for customers with food allergies. Some restaurants designate a person on each shift who knows this information well to be in charge of discussing allergy concerns with customers. To be useful, this information must be accurate and updated regularly. A manufacturer may change the ingredients in salsa, for example, and these changes are important to note.
2. If you are not sure what is in a menu item, don't give the customer false reassurances. It's better to say "I don't know—why not pick something else" than to give false information that could result in the customer's death and hundreds of thousands of dollars in fines and lawsuits.
3. Staff need training in the nature of food allergies, the foods commonly involved in anaphylactic shock, the restaurant's procedure on identifying and handling customers with food allergies, and emergency procedures.
4. Kitchen staff need training in avoiding ingredient substitutions. They also need to be trained to prepare and serve foods without contacting the foods most likely to cause anaphylactic shock: nuts, peanuts, fish, and shellfish. This means that all preparation and cooking equipment

should be thoroughly cleaned after working with these foods. Remember that even minute amounts of the offending food, sometimes even a strong smell of the food, can cause anaphylactic reactions.

The prevalence of food allergy is growing and probably will continue to grow. Take it seriously when a customer asks whether there are walnuts in your Waldorf salad. Some customers who ask such questions are probably just trying to avoid an upset stomach, but for some it's a much more serious, and possibly life-threatening, matter. Since you don't know which customer really suffers from food allergies, take every customer seriously.

Table 13-14 lists foods to omit for specific allergies.

TABLE 13-14 Foods to Omit for Specific Allergies

Food Allergen	Foods to Omit	Check Food Labels for
Milk	All fluid milk including buttermilk, evaporated or condensed milk, nonfat dry milk, all cheeses, all yogurts, ice cream and ice milk, butter, many margarines, most nondairy creamers and whipped toppings, hot cocoa mixes, creamed soups, many breads, crackers and cereals, pancakes, waffles, many baked goods such as cakes and cookies (check the label), fudge, instant potatoes, custards, puddings, some hot dogs and luncheon meats	Instant nonfat dry milk, nonfat milk, milk solids, whey, curds, casein, caseinate, milk, lactose-free milk, lactalbumin, lactoglobulin, sour cream, butter, cheese, cheese food, butter, milk chocolate, buttermilk
Eggs	All forms of eggs, most egg substitutes, eggnog, any baked goods made with eggs such as muffins and cookies or glazed with eggs such as sweet rolls, ice cream, sherbet, custards, meringues, cream pies, puddings, French toast, pancakes, waffles, some candies, some salad dressings and sandwich spreads such as mayonnaise, any sauce made with egg such as hollandaise, souffles, any meat or potato made with egg, all pastas unless egg free, soups made with eggs of noodles, soups made with stocks that were cleared with eggs, marshmallows	Eggs, albumin, globulin, livetin, ovalbumin, ovomucin, ovomucoid, ovoglobulin egg albumin, ovovitellin, vitellin
Gluten	All foods containing wheat, oats, barley or rye as flour or in any other form, salad dressings, gravies, malted beverages, postum, soy sauce, instant puddings, distilled vinegar, beer, ale, some wines, gin, whiskey, vodka	Flour* (unless from sources noted below), modified food starch, monosodium glutamate, hydrolyzed vegetable protein, cereals, malt or cereal extracts, food starch, vegetable gum, wheat germ, wheat bran, bran, semolina, malt flavoring, distilled vinegar, emulsifiers, stabilizers

* Corn, rice, soy, arrowroot, tapioca, and potato do not contain gluten, so they are safe to use.

Appendix A
Nutritive Value of Foods

Appendix A lists the nutritive values of foods commonly consumed in the United States and makes up the bulk of this publication. The data source is USDA Nutrient Database for Standard Reference, Release 13 (U.S. Department of Agriculture, Agricultural Research Service 2000). Most differences in values between this table and the Standard Reference are due to rounding.

Foods are grouped under the following headings:

Beverages
Dairy products
Eggs
Fats and oils
Fish and shellfish
Fruits and fruit juices
Grain products
Legumes, nuts, and seeds
Meat and meat products
Mixed dishes and fast foods
Poultry and poultry products
Soups, sauces, and gravies
Sugars and sweets
Vegetables and vegetable products
Miscellaneous items

Most of the foods listed are in ready-to-eat form. Some are basic products widely used in food preparation, such as flour, oil, and cornmeal. Most snack foods are found under Grain Products.

Measures and weights. The approximate measure given for each food is in cups, ounces, pounds, some other well-known unit, or a piece of a specified size. The measures do not necessarily represent a serving, but the unit given may be used to calculate a variety of serving sizes. For example, nutrient values are given for 1 cup of applesauce. If the serving you consume is 1/2 cup, divide the values by 2 or multiply by 0.5.

For fluids, the cup measure refers to the standard measuring cup of 8 fluid ounces. The ounce is one-sixteenth of a pound, unless "fluid ounce" is indicated. The weight of a fluid ounce varies according to the food. If the household measure of a food is listed as 1 ounce, the nutrients are based on a weight of 28.35 grams, rounded to 28 grams in the table. All measure weights are actual weights or rounded to the nearest whole number.

The table gives the weight in grams for an approximate measure of each food. The weight applies to only the edible portion (part of food normally eaten), such as the banana pulp without the peel. Some poultry descriptions provide weights for the whole part, such as a drumstick, including skin and/or bone. Keep in mind that the nutritive values are only for the edible portions indicated in the description. For example, item 877, roasted chicken drumstick, indicates a weight of 2.9 oz (82 grams) with the bone and skin. But note that the weight of one drumstick, meat only, is listed as 44 grams (about 1-1/2 oz). So the skin and bone equal 38 grams (82 minus 44). Nutrient values are always given for the gram weight listed in the column Weight—in this case, 44 grams.

Food values. Values are listed for water; calories; protein; total fat; saturated, monosaturated, and polyunsaturated fatty acids; cholesterol; carbohydrate; total dietary fiber; four minerals (calcium, iron, potassium, and sodium); and five vitamins (vitamin A, thiamin, riboflavin, niacin, and ascorbic acid, or vitamin C). Water content is included because the percentage of moisture is helpful for identification and comparison of many food items. For example, to identify whether the cocoa listed is powder or prepared, you could check the water value, which is much less for cocoa powder. Values are in grams or milligrams except for water, calories, and vitamin A.

Food energy is reported as calories. A calorie is the unit of measure for the amount of energy that protein, fat, and carbohydrate furnish the body. Alcohol also contributes to the calorie content of alcoholic beverages. The official unit of measurement for food energy is actually kilocalories (kcal), but the term "calories" is commonly used in its place. In fact, "calories" is used on the food label.

Vitamin A is reported in two different units: International Units (IU) are used on food labels and in the past for expressing vitamin A activity; Retinol Equivalents (RE) are the units released in 1989 by the Food and Nutrition Board for expressing the RDAs for vitamin A.

Values for calories and nutrients shown in the table are the amounts in the part of the item that is customarily eaten—corn without cob, meat without bones, and peaches without pits. Nutrient values are averages for products presented here. Values for some nutrients may vary more widely for specific food items. For example, the vitamin A content of beef liver varies widely, but the values listed in the table represent an average for that food.

In some cases, as with many vegetables, values for fat may be trace (Tr), yet there will be numerical values listed for some of the fatty acids. The values for fat have been rounded to whole numbers, unless they are between 0 and 0.5; then they are listed as trace. This definition of trace also applies to the other nutrients in the table that are rounded to whole numbers.

Other uses of "trace" in the table are:

- For nutrients rounded to one decimal place, values falling between 0 and 0.05 are trace.
- For nutrients rounded to two decimal places, values falling between 0 and 0.005 are trace.

Thiamin, riboflavin, niacin, and iron values in enriched white flours, white bread and rolls, cornmeals, pastas, farina, and rice are based on the current enrichment levels established by the Food and Drug Administration. Enrichment levels for riboflavin in rice were not in effect at press time and are not used in the table. Enriched flour is used in most home-prepared and commercially prepared baked goods.

Niacin values given are for preformed niacin that occurs naturally in foods. The values do not include additional niacin that may be formed in the body from tryptophan, an essential amino acid in the protein of most foods.

Nutrient values for many prepared items were calculated from the ingredients in typical recipes. Examples are biscuits, cornbread, mashed potatoes, white sauce, and many dessert foods. Adjustments were made for nutrient losses during cooking.

Nutrient values for toast and cooked vegetables do not include any added fat, either during preparation or at the table. Cutting or shredding vegetables may destroy part of some vitamins, especially ascorbic acid. Since much losses are variable, no deduction has been made.

Values for cooked dry beans, vegetables, pasta, noodles, rice, cereal, meat, poultry, and fish are without salt added. If hot cereals are prepared with salt, the sodium content ranges from about 324-374 mg for Malt-O-Meal, Cream of Wheat, and rolled oats. The sodium value for corn grits is about 540 mg; sodium for Wheatena is about 238 mg. Sodium values for canned vegetables labeled as "no salt added" are similar to those listed for the cooked vegetables.

The mineral contribution of water was not considered for coffee, tea, soups, sauces, or concentrated fruit juices prepared with water. Sweetened items contain sugar unless identified as artificially sweetened.

Several manufactured items—including some milk products, ready-to-eat breakfast, imitation cream products, fruit drinks, and various mixes—

are included in the table. Such foods may be fortified with one or more nutrients; the label will describe any fortification. Values for these foods may be based on products from several manufacturers, so they may differ from the values provided by any one source. Nutrient values listed on food labels may also differ from those in the table because of rounding on labels.

Nutrient values represent meats after they have been cooked and drained of the drippings. For many cuts, two sets of values are shown: meat including lean and fat parts, and lean meat from which the outer fat layer and large fat pads have been removed either before or after cooking.

In the entries for cheeseburger and hamburger in Mixed Dishes and Fast Foods, "condiments" refers to catsup, mustard, salt, and pepper; "vegetetables" refers to lettuce, tomato, onion, and pickle; "regular" is a 2-oz patty, and large is a 4-oz patty (precooked weight).

Source: The Nutritive Value of Foods. USDA Home and Garden Bulletin no. 72, 2002.

Food No.	Food Description	Measure of edible portion	Weight (g)	Water (%)	Calories (kcal)	Protein (g)	Total fat (g)	Fatty acids Saturated (g)	Monounsaturated (g)	Polyunsaturated (g)

Beverages

Food No.	Food Description	Measure of edible portion	Weight (g)	Water (%)	Calories (kcal)	Protein (g)	Total fat (g)	Saturated (g)	Monounsaturated (g)	Polyunsaturated (g)
	Alcoholic									
	Beer									
1	Regular	12 fl oz	355	92	146	1	0	0.0	0.0	0.0
2	Light	12 fl oz	354	95	99	1	0	0.0	0.0	0.0
	Gin, rum, vodka, whiskey									
3	80 proof	1.5 fl oz	42	67	97	0	0	0.0	0.0	0.0
4	86 proof	1.5 fl oz	42	64	105	0	0	0.0	0.0	0.0
5	90 proof	1.5 fl oz	42	62	110	0	0	0.0	0.0	0.0
6	Liqueur, coffee, 53 proof	1.5 fl oz	52	31	175	Tr	Tr	0.1	Tr	0.1
	Mixed drinks, prepared from recipe									
7	Daiquiri	2 fl oz	60	70	112	Tr	Tr	Tr	Tr	Tr
8	Pina colada	4.5 fl oz	141	65	262	1	3	1.2	0.2	0.5
	Wine									
	Dessert									
9	Dry	3.5 fl oz	103	80	130	Tr	0	0.0	0.0	0.0
10	Sweet	3.5 fl oz	103	73	158	Tr	0	0.0	0.0	0.0
	Table									
11	Red	3.5 fl oz	103	89	74	Tr	0	0.0	0.0	0.0
12	White	3.5 fl oz	103	90	70	Tr	0	0.0	0.0	0.0
	Carbonated*									
13	Club soda	12 fl oz	355	100	0	0	0	0.0	0.0	0.0
14	Cola type	12 fl oz	370	89	152	0	0	0.0	0.0	0.0
	Diet, sweetened with aspartame									
15	Cola	12 fl oz	355	100	4	Tr	0	0.0	0.0	0.0
16	Other than cola or pepper type	12 fl oz	355	100	0	Tr	0	0.0	0.0	0.0
17	Ginger ale	12 fl oz	366	91	124	0	0	0.0	0.0	0.0
18	Grape	12 fl oz	372	89	160	0	0	0.0	0.0	0.0
19	Lemon lime	12 fl oz	368	90	147	0	0	0.0	0.0	0.0
20	Orange	12 fl oz	372	88	179	0	0	0.0	0.0	0.0
21	Pepper type	12 fl oz	368	89	151	0	Tr	0.3	0.0	0.0
22	Root beer	12 fl oz	370	89	152	0	0	0.0	0.0	0.0
	Chocolate flavored beverage mix									
23	Powder	2-3 heaping tsp	22	1	75	1	1	0.4	0.2	Tr
24	Prepared with milk	1 cup	266	81	226	9	9	5.5	2.6	0.3
	Cocoa									
	Powder containing nonfat dry milk									
25	Powder	3 heaping tsp	28	2	102	3	1	0.7	0.4	Tr
26	Prepared (6 oz water plus 1 oz powder)	1 serving	206	86	103	3	1	0.7	0.4	Tr
	Powder containing nonfat dry milk and aspartame									
27	Powder	½-oz envelope	15	3	48	4	Tr	0.3	0.1	Tr
28	Prepared (6 oz water plus 1 envelope mix)	1 serving	192	92	48	4	Tr	0.3	0.1	Tr
	Coffee									
29	Brewed	6 fl oz	178	99	4	Tr	0	Tr	0.0	Tr
30	Espresso	2 fl oz	60	98	5	Tr	Tr	0.1	0.0	0.1
31	Instant, prepared (1 rounded tsp powder plus 6 fl oz water)	6 fl oz	179	99	4	Tr	0	Tr	0.0	Tr

*Mineral content varies depending on water source.

Cholesterol (mg)	Carbohydrate (g)	Total dietary fiber (g)	Calcium (mg)	Iron (mg)	Potassium (mg)	Sodium (mg)	Vitamin A		Thiamin (mg)	Riboflavin (mg)	Niacin (mg)	Ascorbic acid (mg)	Food No.
							(IU)	(RE)					
0	13	0.7	18	0.1	89	18	0	0	0.02	0.09	1.6	0	1
0	5	0.0	18	0.1	64	11	0	0	0.03	0.11	1.4	0	2
0	0	0.0	0	Tr	1	Tr	0	0	Tr	Tr	Tr	0	3
0	Tr	0.0	0	Tr	1	Tr	0	0	Tr	Tr	Tr	0	4
0	0	0.0	0	Tr	1	Tr	0	0	Tr	Tr	Tr	0	5
0	24	0.0	1	Tr	16	4	0	0	Tr	0.01	0.1	0	6
0	4	0.0	2	0.1	13	3	2	0	0.01	Tr	Tr	1	7
0	40	0.8	11	0.3	100	8	3	0	0.04	0.02	0.2	7	8
0	4	0.0	8	0.2	95	9	0	0	0.02	0.02	0.2	0	9
0	12	0.0	8	0.2	95	9	0	0	0.02	0.02	0.2	0	10
0	2	0.0	8	0.4	115	5	0	0	0.01	0.03	0.1	0	11
0	1	0.0	9	0.3	82	5	0	0	Tr	0.01	0.1	0	12
0	0	0.0	18	Tr	7	75	0	0	0.00	0.00	0.0	0	13
0	38	0.0	11	0.1	4	15	0	0	0.00	0.00	0.0	0	14
0	Tr	0.0	14	0.1	0	21	0	0	0.02	0.08	0.0	0	15
0	0	0.0	14	0.1	7	21	0	0	0.00	0.00	0.0	0	16
0	32	0.0	11	0.7	4	26	0	0	0.00	0.00	0.0	0	17
0	42	0.0	11	0.3	4	56	0	0	0.00	0.00	0.0	0	18
0	38	0.0	7	0.3	4	40	0	0	0.00	0.00	0.1	0	19
0	46	0.0	19	0.2	7	45	0	0	0.00	0.00	0.0	0	20
0	38	0.0	11	0.1	4	37	0	0	0.00	0.00	0.0	0	21
0	39	0.0	19	0.2	4	48	0	0	0.00	0.00	0.0	0	22
0	20	1.3	8	0.7	128	45	4	Tr	0.01	0.03	0.1	Tr	23
32	31	1.3	301	0.8	497	165	311	77	0.10	0.43	0.3	2	24
1	22	0.3	92	0.3	202	143	4	1	0.03	0.16	0.2	1	25
2	22	2.5	97	0.4	202	148	4	0	0.03	0.16	0.2	Tr	26
1	9	0.4	86	0.7	405	168	5	1	0.04	0.21	0.2	0	27
2	8	0.4	90	0.7	405	173	4	0	0.04	0.21	0.2	0	28
0	1	0.0	4	0.1	96	4	0	0	0.00	0.00	0.4	0	29
0	1	0.0	1	0.1	69	8	0	0	Tr	0.11	3.1	Tr	30
0	1	0.0	5	0.1	64	5	0	0	0.00	Tr	0.5	0	31

Food No.	Food Description	Measure of edible portion	Weight (g)	Water (%)	Calories (kcal)	Pro-tein (g)	Total fat (g)	Fatty acids Satu-rated (g)	Mono-unsatu-rated (g)	Poly-unsatu-rated (g)
	Beverages (continued)									
	Fruit drinks, noncarbonated, canned or bottled, with added ascorbic acid									
32	Cranberry juice cocktail	8 fl oz	253	86	144	0	Tr	Tr	Tr	0.1
33	Fruit punch drink	8 fl oz	248	88	117	0	0	Tr	Tr	Tr
34	Grape drink	8 fl oz	250	88	113	0	0	Tr	0.0	Tr
35	Pineapple grapefruit juice drink	8 fl oz	250	88	118	1	Tr	Tr	Tr	0.1
36	Pineapple orange juice drink	8 fl oz	250	87	125	3	0	0.0	0.0	0.0
	Lemonade									
37	Frozen concentrate, prepared	8 fl oz	248	89	99	Tr	0	Tr	Tr	Tr
	Powder, prepared with water									
38	Regular	8 fl oz	266	89	112	0	0	Tr	Tr	Tr
39	Low calorie, sweetened with aspartame	8 fl oz	237	99	5	0	0	0.0	0.0	0.0
	Malted milk, with added nutrients									
	Chocolate									
40	Powder	3 heaping tsp	21	3	75	1	1	0.4	0.2	0.1
41	Prepared	1 cup	265	81	225	9	9	5.5	2.6	0.4
	Natural									
42	Powder	4-5 heaping tsp	21	3	80	2	1	0.3	0.2	0.1
43	Prepared	1 cup	265	81	231	10	9	5.4	2.5	0.4
	Milk and milk beverages. See Dairy Products.									
44	Rice beverage, canned (RICE DREAM)	1 cup	245	89	120	Tr	2	0.2	1.3	0.3
	Soy milk. See Legumes, Nuts, and Seeds.									
	Tea									
	Brewed									
45	Black	6 fl oz	178	100	2	0	0	Tr	Tr	Tr
	Herb									
46	Chamomile	6 fl oz	178	100	2	0	0	Tr	Tr	Tr
47	Other than chamomile	6 fl oz	178	100	2	0	0	Tr	Tr	Tr
	Instant, powder, prepared									
48	Unsweetened	8 fl oz	237	100	2	0	0	0.0	0.0	0.0
49	Sweetened, lemon flavor	8 fl oz	259	91	88	Tr	0	Tr	Tr	Tr
50	Sweetened with saccharin, lemon flavor	8 fl oz	237	99	5	0	0	0.0	0.0	Tr
51	Water, tap	8 fl oz	237	100	0	0	0	0.0	0.0	0.0
	Dairy Products									
	Butter. See Fats and Oils.									
	Cheese									
	Natural									
52	Blue	1 oz	28	42	100	6	8	5.3	2.2	0.2
53	Camembert (3 wedges per 4-oz container)	1 wedge	38	52	114	8	9	5.8	2.7	0.3
	Cheddar									
54	Cut pieces	1 oz	28	37	114	7	9	6.0	2.7	0.3
55		1 cubic inch	17	37	68	4	6	3.6	1.6	0.2
56	Shredded	1 cup	113	37	455	28	37	23.8	10.6	1.1

Choles-terol (mg)	Carbo-hydrate (g)	Total dietary fiber (g)	Calcium (mg)	Iron (mg)	Potas-sium (mg)	Sodium (mg)	Vitamin A		Thiamin (mg)	Ribo-flavin (mg)	Niacin (mg)	Ascor-bic acid (mg)	Food No.
							(IU)	(RE)					
0	36	0.3	8	0.4	46	5	10	0	0.02	0.02	0.1	90	32
0	30	0.2	20	0.5	62	55	35	2	0.05	0.06	0.1	73	33
0	29	0.0	8	0.4	13	15	3	0	0.01	0.01	0.1	85	34
0	29	0.3	18	0.8	153	35	88	10	0.08	0.04	0.7	115	35
0	30	0.3	13	0.7	115	8	1,328	133	0.08	0.05	0.5	56	36
0	26	0.2	7	0.4	37	7	52	5	0.01	0.05	Tr	10	37
0	29	0.0	29	0.1	3	19	0	0	0.00	Tr	0.0	34	38
0	1	0.0	50	0.1	0	7	0	0	0.00	0.00	0.0	6	39
1	18	0.2	93	3.6	251	125	2,751	824	0.64	0.86	10.7	32	40
34	29	0.3	384	3.8	620	244	3,058	901	0.73	1.26	10.9	34	41
4	17	0.1	79	3.5	203	85	2,222	668	0.62	0.75	10.2	27	42
34	28	0.0	371	3.6	572	204	2,531	742	0.71	1.14	10.4	29	43
0	25	0.0	20	0.2	69	86	5	0	0.08	0.01	1.9	1	44
0	1	0.0	0	Tr	66	5	0	0	0.00	0.02	0.0	0	45
0	Tr	0.0	4	0.1	16	2	36	4	0.02	0.01	0.0	0	46
0	Tr	0.0	4	0.1	16	2	0	0	0.02	0.01	0.0	0	47
0	Tr	0.0	5	Tr	47	7	0	0	0.00	Tr	0.1	0	48
0	22	0.0	5	0.1	49	8	0	0	0.00	0.05	0.1	0	49
0	1	0.0	5	0.1	40	24	0	0	0.00	0.01	0.1	0	50
0	0	0.0	5	Tr	0	7	0	0	0.00	0.00	0.0	0	51
21	1	0.0	150	0.1	73	396	204	65	0.01	0.11	0.3	0	52
27	Tr	0.0	147	0.1	71	320	351	96	0.01	0.19	0.2	0	53
30	Tr	0.0	204	0.2	28	176	300	79	0.01	0.11	Tr	0	54
18	Tr	0.0	123	0.1	17	105	180	47	Tr	0.06	Tr	0	55
119	1	0.0	815	0.8	111	701	1,197	314	0.03	0.42	0.1	0	56

Food No.	Food Description	Measure of edible portion	Weight (g)	Water (%)	Calories (kcal)	Protein (g)	Total fat (g)	Fatty acids: Saturated (g)	Fatty acids: Monounsaturated (g)	Fatty acids: Polyunsaturated (g)

Dairy Products (continued)

Food No.	Food Description	Measure of edible portion	Weight (g)	Water (%)	Calories (kcal)	Protein (g)	Total fat (g)	Saturated (g)	Monounsaturated (g)	Polyunsaturated (g)
	Cheese (continued)									
	Natural (continued)									
	Cottage									
	Creamed (4% fat)									
57	Large curd	1 cup	225	79	233	28	10	6.4	2.9	0.3
58	Small curd	1 cup	210	79	217	26	9	6.0	2.7	0.3
59	With fruit	1 cup	226	72	279	22	8	4.9	2.2	0.2
60	Low fat (2%)	1 cup	226	79	203	31	4	2.8	1.2	0.1
61	Low fat (1%)	1 cup	226	82	164	28	2	1.5	0.7	0.1
62	Uncreamed (dry curd, less than ½% fat)	1 cup	145	80	123	25	1	0.4	0.2	Tr
	Cream									
63	Regular	1 oz	28	54	99	2	10	6.2	2.8	0.4
64		1 tbsp	15	54	51	1	5	3.2	1.4	0.2
65	Low fat	1 tbsp	15	64	35	2	3	1.7	0.7	0.1
66	Fat free	1 tbsp	16	76	15	2	Tr	0.1	0.1	Tr
67	Feta	1 oz	28	55	75	4	6	4.2	1.3	0.2
68	Low fat, cheddar or colby	1 oz	28	63	49	7	2	1.2	0.6	0.1
	Mozzarella, made with									
69	Whole milk	1 oz	28	54	80	6	6	3.7	1.9	0.2
70	Part skim milk (low moisture)	1 oz	28	49	79	8	5	3.1	1.4	0.1
71	Muenster	1 oz	28	42	104	7	9	5.4	2.5	0.2
72	Neufchatel	1 oz	28	62	74	3	7	4.2	1.9	0.2
73	Parmesan, grated	1 cup	100	18	456	42	30	19.1	8.7	0.7
74		1 tbsp	5	18	23	2	2	1.0	0.4	Tr
75		1 oz	28	18	129	12	9	5.4	2.5	0.2
76	Provolone	1 oz	28	41	100	7	8	4.8	2.1	0.2
	Ricotta, made with									
77	Whole milk	1 cup	246	72	428	28	32	20.4	8.9	0.9
78	Part skim milk	1 cup	246	74	340	28	19	12.1	5.7	0.6
79	Swiss	1 oz	28	37	107	8	8	5.0	2.1	0.3
	Pasteurized process cheese									
	American									
80	Regular	1 oz	28	39	106	6	9	5.6	2.5	0.3
81	Fat free	1 slice	21	57	31	5	Tr	0.1	Tr	Tr
82	Swiss	1 oz	28	42	95	7	7	4.5	2.0	0.2
83	Pasteurized process cheese food, American	1 oz	28	43	93	6	7	4.4	2.0	0.2
84	Pasteurized process cheese spread, American	1 oz	28	48	82	5	6	3.8	1.8	0.2
	Cream, sweet									
85	Half and half (cream and milk)	1 cup	242	81	315	7	28	17.3	8.0	1.0
86		1 tbsp	15	81	20	Tr	2	1.1	0.5	0.1
87	Light, coffee, or table	1 cup	240	74	469	6	46	28.8	13.4	1.7
88		1 tbsp	15	74	29	Tr	3	1.8	0.8	0.1
	Whipping, unwhipped (volume about double when whipped)									
89	Light	1 cup	239	64	699	5	74	46.2	21.7	2.1
90		1 tbsp	15	64	44	Tr	5	2.9	1.4	0.1
91	Heavy	1 cup	238	58	821	5	88	54.8	25.4	3.3
92		1 tbsp	15	58	52	Tr	6	3.5	1.6	0.2
93	Whipped topping (pressurized)	1 cup	60	61	154	2	13	8.3	3.9	0.5
94		1 tbsp	3	61	8	Tr	1	0.4	0.2	Tr

Choles-terol (mg)	Carbo-hydrate (g)	Total dietary fiber (g)	Calcium (mg)	Iron (mg)	Potas-sium (mg)	Sodium (mg)	Vitamin A (IU)	Vitamin A (RE)	Thiamin (mg)	Ribo-flavin (mg)	Niacin (mg)	Ascor-bic acid (mg)	Food No.
34	6	0.0	135	0.3	190	911	367	108	0.05	0.37	0.3	0	57
31	6	0.0	126	0.3	177	850	342	101	0.04	0.34	0.3	0	58
25	30	0.0	108	0.2	151	915	278	81	0.04	0.29	0.2	0	59
19	8	0.0	155	0.4	217	918	158	45	0.05	0.42	0.3	0	60
10	6	0.0	138	0.3	193	918	84	25	0.05	0.37	0.3	0	61
10	3	0.0	46	0.3	47	19	44	12	0.04	0.21	0.2	0	62
31	1	0.0	23	0.3	34	84	405	108	Tr	0.06	Tr	0	63
16	Tr	0.0	12	0.2	17	43	207	55	Tr	0.03	Tr	0	64
8	1	0.0	17	0.3	25	44	108	33	Tr	0.04	Tr	0	65
1	1	0.0	29	Tr	25	85	145	44	0.01	0.03	Tr	0	66
25	1	0.0	140	0.2	18	316	127	36	0.04	0.24	0.3	0	67
6	1	0.0	118	0.1	19	174	66	18	Tr	0.06	Tr	0	68
22	1	0.0	147	0.1	19	106	225	68	Tr	0.07	Tr	0	69
15	1	0.0	207	0.1	27	150	199	54	0.01	0.10	Tr	0	70
27	Tr	0.0	203	0.1	38	178	318	90	Tr	0.09	Tr	0	71
22	1	0.0	21	0.1	32	113	321	85	Tr	0.06	Tr	0	72
79	4	0.0	1,376	1.0	107	1,862	701	173	0.05	0.39	0.3	0	73
4	Tr	0.0	69	Tr	5	93	35	9	Tr	0.02	Tr	0	74
22	1	0.0	390	0.3	30	528	199	49	0.01	0.11	0.1	0	75
20	1	0.0	214	0.1	39	248	231	75	0.01	0.09	Tr	0	76
124	7	0.0	509	0.9	257	207	1,205	330	0.03	0.48	0.3	0	77
76	13	0.0	669	1.1	308	307	1,063	278	0.05	0.46	0.2	0	78
26	1	0.0	272	Tr	31	74	240	72	0.01	0.10	Tr	0	79
27	Tr	0.0	174	0.1	46	406	343	82	0.01	0.10	Tr	0	80
2	3	0.0	145	0.1	60	321	308	92	0.01	0.10	Tr	0	81
24	1	0.0	219	0.2	61	388	229	65	Tr	0.08	Tr	0	82
18	2	0.0	163	0.2	79	337	259	62	0.01	0.13	Tr	0	83
16	2	0.0	159	0.1	69	381	223	54	0.01	0.12	Tr	0	84
89	10	0.0	254	0.2	314	98	1,050	259	0.08	0.36	0.2	2	85
6	1	0.0	16	Tr	19	6	65	16	0.01	0.02	Tr	Tr	86
159	9	0.0	231	0.1	292	95	1,519	437	0.08	0.36	0.1	2	87
10	1	0.0	14	Tr	18	6	95	27	Tr	0.02	Tr	Tr	88
265	7	0.0	166	0.1	231	82	2,694	705	0.06	0.30	0.1	1	89
17	Tr	0.0	10	Tr	15	5	169	44	Tr	0.02	Tr	Tr	90
326	7	0.0	154	0.1	179	89	3,499	1,002	0.05	0.26	0.1	1	91
21	Tr	0.0	10	Tr	11	6	221	63	Tr	0.02	Tr	Tr	92
46	7	0.0	61	Tr	88	78	506	124	0.02	0.04	Tr	0	93
2	Tr	0.0	3	Tr	4	4	25	6	Tr	Tr	Tr	0	94

Dairy Products (continued)

Food No.	Food Description	Measure of edible portion	Weight (g)	Water (%)	Calories (kcal)	Protein (g)	Total fat (g)	Fatty acids: Saturated (g)	Fatty acids: Monounsaturated (g)	Fatty acids: Polyunsaturated (g)
	Cream, sour									
95	Regular	1 cup	230	71	493	7	48	30.0	13.9	1.8
96		1 tbsp	12	71	26	Tr	3	1.6	0.7	0.1
97	Reduced fat	1 tbsp	15	80	20	Tr	2	1.1	0.5	0.1
98	Fat free	1 tbsp	16	81	12	Tr	0	0.0	0.0	0.0
	Cream product, imitation (made with vegetable fat)									
	Sweet									
	Creamer									
99	Liquid (frozen)	1 tbsp	15	77	20	Tr	1	0.3	1.1	Tr
100	Powdered	1 tsp	2	2	11	Tr	1	0.7	Tr	Tr
	Whipped topping									
101	Frozen	1 cup	75	50	239	1	19	16.3	1.2	0.4
102		1 tbsp	4	50	13	Tr	1	0.9	0.1	Tr
103	Powdered, prepared with whole milk	1 cup	80	67	151	3	10	8.5	0.7	0.2
104		1 tbsp	4	67	8	Tr	Tr	0.4	Tr	Tr
105	Pressurized	1 cup	70	60	184	1	16	13.2	1.3	0.2
106		1 tbsp	4	60	11	Tr	1	0.8	0.1	Tr
107	Sour dressing (filled cream type, nonbutterfat)	1 cup	235	75	417	8	39	31.2	4.6	1.1
108		1 tbsp	12	75	21	Tr	2	1.6	0.2	0.1
	Frozen dessert									
	Frozen yogurt, soft serve									
109	Chocolate	½ cup	72	64	115	3	4	2.6	1.3	0.2
110	Vanilla	½ cup	72	65	114	3	4	2.5	1.1	0.2
	Ice cream									
	Regular									
111	Chocolate	½ cup	66	56	143	3	7	4.5	2.1	0.3
112	Vanilla	½ cup	66	61	133	2	7	4.5	2.1	0.3
113	Light (50% reduced fat), vanilla	½ cup	66	68	92	3	3	1.7	0.8	0.1
114	Premium low fat, chocolate	½ cup	72	61	113	3	2	1.0	0.6	0.1
115	Rich, vanilla	½ cup	74	57	178	3	12	7.4	3.4	0.4
116	Soft serve, french vanilla	½ cup	86	60	185	4	11	6.4	3.0	0.4
117	Sherbet, orange	½ cup	74	66	102	1	1	0.9	0.4	0.1
	Milk									
	Fluid, no milk solids added									
118	Whole (3.3% fat)	1 cup	244	88	150	8	8	5.1	2.4	0.3
119	Reduced fat (2%)	1 cup	244	89	121	8	5	2.9	1.4	0.2
120	Lowfat (1%)	1 cup	244	90	102	8	3	1.6	0.7	0.1
121	Nonfat (skim)	1 cup	245	91	86	8	Tr	0.3	0.1	Tr
122	Buttermilk	1 cup	245	90	99	8	2	1.3	0.6	0.1
	Canned									
123	Condensed, sweetened	1 cup	306	27	982	24	27	16.8	7.4	1.0
	Evaporated									
124	Whole milk	1 cup	252	74	339	17	19	11.6	5.9	0.6
125	Skim milk	1 cup	256	79	199	19	1	0.3	0.2	Tr
	Dried									
126	Buttermilk	1 cup	120	3	464	41	7	4.3	2.0	0.3
127	Nonfat, instant, with added vitamin A	1 cup	68	4	244	24	Tr	0.3	0.1	Tr
	Milk beverage									
	Chocolate milk (commercial)									
128	Whole	1 cup	250	82	208	8	8	5.3	2.5	0.3
129	Reduced fat (2%)	1 cup	250	84	179	8	5	3.1	1.5	0.2
130	Lowfat (1%)	1 cup	250	85	158	8	3	1.5	0.8	0.1

*The vitamin A values listed for imitation sweet cream products are mostly from beta-carotene added for coloring.

Cholesterol (mg)	Carbohydrate (g)	Total dietary fiber (g)	Calcium (mg)	Iron (mg)	Potassium (mg)	Sodium (mg)	Vitamin A (IU)	Vitamin A (RE)	Thiamin (mg)	Riboflavin (mg)	Niacin (mg)	Ascorbic acid (mg)	Food No.
102	10	0.0	268	0.1	331	123	1,817	449	0.08	0.34	0.2	2	95
5	1	0.0	14	Tr	17	6	95	23	Tr	0.02	Tr	Tr	96
6	1	0.0	16	Tr	19	6	68	17	0.01	0.02	Tr	Tr	97
1	2	0.0	20	0.0	21	23	100	13	0.01	0.02	Tr	0	98
0	2	0.0	1	Tr	29	12	13*	1*	0.00	0.00	0.0	0	99
0	1	0.0	Tr	Tr	16	4	4	Tr	0.00	Tr	0.0	0	100
0	17	0.0	5	0.1	14	19	646*	65*	0.00	0.00	0.0	0	101
0	1	0.0	Tr	Tr	1	1	34*	3*	0.00	0.00	0.0	0	102
8	13	0.0	72	Tr	121	53	289*	39*	0.02	0.09	Tr	1	103
Tr	1	0.0	4	Tr	6	3	14*	2*	Tr	Tr	Tr	Tr	104
0	11	0.0	4	Tr	13	43	331*	33*	0.00	0.00	0.0	0	105
0	1	0.0	Tr	Tr	1	2	19*	2*	0.00	0.00	0.0	0	106
13	11	0.0	266	0.1	380	113	24	5	0.09	0.38	0.2	2	107
1	1	0.0	14	Tr	19	6	1	Tr	Tr	0.02	Tr	Tr	108
4	18	1.6	106	0.9	188	71	115	31	0.03	0.15	0.2	Tr	109
1	17	0.0	103	0.2	152	63	153	41	0.03	0.16	0.2	1	110
22	19	0.8	72	0.6	164	50	275	79	0.03	0.13	0.1	Tr	111
29	16	0.0	84	0.1	131	53	270	77	0.03	0.16	0.1	Tr	112
9	15	0.0	92	0.1	139	56	109	31	0.04	0.17	0.1	1	113
7	22	0.7	107	0.4	179	50	163	47	0.02	0.13	0.1	1	114
45	17	0.0	87	Tr	118	41	476	136	0.03	0.12	0.1	1	115
78	19	0.0	113	0.2	152	52	464	132	0.04	0.16	0.1	1	116
4	22	0.0	40	0.1	71	34	56	10	0.02	0.06	Tr	2	117
33	11	0.0	291	0.1	370	120	307	76	0.09	0.40	0.2	2	118
18	12	0.0	297	0.1	377	122	500	139	0.10	0.40	0.2	2	119
10	12	0.0	300	0.1	381	123	500	144	0.10	0.41	0.2	2	120
4	12	0.0	302	0.1	406	126	500	149	0.09	0.34	0.2	2	121
9	12	0.0	285	0.1	371	257	81	20	0.08	0.38	0.1	2	122
104	166	0.0	868	0.6	1,136	389	1,004	248	0.28	1.27	0.6	8	123
74	25	0.0	657	0.5	764	267	612	136	0.12	0.80	0.5	5	124
9	29	0.0	741	0.7	849	294	1,004	300	0.12	0.79	0.4	3	125
83	59	0.0	1,421	0.4	1,910	621	262	65	0.47	1.89	1.1	7	126
12	35	0.0	837	0.2	1,160	373	1,612	483	0.28	1.19	0.6	4	127
31	26	2.0	280	0.6	417	149	303	73	0.09	0.41	0.3	2	128
17	26	1.3	284	0.6	422	151	500	143	0.09	0.41	0.3	2	129
7	26	1.3	287	0.6	426	152	500	148	0.10	0.42	0.3	2	130

Food No.	Food Description	Measure of edible portion	Weight (g)	Water (%)	Calories (kcal)	Protein (g)	Total fat (g)	Fatty acids		
								Saturated (g)	Monounsaturated (g)	Polyunsaturated (g)

Dairy Products (continued)

Food No.	Food Description	Measure of edible portion	Weight (g)	Water (%)	Calories (kcal)	Protein (g)	Total fat (g)	Saturated (g)	Monounsaturated (g)	Polyunsaturated (g)
	Milk beverage (continued)									
131	Eggnog (commercial)	1 cup	254	74	342	10	19	11.3	5.7	0.9
	Milk shake, thick									
132	Chocolate	10.6 fl oz	300	72	356	9	8	5.0	2.3	0.3
133	Vanilla	11 fl oz	313	74	350	12	9	5.9	2.7	0.4
	Sherbet. See Dairy Products, frozen dessert.									
	Yogurt									
	With added milk solids									
	Made with lowfat milk									
134	Fruit flavored	8-oz container	227	74	231	10	2	1.6	0.7	0.1
135	Plain	8-oz container	227	85	144	12	4	2.3	1.0	0.1
	Made with nonfat milk									
136	Fruit flavored	8-oz container	227	75	213	10	Tr	0.3	0.1	Tr
137	Plain	8-oz container	227	85	127	13	Tr	0.3	0.1	Tr
	Without added milk solids									
138	Made with whole milk, plain	8-oz container	227	88	139	8	7	4.8	2.0	0.2
139	Made with nonfat milk, low calorie sweetener, vanilla or lemon flavor	8-oz container	227	87	98	9	Tr	0.3	0.1	Tr

Eggs

Food No.	Food Description	Measure of edible portion	Weight (g)	Water (%)	Calories (kcal)	Protein (g)	Total fat (g)	Saturated (g)	Monounsaturated (g)	Polyunsaturated (g)
	Egg									
	Raw									
140	Whole	1 medium	44	75	66	5	4	1.4	1.7	0.6
141		1 large	50	75	75	6	5	1.6	1.9	0.7
142		1 extra large	58	75	86	7	6	1.8	2.2	0.8
143	White	1 large	33	88	17	4	0	0.0	0.0	0.0
144	Yolk	1 large	17	49	59	3	5	1.6	1.9	0.7
	Cooked, whole									
145	Fried, in margarine, with salt	1 large	46	69	92	6	7	1.9	2.7	1.3
146	Hard cooked, shell removed	1 large	50	75	78	6	5	1.6	2.0	0.7
147		1 cup, chopped	136	75	211	17	14	4.4	5.5	1.9
148	Poached, with salt	1 large	50	75	75	6	5	1.5	1.9	0.7
149	Scrambled, in margarine, with whole milk, salt	1 large	61	73	101	7	7	2.2	2.9	1.3
150	Egg substitute, liquid	¼ cup	63	83	53	8	2	0.4	0.6	1.0

Fats and Oils

Food No.	Food Description	Measure of edible portion	Weight (g)	Water (%)	Calories (kcal)	Protein (g)	Total fat (g)	Saturated (g)	Monounsaturated (g)	Polyunsaturated (g)
	Butter (4 sticks per lb)									
151	Salted	1 stick	113	16	813	1	92	57.3	26.6	3.4
152		1 tbsp	14	16	102	Tr	12	7.2	3.3	0.4
153		1 tsp	5	16	36	Tr	4	2.5	1.2	0.2
154	Unsalted	1 stick	113	18	813	1	92	57.3	26.6	3.4
155	Lard	1 cup	205	0	1,849	0	205	80.4	92.5	23.0
156		1 tbsp	13	0	115	0	13	5.0	5.8	1.4
	Margarine, vitamin A-fortified, salt added									
	Regular (about 80% fat)									
157	Hard (4 sticks per lb)	1 stick	113	16	815	1	91	17.9	40.6	28.8
158		1 tbsp	14	16	101	Tr	11	2.2	5.0	3.6
159		1 tsp	5	16	34	Tr	4	0.7	1.7	1.2
160	Soft	1 cup	227	16	1,626	2	183	31.3	64.7	78.5
161		1 tsp	5	16	34	Tr	4	0.6	1.3	1.6

Cholesterol (mg)	Carbohydrate (g)	Total dietary fiber (g)	Calcium (mg)	Iron (mg)	Potassium (mg)	Sodium (mg)	Vitamin A (IU)	Vitamin A (RE)	Thiamin (mg)	Riboflavin (mg)	Niacin (mg)	Ascorbic acid (mg)	Food No.
149	34	0.0	330	0.5	420	138	894	203	0.09	0.48	0.3	4	131
32	63	0.9	396	0.9	672	333	258	63	0.14	0.67	0.4	0	132
37	56	0.0	457	0.3	572	299	357	88	0.09	0.61	0.5	0	133
10	43	0.0	345	0.2	442	133	104	25	0.08	0.40	0.2	1	134
14	16	0.0	415	0.2	531	159	150	36	0.10	0.49	0.3	2	135
5	43	0.0	345	0.2	440	132	16	5	0.09	0.41	0.2	2	136
4	17	0.0	452	0.2	579	174	16	5	0.11	0.53	0.3	2	137
29	11	0.0	274	0.1	351	105	279	68	0.07	0.32	0.2	1	138
5	17	0.0	325	0.3	402	134	0	0	0.08	0.37	0.2	2	139
187	1	0.0	22	0.6	53	55	279	84	0.03	0.22	Tr	0	140
213	1	0.0	25	0.7	61	63	318	96	0.03	0.25	Tr	0	141
247	1	0.0	28	0.8	70	73	368	111	0.04	0.29	Tr	0	142
0	Tr	0.0	2	Tr	48	55	0	0	Tr	0.15	Tr	0	143
213	Tr	0.0	23	0.6	16	7	323	97	0.03	0.11	Tr	0	144
211	1	0.0	25	0.7	61	162	394	114	0.03	0.24	Tr	0	145
212	1	0.0	25	0.6	63	62	280	84	0.03	0.26	Tr	0	146
577	2	0.0	68	1.6	171	169	762	228	0.09	0.70	0.1	0	147
212	1	0.0	25	0.7	60	140	316	95	0.02	0.22	Tr	0	148
215	1	0.0	43	0.7	84	171	416	119	0.03	0.27	Tr	Tr	149
1	Tr	0.0	33	1.3	208	112	1,361	136	0.07	0.19	0.1	0	150
248	Tr	0.0	27	0.2	29	937	3,468	855	0.01	0.04	Tr	0	151
31	Tr	0.0	3	Tr	4	117	434	107	Tr	Tr	Tr	0	152
11	Tr	0.0	1	Tr	1	41	153	38	Tr	Tr	Tr	0	153
248	Tr	0.0	27	0.2	29	12	3,468	855	0.01	0.04	Tr	0	154
195	0	0.0	Tr	0.0	Tr	Tr	0	0	0.00	0.00	0.0	0	155
12	0	0.0	Tr	0.0	Tr	Tr	0	0	0.00	0.00	0.0	0	156
0	1	0.0	34	0.1	48	1,070	4,050	906	0.01	0.04	Tr	Tr	157
0	Tr	0.0	4	Tr	6	132	500	112	Tr	0.01	Tr	Tr	158
0	Tr	0.0	1	Tr	2	44	168	38	Tr	Tr	Tr	Tr	159
0	1	0.0	60	0.0	86	2,449	8,106	1,814	0.02	0.07	Tr	Tr	160
0	Tr	0.0	1	0.0	2	51	168	38	Tr	Tr	Tr	Tr	161

Food No.	Food Description	Measure of edible portion	Weight (g)	Water (%)	Calories (kcal)	Protein (g)	Total fat (g)	Fatty acids: Saturated (g)	Fatty acids: Monounsaturated (g)	Fatty acids: Polyunsaturated (g)

Fats and Oils (continued)

Food No.	Food Description	Measure of edible portion	Weight (g)	Water (%)	Calories (kcal)	Protein (g)	Total fat (g)	Saturated (g)	Monounsaturated (g)	Polyunsaturated (g)
	Margarine, vitamin A-fortified, salt added (continued)									
	Spread (about 60% fat)									
162	Hard (4 sticks per lb)	1 stick	115	37	621	1	70	16.2	29.9	20.8
163		1 tbsp	14	37	76	Tr	9	2.0	3.6	2.5
164		1 tsp	5	37	26	Tr	3	0.7	1.2	0.9
165	Soft	1 cup	229	37	1,236	1	139	29.3	72.1	31.6
166		1 tsp	5	37	26	Tr	3	0.6	1.5	0.7
167	Spread (about 40% fat)	1 cup	232	58	801	1	90	17.9	36.4	32.0
168		1 tsp	5	58	17	Tr	2	0.4	0.8	0.7
169	Margarine butter blend	1 stick	113	16	811	1	91	32.1	37.0	18.0
170		1 tbsp	14	16	102	Tr	11	4.0	4.7	2.3
	Oils, salad or cooking									
171	Canola	1 cup	218	0	1,927	0	218	15.5	128.4	64.5
172		1 tbsp	14	0	124	0	14	1.0	8.2	4.1
173	Corn	1 cup	218	0	1,927	0	218	27.7	52.8	128.0
174		1 tbsp	14	0	120	0	14	1.7	3.3	8.0
175	Olive	1 cup	216	0	1,909	0	216	29.2	159.2	18.1
176		1 tbsp	14	0	119	0	14	1.8	9.9	1.1
177	Peanut	1 cup	216	0	1,909	0	216	36.5	99.8	69.1
178		1 tbsp	14	0	119	0	14	2.3	6.2	4.3
179	Safflower, high oleic	1 cup	218	0	1,927	0	218	13.5	162.7	31.3
180		1 tbsp	14	0	120	0	14	0.8	10.2	2.0
181	Sesame	1 cup	218	0	1,927	0	218	31.0	86.5	90.9
182		1 tbsp	14	0	120	0	14	1.9	5.4	5.7
183	Soybean, hydrogenated	1 cup	218	0	1,927	0	218	32.5	93.7	82.0
184		1 tbsp	14	0	120	0	14	2.0	5.8	5.1
185	Soybean, hydrogenated and cottonseed oil blend	1 cup	218	0	1,927	0	218	39.2	64.3	104.9
186		1 tbsp	14	0	120	0	14	2.4	4.0	6.5
187	Sunflower	1 cup	218	0	1,927	0	218	22.5	42.5	143.2
188		1 tbsp	14	0	120	0	14	1.4	2.7	8.9
	Salad dressings									
	Commercial									
	Blue cheese									
189	Regular	1 tbsp	15	32	77	1	8	1.5	1.9	4.3
190	Low calorie	1 tbsp	15	80	15	1	1	0.4	0.3	0.4
	Caesar									
191	Regular	1 tbsp	15	34	78	Tr	8	1.3	2.0	4.8
192	Low calorie	1 tbsp	15	73	17	Tr	1	0.1	0.2	0.4
	French									
193	Regular	1 tbsp	16	38	67	Tr	6	1.5	1.2	3.4
194	Low calorie	1 tbsp	16	69	22	Tr	1	0.1	0.2	0.6
	Italian									
195	Regular	1 tbsp	15	38	69	Tr	7	1.0	1.6	4.1
196	Low calorie	1 tbsp	15	82	16	Tr	1	0.2	0.3	0.9
	Mayonnaise									
197	Regular	1 tbsp	14	15	99	Tr	11	1.6	3.1	5.7
198	Light, cholesterol free	1 tbsp	15	56	49	Tr	5	0.7	1.1	2.8
199	Fat free	1 tbsp	16	84	12	0	Tr	0.1	0.1	0.2
	Russian									
200	Regular	1 tbsp	15	35	76	Tr	8	1.1	1.8	4.5
201	Low calorie	1 tbsp	16	65	23	Tr	1	0.1	0.1	0.4
	Thousand island									
202	Regular	1 tbsp	16	46	59	Tr	6	0.9	1.3	3.1
203	Low calorie	1 tbsp	15	69	24	Tr	2	0.2	0.4	0.9

Cholesterol (mg)	Carbohydrate (g)	Total dietary fiber (g)	Calcium (mg)	Iron (mg)	Potassium (mg)	Sodium (mg)	Vitamin A (IU)	Vitamin A (RE)	Thiamin (mg)	Riboflavin (mg)	Niacin (mg)	Ascorbic acid (mg)	Food No.
0	0	0.0	24	0.0	34	1,143	4,107	919	0.01	0.03	Tr	Tr	162
0	0	0.0	3	0.0	4	139	500	112	Tr	Tr	Tr	Tr	163
0	0	0.0	1	0.0	1	48	171	38	Tr	Tr	Tr	Tr	164
0	0	0.0	48	0.0	68	2,276	8,178	1,830	0.02	0.06	Tr	Tr	165
0	0	0.0	1	0.0	1	48	171	38	Tr	Tr	Tr	Tr	166
0	1	0.0	41	0.0	59	2,226	8,285	1,854	0.01	0.05	Tr	Tr	167
0	Tr	0.0	1	0.0	1	46	171	38	Tr	Tr	Tr	Tr	168
99	1	0.0	32	0.1	41	1,014	4,035	903	0.01	0.04	Tr	Tr	169
12	Tr	0.0	4	Tr	5	127	507	113	Tr	Tr	Tr	Tr	170
0	0	0.0	0	0.0	0	0	0	0	0.00	0.00	0.0	0	171
0	0	0.0	0	0.0	0	0	0	0	0.00	0.00	0.0	0	172
0	0	0.0	0	0.0	0	0	0	0	0.00	0.00	0.0	0	173
0	0	0.0	0	0.0	0	0	0	0	0.00	0.00	0.0	0	174
0	0	0.0	Tr	0.8	0	Tr	0	0	0.00	0.00	0.0	0	175
0	0	0.0	Tr	0.1	0	Tr	0	0	0.00	0.00	0.0	0	176
0	0	0.0	Tr	0.1	Tr	Tr	0	0	0.00	0.00	0.0	0	177
0	0	0.0	Tr	Tr	Tr	Tr	0	0	0.00	0.00	0.0	0	178
0	0	0.0	0	0.0	0	0	0	0	0.00	0.00	0.0	0	179
0	0	0.0	0	0.0	0	0	0	0	0.00	0.00	0.0	0	180
0	0	0.0	0	0.0	0	0	0	0	0.00	0.00	0.0	0	181
0	0	0.0	0	0.0	0	0	0	0	0.00	0.00	0.0	0	182
0	0	0.0	0	0.0	0	0	0	0	0.00	0.00	0.0	0	183
0	0	0.0	0	0.0	0	0	0	0	0.00	0.00	0.0	0	184
0	0	0.0	0	0.0	0	0	0	0	0.00	0.00	0.0	0	185
0	0	0.0	0	0.0	0	0	0	0	0.00	0.00	0.0	0	186
0	0	0.0	0	0.0	0	0	0	0	0.00	0.00	0.0	0	187
0	0	0.0	0	0.0	0	0	0	0	0.00	0.00	0.0	0	188
3	1	0.0	12	Tr	6	167	32	10	Tr	0.02	Tr	Tr	189
Tr	Tr	0.0	14	0.1	1	184	2	Tr	Tr	0.02	Tr	Tr	190
Tr	Tr	Tr	4	Tr	4	158	3	Tr	Tr	Tr	Tr	0	191
Tr	3	Tr	4	Tr	4	162	3	Tr	Tr	Tr	Tr	0	192
0	3	0.0	2	0.1	12	214	203	20	Tr	Tr	Tr	0	193
0	4	0.0	2	0.1	13	128	212	21	0.00	0.00	0.0	0	194
0	1	0.0	1	Tr	2	116	11	4	Tr	Tr	Tr	0	195
1	1	Tr	Tr	Tr	2	118	0	0	0.00	0.00	0.0	0	196
8	Tr	0.0	2	0.1	5	78	39	12	0.00	0.00	Tr	0	197
0	1	0.0	0	0.0	10	107	18	2	0.00	0.00	0.0	0	198
0	2	0.6	0	0.0	15	190	0	0	0.00	0.00	0.0	0	199
3	2	0.0	3	0.1	24	133	106	32	0.01	0.01	0.1	1	200
1	4	Tr	3	0.1	26	141	9	3	Tr	Tr	Tr	1	201
4	2	0.0	2	0.1	18	109	50	15	Tr	Tr	Tr	0	202
2	2	0.2	2	0.1	17	153	49	15	Tr	Tr	Tr	0	203

Food No.	Food Description	Measure of edible portion	Weight (g)	Water (%)	Calories (kcal)	Protein (g)	Total fat (g)	Fatty acids Saturated (g)	Fatty acids Monounsaturated (g)	Fatty acids Polyunsaturated (g)
	Fats and Oils (continued)									
	Salad dressings (continued)									
	Prepared from home recipe									
204	Cooked, made with margarine	1 tbsp	16	69	25	1	2	0.5	0.6	0.3
205	French	1 tbsp	14	24	88	Tr	10	1.8	2.9	4.7
206	Vinegar and oil	1 tbsp	16	47	70	0	8	1.4	2.3	3.8
207	Shortening (hydrogenated soybean and cottonseed oils)	1 cup	205	0	1,812	0	205	51.3	91.2	53.5
208		1 tbsp	13	0	113	0	13	3.2	5.7	3.3
	Fish and Shellfish									
209	Catfish, breaded, fried	3 oz	85	59	195	15	11	2.8	4.8	2.8
	Clam									
210	Raw, meat only	3 oz	85	82	63	11	1	0.1	0.1	0.2
211		1 medium	15	82	11	2	Tr	Tr	Tr	Tr
212	Breaded, fried	¾ cup	115	29	451	13	26	6.6	11.4	6.8
213	Canned, drained solids	3 oz	85	64	126	22	2	0.2	0.1	0.5
214		1 cup	160	64	237	41	3	0.3	0.3	0.9
	Cod									
215	Baked or broiled	3 oz	85	76	89	20	1	0.1	0.1	0.3
216		1 fillet	90	76	95	21	1	0.1	0.1	0.3
217	Canned, solids and liquid	3 oz	85	76	89	19	1	0.1	0.1	0.2
	Crab									
	Alaska king									
218	Steamed	1 leg	134	78	130	26	2	0.2	0.2	0.7
219		3 oz	85	78	82	16	1	0.1	0.2	0.5
220	Imitation, from surimi	3 oz	85	74	87	10	1	0.2	0.2	0.6
	Blue									
221	Steamed	3 oz	85	77	87	17	2	0.2	0.2	0.6
222	Canned crabmeat	1 cup	135	76	134	28	2	0.3	0.3	0.6
223	Crab cake, with egg, onion, fried in margarine	1 cake	60	71	93	12	5	0.9	1.7	1.4
224	Fish fillet, battered or breaded, fried	1 fillet	91	54	211	13	11	2.6	2.3	5.7
225	Fish stick and portion, breaded, frozen, reheated	1 stick (4" x 1" x ½")	28	46	76	4	3	0.9	1.4	0.9
226		1 portion (4" x 2" x ½")	57	46	155	9	7	1.8	2.9	1.8
227	Flounder or sole, baked or broiled	3 oz	85	73	99	21	1	0.3	0.2	0.5
228		1 fillet	127	73	149	31	2	0.5	0.3	0.8
229	Haddock, baked or broiled	3 oz	85	74	95	21	1	0.1	0.1	0.3
230		1 fillet	150	74	168	36	1	0.3	0.2	0.5
231	Halibut, baked or broiled	3 oz	85	72	119	23	2	0.4	0.8	0.8
232		½ fillet	159	72	223	42	5	0.7	1.5	1.5
233	Herring, pickled	3 oz	85	55	223	12	15	2.0	10.2	1.4
234	Lobster, steamed	3 oz	85	76	83	17	1	0.1	0.1	0.1
235	Ocean perch, baked or broiled	3 oz	85	73	103	20	2	0.3	0.7	0.5
236		1 fillet	50	73	61	12	1	0.2	0.4	0.3
	Oyster									
237	Raw, meat only	1 cup	248	85	169	17	6	1.9	0.8	2.4
238		6 medium	84	85	57	6	2	0.6	0.3	0.8
239	Breaded, fried	3 oz	85	65	167	7	11	2.7	4.0	2.8
240	Pollock, baked or broiled	3 oz	85	74	96	20	1	0.2	0.1	0.4
241		1 fillet	60	74	68	14	1	0.1	0.1	0.3
242	Rockfish, baked or broiled	3 oz	85	73	103	20	2	0.4	0.4	0.5
243		1 fillet	149	73	180	36	3	0.7	0.7	0.9

Cholesterol (mg)	Carbohydrate (g)	Total dietary fiber (g)	Calcium (mg)	Iron (mg)	Potassium (mg)	Sodium (mg)	Vitamin A (IU)	Vitamin A (RE)	Thiamin (mg)	Riboflavin (mg)	Niacin (mg)	Ascorbic acid (mg)	Food No.
9	2	0.0	13	0.1	19	117	66	20	0.01	0.02	Tr	Tr	204
0	Tr	0.0	1	Tr	3	92	72	22	Tr	Tr	Tr	Tr	205
0	Tr	0.0	0	0.0	1	Tr	0	0	0.00	0.00	0.0	0	206
0	0	0.0	0	0.0	0	0	0	0	0.00	0.00	0.0	0	207
0	0	0.0	0	0.0	0	0	0	0	0.00	0.00	0.0	0	208
69	7	0.6	37	1.2	289	238	24	7	0.06	0.11	1.9	0	209
29	2	0.0	39	11.9	267	48	255	77	0.07	0.18	1.5	11	210
5	Tr	0.0	7	2.0	46	8	44	13	0.01	0.03	0.3	2	211
87	39	0.3	21	3.0	266	834	122	37	0.21	0.26	2.9	0	212
57	4	0.0	78	23.8	534	95	485	145	0.13	0.36	2.9	19	213
107	8	0.0	147	44.7	1,005	179	912	274	0.24	0.68	5.4	35	214
40	0	0.0	8	0.3	439	77	27	9	0.02	0.04	2.1	3	215
42	0	0.0	8	0.3	465	82	29	9	0.02	0.05	2.2	3	216
47	0	0.0	18	0.4	449	185	39	12	0.07	0.07	2.1	1	217
71	0	0.0	79	1.0	351	1,436	39	12	0.07	0.07	1.8	10	218
45	0	0.0	50	0.6	223	911	25	8	0.05	0.05	1.1	6	219
17	9	0.0	11	0.3	77	715	56	17	0.03	0.02	0.2	0	220
85	0	0.0	88	0.8	275	237	5	2	0.09	0.04	2.8	3	221
120	0	0.0	136	1.1	505	450	7	3	0.11	0.11	1.8	4	222
90	Tr	0.0	63	0.6	194	198	151	49	0.05	0.05	1.7	2	223
31	15	0.5	16	1.9	291	484	35	11	0.10	0.10	1.9	0	224
31	7	0.0	6	0.2	73	163	30	9	0.04	0.05	0.6	0	225
64	14	0.0	11	0.4	149	332	60	18	0.07	0.10	1.2	0	226
58	0	0.0	15	0.3	292	89	32	9	0.07	0.10	1.9	0	227
86	0	0.0	23	0.4	437	133	48	14	0.10	0.14	2.8	0	228
63	0	0.0	36	1.1	339	74	54	16	0.03	0.04	3.9	0	229
111	0	0.0	63	2.0	599	131	95	29	0.06	0.07	6.9	0	230
35	0	0.0	51	0.9	490	59	152	46	0.06	0.08	6.1	0	231
65	0	0.0	95	1.7	916	110	285	86	0.11	0.14	11.3	0	232
11	8	0.0	65	1.0	59	740	732	219	0.03	0.12	2.8	0	233
61	1	0.0	52	0.3	299	323	74	22	0.01	0.06	0.9	0	234
46	0	0.0	116	1.0	298	82	39	12	0.11	0.11	2.1	1	235
27	0	0.0	69	0.6	175	48	23	7	0.07	0.07	1.2	Tr	236
131	10	0.0	112	16.5	387	523	248	74	0.25	0.24	3.4	9	237
45	3	0.0	38	5.6	131	177	84	25	0.08	0.08	1.2	3	238
69	10	0.2	53	5.9	207	354	257	77	0.13	0.17	1.4	3	239
82	0	0.0	5	0.2	329	99	65	20	0.06	0.06	1.4	0	240
58	0	0.0	4	0.2	232	70	46	14	0.04	0.05	1.0	0	241
37	0	0.0	10	0.5	442	65	186	56	0.04	0.07	3.3	0	242
66	0	0.0	18	0.8	775	115	326	98	0.07	0.13	5.8	0	243

Food No.	Food Description	Measure of edible portion	Weight (g)	Water (%)	Calories (kcal)	Protein (g)	Total fat (g)	Fatty acids: Saturated (g)	Fatty acids: Monounsaturated (g)	Fatty acids: Polyunsaturated (g)

Fish and Shellfish (continued)

Food No.	Food Description	Measure of edible portion	Weight (g)	Water (%)	Calories (kcal)	Protein (g)	Total fat (g)	Saturated (g)	Monounsaturated (g)	Polyunsaturated (g)
244	Roughy, orange, baked or broiled	3 oz	85	69	76	16	1	Tr	0.5	Tr
	Salmon									
245	Baked or broiled (red)	3 oz	85	62	184	23	9	1.6	4.5	2.0
246		½ fillet	155	62	335	42	17	3.0	8.2	3.7
247	Canned (pink), solids and liquid (includes bones)	3 oz	85	69	118	17	5	1.3	1.5	1.7
248	Smoked (chinook)	3 oz	85	72	99	16	4	0.8	1.7	0.8
249	Sardine, Atlantic, canned in oil, drained solids (includes bones)	3 oz	85	60	177	21	10	1.3	3.3	4.4
	Scallop, cooked									
250	Breaded, fried	6 large	93	58	200	17	10	2.5	4.2	2.7
251	Steamed	3 oz	85	73	95	20	1	0.1	0.1	0.4
	Shrimp									
252	Breaded, fried	3 oz	85	53	206	18	10	1.8	3.2	4.3
253		6 large	45	53	109	10	6	0.9	1.7	2.3
254	Canned, drained solids	3 oz	85	73	102	20	2	0.3	0.2	0.6
255	Swordfish, baked or broiled	3 oz	85	69	132	22	4	1.2	1.7	1.0
256		1 piece	106	69	164	27	5	1.5	2.1	1.3
257	Trout, baked or broiled	3 oz	85	68	144	21	6	1.8	1.8	2.0
258		1 fillet	71	68	120	17	5	1.5	1.5	1.7
	Tuna									
259	Baked or broiled	3 oz	85	63	118	25	1	0.3	0.2	0.3
	Canned, drained solids									
260	Oil pack, chunk light	3 oz	85	60	168	25	7	1.3	2.5	2.5
261	Water pack, chunk light	3 oz	85	75	99	22	1	0.2	0.1	0.3
262	Water pack, solid white	3 oz	85	73	109	20	3	0.7	0.7	0.9
263	Tuna salad: light tuna in oil, pickle relish, mayo type salad dressing	1 cup	205	63	383	33	19	3.2	5.9	8.5

Fruits and Fruit Juices

Food No.	Food Description	Measure of edible portion	Weight (g)	Water (%)	Calories (kcal)	Protein (g)	Total fat (g)	Saturated (g)	Monounsaturated (g)	Polyunsaturated (g)
	Apples									
	Raw									
264	Unpeeled, 2¾" dia (about 3 per lb)	1 apple	138	84	81	Tr	Tr	0.1	Tr	0.1
265	Peeled, sliced	1 cup	110	84	63	Tr	Tr	0.1	Tr	0.1
266	Dried (sodium bisulfite used to preserve color)	5 rings	32	32	78	Tr	Tr	Tr	Tr	Tr
267	Apple juice, bottled or canned	1 cup	248	88	117	Tr	Tr	Tr	Tr	0.1
268	Apple pie filling, canned	⅛ of 21-oz can	74	73	75	Tr	Tr	Tr	0.0	Tr
	Applesauce, canned									
269	Sweetened	1 cup	255	80	194	Tr	Tr	0.1	Tr	0.1
270	Unsweetened	1 cup	244	88	105	Tr	Tr	Tr	Tr	Tr
	Apricots									
271	Raw, without pits (about 12 per lb with pits)	1 apricot	35	86	17	Tr	Tr	Tr	0.1	Tr
	Canned, halves, fruit and liquid									
272	Heavy syrup pack	1 cup	258	78	214	1	Tr	Tr	0.1	Tr
273	Juice pack	1 cup	244	87	117	2	Tr	Tr	Tr	Tr
274	Dried, sulfured	10 halves	35	31	83	1	Tr	Tr	0.1	Tr
275	Apricot nectar, canned, with added ascorbic acid	1 cup	251	85	141	1	Tr	Tr	0.1	Tr
	Asian pear, raw									
276	2¼" high x 2½" dia	1 pear	122	88	51	1	Tr	Tr	0.1	0.1
277	3⅜" high x 3" dia	1 pear	275	88	116	1	1	Tr	0.1	0.2

Choles-terol (mg)	Carbo-hydrate (g)	Total dietary fiber (g)	Calcium (mg)	Iron (mg)	Potas-sium (mg)	Sodium (mg)	Vitamin A (IU)	Vitamin A (RE)	Thiamin (mg)	Ribo-flavin (mg)	Niacin (mg)	Ascor-bic acid (mg)	Food No.
22	0	0.0	32	0.2	327	69	69	20	0.10	0.16	3.1	0	244
74	0	0.0	6	0.5	319	56	178	54	0.18	0.15	5.7	0	245
135	0	0.0	11	0.9	581	102	324	98	0.33	0.27	10.3	0	246
47	0	0.0	181	0.7	277	471	47	14	0.02	0.16	5.6	0	247
20	0	0.0	9	0.7	149	666	75	22	0.02	0.09	4.0	0	248
121	0	0.0	325	2.5	337	429	190	57	0.07	0.19	4.5	0	249
57	9	0.2	39	0.8	310	432	70	20	0.04	0.10	1.4	2	250
45	3	0.0	98	2.6	405	225	85	26	0.09	0.05	1.1	0	251
150	10	0.3	57	1.1	191	292	161	48	0.11	0.12	2.6	1	252
80	5	0.2	30	0.6	101	155	85	25	0.06	0.06	1.4	1	253
147	1	0.0	50	2.3	179	144	51	15	0.02	0.03	2.3	2	254
43	0	0.0	5	0.9	314	98	116	35	0.04	0.10	10.0	1	255
53	0	0.0	6	1.1	391	122	145	43	0.05	0.12	12.5	1	256
58	0	0.0	73	0.3	375	36	244	73	0.20	0.07	7.5	3	257
48	0	0.0	61	0.2	313	30	204	61	0.17	0.06	6.2	2	258
49	0	0.0	18	0.8	484	40	58	17	0.43	0.05	10.1	1	259
15	0	0.0	11	1.2	176	301	66	20	0.03	0.10	10.5	0	260
26	0	0.0	9	1.3	201	287	48	14	0.03	0.06	11.3	0	261
36	0	0.0	12	0.8	201	320	16	5	0.01	0.04	4.9	0	262
27	19	0.0	35	2.1	365	824	199	55	0.06	0.14	13.7	5	263
0	21	3.7	10	0.2	159	0	73	7	0.02	0.02	0.1	8	264
0	16	2.1	4	0.1	124	0	48	4	0.02	0.01	0.1	4	265
0	21	2.8	4	0.4	144	28	0	0	0.00	0.05	0.3	1	266
0	29	0.2	17	0.9	295	7	2	0	0.05	0.04	0.2	2	267
0	19	0.7	3	0.2	33	33	10	1	0.01	0.01	Tr	1	268
0	51	3.1	10	0.9	156	8	28	3	0.03	0.07	0.5	4	269
0	28	2.9	7	0.3	183	5	71	7	0.03	0.06	0.5	3	270
0	4	0.8	5	0.2	104	Tr	914	91	0.01	0.01	0.2	4	271
0	55	4.1	23	0.8	361	10	3,173	317	0.05	0.06	1.0	8	272
0	30	3.9	29	0.7	403	10	4,126	412	0.04	0.05	0.8	12	273
0	22	3.2	16	1.6	482	4	2,534	253	Tr	0.05	1.0	1	274
0	36	1.5	18	1.0	286	8	3,303	331	0.02	0.04	0.7	137	275
0	13	4.4	5	0.0	148	0	0	0	0.01	0.01	0.3	5	276
0	29	9.9	11	0.0	333	0	0	0	0.02	0.03	0.6	10	277

								Fatty acids		
Food No.	Food Description	Measure of edible portion	Weight (g)	Water (%)	Calories (kcal)	Protein (g)	Total fat (g)	Saturated (g)	Monounsaturated (g)	Polyunsaturated (g)

Fruits and Fruit Juices (continued)

Food No.	Food Description	Measure of edible portion	Weight (g)	Water (%)	Calories (kcal)	Protein (g)	Total fat (g)	Saturated (g)	Monounsaturated (g)	Polyunsaturated (g)
	Avocados, raw, without skin and seed									
278	California (about 1/5 whole)	1 oz	28	73	50	1	5	0.7	3.2	0.6
279	Florida (about 1/10 whole)	1 oz	28	80	32	Tr	3	0.5	1.4	0.4
	Bananas, raw									
280	Whole, medium (7" to 7 7/8" long)	1 banana	118	74	109	1	1	0.2	Tr	0.1
281	Sliced	1 cup	150	74	138	2	1	0.3	0.1	0.1
282	Blackberries, raw	1 cup	144	86	75	1	1	Tr	0.1	0.3
	Blueberries									
283	Raw	1 cup	145	85	81	1	1	Tr	0.1	0.2
284	Frozen, sweetened, thawed	1 cup	230	77	186	1	Tr	Tr	Tr	0.1
	Cantaloupe. See Melons.									
	Carambola (starfruit), raw									
285	Whole (3 5/8" long)	1 fruit	91	91	30	Tr	Tr	Tr	Tr	0.2
286	Sliced	1 cup	108	91	36	1	Tr	Tr	Tr	0.2
	Cherries									
287	Sour, red, pitted, canned, water pack	1 cup	244	90	88	2	Tr	0.1	0.1	0.1
288	Sweet, raw, without pits and stems	10 cherries	68	81	49	1	1	0.1	0.2	0.2
289	Cherry pie filling, canned	1/5 of 21-oz can	74	71	85	Tr	Tr	Tr	Tr	Tr
290	Cranberries, dried, sweetened	1/4 cup	28	12	92	Tr	Tr	Tr	Tr	0.1
291	Cranberry sauce, sweetened, canned (about 8 slices per can)	1 slice	57	61	86	Tr	Tr	Tr	Tr	Tr
	Dates, without pits									
292	Whole	5 dates	42	23	116	1	Tr	0.1	0.1	Tr
293	Chopped	1 cup	178	23	490	4	1	0.3	0.3	0.1
294	Figs, dried	2 figs	38	28	97	1	Tr	0.1	0.1	0.2
	Fruit cocktail, canned, fruit and liquid									
295	Heavy syrup pack	1 cup	248	80	181	1	Tr	Tr	Tr	0.1
296	Juice pack	1 cup	237	87	109	1	Tr	Tr	Tr	Tr
	Grapefruit									
	Raw, without peel, membrane and seeds (3 3/4" dia)									
297	Pink or red	1/2 grapefruit	123	91	37	1	Tr	Tr	Tr	Tr
298	White	1/2 grapefruit	118	90	39	1	Tr	Tr	Tr	Tr
299	Canned, sections with light syrup	1 cup	254	84	152	1	Tr	Tr	Tr	0.1
	Grapefruit juice									
	Raw									
300	Pink	1 cup	247	90	96	1	Tr	Tr	Tr	0.1
301	White	1 cup	247	90	96	1	Tr	Tr	Tr	0.1
	Canned									
302	Unsweetened	1 cup	247	90	94	1	Tr	Tr	Tr	0.1
303	Sweetened	1 cup	250	87	115	1	Tr	Tr	Tr	0.1
	Frozen concentrate, unsweetened									
304	Undiluted	6-fl-oz can	207	62	302	4	1	0.1	0.1	0.2
305	Diluted with 3 parts water by volume	1 cup	247	89	101	1	Tr	Tr	Tr	0.1
306	Grapes, seedless, raw	10 grapes	50	81	36	Tr	Tr	0.1	Tr	0.1
307		1 cup	160	81	114	1	1	0.3	Tr	0.3

Cholesterol (mg)	Carbohydrate (g)	Total dietary fiber (g)	Calcium (mg)	Iron (mg)	Potassium (mg)	Sodium (mg)	Vitamin A (IU)	Vitamin A (RE)	Thiamin (mg)	Riboflavin (mg)	Niacin (mg)	Ascorbic acid (mg)	Food No.
0	2	1.4	3	0.3	180	3	174	17	0.03	0.03	0.5	2	278
0	3	1.5	3	0.2	138	1	174	17	0.03	0.03	0.5	2	279
0	28	2.8	7	0.4	467	1	96	9	0.05	0.12	0.6	11	280
0	35	3.6	9	0.5	594	2	122	12	0.07	0.15	0.8	14	281
0	18	7.6	46	0.8	282	0	238	23	0.04	0.06	0.6	30	282
0	20	3.9	9	0.2	129	9	145	15	0.07	0.07	0.5	19	283
0	50	4.8	14	0.9	138	2	101	9	0.05	0.12	0.6	2	284
0	7	2.5	4	0.2	148	2	449	45	0.03	0.02	0.4	19	285
0	8	2.9	4	0.3	176	2	532	53	0.03	0.03	0.4	23	286
0	22	2.7	27	3.3	239	17	1,840	183	0.04	0.10	0.4	5	287
0	11	1.6	10	0.3	152	0	146	14	0.03	0.04	0.3	5	288
0	21	0.4	8	0.2	78	13	152	16	0.02	0.01	0.1	3	289
0	24	2.5	5	0.1	24	1	0	0	0.01	0.03	Tr	Tr	290
0	22	0.6	2	0.1	15	17	11	1	0.01	0.01	0.1	1	291
0	31	3.2	13	0.5	274	1	21	2	0.04	0.04	0.9	0	292
0	131	13.4	57	2.0	1,161	5	89	9	0.16	0.18	3.9	0	293
0	25	4.6	55	0.8	271	4	51	5	0.03	0.03	0.3	Tr	294
0	47	2.5	15	0.7	218	15	508	50	0.04	0.05	0.9	5	295
0	28	2.4	19	0.5	225	9	723	73	0.03	0.04	1.0	6	296
0	9	1.4	14	0.1	159	0	319	32	0.04	0.02	0.2	47	297
0	10	1.3	14	0.1	175	0	12	1	0.04	0.02	0.3	39	298
0	39	1.0	36	1.0	328	5	0	0	0.10	0.05	0.6	54	299
0	23	0.2	22	0.5	400	2	1,087	109	0.10	0.05	0.5	94	300
0	23	0.2	22	0.5	400	2	25	2	0.10	0.05	0.5	94	301
0	22	0.2	17	0.5	378	2	17	2	0.10	0.05	0.6	72	302
0	28	0.3	20	0.9	405	5	0	0	0.10	0.06	0.8	67	303
0	72	0.8	56	1.0	1,002	6	64	6	0.30	0.16	1.6	248	304
0	24	0.2	20	0.3	336	2	22	2	0.10	0.05	0.5	83	305
0	9	0.5	6	0.1	93	1	37	4	0.05	0.03	0.2	5	306
0	28	1.6	18	0.4	296	3	117	11	0.15	0.09	0.5	17	307

Food No.	Food Description	Measure of edible portion	Weight (g)	Water (%)	Calories (kcal)	Protein (g)	Total fat (g)	Fatty acids: Saturated (g)	Fatty acids: Monounsaturated (g)	Fatty acids: Polyunsaturated (g)

Fruits and Fruit Juices (continued)

Food No.	Food Description	Measure of edible portion	Weight (g)	Water (%)	Calories (kcal)	Protein (g)	Total fat (g)	Saturated (g)	Monounsaturated (g)	Polyunsaturated (g)
	Grape juice									
308	Canned or bottled	1 cup	253	84	154	1	Tr	0.1	Tr	0.1
	Frozen concentrate, sweetened, with added vitamin C									
309	Undiluted	6-fl-oz can	216	54	387	1	1	0.2	Tr	0.2
310	Diluted with 3 parts water by volume	1 cup	250	87	128	Tr	Tr	0.1	Tr	0.1
311	Kiwi fruit, raw, without skin (about 5 per lb with skin)	1 medium	76	83	46	1	Tr	Tr	Tr	0.2
312	Lemons, raw, without peel (2⅛" dia with peel)	1 lemon	58	89	17	1	Tr	Tr	Tr	0.1
	Lemon juice									
313	Raw (from 2⅛"-dia lemon)	juice of 1 lemon	47	91	12	Tr	0	0.0	0.0	0.0
314	Canned or bottled, unsweetened	1 cup	244	92	51	1	1	0.1	Tr	0.2
315		1 tbsp	15	92	3	Tr	Tr	Tr	Tr	Tr
	Lime juice									
316	Raw (from 2"-dia lime)	juice of 1 lime	38	90	10	Tr	Tr	Tr	Tr	Tr
317	Canned, unsweetened	1 cup	246	93	52	1	1	0.1	0.1	0.2
318		1 tbsp	15	93	3	Tr	Tr	Tr	Tr	Tr
	Mangos, raw, without skin and seed (about 1½ per lb with skin and seed)									
319	Whole	1 mango	207	82	135	1	1	0.1	0.2	0.1
320	Sliced	1 cup	165	82	107	1	Tr	0.1	0.2	0.1
	Melons, raw, without rind and cavity contents									
	Cantaloupe (5" dia)									
321	Wedge	⅛ melon	69	90	24	1	Tr	Tr	Tr	0.1
322	Cubes	1 cup	160	90	56	1	Tr	0.1	Tr	0.2
	Honeydew (6"-7" dia)									
323	Wedge	⅛ melon	160	90	56	1	Tr	Tr	Tr	0.1
324	Diced (about 20 pieces per cup)	1 cup	170	90	60	1	Tr	Tr	Tr	0.1
325	Mixed fruit, frozen, sweetened, thawed (peach, cherry, raspberry, grape and boysenberry)	1 cup	250	74	245	4	Tr	0.1	0.1	0.2
326	Nectarines, raw (2½" dia)	1 nectarine	136	86	67	1	1	0.1	0.2	0.3
	Oranges, raw									
327	Whole, without peel and seeds (2⅝" dia)	1 orange	131	87	62	1	Tr	Tr	Tr	Tr
328	Sections without membranes	1 cup	180	87	85	2	Tr	Tr	Tr	Tr
	Orange juice									
329	Raw, all varieties	1 cup	248	88	112	2	Tr	0.1	0.1	0.1
330		juice from 1 orange	86	88	39	1	Tr	Tr	Tr	Tr
331	Canned, unsweetened	1 cup	249	89	105	1	Tr	Tr	0.1	0.1
332	Chilled (refrigerator case)	1 cup	249	88	110	2	1	0.1	0.1	0.2
	Frozen concentrate									
333	Undiluted	6-fl-oz can	213	58	339	5	Tr	0.1	0.1	0.1
334	Diluted with 3 parts water by volume	1 cup	249	88	112	2	Tr	Tr	Tr	Tr
	Papayas, raw									
335	½" cubes	1 cup	140	89	55	1	Tr	0.1	0.1	Tr
336	Whole (5⅛" long x 3" dia)	1 papaya	304	89	119	2	Tr	0.1	0.1	0.1

*Sodium benzoate and sodium bisulfite added as preservatives.

Choles-terol (mg)	Carbo-hydrate (g)	Total dietary fiber (g)	Calcium (mg)	Iron (mg)	Potas-sium (mg)	Sodium (mg)	Vitamin A (IU)	Vitamin A (RE)	Thiamin (mg)	Ribo-flavin (mg)	Niacin (mg)	Ascor-bic acid (mg)	Food No.
0	38	0.3	23	0.6	334	8	20	3	0.07	0.09	0.7	Tr	308
0	96	0.6	28	0.8	160	15	58	6	0.11	0.20	0.9	179	309
0	32	0.3	10	0.3	53	5	20	3	0.04	0.07	0.3	60	310
0	11	2.6	20	0.3	252	4	133	14	0.02	0.04	0.4	74	311
0	5	1.6	15	0.3	80	1	17	2	0.02	0.01	0.1	31	312
0	4	0.2	3	Tr	58	Tr	9	1	0.01	Tr	Tr	22	313
0	16	1.0	27	0.3	249	51*	37	5	0.10	0.02	0.5	61	314
0	1	0.1	2	Tr	16	3*	2	Tr	0.01	Tr	Tr	4	315
0	3	0.2	3	Tr	41	Tr	4	Tr	0.01	Tr	Tr	11	316
0	16	1.0	30	0.6	185	39*	39	5	0.08	0.01	0.4	16	317
0	1	0.1	2	Tr	11	2*	2	Tr	Tr	Tr	Tr	1	318
0	35	3.7	21	0.3	323	4	8,061	805	0.12	0.12	1.2	57	319
0	28	3.0	17	0.2	257	3	6,425	642	0.10	0.09	1.0	46	320
0	6	0.6	8	0.1	213	6	2,225	222	0.02	0.01	0.4	29	321
0	13	1.3	18	0.3	494	14	5,158	515	0.06	0.03	0.9	68	322
0	15	1.0	10	0.1	434	16	64	6	0.12	0.03	1.0	40	323
0	16	1.0	10	0.1	461	17	68	7	0.13	0.03	1.0	42	324
0	61	4.8	18	0.7	328	8	805	80	0.04	0.09	1.0	188	325
0	16	2.2	7	0.2	288	0	1,001	101	0.02	0.06	1.3	7	326
0	15	3.1	52	0.1	237	0	269	28	0.11	0.05	0.4	70	327
0	21	4.3	72	0.2	326	0	369	38	0.16	0.07	0.5	96	328
0	26	0.5	27	0.5	496	2	496	50	0.22	0.07	1.0	124	329
0	9	0.2	9	0.2	172	1	172	17	0.08	0.03	0.3	43	330
0	25	0.5	20	1.1	436	5	436	45	0.15	0.07	0.8	86	331
0	25	0.5	25	0.4	473	2	194	20	0.28	0.05	0.7	82	332
0	81	1.7	68	0.7	1,436	6	588	60	0.60	0.14	1.5	294	333
0	27	0.5	22	0.2	473	2	194	20	0.20	0.04	0.5	97	334
0	14	2.5	34	0.1	360	4	398	39	0.04	0.04	0.5	87	335
0	30	5.5	73	0.3	781	9	863	85	0.08	0.10	1.0	188	336

Food No.	Food Description	Measure of edible portion	Weight (g)	Water (%)	Calories (kcal)	Protein (g)	Total fat (g)	Fatty acids: Saturated (g)	Fatty acids: Monounsaturated (g)	Fatty acids: Polyunsaturated (g)

Fruits and Fruit Juices (continued)

Food No.	Food Description	Measure of edible portion	Weight (g)	Water (%)	Calories (kcal)	Protein (g)	Total fat (g)	Saturated (g)	Monounsaturated (g)	Polyunsaturated (g)
	Peaches									
	Raw									
337	Whole, 2½" dia, pitted (about 4 per lb)	1 peach	98	88	42	1	Tr	Tr	Tr	Tr
338	Sliced	1 cup	170	88	73	1	Tr	Tr	0.1	0.1
	Canned, fruit and liquid									
339	Heavy syrup pack	1 cup	262	79	194	1	Tr	Tr	0.1	0.1
340		1 half	98	79	73	Tr	Tr	Tr	Tr	Tr
341	Juice pack	1 cup	248	87	109	2	Tr	Tr	Tr	Tr
342		1 half	98	87	43	1	Tr	Tr	Tr	Tr
343	Dried, sulfured	3 halves	39	32	93	1	Tr	Tr	0.1	0.1
344	Frozen, sliced, sweetened, with added ascorbic acid, thawed	1 cup	250	75	235	2	Tr	Tr	0.1	0.2
	Pears									
345	Raw, with skin, cored, 2½" dia	1 pear	166	84	98	1	1	Tr	0.1	0.2
	Canned, fruit and liquid									
346	Heavy syrup pack	1 cup	266	80	197	1	Tr	Tr	0.1	0.1
347		1 half	76	80	56	Tr	Tr	Tr	Tr	Tr
348	Juice pack	1 cup	248	86	124	1	Tr	Tr	Tr	Tr
349		1 half	76	86	38	Tr	Tr	Tr	Tr	Tr
	Pineapple									
350	Raw, diced	1 cup	155	87	76	1	1	Tr	0.1	0.2
	Canned, fruit and liquid									
	Heavy syrup pack									
351	Crushed, sliced, or chunks	1 cup	254	79	198	1	Tr	Tr	Tr	0.1
352	Slices (3" dia)	1 slice	49	79	38	Tr	Tr	Tr	Tr	Tr
	Juice pack									
353	Crushed, sliced, or chunks	1 cup	249	84	149	1	Tr	Tr	Tr	0.1
354	Slice (3" dia)	1 slice	47	84	28	Tr	Tr	Tr	Tr	Tr
355	Pineapple juice, unsweetened, canned	1 cup	250	86	140	1	Tr	Tr	Tr	0.1
	Plantain, without peel									
356	Raw	1 medium	179	65	218	2	1	0.3	0.1	0.1
357	Cooked, slices	1 cup	154	67	179	1	Tr	0.1	Tr	0.1
	Plums									
358	Raw (2⅛" dia)	1 plum	66	85	36	1	Tr	Tr	0.3	0.1
	Canned, purple, fruit and liquid									
359	Heavy syrup pack	1 cup	258	76	230	1	Tr	Tr	0.2	0.1
360		1 plum	46	76	41	Tr	Tr	Tr	Tr	Tr
361	Juice pack	1 cup	252	84	146	1	Tr	Tr	Tr	Tr
362		1 plum	46	84	27	Tr	Tr	Tr	Tr	Tr
	Prunes, dried, pitted									
363	Uncooked	5 prunes	42	32	100	1	Tr	Tr	0.1	Tr
364	Stewed, unsweetened, fruit and liquid	1 cup	248	70	265	3	1	Tr	0.4	0.1
365	Prune juice, canned or bottled	1 cup	256	81	182	2	Tr	Tr	0.1	Tr
	Raisins, seedless									
366	Cup, not packed	1 cup	145	15	435	5	1	0.2	Tr	0.2
367	Packet, ½ oz (1½ tbsp)	1 packet	14	15	42	Tr	Tr	Tr	Tr	Tr
	Raspberries									
368	Raw	1 cup	123	87	60	1	1	Tr	0.1	0.4
369	Frozen, sweetened, thawed	1 cup	250	73	258	2	Tr	Tr	Tr	0.2
370	Rhubarb, frozen, cooked, with sugar	1 cup	240	68	278	1	Tr	Tr	Tr	0.1

Choles-terol (mg)	Carbo-hydrate (g)	Total dietary fiber (g)	Calcium (mg)	Iron (mg)	Potas-sium (mg)	Sodium (mg)	Vitamin A		Thiamin (mg)	Ribo-flavin (mg)	Niacin (mg)	Ascor-bic acid (mg)	Food No.
							(IU)	(RE)					
0	11	2.0	5	0.1	193	0	524	53	0.02	0.04	1.0	6	337
0	19	3.4	9	0.2	335	0	910	92	0.03	0.07	1.7	11	338
0	52	3.4	8	0.7	241	16	870	86	0.03	0.06	1.6	7	339
0	20	1.3	3	0.3	90	6	325	32	0.01	0.02	0.6	3	340
0	29	3.2	15	0.7	317	10	945	94	0.02	0.04	1.4	9	341
0	11	1.3	6	0.3	125	4	373	37	0.01	0.02	0.6	4	342
0	24	3.2	11	1.6	388	3	844	84	Tr	0.08	1.7	2	343
0	60	4.5	8	0.9	325	15	710	70	0.03	0.09	1.6	236	344
0	25	4.0	18	0.4	208	0	33	3	0.03	0.07	0.2	7	345
0	51	4.3	13	0.6	173	13	0	0	0.03	0.06	0.6	3	346
0	15	1.2	4	0.2	49	4	0	0	0.01	0.02	0.2	1	347
0	32	4.0	22	0.7	238	10	15	2	0.03	0.03	0.5	4	348
0	10	1.2	7	0.2	73	3	5	1	0.01	0.01	0.2	1	349
0	19	1.9	11	0.6	175	2	36	3	0.14	0.06	0.7	24	350
0	51	2.0	36	1.0	264	3	36	3	0.23	0.06	0.7	19	351
0	10	0.4	7	0.2	51	Tr	7	Tr	0.04	0.01	0.1	4	352
0	39	2.0	35	0.7	304	2	95	10	0.24	0.05	0.7	24	353
0	7	0.4	7	0.1	57	Tr	18	2	0.04	0.01	0.1	4	354
0	34	0.5	43	0.7	335	3	13	0	0.14	0.06	0.6	27	355
0	57	4.1	5	1.1	893	7	2,017	202	0.09	0.10	1.2	33	356
0	48	3.5	3	0.9	716	8	1,400	140	0.07	0.08	1.2	17	357
0	9	1.0	3	0.1	114	0	213	21	0.03	0.06	0.3	6	358
0	60	2.6	23	2.2	235	49	668	67	0.04	0.10	0.8	1	359
0	11	0.5	4	0.4	42	9	119	12	0.01	0.02	0.1	Tr	360
0	38	2.5	25	0.9	388	3	2,543	255	0.06	0.15	1.2	7	361
0	7	0.5	5	0.2	71	Tr	464	46	0.01	0.03	0.2	1	362
0	26	3.0	21	1.0	313	2	835	84	0.03	0.07	0.8	1	363
0	70	16.4	57	2.8	828	5	759	77	0.06	0.25	1.8	7	364
0	45	2.6	31	3.0	707	10	8	0	0.04	0.18	2.0	10	365
0	115	5.8	71	3.0	1,089	17	12	1	0.23	0.13	1.2	5	366
0	11	0.6	7	0.3	105	2	1	Tr	0.02	0.01	0.1	Tr	367
0	14	8.4	27	0.7	187	0	160	16	0.04	0.11	1.1	31	368
0	65	11.0	38	1.6	285	3	150	15	0.05	0.11	0.6	41	369
0	75	4.8	348	0.5	230	2	166	17	0.04	0.06	0.5	8	370

Food No.	Food Description	Measure of edible portion	Weight (g)	Water (%)	Calories (kcal)	Protein (g)	Total fat (g)	Fatty acids: Saturated (g)	Fatty acids: Monounsaturated (g)	Fatty acids: Polyunsaturated (g)
	Fruits and Fruit Juices (continued)									
	Strawberries									
	Raw, capped									
371	Large (1⅛" dia)	1 strawberry	18	92	5	Tr	Tr	Tr	Tr	Tr
372	Medium (1¼" dia)	1 strawberry	12	92	4	Tr	Tr	Tr	Tr	Tr
373	Sliced	1 cup	166	92	50	1	1	Tr	0.1	0.3
374	Frozen, sweetened, sliced, thawed	1 cup	255	73	245	1	Tr	Tr	Tr	0.2
	Tangerines									
375	Raw, without peel and seeds (2⅜" dia)	1 tangerine	84	88	37	1	Tr	Tr	Tr	Tr
376	Canned (mandarin oranges), light syrup, fruit and liquid	1 cup	252	83	154	1	Tr	Tr	Tr	0.1
377	Tangerine juice, canned, sweetened	1 cup	249	87	125	1	Tr	Tr	Tr	0.1
	Watermelon, raw (15" long x 7½" dia)									
378	Wedge (about 1/16 of melon)	1 wedge	286	92	92	2	1	0.1	0.3	0.4
379	Diced	1 cup	152	92	49	1	1	0.1	0.2	0.2
	Grain Products									
	Bagels, enriched									
380	Plain	3½" bagel	71	33	195	7	1	0.2	0.1	0.5
381		4" bagel	89	33	245	9	1	0.2	0.1	0.6
382	Cinnamon raisin	3½" bagel	71	32	195	7	1	0.2	0.1	0.5
383		4" bagel	89	32	244	9	2	0.2	0.2	0.6
384	Egg	3½" bagel	71	33	197	8	1	0.3	0.3	0.5
385		4" bagel	89	33	247	9	2	0.4	0.4	0.6
386	Banana bread, prepared from recipe, with margarine	1 slice	60	29	196	3	6	1.3	2.7	1.9
	Barley, pearled									
387	Uncooked	1 cup	200	10	704	20	2	0.5	0.3	1.1
388	Cooked	1 cup	157	69	193	4	1	0.1	0.1	0.3
	Biscuits, plain or buttermilk, enriched									
389	Prepared from recipe, with 2% milk	2½" biscuit	60	29	212	4	10	2.6	4.2	2.5
390		4" biscuit	101	29	358	7	16	4.4	7.0	4.2
	Refrigerated dough, baked									
391	Regular	2½" biscuit	27	28	93	2	4	1.0	2.2	0.5
392	Lower fat	2¼" biscuit	21	28	63	2	1	0.3	0.6	0.2
	Breads, enriched									
393	Cracked wheat	1 slice	25	36	65	2	1	0.2	0.5	0.2
394	Egg bread (challah)	½" slice	40	35	115	4	2	0.6	0.9	0.4
395	French or vienna (includes sourdough)	½" slice	25	34	69	2	1	0.2	0.3	0.2
396	Indian fry (navajo) bread	5" bread	90	27	296	6	9	2.1	3.6	2.3
397		10½" bread	160	27	526	11	15	3.7	6.4	4.1
398	Italian	1 slice	20	36	54	2	1	0.2	0.2	0.3
	Mixed grain									
399	Untoasted	1 slice	26	38	65	3	1	0.2	0.4	0.2
400	Toasted	1 slice	24	32	65	3	1	0.2	0.4	0.2
	Oatmeal									
401	Untoasted	1 slice	27	37	73	2	1	0.2	0.4	0.5
402	Toasted	1 slice	25	31	73	2	1	0.2	0.4	0.5
403	Pita	4" pita	28	32	77	3	Tr	Tr	Tr	0.1
404		6½" pita	60	32	165	5	1	0.1	0.1	0.3

Cholesterol (mg)	Carbohydrate (g)	Total dietary fiber (g)	Calcium (mg)	Iron (mg)	Potassium (mg)	Sodium (mg)	Vitamin A (IU)	Vitamin A (RE)	Thiamin (mg)	Riboflavin (mg)	Niacin (mg)	Ascorbic acid (mg)	Food No.
0	1	0.4	3	0.1	30	Tr	5	1	Tr	0.01	Tr	10	371
0	1	0.3	2	Tr	20	Tr	3	Tr	Tr	0.01	Tr	7	372
0	12	3.8	23	0.6	276	2	45	5	0.03	0.11	0.4	94	373
0	66	4.8	28	1.5	250	8	61	5	0.04	0.13	1.0	106	374
0	9	1.9	12	0.1	132	1	773	77	0.09	0.02	0.1	26	375
0	41	1.8	18	0.9	197	15	2,117	212	0.13	0.11	1.1	50	376
0	30	0.5	45	0.5	443	2	1,046	105	0.15	0.05	0.2	55	377
0	21	1.4	23	0.5	332	6	1,047	106	0.23	0.06	0.6	27	378
0	11	0.8	12	0.3	176	3	556	56	0.12	0.03	0.3	15	379
0	38	1.6	53	2.5	72	379	0	0	0.38	0.22	3.2	0	380
0	48	2.0	66	3.2	90	475	0	0	0.48	0.28	4.1	0	381
0	39	1.6	13	2.7	105	229	52	0	0.27	0.20	2.2	Tr	382
0	49	2.0	17	3.4	132	287	65	0	0.34	0.25	2.7	1	383
17	38	1.6	9	2.8	48	359	77	23	0.38	0.17	2.4	Tr	384
21	47	2.0	12	3.5	61	449	97	29	0.48	0.21	3.1	1	385
26	33	0.7	13	0.8	80	181	278	72	0.10	0.12	0.9	1	386
0	155	31.2	58	5.0	560	18	44	4	0.38	0.23	9.2	0	387
0	44	6.0	17	2.1	146	5	11	2	0.13	0.10	3.2	0	388
2	27	0.9	141	1.7	73	348	49	14	0.21	0.19	1.8	Tr	389
3	45	1.5	237	2.9	122	586	83	23	0.36	0.31	3.0	Tr	390
0	13	0.4	5	0.7	42	325	0	0	0.09	0.06	0.8	0	391
0	12	0.4	4	0.6	39	305	0	0	0.09	0.05	0.7	0	392
0	12	1.4	11	0.7	44	135	0	0	0.09	0.06	0.9	0	393
20	19	0.9	37	1.2	46	197	30	9	0.18	0.17	1.9	0	394
0	13	0.8	19	0.6	28	152	0	0	0.13	0.08	1.2	0	395
0	48	1.6	210	3.2	67	626	0	0	0.39	0.27	3.3	0	396
0	85	2.9	373	5.8	118	1,112	0	0	0.69	0.49	5.8	0	397
0	10	0.5	16	0.6	22	117	0	0	0.09	0.06	0.9	0	398
0	12	1.7	24	0.9	53	127	0	0	0.11	0.09	1.1	Tr	399
0	12	1.6	24	0.9	53	127	0	0	0.08	0.08	1.0	Tr	400
0	13	1.1	18	0.7	38	162	4	1	0.11	0.06	0.8	0	401
0	13	1.1	18	0.7	39	163	4	1	0.09	0.06	0.8	Tr	402
0	16	0.6	24	0.7	34	150	0	0	0.17	0.09	1.3	0	403
0	33	1.3	52	1.6	72	322	0	0	0.36	0.20	2.8	0	404

Food No.	Food Description	Measure of edible portion	Weight (g)	Water (%)	Calories (kcal)	Protein (g)	Total fat (g)	Fatty acids Saturated (g)	Fatty acids Monounsaturated (g)	Fatty acids Polyunsaturated (g)

Grain Products (continued)

Food No.	Food Description	Measure of edible portion	Weight (g)	Water (%)	Calories (kcal)	Protein (g)	Total fat (g)	Saturated (g)	Monounsaturated (g)	Polyunsaturated (g)
	Breads, enriched (continued)									
	Pumpernickel									
405	Untoasted	1 slice	32	38	80	3	1	0.1	0.3	0.4
406	Toasted	1 slice	29	32	80	3	1	0.1	0.3	0.4
	Raisin									
407	Untoasted	1 slice	26	34	71	2	1	0.3	0.6	0.2
408	Toasted	1 slice	24	28	71	2	1	0.3	0.6	0.2
	Rye									
409	Untoasted	1 slice	32	37	83	3	1	0.2	0.4	0.3
410	Toasted	1 slice	24	31	68	2	1	0.2	0.3	0.2
411	Rye, reduced calorie	1 slice	23	46	47	2	1	0.1	0.2	0.2
	Wheat									
412	Untoasted	1 slice	25	37	65	2	1	0.2	0.4	0.2
413	Toasted	1 slice	23	32	65	2	1	0.2	0.4	0.2
414	Wheat, reduced calorie	1 slice	23	43	46	2	1	0.1	0.1	0.2
	White									
415	Untoasted	1 slice	25	37	67	2	1	0.1	0.2	0.5
416	Toasted	1 slice	22	30	64	2	1	0.1	0.2	0.5
417	Soft crumbs	1 cup	45	37	120	4	2	0.2	0.3	0.9
418	White, reduced calorie	1 slice	23	43	48	2	1	0.1	0.2	0.1
	Bread, whole wheat									
419	Untoasted	1 slice	28	38	69	3	1	0.3	0.5	0.3
420	Toasted	1 slice	25	30	69	3	1	0.3	0.5	0.3
	Bread crumbs, dry, grated									
421	Plain, enriched	1 cup	108	6	427	14	6	1.3	2.6	1.2
422		1 oz	28	6	112	4	2	0.3	0.7	0.3
423	Seasoned, unenriched	1 cup	120	6	440	17	3	0.9	1.2	0.8
	Bread crumbs, soft. See White bread.									
424	Bread stuffing, prepared from dry mix	½ cup	100	65	178	3	9	1.7	3.8	2.6
425	Breakfast bar, cereal crust withfruit filling, fat free	1 bar	37	14	121	2	Tr	Tr	Tr	0.1
	Breakfast Cereals									
	Hot type, cooked									
	Corn (hominy) grits									
	Regular or quick, enriched									
426	White	1 cup	242	85	145	3	Tr	0.1	0.1	0.2
427	Yellow	1 cup	242	85	145	3	Tr	0.1	0.1	0.2
428	Instant, plain	1 packet	137	82	89	2	Tr	Tr	Tr	0.1
	CREAM OF WHEAT									
429	Regular	1 cup	251	87	133	4	1	0.1	0.1	0.3
430	Quick	1 cup	239	87	129	4	Tr	0.1	0.1	0.3
431	Mix'n Eat, plain	1 packet	142	82	102	3	Tr	Tr	Tr	0.2
432	MALT O MEAL	1 cup	240	88	122	4	Tr	0.1	0.1	Tr
	Oatmeal									
433	Regular, quick or instant, plain, nonfortified	1 cup	234	85	145	6	2	0.4	0.7	0.9
434	Instant, fortified, plain	1 packet	177	86	104	4	2	0.3	0.6	0.7
	QUAKER instant									
435	Apples and cinnamon	1 packet	149	79	125	3	1	0.3	0.5	0.6
436	Maple and brown sugar	1 packet	155	75	153	4	2	0.4	0.6	0.7
437	WHEATENA	1 cup	243	85	136	5	1	0.2	0.2	0.6
	Ready to eat									
438	ALL BRAN	½ cup	30	3	79	4	1	0.2	0.2	0.5
439	APPLE CINNAMON CHEERIOS	¾ cup	30	3	118	2	2	0.3	0.6	0.2
440	APPLE JACKS	1 cup	30	3	116	1	Tr	0.1	0.1	0.2

Cholesterol (mg)	Carbohydrate (g)	Total dietary fiber (g)	Calcium (mg)	Iron (mg)	Potassium (mg)	Sodium (mg)	Vitamin A (IU)	Vitamin A (RE)	Thiamin (mg)	Riboflavin (mg)	Niacin (mg)	Ascorbic acid (mg)	Food No.
0	15	2.1	22	0.9	67	215	0	0	0.10	0.10	1.0	0	405
0	15	2.1	21	0.9	66	214	0	0	0.08	0.09	0.9	0	406
0	14	1.1	17	0.8	59	101	0	0	0.09	0.10	0.9	Tr	407
0	14	1.1	17	0.8	59	102	Tr	0	0.07	0.09	0.8	Tr	408
0	15	1.9	23	0.9	53	211	2	Tr	0.14	0.11	1.2	Tr	409
0	13	1.5	19	0.7	44	174	1	0	0.09	0.08	0.9	Tr	410
0	9	2.8	17	0.7	23	93	1	0	0.08	0.06	0.6	Tr	411
0	12	1.1	26	0.8	50	133	0	0	0.10	0.07	1.0	0	412
0	12	1.2	26	0.8	50	132	0	0	0.08	0.06	0.9	0	413
0	10	2.8	18	0.7	28	118	0	0	0.10	0.07	0.9	Tr	414
Tr	12	0.6	27	0.8	30	135	0	0	0.12	0.09	1.0	0	415
Tr	12	0.6	26	0.7	29	130	0	0	0.09	0.07	0.9	0	416
Tr	22	1.0	49	1.4	54	242	0	0	0.21	0.15	1.8	0	417
0	10	2.2	22	0.7	17	104	1	Tr	0.09	0.07	0.8	Tr	418
0	13	1.9	20	0.9	71	148	0	0	0.10	0.06	1.1	0	419
0	13	1.9	20	0.9	71	148	0	0	0.08	0.05	1.0	0	420
0	78	2.6	245	6.6	239	931	1	0	0.83	0.47	7.4	0	421
0	21	0.7	64	1.7	63	244	Tr	0	0.22	0.12	1.9	0	422
1	84	5.0	119	3.8	324	3,180	16	4	0.19	0.20	3.3	Tr	423
0	22	2.9	32	1.1	74	543	313	81	0.14	0.11	1.5	0	424
Tr	28	0.8	49	4.5	92	203	1,249	125	1.01	0.42	5.0	1	425
0	31	0.5	0	1.5	53	0	0	0	0.24	0.15	2.0	0	426
0	31	0.5	0	1.5	53	0	145	15	0.24	0.15	2.0	0	427
0	21	1.2	8	8.2	38	289	0	0	0.15	0.08	1.4	0	428
0	28	1.8	50	10.3	43	3	0	0	0.25	0.00	1.5	0	429
0	27	1.2	50	10.3	45	139	0	0	0.24	0.00	1.4	0	430
0	21	0.4	20	8.1	38	241	1,252	376	0.43	0.28	5.0	0	431
0	26	1.0	5	9.6	31	2	0	0	0.48	0.24	5.8	0	432
0	25	4.0	19	1.6	131	2	37	5	0.26	0.05	0.3	0	433
0	18	3.0	163	6.3	99	285	1,510	453	0.53	0.28	5.5	0	434
0	26	2.5	104	3.9	106	121	1,019	305	0.30	0.35	4.1	Tr	435
0	31	2.6	105	3.9	112	234	1,008	302	0.30	0.34	4.0	0	436
0	29	6.6	10	1.4	187	5	0	0	0.02	0.05	1.3	0	437
0	23	9.7	106	4.5	342	61	750	225	0.39	0.42	5.0	15	438
0	25	1.6	35	4.5	60	150	750	225	0.38	0.43	5.0	15	439
0	27	0.6	3	4.5	32	134	750	225	0.39	0.42	5.0	15	440

Food No.	Food Description	Measure of edible portion	Weight (g)	Water (%)	Calories (kcal)	Protein (g)	Total fat (g)	Fatty acids Saturated (g)	Monounsaturated (g)	Polyunsaturated (g)

Grain Products (continued)

Food No.	Food Description	Measure of edible portion	Weight (g)	Water (%)	Calories (kcal)	Protein (g)	Total fat (g)	Saturated (g)	Monounsaturated (g)	Polyunsaturated (g)
	Breakfast Cereals (continued)									
	Ready to eat (continued)									
441	BASIC 4	1 cup	55	7	201	4	3	0.4	1.0	1.1
442	BERRY BERRY KIX	¾ cup	30	2	120	1	1	0.2	0.5	0.1
443	CAP'N CRUNCH	¾ cup	27	2	107	1	1	0.4	0.3	0.2
444	CAP'N CRUNCH'S CRUNCHBERRIES	¾ cup	26	2	104	1	1	0.3	0.3	0.2
445	CAP'N CRUNCH'S PEANUT BUTTER CRUNCH	¾ cup	27	2	112	2	2	0.5	0.8	0.5
446	CHEERIOS	1 cup	30	3	110	3	2	0.4	0.6	0.2
	CHEX									
447	Corn	1 cup	30	3	113	2	Tr	0.1	0.1	0.2
448	Honey nut	¾ cup	30	2	117	2	1	0.1	0.4	0.2
449	Multi bran	1 cup	49	3	165	4	1	0.2	0.3	0.5
450	Rice	1¼ cup	31	3	117	2	Tr	Tr	Tr	Tr
451	Wheat	1 cup	30	3	104	3	1	0.1	0.1	0.3
452	CINNAMON LIFE	1 cup	50	4	190	4	2	0.3	0.6	0.8
453	CINNAMON TOAST CRUNCH	¾ cup	30	2	124	2	3	0.5	0.9	0.5
454	COCOA KRISPIES	¾ cup	31	2	120	2	1	0.6	0.1	0.1
455	COCOA PUFFS	1 cup	30	2	119	1	1	0.2	0.3	Tr
	Corn Flakes									
456	GENERAL MILLS, TOTAL	1⅓ cup	30	3	112	2	Tr	0.2	0.1	Tr
457	KELLOGG'S	1 cup	28	3	102	2	Tr	0.1	Tr	0.1
458	CORN POPS	1 cup	31	3	118	1	Tr	0.1	0.1	Tr
459	CRISPIX	1 cup	29	3	108	2	Tr	0.1	0.1	0.1
460	Complete Wheat Bran Flakes	¾ cup	29	4	95	3	1	0.1	0.1	0.4
461	FROOT LOOPS	1 cup	30	2	117	1	1	0.4	0.2	0.3
462	FROSTED FLAKES	¾ cup	31	3	119	1	Tr	0.1	Tr	0.1
	FROSTED MINI WHEATS									
463	Regular	1 cup	51	5	173	5	1	0.2	0.1	0.6
464	Bite size	1 cup	55	5	187	5	1	0.2	0.2	0.6
465	GOLDEN GRAHAMS	¾ cup	30	3	116	2	1	0.2	0.3	0.2
466	HONEY FROSTED WHEATIES	¾ cup	30	3	110	2	Tr	0.1	Tr	Tr
467	HONEY NUT CHEERIOS	1 cup	30	2	115	3	1	0.2	0.5	0.2
468	HONEY NUT CLUSTERS	1 cup	55	3	213	5	3	0.4	1.8	0.4
469	KIX	1⅓ cup	30	2	114	2	1	0.2	0.1	Tr
470	LIFE	¾ cup	32	4	121	3	1	0.2	0.4	0.6
471	LUCKY CHARMS	1 cup	30	2	116	2	1	0.2	0.4	0.2
472	NATURE VALLEY Granola	¾ cup	55	4	248	6	10	1.3	6.5	1.9
	100% Natural Cereal									
473	With oats, honey, and raisins	½ cup	51	4	218	5	7	3.2	3.2	0.8
474	With raisins, low fat	½ cup	50	4	195	4	3	0.8	1.3	0.5
475	PRODUCT 19	1 cup	30	3	110	3	Tr	Tr	0.2	0.2
476	Puffed Rice	1 cup	14	3	56	1	Tr	Tr	Tr	Tr
477	Puffed Wheat	1 cup	12	3	44	2	Tr	Tr	Tr	Tr
	Raisin Bran									
478	GENERAL MILLS, TOTAL	1 cup	55	9	178	4	1	0.2	0.2	0.2
479	KELLOGG'S	1 cup	61	8	186	6	1	0.0	0.2	0.8
480	RAISIN NUT BRAN	1 cup	55	5	209	5	4	0.7	1.9	0.5
481	REESE'S PEANUT BUTTER PUFFS	¾ cup	30	2	129	3	3	0.6	1.4	0.6
482	RICE KRISPIES	1¼ cup	33	3	124	2	Tr	0.1	0.1	0.2

Cholesterol (mg)	Carbohydrate (g)	Total dietary fiber (g)	Calcium (mg)	Iron (mg)	Potassium (mg)	Sodium (mg)	Vitamin A (IU)	Vitamin A (RE)	Thiamin (mg)	Riboflavin (mg)	Niacin (mg)	Ascorbic acid (mg)	Food No.
0	42	3.4	310	4.5	162	323	1,250	375	0.37	0.42	5.0	15	441
0	26	0.2	66	4.5	24	185	750	225	0.38	0.43	5.0	15	442
0	23	0.9	5	4.5	35	208	36	4	0.38	0.42	5.0	0	443
0	22	0.6	7	4.5	37	190	33	5	0.37	0.42	5.0	Tr	444
0	22	0.8	3	4.5	62	204	37	4	0.38	0.42	5.0	0	445
0	23	2.6	55	8.1	89	284	1,250	375	0.38	0.43	5.0	15	446
0	26	0.5	100	9.0	32	289	0	0	0.38	0.00	5.0	6	447
0	26	0.4	102	9.0	27	224	0	0	0.38	0.44	5.0	6	448
0	41	6.4	95	13.7	191	325	0	0	0.32	0.00	4.4	5	449
0	27	0.3	104	9.0	36	291	0	0	0.38	0.02	5.0	6	450
0	24	3.3	60	9.0	116	269	0	0	0.23	0.04	3.0	4	451
0	40	3.0	135	7.5	113	220	16	2	0.63	0.71	8.4	Tr	452
0	24	1.5	42	4.5	44	210	750	225	0.38	0.43	5.0	15	453
0	27	0.4	4	1.8	60	210	750	225	0.37	0.43	5.0	15	454
0	27	0.2	33	4.5	52	181	0	0	0.38	0.43	5.0	15	455
0	26	0.8	237	18.0	34	203	1,250	375	1.50	1.70	20.1	60	456
0	24	0.8	1	8.7	25	298	700	210	0.36	0.39	4.7	14	457
0	28	0.4	2	1.9	23	123	775	233	0.40	0.43	5.2	16	458
0	25	0.6	3	1.8	35	240	750	225	0.38	0.44	5.0	15	459
0	23	4.6	14	8.1	175	226	1,208	363	0.38	0.44	5.0	15	460
0	26	0.6	3	4.2	32	141	703	211	0.39	0.42	5.0	14	461
0	28	0.6	1	4.5	20	200	750	225	0.37	0.43	5.0	15	462
0	42	5.5	18	14.3	170	2	0	0	0.36	0.41	5.0	0	463
0	45	5.9	0	15.4	186	2	0	0	0.33	0.39	4.7	0	464
0	26	0.9	14	4.5	53	275	750	225	0.38	0.43	5.0	15	465
0	26	1.5	8	4.5	56	211	750	225	0.38	0.43	5.0	15	466
0	24	1.6	20	4.5	85	259	750	225	0.38	0.43	5.0	15	467
0	43	4.2	72	4.5	171	239	0	0	0.37	0.42	5.0	9	468
0	26	0.8	44	8.1	41	263	1,250	375	0.38	0.43	5.0	15	469
0	25	2.0	98	9.0	79	174	12	1	0.40	0.45	5.3	0	470
0	25	1.2	32	4.5	54	203	750	225	0.38	0.43	5.0	15	471
0	36	3.5	41	1.7	183	89	0	0	0.17	0.06	0.6	0	472
1	36	3.7	39	1.7	214	11	4	1	0.14	0.09	0.8	Tr	473
1	40	3.0	30	1.3	169	129	9	1	0.15	0.06	0.9	Tr	474
0	25	1.0	3	18.0	41	216	750	225	1.50	1.71	20.0	60	475
0	13	0.2	1	4.4	16	Tr	0	0	0.36	0.25	4.9	0	476
0	10	0.5	3	3.8	42	Tr	0	0	0.31	0.22	4.2	0	477
0	43	5.0	238	18.0	287	240	1,250	375	1.50	1.70	20.0	0	478
0	47	8.2	35	5.0	437	354	832	250	0.43	0.49	5.6	0	479
0	41	5.1	74	4.5	218	246	0	0	0.37	0.42	5.0	0	480
0	23	0.4	21	4.5	62	177	750	225	0.38	0.43	5.0	15	481
0	29	0.4	3	2.0	42	354	825	248	0.43	0.46	5.5	17	482

Food No.	Food Description	Measure of edible portion	Weight (g)	Water (%)	Calories (kcal)	Protein (g)	Total fat (g)	Fatty acids: Saturated (g)	Fatty acids: Monounsaturated (g)	Fatty acids: Polyunsaturated (g)

Grain Products (continued)

Food No.	Food Description	Measure of edible portion	Weight (g)	Water (%)	Calories (kcal)	Protein (g)	Total fat (g)	Saturated (g)	Monounsaturated (g)	Polyunsaturated (g)
	Breakfast Cereals (continued)									
	Ready to eat (continued)									
483	RICE KRISPIES TREATS cereal	¾ cup	30	4	120	1	2	0.4	1.0	0.2
484	SHREDDED WHEAT	2 biscuits	46	4	156	5	1	0.1	NA	NA
485	SMACKS	¾ cup	27	3	103	2	1	0.3	0.1	0.2
486	SPECIAL K	1 cup	31	3	115	6	Tr	0.0	0.0	0.2
487	QUAKER Toasted Oatmeal, Honey Nut	1 cup	49	3	191	5	3	0.5	1.2	0.7
488	TOTAL, Whole Grain	¾ cup	30	3	105	3	1	0.2	0.1	0.1
489	TRIX	1 cup	30	2	122	1	2	0.4	0.9	0.3
490	WHEATIES	1 cup	30	3	110	3	1	0.2	0.2	0.2
	Brownies, without icing									
	Commercially prepared									
491	Regular, large (2¾" sq x ⅞")	1 brownie	56	14	227	3	9	2.4	5.0	1.3
492	Fat free, 2" sq	1 brownie	28	12	89	1	Tr	0.2	0.1	Tr
493	Prepared from dry mix, reduced calorie, 2" sq	1 brownie	22	13	84	1	2	1.1	1.0	0.2
494	Buckwheat flour, whole groat	1 cup	120	11	402	15	4	0.8	1.1	1.1
495	Buckwheat groats, roasted (kasha), cooked	1 cup	168	76	155	6	1	0.2	0.3	0.3
	Bulgur									
496	Uncooked	1 cup	140	9	479	17	2	0.3	0.2	0.8
497	Cooked	1 cup	182	78	151	6	Tr	0.1	0.1	0.2
	Cakes, prepared from dry mix									
498	Angelfood (1/12 of 10" dia)	1 piece	50	33	129	3	Tr	Tr	Tr	0.1
499	Yellow, light, with water, egg whites, no frosting (1/12 of 9" dia)	1 piece	69	37	181	3	2	1.1	0.9	0.2
	Cakes, prepared from recipe									
500	Chocolate, without frosting (1/12 of 9" dia)	1 piece	95	24	340	5	14	5.2	5.7	2.6
501	Gingerbread (1/9 of 8" square)	1 piece	74	28	263	3	12	3.1	5.3	3.1
502	Pineapple upside down (1/9 of 8" square)	1 piece	115	32	367	4	14	3.4	6.0	3.8
503	Shortcake, biscuit type (about 3" dia)	1 shortcake	65	28	225	4	9	2.5	3.9	2.4
504	Sponge (1/12 of 16-oz cake)	1 piece	63	29	187	5	3	0.8	1.0	0.4
	White									
505	With coconut frosting (1/12 of 9" dia)	1 piece	112	21	399	5	12	4.4	4.1	2.4
506	Without frosting (1/12 of 9" dia)	1 piece	74	23	264	4	9	2.4	3.9	2.3
	Cakes, commercially prepared									
507	Angelfood (1/12 of 12-oz cake)	1 piece	28	33	72	2	Tr	Tr	Tr	0.1
508	Boston cream (⅙ of pie)	1 piece	92	45	232	2	8	2.2	4.2	0.9
509	Chocolate with chocolate frosting (⅛ of 18-oz cake)	1 piece	64	23	235	3	10	3.1	5.6	1.2
510	Coffeecake, crumb (1/9 of 20-oz cake)	1 piece	63	22	263	4	15	3.7	8.2	2.0
511	Fruitcake	1 piece	43	25	139	1	4	0.5	1.8	1.4
	Pound									
512	Butter (1/12 of 12-oz cake)	1 piece	28	25	109	2	6	3.2	1.7	0.3
513	Fat free (3¼" x 2¾" x ⅝" slice)	1 slice	28	31	79	2	Tr	0.1	Tr	0.1

Choles-terol (mg)	Carbo-hydrate (g)	Total dietary fiber (g)	Calcium (mg)	Iron (mg)	Potas-sium (mg)	Sodium (mg)	Vitamin A (IU)	Vitamin A (RE)	Thiamin (mg)	Ribo-flavin (mg)	Niacin (mg)	Ascor-bic acid (mg)	Food No.
0	26	0.3	2	1.8	19	190	750	225	0.39	0.42	5.0	15	483
0	38	5.3	20	1.4	196	3	0	NA	0.12	0.05	2.6	0	484
0	24	0.9	3	1.8	42	51	750	225	0.38	0.43	5.0	15	485
0	22	1.0	5	8.7	55	250	750	225	0.53	0.59	7.0	15	486
Tr	39	3.3	27	4.5	185	166	500	150	0.37	0.42	5.0	6	487
0	24	2.6	258	18.0	97	199	1,250	375	1.50	1.70	20.1	60	488
0	26	0.7	32	4.5	18	197	750	225	0.38	0.43	5.0	15	489
0	24	2.1	55	8.1	104	222	750	225	0.38	0.43	5.0	15	490
10	36	1.2	16	1.3	83	175	39	3	0.14	0.12	1.0	0	491
0	22	1.0	17	0.7	89	90	1	Tr	0.03	0.04	0.3	Tr	492
0	16	0.8	3	0.3	69	21	0	0	0.02	0.03	0.2	0	493
0	85	12.0	49	4.9	692	13	0	0	0.50	0.23	7.4	0	494
0	33	4.5	12	1.3	148	7	0	0	0.07	0.07	1.6	0	495
0	106	25.6	49	3.4	574	24	0	0	0.32	0.16	7.2	0	496
0	34	8.2	18	1.7	124	9	0	0	0.10	0.05	1.8	0	497
0	29	0.1	42	0.1	68	255	0	0	0.05	0.10	0.1	0	498
0	37	0.6	69	0.6	41	279	6	1	0.06	0.12	0.6	0	499
55	51	1.5	57	1.5	133	299	133	38	0.13	0.20	1.1	Tr	500
24	36	0.7	53	2.1	325	242	36	10	0.14	0.12	1.3	Tr	501
25	58	0.9	138	1.7	129	367	291	75	0.18	0.18	1.4	1	502
2	32	0.8	133	1.7	69	329	47	12	0.20	0.18	1.7	Tr	503
107	36	0.4	26	1.0	89	144	163	49	0.10	0.19	0.8	0	504
1	71	1.1	101	1.3	111	318	43	12	0.14	0.21	1.2	Tr	505
1	42	0.6	96	1.1	70	242	41	12	0.14	0.18	1.1	Tr	506
0	16	0.4	39	0.1	26	210	0	0	0.03	0.14	0.2	0	507
34	39	1.3	21	0.3	36	132	74	21	0.38	0.25	0.2	Tr	508
27	35	1.8	28	1.4	128	214	54	16	0.02	0.09	0.4	Tr	509
20	29	1.3	34	1.2	77	221	70	21	0.13	0.14	1.1	Tr	510
2	26	1.6	14	0.9	66	116	9	2	0.02	0.04	0.3	Tr	511
62	14	0.1	10	0.4	33	111	170	44	0.04	0.06	0.4	0	512
0	17	0.3	12	0.6	31	95	27	8	0.04	0.08	0.2	0	513

Food No.	Food Description	Measure of edible portion	Weight (g)	Water (%)	Calories (kcal)	Protein (g)	Total fat (g)	Fatty acids Saturated (g)	Fatty acids Monounsaturated (g)	Fatty acids Polyunsaturated (g)

Grain Products (continued)

Food No.	Food Description	Measure of edible portion	Weight (g)	Water (%)	Calories (kcal)	Protein (g)	Total fat (g)	Saturated (g)	Monounsaturated (g)	Polyunsaturated (g)
	Cakes, commercially prepared (continued)									
	Snack cakes									
514	Chocolate, creme filled, with frosting	1 cupcake	50	20	188	2	7	1.4	2.8	2.6
515	Chocolate, with frosting, low fat	1 cupcake	43	23	131	2	2	0.5	0.8	0.2
516	Sponge, creme filled	1 cake	43	20	155	1	5	1.1	1.7	1.4
517	Sponge, individual shortcake	1 shortcake	30	30	87	2	1	0.2	0.3	0.1
	Yellow									
518	With chocolate frosting	1 piece	64	22	243	2	11	3.0	6.1	1.4
519	With vanilla frosting	1 piece	64	22	239	2	9	1.5	3.9	3.3
520	Cheesecake (1/6 of 17-oz cake)	1 piece	80	46	257	4	18	7.9	6.9	1.3
521	Cheese flavor puffs or twists	1 oz	28	2	157	2	10	1.9	5.7	1.3
522	CHEX mix	1 oz (about 2/3 cup)	28	4	120	3	5	1.6	NA	NA
	Cookies									
523	Butter, commercially prepared	1 cookie	5	5	23	Tr	1	0.6	0.3	Tr
	Chocolate chip, medium (2¼"-2½" dia)									
	Commercially prepared									
524	Regular	1 cookie	10	4	48	1	2	0.7	1.2	0.2
525	Reduced fat	1 cookie	10	4	45	1	2	0.4	0.6	0.5
526	From refrigerated dough (spooned from roll)	1 cookie	26	3	128	1	6	2.0	2.9	0.6
527	Prepared from recipe, with margarine	1 cookie	16	6	78	1	5	1.3	1.7	1.3
528	Devil's food, commercially prepared, fat free	1 cookie	16	18	49	1	Tr	0.1	Tr	Tr
529	Fig bar	1 cookie	16	17	56	1	1	0.2	0.5	0.4
	Molasses									
530	Medium	1 cookie	15	6	65	1	2	0.5	1.1	0.3
531	Large (3½"-4" dia)	1 cookie	32	6	138	2	4	1.0	2.3	0.6
	Oatmeal									
	Commercially prepared, with or without raisins									
532	Regular, large	1 cookie	25	6	113	2	5	1.1	2.5	0.6
533	Soft type	1 cookie	15	11	61	1	2	0.5	1.2	0.3
534	Fat free	1 cookie	11	13	36	1	Tr	Tr	Tr	0.1
535	Prepared from recipe, with raisins (2⅝" dia)	1 cookie	15	6	65	1	2	0.5	1.0	0.8
	Peanut butter									
536	Commercially prepared	1 cookie	15	6	72	1	4	0.7	1.9	0.8
537	Prepared from recipe, with margarine (3" dia)	1 cookie	20	6	95	2	5	0.9	2.2	1.4
	Sandwich type, with creme filling									
538	Chocolate cookie	1 cookie	10	2	47	Tr	2	0.4	0.9	0.7
	Vanilla cookie									
539	Oval	1 cookie	15	2	72	1	3	0.4	1.3	1.1
540	Round	1 cookie	10	2	48	Tr	2	0.3	0.8	0.8
	Shortbread, commercially prepared									
541	Plain (1⅝" sq)	1 cookie	8	4	40	Tr	2	0.5	1.1	0.3
	Pecan									
542	Regular (2" dia)	1 cookie	14	3	76	1	5	1.1	2.6	0.6
543	Reduced fat	1 cookie	16	5	73	1	3	0.6	1.6	0.4

Cholesterol (mg)	Carbohydrate (g)	Total dietary fiber (g)	Calcium (mg)	Iron (mg)	Potassium (mg)	Sodium (mg)	Vitamin A (IU)	Vitamin A (RE)	Thiamin (mg)	Riboflavin (mg)	Niacin (mg)	Ascorbic acid (mg)	Food No.
9	30	0.4	37	1.7	61	213	9	3	0.11	0.15	1.2	0	514
0	29	1.8	15	0.7	96	178	0	0	0.02	0.06	0.3	0	515
7	27	0.2	19	0.5	37	155	7	2	0.07	0.06	0.5	Tr	516
31	18	0.2	21	0.8	30	73	46	14	0.07	0.08	0.6	0	517
35	35	1.2	24	1.3	114	216	70	21	0.08	0.10	0.8	0	518
35	38	0.2	40	0.7	34	220	40	12	0.06	0.04	0.3	0	519
44	20	0.3	41	0.5	72	166	438	117	0.02	0.15	0.2	Tr	520
1	15	0.3	16	0.7	47	298	75	10	0.07	0.10	0.9	Tr	521
0	18	1.6	10	7.0	76	288	41	4	0.44	0.14	4.8	13	522
6	3	Tr	1	0.1	6	18	34	8	0.02	0.02	0.2	0	523
0	7	0.3	3	0.3	14	32	Tr	0	0.02	0.03	0.3	0	524
0	7	0.4	2	0.3	12	38	Tr	0	0.03	0.03	0.3	0	525
7	18	0.4	7	0.7	52	60	15	4	0.04	0.05	0.5	0	526
5	9	0.4	6	0.4	36	58	102	26	0.03	0.03	0.2	Tr	527
0	12	0.3	5	0.4	18	28	Tr	NA	0.01	0.03	0.2	Tr	528
0	11	0.7	10	0.5	33	56	5	1	0.03	0.03	0.3	Tr	529
0	11	0.1	11	1.0	52	69	0	0	0.05	0.04	0.5	0	530
0	24	0.3	24	2.1	111	147	0	0	0.11	0.08	1.0	0	531
0	17	0.7	9	0.6	36	96	5	1	0.07	0.06	0.6	Tr	532
1	10	0.4	14	0.4	20	52	5	1	0.03	0.03	0.3	Tr	533
0	9	0.8	4	0.2	23	33	0	0	0.02	0.03	0.1	0	534
5	10	0.5	15	0.4	36	81	96	25	0.04	0.02	0.2	Tr	535
Tr	9	0.3	5	0.4	25	62	1	Tr	0.03	0.03	0.6	0	536
6	12	0.4	8	0.4	46	104	120	31	0.04	0.04	0.7	Tr	537
0	7	0.3	3	0.4	18	60	Tr	0	0.01	0.02	0.2	0	538
0	11	0.2	4	0.3	14	52	0	0	0.04	0.04	0.4	0	539
0	7	0.2	3	0.2	9	35	0	0	0.03	0.02	0.3	0	540
2	5	0.1	3	0.2	8	36	7	1	0.03	0.03	0.3	0	541
5	8	0.3	4	0.3	10	39	Tr	Tr	0.04	0.03	0.3	0	542
0	11	0.2	8	0.5	15	55	1	Tr	0.05	0.03	0.4	Tr	543

Food No.	Food Description	Measure of edible portion	Weight (g)	Water (%)	Calories (kcal)	Protein (g)	Total fat (g)	Fatty acids Saturated (g)	Monounsaturated (g)	Polyunsaturated (g)

Grain Products (continued)

Food No.	Food Description	Measure of edible portion	Weight (g)	Water (%)	Calories (kcal)	Protein (g)	Total fat (g)	Saturated (g)	Monounsaturated (g)	Polyunsaturated (g)
	Cookies (continued)									
	Sugar									
544	Commercially prepared	1 cookie	15	5	72	1	3	0.8	1.8	0.4
545	From refrigerated dough	1 cookie	15	5	73	1	3	0.9	2.0	0.4
546	Prepared from recipe, with margarine (3" dia)	1 cookie	14	9	66	1	3	0.7	1.4	1.0
547	Vanilla wafer, lower fat, medium size	1 cookie	4	5	18	Tr	1	0.2	0.3	0.2
	Corn chips									
548	Plain	1 oz	28	1	153	2	9	1.3	2.7	4.7
549	Barbecue flavor	1 oz	28	1	148	2	9	1.3	2.7	4.6
	Cornbread									
550	Prepared from mix, piece 3¾" x 2½" x ¾"	1 piece	60	32	188	4	6	1.6	3.1	0.7
551	Prepared from recipe, with 2% milk, piece 2½" sq x 1½"	1 piece	65	39	173	4	5	1.0	1.2	2.1
	Cornmeal, yellow, dry form									
552	Whole grain	1 cup	122	10	442	10	4	0.6	1.2	2.0
553	Degermed, enriched	1 cup	138	12	505	12	2	0.3	0.6	1.0
554	Self rising, degermed, enriched	1 cup	138	10	490	12	2	0.3	0.6	1.0
555	Cornstarch	1 tbsp	8	8	30	Tr	Tr	Tr	Tr	Tr
	Couscous									
556	Uncooked	1 cup	173	9	650	22	1	0.2	0.2	0.4
557	Cooked	1 cup	157	73	176	6	Tr	Tr	Tr	0.1
	Crackers									
558	Cheese, 1" sq	10 crackers	10	3	50	1	3	0.9	1.2	0.2
	Graham, plain									
559	2½" sq	2 squares	14	4	59	1	1	0.2	0.6	0.5
560	Crushed	1 cup	84	4	355	6	8	1.3	3.4	3.2
561	Melba toast, plain	4 pieces	20	5	78	2	1	0.1	0.2	0.3
562	Rye wafer, whole grain, plain	1 wafer	11	5	37	1	Tr	Tr	Tr	Tr
	Saltine									
563	Square	4 crackers	12	4	52	1	1	0.4	0.8	0.2
564	Oyster type	1 cup	45	4	195	4	5	1.3	2.9	0.8
	Sandwich type									
565	Wheat with cheese	1 sandwich	7	4	33	1	1	0.4	0.8	0.2
566	Cheese with peanut butter	1 sandwich	7	4	34	1	2	0.4	0.8	0.3
	Standard snack type									
567	Bite size	1 cup	62	4	311	5	16	2.3	6.6	5.9
568	Round	4 crackers	12	4	60	1	3	0.5	1.3	1.1
569	Wheat, thin square	4 crackers	8	3	38	1	2	0.4	0.9	0.2
570	Whole wheat	4 crackers	16	3	71	1	3	0.5	0.9	1.1
571	Croissant, butter	1 croissant	57	23	231	5	12	6.6	3.1	0.6
572	Croutons, seasoned	1 cup	40	4	186	4	7	2.1	3.8	0.9
	Danish pastry, enriched									
573	Cheese filled	1 danish	71	31	266	6	16	4.8	8.0	1.8
574	Fruit filled	1 danish	71	27	263	4	13	3.5	7.1	1.7
	Doughnuts									
575	Cake type	1 hole	14	21	59	1	3	0.5	1.3	1.1
576		1 medium	47	21	198	2	11	1.7	4.4	3.7
577	Yeast leavened, glazed	1 hole	13	25	52	1	3	0.8	1.7	0.4
578		1 medium	60	25	242	4	14	3.5	7.7	1.7
579	Eclair, prepared from recipe, 5" x 2" x 1¾"	1 eclair	100	52	262	6	16	4.1	6.5	3.9
	English muffin, plain, enriched									
580	Untoasted	1 muffin	57	42	134	4	1	0.1	0.2	0.5
581	Toasted	1 muffin	52	37	133	4	1	0.1	0.2	0.5

Cholesterol (mg)	Carbohydrate (g)	Total dietary fiber (g)	Calcium (mg)	Iron (mg)	Potassium (mg)	Sodium (mg)	Vitamin A (IU)	Vitamin A (RE)	Thiamin (mg)	Riboflavin (mg)	Niacin (mg)	Ascorbic acid (mg)	Food No.
8	10	0.1	3	0.3	9	54	14	4	0.03	0.03	0.4	Tr	544
5	10	0.1	14	0.3	24	70	6	2	0.03	0.02	0.4	0	545
4	8	0.2	10	0.3	11	69	135	35	0.04	0.04	0.3	Tr	546
2	3	0.1	2	0.1	4	12	1	Tr	0.01	0.01	0.1	0	547
0	16	1.4	36	0.4	40	179	27	3	0.01	0.04	0.3	0	548
0	16	1.5	37	0.4	67	216	173	17	0.02	0.06	0.5	Tr	549
37	29	1.4	44	1.1	77	467	123	26	0.15	0.16	1.2	Tr	550
26	28	1.9	162	1.6	96	428	180	35	0.19	0.19	1.5	Tr	551
0	94	8.9	7	4.2	350	43	572	57	0.47	0.25	4.4	0	552
0	107	10.2	7	5.7	224	4	570	57	0.99	0.56	6.9	0	553
0	103	9.8	483	6.5	235	1,860	570	57	0.94	0.53	6.3	0	554
0	7	0.1	Tr	Tr	Tr	1	0	0	0.00	0.00	0.0	0	555
0	134	8.7	42	1.9	287	17	0	0	0.28	0.13	6.0	0	556
0	36	2.2	13	0.6	91	8	0	0	0.10	0.04	1.5	0	557
1	6	0.2	15	0.5	15	100	16	3	0.06	0.04	0.5	0	558
0	11	0.4	3	0.5	19	85	0	0	0.03	0.04	0.6	0	559
0	65	2.4	20	3.1	113	508	0	0	0.19	0.26	3.5	0	560
0	15	1.3	19	0.7	40	166	0	0	0.08	0.05	0.8	0	561
0	9	2.5	4	0.7	54	87	1	0	0.05	0.03	0.2	Tr	562
0	9	0.4	14	0.6	15	156	0	0	0.07	0.06	0.6	0	563
0	32	1.4	54	2.4	58	586	0	0	0.25	0.21	2.4	0	564
Tr	4	0.1	18	0.2	30	98	5	1	0.03	0.05	0.3	Tr	565
Tr	4	0.2	6	0.2	17	69	22	2	0.03	0.02	0.5	Tr	566
0	38	1.0	74	2.2	82	525	0	0	0.25	0.21	2.5	0	567
0	7	0.2	14	0.4	16	102	0	0	0.05	0.04	0.5	0	568
0	5	0.4	4	0.4	15	64	0	0	0.04	0.03	0.4	0	569
0	11	1.7	8	0.5	48	105	0	0	0.03	0.02	0.7	0	570
38	26	1.5	21	1.2	67	424	424	106	0.22	0.14	1.2	Tr	571
3	25	2.0	38	1.1	72	495	16	4	0.20	0.17	1.9	0	572
11	26	0.7	25	1.1	70	320	104	32	0.13	0.18	1.4	Tr	573
81	34	1.3	33	1.3	59	251	53	16	0.19	0.16	1.4	3	574
5	7	0.2	6	0.3	18	76	8	2	0.03	0.03	0.3	Tr	575
17	23	0.7	21	0.9	60	257	27	8	0.10	0.11	0.9	Tr	576
1	6	0.2	6	0.3	14	44	2	1	0.05	0.03	0.4	Tr	577
4	27	0.7	26	1.2	65	205	8	2	0.22	0.13	1.7	Tr	578
127	24	0.6	63	1.2	117	337	718	191	0.12	0.27	0.8	Tr	579
0	26	1.5	99	1.4	75	264	0	0	0.25	0.16	2.2	0	580
0	26	1.5	98	1.4	74	262	0	0	0.20	0.14	2.0	Tr	581

Food No.	Food Description	Measure of edible portion	Weight (g)	Water (%)	Calories (kcal)	Protein (g)	Total fat (g)	Fatty acids: Saturated (g)	Fatty acids: Monounsaturated (g)	Fatty acids: Polyunsaturated (g)

Grain Products (continued)

Food No.	Food Description	Measure of edible portion	Weight (g)	Water (%)	Calories (kcal)	Protein (g)	Total fat (g)	Saturated (g)	Monounsaturated (g)	Polyunsaturated (g)
	French toast									
582	Prepared from recipe, with 2% milk, fried in margarine	1 slice	65	55	149	5	7	1.8	2.9	1.7
583	Frozen, ready to heat	1 slice	59	53	126	4	4	0.9	1.2	0.7
	Granola bar									
584	Hard, plain	1 bar	28	4	134	3	6	0.7	1.2	3.4
	Soft, uncoated									
585	Chocolate chip	1 bar	28	5	119	2	5	2.9	1.0	0.6
586	Raisin	1 bar	28	6	127	2	5	2.7	0.8	0.9
587	Soft, chocolate-coated, peanut butter	1 bar	28	3	144	3	9	4.8	1.9	0.5
588	Macaroni (elbows), enriched, cooked	1 cup	140	66	197	7	1	0.1	0.1	0.4
589	Matzo, plain	1 matzo	28	4	112	3	Tr	0.1	Tr	0.2
	Muffins									
	Blueberry									
590	Commercially prepared (2¾" dia x 2")	1 muffin	57	38	158	3	4	0.8	1.1	1.4
591	Prepared from mix (2¼" dia x 1¾")	1 muffin	50	36	150	3	4	0.7	1.8	1.5
592	Prepared from recipe, with 2% milk	1 muffin	57	40	162	4	6	1.2	1.5	3.1
593	Bran with raisins, toaster type, toasted	1 muffin	34	27	106	2	3	0.5	0.8	1.7
	Corn									
594	Commercially prepared (2½" dia x 2¼")	1 muffin	57	33	174	3	5	0.8	1.2	1.8
595	Prepared from mix (2¼" dia x 1½")	1 muffin	50	31	161	4	5	1.4	2.6	0.6
596	Oat bran, commercially prepared (2½" dia x 2¼")	1 muffin	57	35	154	4	4	0.6	1.0	2.4
597	Noodles, chow mein, canned	1 cup	45	1	237	4	14	2.0	3.5	7.8
	Noodles (egg noodles), enriched, cooked									
598	Regular	1 cup	160	69	213	8	2	0.5	0.7	0.7
599	Spinach	1 cup	160	69	211	8	3	0.6	0.8	0.6
600	NUTRI GRAIN Cereal Bar, fruit filled	1 bar	37	15	136	2	3	0.6	1.9	0.3
	Oat bran									
601	Uncooked	1 cup	94	7	231	16	7	1.2	2.2	2.6
602	Cooked	1 cup	219	84	88	7	2	0.4	0.6	0.7
603	Oriental snack mix	1 oz (about ¼ cup)	28	3	156	5	7	1.1	2.8	3.0
	Pancakes, plain (4" dia)									
604	Frozen, ready to heat	1 pancake	36	45	82	2	1	0.3	0.4	0.3
605	Prepared from complete mix	1 pancake	38	53	74	2	1	0.2	0.3	0.3
606	Prepared from incomplete mix, with 2% milk, egg and oil	1 pancake	38	53	83	3	3	0.8	0.8	1.1
	Pie crust, baked									
	Standard type									
607	From recipe	1 pie shell	180	10	949	12	62	15.5	27.3	16.4
608	From frozen	1 pie shell	126	11	648	6	41	13.3	19.8	5.1
609	Graham cracker	1 pie shell	239	4	1,181	10	60	12.4	27.2	16.5

Cholesterol (mg)	Carbohydrate (g)	Total dietary fiber (g)	Calcium (mg)	Iron (mg)	Potassium (mg)	Sodium (mg)	Vitamin A (IU)	Vitamin A (RE)	Thiamin (mg)	Riboflavin (mg)	Niacin (mg)	Ascorbic acid (mg)	Food No.
75	16	0.7	65	1.1	87	311	315	86	0.13	0.21	1.1	Tr	582
48	19	0.7	63	1.3	79	292	110	32	0.16	0.22	1.6	Tr	583
0	18	1.5	17	0.8	95	83	43	4	0.07	0.03	0.4	Tr	584
Tr	20	1.4	26	0.7	96	77	12	1	0.06	0.04	0.3	0	585
Tr	19	1.2	29	0.7	103	80	0	0	0.07	0.05	0.3	0	586
3	15	0.8	31	0.4	96	55	37	10	0.03	0.06	0.9	Tr	587
0	40	1.8	10	2.0	43	1	0	0	0.29	0.14	2.3	0	588
0	24	0.9	4	0.9	32	1	0	0	0.11	0.08	1.1	0	589
17	27	1.5	32	0.9	70	255	19	5	0.08	0.07	0.6	1	590
23	24	0.6	13	0.6	39	219	39	11	0.07	0.16	1.1	1	591
21	23	1.1	108	1.3	70	251	80	22	0.16	0.16	1.3	1	592
3	19	2.8	13	1.0	60	179	58	16	0.07	0.10	0.8	0	593
15	29	1.9	42	1.6	39	297	119	21	0.16	0.19	1.2	0	594
31	25	1.2	38	1.0	66	398	105	23	0.12	0.14	1.1	Tr	595
0	28	2.6	36	2.4	289	224	0	0	0.15	0.05	0.2	0	596
0	26	1.8	9	2.1	54	198	38	4	0.26	0.19	2.7	0	597
53	40	1.8	19	2.5	45	11	32	10	0.30	0.13	2.4	0	598
53	39	3.7	30	1.7	59	19	165	22	0.39	0.20	2.4	0	599
0	27	0.8	15	1.8	73	110	750	227	0.37	0.41	5.0	0	600
0	62	14.5	55	5.1	532	4	0	0	1.10	0.21	0.9	0	601
0	25	5.7	22	1.9	201	2	0	0	0.35	0.07	0.3	0	602
0	15	3.7	15	0.7	93	117	1	0	0.09	0.04	0.9	Tr	603
3	16	0.6	22	1.3	26	183	36	10	0.14	0.17	1.4	Tr	604
5	14	0.5	48	0.6	67	239	12	3	0.08	0.08	0.7	Tr	605
27	11	0.7	82	0.5	76	192	95	27	0.08	0.12	0.5	Tr	606
0	86	3.0	18	5.2	121	976	0	0	0.70	0.50	6.0	0	607
0	62	1.3	26	2.8	139	815	0	0	0.35	0.48	3.1	0	608
0	156	3.6	50	5.2	210	1,365	1,876	483	0.25	0.42	5.1	0	609

Food No.	Food Description	Measure of edible portion	Weight (g)	Water (%)	Calories (kcal)	Protein (g)	Total fat (g)	Fatty acids Saturated (g)	Fatty acids Monounsaturated (g)	Fatty acids Polyunsaturated (g)

Grain Products (continued)

Food No.	Food Description	Measure of edible portion	Weight (g)	Water (%)	Calories (kcal)	Protein (g)	Total fat (g)	Saturated (g)	Monounsaturated (g)	Polyunsaturated (g)
	Pies									
	Commercially prepared (⅛ of 8" dia)									
610	Apple	1 piece	117	52	277	2	13	4.4	5.1	2.6
611	Blueberry	1 piece	117	53	271	2	12	2.0	5.0	4.1
612	Cherry	1 piece	117	46	304	2	13	3.0	6.8	2.4
613	Chocolate creme	1 piece	113	44	344	3	22	5.6	12.6	2.7
614	Coconut custard	1 piece	104	49	270	6	14	6.1	5.7	1.2
615	Lemon meringue	1 piece	113	42	303	2	10	2.0	3.0	4.1
616	Pecan	1 piece	113	19	452	5	21	4.0	12.1	3.6
617	Pumpkin	1 piece	109	58	229	4	10	1.9	4.4	3.4
	Prepared from recipe (⅛ of 9" dia)									
618	Apple	1 piece	155	47	411	4	19	4.7	8.4	5.2
619	Blueberry	1 piece	147	51	360	4	17	4.3	7.5	4.5
620	Cherry	1 piece	180	46	486	5	22	5.4	9.6	5.8
621	Lemon meringue	1 piece	127	43	362	5	16	4.0	7.1	4.2
622	Pecan	1 piece	122	20	503	6	27	4.9	13.6	7.0
623	Pumpkin	1 piece	155	59	316	7	14	4.9	5.7	2.8
624	Fried, cherry	1 pie	128	38	404	4	21	3.1	9.5	6.9
	Popcorn									
625	Air popped, unsalted	1 cup	8	4	31	1	Tr	Tr	0.1	0.2
626	Oil popped, salted	1 cup	11	3	55	1	3	0.5	0.9	1.5
	Caramel coated									
627	With peanuts	1 cup	42	3	168	3	3	0.4	1.1	1.4
628	Without peanuts	1 cup	35	3	152	1	5	1.3	1.0	1.6
629	Cheese flavor	1 cup	11	3	58	1	4	0.7	1.1	1.7
630	Popcorn cake	1 cake	10	5	38	1	Tr	Tr	0.1	0.1
	Pretzels, made with enriched flour									
631	Stick, 2¼" long	10 pretzels	3	3	11	Tr	Tr	Tr	Tr	Tr
632	Twisted, regular	10 pretzels	60	3	229	5	2	0.5	0.8	0.7
633	Twisted, dutch, 2¾" x 2⅝"	1 pretzel	16	3	61	1	1	0.1	0.2	0.2
	Rice									
634	Brown, long grain, cooked	1 cup	195	73	216	5	2	0.4	0.6	0.6
	White, long grain, enriched									
	Regular									
635	Raw	1 cup	185	12	675	13	1	0.3	0.4	0.3
636	Cooked	1 cup	158	68	205	4	Tr	0.1	0.1	0.1
637	Instant, prepared	1 cup	165	76	162	3	Tr	0.1	0.1	0.1
	Parboiled									
638	Raw	1 cup	185	10	686	13	1	0.3	0.3	0.3
639	Cooked	1 cup	175	72	200	4	Tr	0.1	0.1	0.1
640	Wild, cooked	1 cup	164	74	166	7	1	0.1	0.1	0.3
641	Rice cake, brown rice, plain	1 cake	9	6	35	1	Tr	0.1	0.1	0.1
642	RICE KRISPIES Treat Squares	1 bar	22	6	91	1	2	0.3	0.6	1.1
	Rolls									
643	Dinner	1 roll	28	32	84	2	2	0.5	1.0	0.3
644	Hamburger or hotdog	1 roll	43	34	123	4	2	0.5	0.4	1.1
645	Hard, kaiser	1 roll	57	31	167	6	2	0.3	0.6	1.0
	Spaghetti, cooked									
646	Enriched	1 cup	140	66	197	7	1	0.1	0.1	0.4
647	Whole wheat	1 cup	140	67	174	7	1	0.1	0.1	0.3
	Sweet rolls, cinnamon									
648	Commercial, with raisins	1 roll	60	25	223	4	10	1.8	2.9	4.5
649	Refrigerated dough, baked, with frosting	1 roll	30	23	109	2	4	1.0	2.2	0.5

Choles-terol (mg)	Carbo-hydrate (g)	Total dietary fiber (g)	Calcium (mg)	Iron (mg)	Potas-sium (mg)	Sodium (mg)	Vitamin A (IU)	Vitamin A (RE)	Thiamin (mg)	Ribo-flavin (mg)	Niacin (mg)	Ascor-bic acid (mg)	Food No.
0	40	1.9	13	0.5	76	311	145	35	0.03	0.03	0.3	4	610
0	41	1.2	9	0.4	59	380	164	40	0.01	0.04	0.4	3	611
0	47	0.9	14	0.6	95	288	329	63	0.03	0.03	0.2	1	612
6	38	2.3	41	1.2	144	154	0	0	0.04	0.12	0.8	0	613
36	31	1.9	84	0.8	182	348	114	28	0.09	0.15	0.4	1	614
51	53	1.4	63	0.7	101	165	198	59	0.07	0.24	0.7	4	615
36	65	4.0	19	1.2	84	479	198	53	0.10	0.14	0.3	1	616
22	30	2.9	65	0.9	168	307	3,743	405	0.06	0.17	0.2	1	617
0	58	3.6	11	1.7	122	327	90	19	0.23	0.17	1.9	3	618
0	49	3.6	10	1.8	74	272	62	6	0.22	0.19	1.8	1	619
0	69	3.5	18	3.3	139	344	736	86	0.27	0.23	2.3	2	620
67	50	0.7	15	1.3	83	307	203	56	0.15	0.20	1.2	4	621
106	64	2.2	39	1.8	162	320	410	109	0.23	0.22	1.0	Tr	622
65	41	2.9	146	2.0	288	349	11,833	1,212	0.14	0.31	1.2	3	623
0	55	3.3	28	1.6	83	479	220	22	0.18	0.14	1.8	2	624
0	6	1.2	1	0.2	24	Tr	16	2	0.02	0.02	0.2	0	625
0	6	1.1	1	0.3	25	97	17	2	0.01	0.01	0.2	Tr	626
0	34	1.6	28	1.6	149	124	27	3	0.02	0.05	0.8	0	627
2	28	1.8	15	0.6	38	73	18	4	0.02	0.02	0.8	0	628
1	6	1.1	12	0.2	29	98	27	5	0.01	0.03	0.2	Tr	629
0	8	0.3	1	0.2	33	29	7	1	0.01	0.02	0.6	0	630
0	2	0.1	1	0.1	4	51	0	0	0.01	0.02	0.2	0	631
0	48	1.9	22	2.6	88	1,029	0	0	0.28	0.37	3.2	0	632
0	13	0.5	6	0.7	23	274	0	0	0.07	0.10	0.8	0	633
0	45	3.5	20	0.8	84	10	0	0	0.19	0.05	3.0	0	634
0	148	2.4	52	8.0	213	9	0	0	1.07	0.09	7.8	0	635
0	45	0.6	16	1.9	55	2	0	0	0.26	0.02	2.3	0	636
0	35	1.0	13	1.0	7	5	0	0	0.12	0.08	1.5	0	637
0	151	3.1	111	6.6	222	9	0	0	1.10	0.13	6.7	0	638
0	43	0.7	33	2.0	65	5	0	0	0.44	0.03	2.5	0	639
0	35	3.0	5	1.0	166	5	0	0	0.09	0.14	2.1	0	640
0	7	0.4	1	0.1	26	29	4	Tr	0.01	0.01	0.7	0	641
0	18	0.1	1	0.5	9	77	200	60	0.15	0.18	2.0	0	642
Tr	14	0.8	33	0.9	37	146	0	0	0.14	0.09	1.1	Tr	643
0	22	1.2	60	1.4	61	241	0	0	0.21	0.13	1.7	Tr	644
0	30	1.3	54	1.9	62	310	0	0	0.27	0.19	2.4	0	645
0	40	2.4	10	2.0	43	1	0	0	0.29	0.14	2.3	0	646
0	37	6.3	21	1.5	62	4	0	0	0.15	0.06	1.0	0	647
40	31	1.4	43	1.0	67	230	129	38	0.19	0.16	1.4	1	648
0	17	0.6	10	0.8	19	250	1	0	0.12	0.07	1.1	Tr	649

Food No.	Food Description	Measure of edible portion	Weight (g)	Water (%)	Calories (kcal)	Protein (g)	Total fat (g)	Fatty acids Saturated (g)	Fatty acids Monounsaturated (g)	Fatty acids Polyunsaturated (g)

Grain Products (continued)

Food No.	Food Description	Measure of edible portion	Weight (g)	Water (%)	Calories (kcal)	Protein (g)	Total fat (g)	Saturated (g)	Monounsaturated (g)	Polyunsaturated (g)
650	Taco shell, baked	1 medium	13	6	62	1	3	0.4	1.2	1.1
651	Tapioca, pearl, dry	1 cup	152	11	544	Tr	Tr	Tr	Tr	Tr
	Toaster pastries									
652	Brown sugar cinnamon	1 pastry	50	11	206	3	7	1.8	4.0	0.9
653	Chocolate with frosting	1 pastry	52	13	201	3	5	1.0	2.7	1.1
654	Fruit filled	1 pastry	52	12	204	2	5	0.8	2.2	2.0
655	Low fat	1 pastry	52	12	193	2	3	0.7	1.7	0.5
	Tortilla chips									
	Plain									
656	Regular	1 oz	28	2	142	2	7	1.4	4.4	1.0
657	Low fat, baked	10 chips	14	2	54	2	1	0.1	0.2	0.4
	Nacho flavor									
658	Regular	1 oz	28	2	141	2	7	1.4	4.3	1.0
659	Light, reduced fat	1 oz	28	1	126	2	4	0.8	2.5	0.6
	Tortillas, ready to cook (about 6" dia)									
660	Corn	1 tortilla	26	44	58	1	1	0.1	0.2	0.3
661	Flour	1 tortilla	32	27	104	3	2	0.6	1.2	0.3
	Waffles, plain									
662	Prepared from recipe, 7" dia	1 waffle	75	42	218	6	11	2.1	2.6	5.1
663	Frozen, toasted, 4" dia	1 waffle	33	42	87	2	3	0.5	1.1	0.9
664	Low fat, 4" dia	1 waffle	35	43	83	2	1	0.3	0.4	0.4
	Wheat flours									
	All purpose, enriched									
665	Sifted, spooned	1 cup	115	12	419	12	1	0.2	0.1	0.5
666	Unsifted, spooned	1 cup	125	12	455	13	1	0.2	0.1	0.5
667	Bread, enriched	1 cup	137	13	495	16	2	0.3	0.2	1.0
668	Cake or pastry flour, enriched, unsifted, spooned	1 cup	137	13	496	11	1	0.2	0.1	0.5
669	Self rising, enriched, unsifted, spooned	1 cup	125	11	443	12	1	0.2	0.1	0.5
670	Whole wheat, from hard wheats, stirred, spooned	1 cup	120	10	407	16	2	0.4	0.3	0.9
671	Wheat germ, toasted, plain	1 tbsp	7	6	27	2	1	0.1	0.1	0.5

Legumes, Nuts, and Seeds

Food No.	Food Description	Measure of edible portion	Weight (g)	Water (%)	Calories (kcal)	Protein (g)	Total fat (g)	Saturated (g)	Monounsaturated (g)	Polyunsaturated (g)
	Almonds, shelled									
672	Sliced	1 cup	95	5	549	20	48	3.7	30.5	11.6
673	Whole	1 oz (24 nuts)	28	5	164	6	14	1.1	9.1	3.5
	Beans, dry									
	Cooked									
674	Black	1 cup	172	66	227	15	1	0.2	0.1	0.4
675	Great Northern	1 cup	177	69	209	15	1	0.2	Tr	0.3
676	Kidney, red	1 cup	177	67	225	15	1	0.1	0.1	0.5
677	Lima, large	1 cup	188	70	216	15	1	0.2	0.1	0.3
678	Pea (navy)	1 cup	182	63	258	16	1	0.3	0.1	0.4
679	Pinto	1 cup	171	64	234	14	1	0.2	0.2	0.3
	Canned, solids and liquid									
	Baked beans									
680	Plain or vegetarian	1 cup	254	73	236	12	1	0.3	0.1	0.5
681	With frankfurters	1 cup	259	69	368	17	17	6.1	7.3	2.2
682	With pork in tomato sauce	1 cup	253	73	248	13	3	1.0	1.1	0.3
683	With pork in sweet sauce	1 cup	253	71	281	13	4	1.4	1.6	0.5
684	Kidney, red	1 cup	256	77	218	13	1	0.1	0.1	0.5
685	Lima, large	1 cup	241	77	190	12	Tr	0.1	Tr	0.2
686	White	1 cup	262	70	307	19	1	0.2	0.1	0.3

Cholesterol (mg)	Carbohydrate (g)	Total dietary fiber (g)	Calcium (mg)	Iron (mg)	Potassium (mg)	Sodium (mg)	Vitamin A (IU)	Vitamin A (RE)	Thiamin (mg)	Riboflavin (mg)	Niacin (mg)	Ascorbic acid (mg)	Food No.
0	8	1.0	21	0.3	24	49	0	0	0.03	0.01	0.2	0	650
0	135	1.4	30	2.4	17	2	0	0	0.01	0.00	0.0	0	651
0	34	0.5	17	2.0	57	212	493	112	0.19	0.29	2.3	Tr	652
0	37	0.6	20	1.8	82	203	500	NA	0.16	0.16	2.0	0	653
0	37	1.1	14	1.8	58	218	501	2	0.15	0.19	2.0	Tr	654
0	40	0.8	23	1.8	34	131	494	49	0.15	0.29	2.0	2	655
0	18	1.8	44	0.4	56	150	56	6	0.02	0.05	0.4	0	656
0	11	0.7	22	0.2	37	57	52	6	0.03	0.04	0.1	Tr	657
1	18	1.5	42	0.4	61	201	105	12	0.04	0.05	0.4	1	658
1	20	1.4	45	0.5	77	284	108	12	0.06	0.08	0.1	Tr	659
0	12	1.4	46	0.4	40	42	0	0	0.03	0.02	0.4	0	660
0	18	1.1	40	1.1	42	153	0	0	0.17	0.09	1.1	0	661
52	25	0.7	191	1.7	119	383	171	49	0.20	0.26	1.6	Tr	662
8	13	0.8	77	1.5	42	260	400	120	0.13	0.16	1.5	0	663
9	15	0.4	20	1.9	50	155	506	NA	0.31	0.26	2.6	0	664
0	88	3.1	17	5.3	123	2	0	0	0.90	0.57	6.8	0	665
0	95	3.4	19	5.8	134	3	0	0	0.98	0.62	7.4	0	666
0	99	3.3	21	6.0	137	3	0	0	1.11	0.70	10.3	0	667
0	107	2.3	19	10.0	144	3	0	0	1.22	0.59	9.3	0	668
0	93	3.4	423	5.8	155	1,588	0	0	0.84	0.52	7.3	0	669
0	87	14.6	41	4.7	486	6	0	0	0.54	0.26	7.6	0	670
0	3	0.9	3	0.6	66	Tr	0	0	0.12	0.06	0.4	Tr	671
0	19	11.2	236	4.1	692	1	10	1	0.23	0.77	3.7	0	672
0	6	3.3	70	1.2	206	Tr	3	Tr	0.07	0.23	1.1	0	673
0	41	15.0	46	3.6	611	2	10	2	0.42	0.10	0.9	0	674
0	37	12.4	120	3.8	692	4	2	0	0.28	0.10	1.2	2	675
0	40	13.1	50	5.2	713	4	0	0	0.28	0.10	1.0	2	676
0	39	13.2	32	4.5	955	4	0	0	0.30	0.10	0.8	0	677
0	48	11.6	127	4.5	670	2	4	0	0.37	0.11	1.0	2	678
0	44	14.7	82	4.5	800	3	3	0	0.32	0.16	0.7	4	679
0	52	12.7	127	0.7	752	1,008	434	43	0.39	0.15	1.1	8	680
16	40	17.9	124	4.5	609	1,114	399	39	0.15	0.15	2.3	6	681
18	49	12.1	142	8.3	759	1,113	314	30	0.13	0.12	1.3	8	682
18	53	13.2	154	4.2	673	850	288	28	0.12	0.15	0.9	8	683
0	40	16.4	61	3.2	658	873	0	0	0.27	0.23	1.2	3	684
0	36	11.6	51	4.4	530	810	0	0	0.13	0.08	0.6	0	685
0	57	12.6	191	7.8	1,189	13	0	0	0.25	0.10	0.3	0	686

Food No.	Food Description	Measure of edible portion	Weight (g)	Water (%)	Calories (kcal)	Protein (g)	Total fat (g)	Fatty acids Saturated (g)	Monounsaturated (g)	Polyunsaturated (g)

Legumes, Nuts, and Seeds (continued)

Food No.	Food Description	Measure of edible portion	Weight (g)	Water (%)	Calories (kcal)	Protein (g)	Total fat (g)	Saturated (g)	Monounsaturated (g)	Polyunsaturated (g)
	Black eyed peas, dry									
687	Cooked	1 cup	172	70	200	13	1	0.2	0.1	0.4
688	Canned, solids and liquid	1 cup	240	80	185	11	1	0.3	0.1	0.6
689	Brazil nuts, shelled	1 oz (6-8 nuts)	28	3	186	4	19	4.6	6.5	6.8
690	Carob flour	1 cup	103	4	229	5	1	0.1	0.2	0.2
	Cashews, salted									
691	Dry roasted	1 oz	28	2	163	4	13	2.6	7.7	2.2
692	Oil roasted	1 cup	130	4	749	21	63	12.4	36.9	10.6
693		1 oz (18 nuts)	28	4	163	5	14	2.7	8.1	2.3
694	Chestnuts, European, roasted, shelled	1 cup	143	40	350	5	3	0.6	1.1	1.2
	Chickpeas, dry									
695	Cooked	1 cup	164	60	269	15	4	0.4	1.0	1.9
696	Canned, solids and liquid	1 cup	240	70	286	12	3	0.3	0.6	1.2
	Coconut									
	Raw									
697	Piece, about 2" x 2" x ½"	1 piece	45	47	159	1	15	13.4	0.6	0.2
698	Shredded, not packed	1 cup	80	47	283	3	27	23.8	1.1	0.3
699	Dried, sweetened, shredded	1 cup	93	13	466	3	33	29.3	1.4	0.4
700	Hazelnuts (filberts), chopped	1 cup	115	5	722	17	70	5.1	52.5	9.1
701		1 oz	28	5	178	4	17	1.3	12.9	2.2
702	Hummus, commercial	1 tbsp	14	67	23	1	1	0.2	0.6	0.5
703	Lentils, dry, cooked	1 cup	198	70	230	18	1	0.1	0.1	0.3
704	Macadamia nuts, dry roasted, salted	1 cup	134	2	959	10	102	16.0	79.4	2.0
705		1 oz (10-12 nuts)	28	2	203	2	22	3.4	16.8	0.4
	Mixed nuts, with peanuts, salted									
706	Dry roasted	1 oz	28	2	168	5	15	2.0	8.9	3.1
707	Oil roasted	1 oz	28	2	175	5	16	2.5	9.0	3.8
	Peanuts									
	Dry roasted									
708	Salted	1 oz (about 28)	28	2	166	7	14	2.0	7.0	4.4
709	Unsalted	1 cup	146	2	854	35	73	10.1	36.0	22.9
710		1 oz (about 28)	28	2	166	7	14	2.0	7.0	4.4
711	Oil roasted, salted	1 cup	144	2	837	38	71	9.9	35.2	22.4
712		1 oz	28	2	165	7	14	1.9	6.9	4.4
	Peanut butter									
	Regular									
713	Smooth style	1 tbsp	16	1	95	4	8	1.7	3.9	2.2
714	Chunk style	1 tbsp	16	1	94	4	8	1.5	3.8	2.3
715	Reduced fat, smooth	1 tbsp	18	1	94	5	6	1.3	2.9	1.8
716	Peas, split, dry, cooked	1 cup	196	69	231	16	1	0.1	0.2	0.3
717	Pecans, halves	1 cup	108	4	746	10	78	6.7	44.0	23.3
718		1 oz (20 halves)	28	4	196	3	20	1.8	11.6	6.1
719	Pine nuts (pignolia), shelled	1 oz	28	7	160	7	14	2.2	5.4	6.1
720		1 tbsp	9	7	49	2	4	0.7	1.6	1.8
721	Pistachio nuts, dry roasted, with salt, shelled	1 oz (47 nuts)	28	2	161	6	13	1.6	6.8	3.9
722	Pumpkin and squash kernels, roasted, with salt	1 oz (142 seeds)	28	7	148	9	12	2.3	3.7	5.4
723	Refried beans, canned	1 cup	252	76	237	14	3	1.2	1.4	0.4
724	Sesame seeds	1 tbsp	8	5	47	2	4	0.6	1.7	1.9
725	Soybeans, dry, cooked	1 cup	172	63	298	29	15	2.2	3.4	8.7
	Soy products									
726	Miso	1 cup	275	41	567	32	17	2.4	3.7	9.4
727	Soy milk	1 cup	245	93	81	7	5	0.5	0.8	2.0

Cholesterol (mg)	Carbohydrate (g)	Total dietary fiber (g)	Calcium (mg)	Iron (mg)	Potassium (mg)	Sodium (mg)	Vitamin A (IU)	Vitamin A (RE)	Thiamin (mg)	Riboflavin (mg)	Niacin (mg)	Ascorbic acid (mg)	Food No.
0	36	11.2	41	4.3	478	7	26	3	0.35	0.09	0.9	1	687
0	33	7.9	48	2.3	413	718	31	2	0.18	0.18	0.8	6	688
0	4	1.5	50	1.0	170	1	0	0	0.28	0.03	0.5	Tr	689
0	92	41.0	358	3.0	852	36	14	1	0.05	0.47	2.0	Tr	690
0	9	0.9	13	1.7	160	181	0	0	0.06	0.06	0.4	0	691
0	37	4.9	53	5.3	689	814	0	0	0.55	0.23	2.3	0	692
0	8	1.1	12	1.2	150	177	0	0	0.12	0.05	0.5	0	693
0	76	7.3	41	1.3	847	3	34	3	0.35	0.25	1.9	37	694
0	45	12.5	80	4.7	477	11	44	5	0.19	0.10	0.9	2	695
0	54	10.6	77	3.2	413	718	58	5	0.07	0.08	0.3	9	696
0	7	4.1	6	1.1	160	9	0	0	0.03	0.01	0.2	1	697
0	12	7.2	11	1.9	285	16	0	0	0.05	0.02	0.4	3	698
0	44	4.2	14	1.8	313	244	0	0	0.03	0.02	0.4	1	699
0	19	11.2	131	5.4	782	0	46	5	0.74	0.13	2.1	7	700
0	5	2.7	32	1.3	193	0	11	1	0.18	0.03	0.5	2	701
0	2	0.8	5	0.3	32	53	4	Tr	0.03	0.01	0.1	0	702
0	40	15.6	38	6.6	731	4	16	2	0.33	0.14	2.1	3	703
0	17	10.7	94	3.6	486	355	0	0	0.95	0.12	3.0	1	704
0	4	2.3	20	0.8	103	75	0	0	0.20	0.02	0.6	Tr	705
0	7	2.6	20	1.0	169	190	4	Tr	0.06	0.06	1.3	Tr	706
0	6	2.6	31	0.9	165	185	5	1	0.14	0.06	1.4	Tr	707
0	6	2.3	15	0.6	187	230	0	0	0.12	0.03	3.8	0	708
0	31	11.7	79	3.3	961	9	0	0	0.64	0.14	19.7	0	709
0	6	2.3	15	0.6	187	2	0	0	0.12	0.03	3.8	0	710
0	27	13.2	127	2.6	982	624	0	0	0.36	0.16	20.6	0	711
0	5	2.6	25	0.5	193	123	0	0	0.07	0.03	4.0	0	712
0	3	0.9	6	0.3	107	75	0	0	0.01	0.02	2.1	0	713
0	3	1.1	7	0.3	120	78	0	0	0.02	0.02	2.2	0	714
0	6	0.9	6	0.3	120	97	0	0	0.05	0.01	2.6	0	715
0	41	16.3	27	2.5	710	4	14	2	0.37	0.11	1.7	1	716
0	15	10.4	76	2.7	443	0	83	9	0.71	0.14	1.3	1	717
0	4	2.7	20	0.7	116	0	22	2	0.19	0.04	0.3	Tr	718
0	4	1.3	7	2.6	170	1	8	1	0.23	0.05	1.0	1	719
0	1	0.4	2	0.8	52	Tr	2	Tr	0.07	0.02	0.3	Tr	720
0	8	2.9	31	1.2	293	121	151	15	0.24	0.04	0.4	1	721
0	4	1.1	12	4.2	229	163	108	11	0.06	0.09	0.5	1	722
20	39	13.4	88	4.2	673	753	0	0	0.07	0.04	0.8	15	723
0	1	0.9	10	0.6	33	3	5	1	0.06	0.01	0.4	0	724
0	17	10.3	175	8.8	886	2	15	2	0.27	0.49	0.7	3	725
0	77	14.9	182	7.5	451	10,029	239	25	0.27	0.69	2.4	0	726
0	4	3.2	10	1.4	345	29	78	7	0.39	0.17	0.4	0	727

Food No.	Food Description	Measure of edible portion	Weight (g)	Water (%)	Calories (kcal)	Protein (g)	Total fat (g)	Fatty acids: Saturated (g)	Fatty acids: Monounsaturated (g)	Fatty acids: Polyunsaturated (g)

Legumes, Nuts, and Seeds (continued)

Food No.	Food Description	Measure of edible portion	Weight (g)	Water (%)	Calories (kcal)	Protein (g)	Total fat (g)	Saturated (g)	Monounsaturated (g)	Polyunsaturated (g)
	Soy products (continued)									
	Tofu									
728	Firm	¼ block	81	84	62	7	4	0.5	0.8	2.0
729	Soft, piece 2½" x 2¾" x 1"	1 piece	120	87	73	8	4	0.6	1.0	2.5
730	Sunflower seed kernels, dry roasted, with salt	¼ cup	32	1	186	6	16	1.7	3.0	10.5
731		1 oz	28	1	165	5	14	1.5	2.7	9.3
732	Tahini	1 tbsp	15	3	89	3	8	1.1	3.0	3.5
733	Walnuts, English	1 cup, chopped	120	4	785	18	78	7.4	10.7	56.6
734		1 oz (14 halves)	28	4	185	4	18	1.7	2.5	13.4

Meat and Meat Products

Food No.	Food Description	Measure of edible portion	Weight (g)	Water (%)	Calories (kcal)	Protein (g)	Total fat (g)	Saturated (g)	Monounsaturated (g)	Polyunsaturated (g)
	Beef, cooked									
	Cuts braised, simmered, or pot roasted									
	Relatively fat, such as chuck blade, piece, 2½" x 2½" x ¾"									
735	Lean and fat	3 oz	85	47	293	23	22	8.7	9.4	0.8
736	Lean only	3 oz	85	55	213	26	11	4.3	4.8	0.4
	Relatively lean, such as bottom round, piece, 4⅛" x 2¼" x ½"									
737	Lean and fat	3 oz	85	52	234	24	14	5.4	6.2	0.5
738	Lean only	3 oz	85	58	178	27	7	2.4	3.1	0.3
	Ground beef, broiled									
739	83% lean	3 oz	85	57	218	22	14	5.5	6.1	0.5
740	79% lean	3 oz	85	56	231	21	16	6.2	6.9	0.6
741	73% lean	3 oz	85	54	246	20	18	6.9	7.7	0.7
742	Liver, fried, slice, 6½" x 2⅜" x ⅜"	3 oz	85	56	184	23	7	2.3	1.4	1.5
	Roast, oven cooked, no liquid added									
	Relatively fat, such as rib, 2 pieces, 4⅛" x 2¼" x ¼"									
743	Lean and fat	3 oz	85	47	304	19	25	9.9	10.6	0.9
744	Lean only	3 oz	85	59	195	23	11	4.2	4.5	0.3
	Relatively lean, such as eye of round, 2 pieces, 2½" x 2½" x ⅜"									
745	Lean and fat	3 oz	85	59	195	23	11	4.2	4.7	0.4
746	Lean only	3 oz	85	65	143	25	4	1.5	1.8	0.1
	Steak, sirloin, broiled, piece, 2½" x 2½" x ¾"									
747	Lean and fat	3 oz	85	57	219	24	13	5.2	5.6	0.5
748	Lean only	3 oz	85	62	166	26	6	2.4	2.6	0.2
749	Beef, canned, corned	3 oz	85	58	213	23	13	5.3	5.1	0.5
750	Beef, dried, chipped	1 oz	28	57	47	8	1	0.5	0.5	0.1
	Lamb, cooked									
	Chops									
	Arm, braised									
751	Lean and fat	3 oz	85	44	294	26	20	8.4	8.7	1.5
752	Lean only	3 oz	85	49	237	30	12	4.3	5.2	0.8
	Loin, broiled									
753	Lean and fat	3 oz	85	52	269	21	20	8.4	8.2	1.4
754	Lean only	3 oz	85	61	184	25	8	3.0	3.6	0.5

Cholesterol (mg)	Carbohydrate (g)	Total dietary fiber (g)	Calcium (mg)	Iron (mg)	Potassium (mg)	Sodium (mg)	Vitamin A (IU)	Vitamin A (RE)	Thiamin (mg)	Riboflavin (mg)	Niacin (mg)	Ascorbic acid (mg)	Food No.
0	2	0.3	131	1.2	143	6	6	1	0.08	0.08	Tr	Tr	728
0	2	0.2	133	1.3	144	10	8	1	0.06	0.04	0.6	Tr	729
0	8	2.9	22	1.2	272	250	0	0	0.03	0.08	2.3	Tr	730
0	7	2.6	20	1.1	241	221	0	0	0.03	0.07	2.0	Tr	731
0	3	1.4	64	1.3	62	17	10	1	0.18	0.07	0.8	0	732
0	16	8.0	125	3.5	529	2	49	5	0.41	0.18	2.3	2	733
0	4	1.9	29	0.8	125	1	12	1	0.10	0.04	0.5	Tr	734
88	0	0.0	11	2.6	196	54	0	0	0.06	0.20	2.1	0	735
90	0	0.0	11	3.1	224	60	0	0	0.07	0.24	2.3	0	736
82	0	0.0	5	2.7	240	43	0	0	0.06	0.20	3.2	0	737
82	0	0.0	4	2.9	262	43	0	0	0.06	0.22	3.5	0	738
71	0	0.0	6	2.0	266	60	0	0	0.05	0.23	4.2	0	739
74	0	0.0	9	1.8	256	65	0	0	0.04	0.18	4.4	0	740
77	0	0.0	9	2.1	248	71	0	0	0.03	0.16	4.9	0	741
410	7	0.0	9	5.3	309	90	30,689	9,120	0.18	3.52	12.3	20	742
71	0	0.0	9	2.0	256	54	0	0	0.06	0.14	2.9	0	743
68	0	0.0	9	2.4	318	61	0	0	0.07	0.18	3.5	0	744
61	0	0.0	5	1.6	308	50	0	0	0.07	0.14	3.0	0	745
59	0	0.0	4	1.7	336	53	0	0	0.08	0.14	3.2	0	746
77	0	0.0	9	2.6	311	54	0	0	0.09	0.23	3.3	0	747
76	0	0.0	9	2.9	343	56	0	0	0.11	0.25	3.6	0	748
73	0	0.0	10	1.8	116	855	0	0	0.02	0.12	2.1	0	749
12	Tr	0.0	2	1.3	126	984	0	0	0.02	0.06	1.5	0	750
102	0	0.0	21	2.0	260	61	0	0	0.06	0.21	5.7	0	751
103	0	0.0	22	2.3	287	65	0	0	0.06	0.23	5.4	0	752
85	0	0.0	17	1.5	278	65	0	0	0.09	0.21	6.0	0	753
81	0	0.0	16	1.7	320	71	0	0	0.09	0.24	5.8	0	754

Food No.	Food Description	Measure of edible portion	Weight (g)	Water (%)	Calories (kcal)	Protein (g)	Total fat (g)	Fatty acids Saturated (g)	Fatty acids Monounsaturated (g)	Fatty acids Polyunsaturated (g)

Meat and Meat Products (continued)

Food No.	Food Description	Measure of edible portion	Weight (g)	Water (%)	Calories (kcal)	Protein (g)	Total fat (g)	Saturated (g)	Monounsaturated (g)	Polyunsaturated (g)
	Lamb (continued)									
	Leg, roasted, 2 pieces, 4⅛" x 2¼" x ¼"									
755	Lean and fat	3 oz	85	57	219	22	14	5.9	5.9	1.0
756	Lean only	3 oz	85	64	162	24	7	2.3	2.9	0.4
	Rib, roasted, 3 pieces, 2½" x 2½" x ¼"									
757	Lean and fat	3 oz	85	48	305	18	25	10.9	10.6	1.8
758	Lean only	3 oz	85	60	197	22	11	4.0	5.0	0.7
	Pork, cured, cooked									
	Bacon									
759	Regular	3 medium slices	19	13	109	6	9	3.3	4.5	1.1
760	Canadian style (6 slices per 6-oz pkg)	2 slices	47	62	86	11	4	1.3	1.9	0.4
	Ham, light cure, roasted, 2 pieces, 4⅛" x 2¼" x ¼"									
761	Lean and fat	3 oz	85	58	207	18	14	5.1	6.7	1.5
762	Lean only	3 oz	85	66	133	21	5	1.6	2.2	0.5
763	Ham, canned, roasted, 2 pieces, 4⅛" x 2¼" x ¼"	3 oz	85	67	142	18	7	2.4	3.5	0.8
	Pork, fresh, cooked									
	Chop, loin (cut 3 per lb with bone)									
	Broiled									
764	Lean and fat	3 oz	85	58	204	24	11	4.1	5.0	0.8
765	Lean only	3 oz	85	61	172	26	7	2.5	3.1	0.5
	Pan fried									
766	Lean and fat	3 oz	85	53	235	25	14	5.1	6.0	1.6
767	Lean only	3 oz	85	57	197	27	9	3.1	3.8	1.1
	Ham (leg), roasted, piece, 2½" x 2½" x ¾"									
768	Lean and fat	3 oz	85	55	232	23	15	5.5	6.7	1.4
769	Lean only	3 oz	85	61	179	25	8	2.8	3.8	0.7
	Rib roast, piece, 2½" x 2½" x ¾"									
770	Lean and fat	3 oz	85	56	217	23	13	5.0	5.9	1.1
771	Lean only	3 oz	85	59	190	24	9	3.7	4.5	0.7
	Ribs, lean and fat, cooked									
772	Backribs, roasted	3 oz	85	45	315	21	25	9.3	11.4	2.0
773	Country style, braised	3 oz	85	54	252	20	18	6.8	7.9	1.6
774	Spareribs, braised	3 oz	85	40	337	25	26	9.5	11.5	2.3
	Shoulder cut, braised, 3 pieces, 2½" x 2½" x ¼"									
775	Lean and fat	3 oz	85	48	280	24	20	7.2	8.8	1.9
776	Lean only	3 oz	85	54	211	27	10	3.5	4.9	1.0
	Sausages and luncheon meats									
777	Bologna, beef and pork (8 slices per 8-oz pkg)	2 slices	57	54	180	7	16	6.1	7.6	1.4
778	Braunschweiger (6 slices per 6-oz pkg)	2 slices	57	48	205	8	18	6.2	8.5	2.1
779	Brown and serve, cooked, link, 4" x ⅞" raw	2 links	26	45	103	4	9	3.4	4.5	1.0
	Canned, minced luncheon meat									
780	Pork, ham, and chicken, reduced sodium (7 slices per 7-oz can)	2 slices	57	56	172	7	15	5.1	7.1	1.5
781	Pork with ham (12 slices per 12-oz can)	2 slices	57	52	188	8	17	5.7	7.7	1.2
782	Pork and chicken (12 slices per 12-oz can)	2 slices	57	64	117	9	8	2.7	3.8	0.8

Cholesterol (mg)	Carbohydrate (g)	Total dietary fiber (g)	Calcium (mg)	Iron (mg)	Potassium (mg)	Sodium (mg)	Vitamin A (IU)	Vitamin A (RE)	Thiamin (mg)	Riboflavin (mg)	Niacin (mg)	Ascorbic acid (mg)	Food No.
79	0	0.0	9	1.7	266	56	0	0	0.09	0.23	5.6	0	755
76	0	0.0	7	1.8	287	58	0	0	0.09	0.25	5.4	0	756
82	0	0.0	19	1.4	230	62	0	0	0.08	0.18	5.7	0	757
75	0	0.0	18	1.5	268	69	0	0	0.08	0.20	5.2	0	758
16	Tr	0.0	2	0.3	92	303	0	0	0.13	0.05	1.4	0	759
27	1	0.0	5	0.4	181	719	0	0	0.38	0.09	3.2	0	760
53	0	0.0	6	0.7	243	1,009	0	0	0.51	0.19	3.8	0	761
47	0	0.0	6	0.8	269	1,128	0	0	0.58	0.22	4.3	0	762
35	Tr	0.0	6	0.9	298	908	0	0	0.82	0.21	4.3	0	763
70	0	0.0	28	0.7	304	49	8	3	0.91	0.24	4.5	Tr	764
70	0	0.0	26	0.7	319	51	7	2	0.98	0.26	4.7	Tr	765
78	0	0.0	23	0.8	361	68	7	2	0.97	0.26	4.8	1	766
78	0	0.0	20	0.8	382	73	7	2	1.06	0.28	5.1	1	767
80	0	0.0	12	0.9	299	51	9	3	0.54	0.27	3.9	Tr	768
80	0	0.0	6	1.0	317	54	8	3	0.59	0.30	4.2	Tr	769
62	0	0.0	24	0.8	358	39	5	2	0.62	0.26	5.2	Tr	770
60	0	0.0	22	0.8	371	40	5	2	0.64	0.27	5.5	Tr	771
100	0	0.0	38	1.2	268	86	8	3	0.36	0.17	3.0	Tr	772
74	0	0.0	25	1.0	279	50	7	2	0.43	0.22	3.3	1	773
103	0	0.0	40	1.6	272	79	9	3	0.35	0.32	4.7	0	774
93	0	0.0	15	1.4	314	75	8	3	0.46	0.26	4.4	Tr	775
97	0	0.0	7	1.7	344	87	7	2	0.51	0.31	5.0	Tr	776
31	2	0.0	7	0.9	103	581	0	0	0.10	0.08	1.5	0	777
89	2	0.0	5	5.3	113	652	8,009	2,405	0.14	0.87	4.8	0	778
18	1	0.0	3	0.3	49	209	0	0	0.09	0.04	0.9	0	779
43	1	0.0	0	0.4	321	539	0	0	0.15	0.10	1.8	18	780
40	1	0.0	0	0.4	233	758	0	0	0.18	0.10	2.0	0	781
43	1	0.0	0	0.7	352	539	0	0	0.10	0.12	2.0	18	782

Food No.	Food Description	Measure of edible portion	Weight (g)	Water (%)	Calories (kcal)	Protein (g)	Total fat (g)	Fatty acids: Saturated (g)	Fatty acids: Monounsaturated (g)	Fatty acids: Polyunsaturated (g)
	Meat and Meat Products (continued)									
	Sausages and luncheon meats (continued)									
783	Chopped ham (8 slices per 6-oz pkg)	2 slices	21	64	48	4	4	1.2	1.7	0.4
	Cooked ham (8 slices per 8-oz pkg)									
784	Regular	2 slices	57	65	104	10	6	1.9	2.8	0.7
785	Extra lean	2 slices	57	71	75	11	3	0.9	1.3	0.3
	Frankfurter (10 per 1-lb pkg), heated									
786	Beef and pork	1 frank	45	54	144	5	13	4.8	6.2	1.2
787	Beef	1 frank	45	55	142	5	13	5.4	6.1	0.6
	Pork sausage, fresh, cooked									
788	Link (4" x ⅞" raw)	2 links	26	45	96	5	8	2.8	3.6	1.0
789	Patty (3⅞" x ¼" raw)	1 patty	27	45	100	5	8	2.9	3.8	1.0
	Salami, beef and pork									
790	Cooked type (8 slices per 8-oz pkg)	2 slices	57	60	143	8	11	4.6	5.2	1.2
791	Dry type, slice, 3⅛" x 1/16"	2 slices	20	35	84	5	7	2.4	3.4	0.6
792	Sandwich spread (pork, beef)	1 tbsp	15	60	35	1	3	0.9	1.1	0.4
793	Vienna sausage (7 per 4-oz can)	1 sausage	16	60	45	2	4	1.5	2.0	0.3
	Veal, lean and fat, cooked									
794	Cutlet, braised, 4⅛" x 2¼" x ½"	3 oz	85	55	179	31	5	2.2	2.0	0.4
795	Rib, roasted, 2 pieces, 4⅛" x 2¼" x ¼"	3 oz	85	60	194	20	12	4.6	4.6	0.8
	Mixed Dishes and Fast Foods									
	Mixed dishes									
796	Beef macaroni, frozen, HEALTHY CHOICE	1 package	240	78	211	14	2	0.7	1.2	0.3
797	Beef stew, canned	1 cup	232	82	218	11	12	5.2	5.5	0.5
798	Chicken pot pie, frozen	1 small pie	217	60	484	13	29	9.7	12.5	4.5
799	Chili con carne with beans, canned	1 cup	222	74	255	20	8	2.1	2.2	1.4
800	Macaroni and cheese, canned, made with corn oil	1 cup	252	82	199	8	6	3.0	NA	1.3
801	Meatless burger crumbles, MORNINGSTAR FARMS	1 cup	110	60	231	22	13	3.3	4.6	4.9
802	Meatless burger patty, frozen, MORNINGSTAR FARMS	1 patty	85	71	91	14	1	0.1	0.3	0.2
803	Pasta with meatballs in tomato sauce, canned	1 cup	252	78	260	11	10	4.0	4.2	0.6
804	Spaghetti bolognese (meat sauce), frozen, HEALTHY CHOICE	1 package	283	78	255	14	3	1.0	0.9	0.9
805	Spaghetti in tomato sauce with cheese, canned	1 cup	252	80	192	6	2	0.7	0.3	0.3
806	Spinach souffle, home-prepared	1 cup	136	74	219	11	18	7.1	6.8	3.1
807	Tortellini, pasta with cheese filling, frozen	¾ cup (yields 1 cup cooked)	81	31	249	11	6	2.9	1.7	0.4

Cholesterol (mg)	Carbohydrate (g)	Total dietary fiber (g)	Calcium (mg)	Iron (mg)	Potassium (mg)	Sodium (mg)	Vitamin A (IU)	Vitamin A (RE)	Thiamin (mg)	Riboflavin (mg)	Niacin (mg)	Ascorbic acid (mg)	Food No.
11	0	0.0	1	0.2	67	288	0	0	0.13	0.04	0.8	0	783
32	2	0.0	4	0.6	189	751	0	0	0.49	0.14	3.0	0	784
27	1	0.0	4	0.4	200	815	0	0	0.53	0.13	2.8	0	785
23	1	0.0	5	0.5	75	504	0	0	0.09	0.05	1.2	0	786
27	1	0.0	9	0.6	75	462	0	0	0.02	0.05	1.1	0	787
22	Tr	0.0	8	0.3	94	336	0	0	0.19	0.07	1.2	1	788
22	Tr	0.0	9	0.3	97	349	0	0	0.20	0.07	1.2	1	789
37	1	0.0	7	1.5	113	607	0	0	0.14	0.21	2.0	0	790
16	1	0.0	2	0.3	76	372	0	0	0.12	0.06	1.0	0	791
6	2	Tr	2	0.1	17	152	13	1	0.03	0.02	0.3	0	792
8	Tr	0.0	2	0.1	16	152	0	0	0.01	0.02	0.3	0	793
114	0	0.0	7	1.1	326	57	0	0	0.05	0.30	9.0	0	794
94	0	0.0	9	0.8	251	78	0	0	0.04	0.23	5.9	0	795
14	33	4.6	46	2.7	365	444	514	50	0.28	0.16	3.1	58	796
37	16	3.5	28	1.6	404	947	3,860	494	0.17	0.14	2.9	10	797
41	43	1.7	33	2.1	256	857	2,285	343	0.25	0.36	4.1	2	798
24	24	8.2	67	3.3	608	1,032	884	93	0.15	0.15	2.1	1	799
8	29	3.0	113	2.0	123	1,058	713	NA	0.28	0.25	2.5	0	800
0	7	5.1	79	6.4	178	476	0	0	9.92	0.35	3.0	0	801
0	8	4.3	87	2.9	434	383	0	0	0.26	0.55	4.1	0	802
20	31	6.8	28	2.3	416	1,053	920	93	0.19	0.16	3.3	8	803
17	43	5.1	51	3.5	408	473	492	48	0.35	3.77	0.5	15	804
8	39	7.8	40	2.8	305	963	932	58	0.35	0.28	4.5	10	805
184	3	NA	230	1.3	201	763	3,461	675	0.09	0.30	0.5	3	806
34	38	1.5	123	1.2	72	279	50	13	0.25	0.25	2.2	0	807

Food No.	Food Description	Measure of edible portion	Weight (g)	Water (%)	Calories (kcal)	Protein (g)	Total fat (g)	Fatty acids: Saturated (g)	Fatty acids: Monounsaturated (g)	Fatty acids: Polyunsaturated (g)

Mixed Dishes and Fast Foods (continued)

Food No.	Food Description	Measure of edible portion	Weight (g)	Water (%)	Calories (kcal)	Protein (g)	Total fat (g)	Saturated (g)	Monounsaturated (g)	Polyunsaturated (g)
	Fast foods									
	Breakfast items									
808	Biscuit with egg and sausage	1 biscuit	180	43	581	19	39	15.0	16.4	4.4
809	Croissant with egg, cheese, bacon	1 croissant	129	44	413	16	28	15.4	9.2	1.8
	Danish pastry									
810	Cheese filled	1 pastry	91	34	353	6	25	5.1	15.6	2.4
811	Fruit filled	1 pastry	94	29	335	5	16	3.3	10.1	1.6
812	English muffin with egg, cheese, Canadian bacon	1 muffin	137	57	289	17	13	4.7	4.7	1.6
813	French toast with butter	2 slices	135	51	356	10	19	7.7	7.1	2.4
814	French toast sticks	5 sticks	141	30	513	8	29	4.7	12.6	9.9
815	Hashed brown potatoes	½ cup	72	60	151	2	9	4.3	3.9	0.5
816	Pancakes with butter, syrup	2 pancakes	232	50	520	8	14	5.9	5.3	2.0
	Burrito									
817	With beans and cheese	1 burrito	93	54	189	8	6	3.4	1.2	0.9
818	With beans and meat	1 burrito	116	52	255	11	9	4.2	3.5	0.6
	Cheeseburger									
	Regular size, with condiments									
819	Double patty with mayo type dressing, vegetables	1 sandwich	166	51	417	21	21	8.7	7.8	2.7
820	Single patty	1 sandwich	113	48	295	16	14	6.3	5.3	1.1
	Regular size, plain									
821	Double patty	1 sandwich	155	42	457	28	28	13.0	11.0	1.9
822	Double patty with 3-piece bun	1 sandwich	160	43	461	22	22	9.5	8.3	1.8
823	Single patty	1 sandwich	102	37	319	15	15	6.5	5.8	1.5
	Large, with condiments									
824	Single patty with mayo type dressing, vegetables	1 sandwich	219	53	563	28	33	15.0	12.6	2.0
825	Single patty with bacon	1 sandwich	195	44	608	32	37	16.2	14.5	2.7
826	Chicken fillet (breaded and fried) sandwich, plain	1 sandwich	182	47	515	24	29	8.5	10.4	8.4
	Chicken, fried. See Poultry and Poultry Products.									
827	Chicken pieces, boneless, breaded and fried, plain	6 pieces	106	47	319	18	21	4.7	10.5	4.6
828	Chili con carne	1 cup	253	77	256	25	8	3.4	3.4	0.5
829	Chimichanga with beef	1 chimichanga	174	51	425	20	20	8.5	8.1	1.1
830	Coleslaw	¾ cup	99	74	147	1	11	1.6	2.4	6.4
	Desserts									
831	Ice milk, soft, vanilla, in cone	1 cone	103	65	164	4	6	3.5	1.8	0.4
832	Pie, fried, with fruit filling (5" x 3¾")	1 pie	128	38	404	4	21	3.1	9.5	6.9
833	Sundae, hot fudge	1 sundae	158	60	284	6	9	5.0	2.3	0.8
834	Enchilada with cheese	1 enchilada	163	63	319	10	19	10.6	6.3	0.8
835	Fish sandwich, with tartar sauce and cheese	1 sandwich	183	45	523	21	29	8.1	8.9	9.4
836	French fries	1 small	85	35	291	4	16	3.3	9.0	2.7
837		1 medium	134	35	458	6	25	5.2	14.3	4.2
838		1 large	169	35	578	7	31	6.5	18.0	5.3
839	Frijoles (refried beans, chili sauce, cheese)	1 cup	167	69	225	11	8	4.1	2.6	0.7

Cholesterol (mg)	Carbohydrate (g)	Total dietary fiber (g)	Calcium (mg)	Iron (mg)	Potassium (mg)	Sodium (mg)	Vitamin A (IU)	Vitamin A (RE)	Thiamin (mg)	Riboflavin (mg)	Niacin (mg)	Ascorbic acid (mg)	Food No.
302	41	0.9	155	4.0	320	1,141	635	164	0.50	0.45	3.6	0	808
215	24	NA	151	2.2	201	889	472	120	0.35	0.34	2.2	2	809
20	29	NA	70	1.8	116	319	155	43	0.26	0.21	2.5	3	810
19	45	NA	22	1.4	110	333	86	24	0.29	0.21	1.8	2	811
234	27	1.5	151	2.4	199	729	586	156	0.49	0.45	3.3	2	812
116	36	NA	73	1.9	177	513	473	146	0.58	0.50	3.9	Tr	813
75	58	2.7	78	3.0	127	499	45	13	0.23	0.25	3.0	0	814
9	16	NA	7	0.5	267	290	18	3	0.08	0.01	1.1	5	815
58	91	NA	128	2.6	251	1,104	281	70	0.39	0.56	3.4	3	816
14	27	NA	107	1.1	248	583	625	119	0.11	0.35	1.8	1	817
24	33	NA	53	2.5	329	670	319	32	0.27	0.42	2.7	1	818
60	35	NA	171	3.4	335	1,051	398	65	0.35	0.28	8.1	2	819
37	27	NA	111	2.4	223	616	462	94	0.25	0.23	3.7	2	820
110	22	NA	233	3.4	308	636	332	79	0.25	0.37	6.0	0	821
80	44	NA	224	3.7	285	891	277	66	0.34	0.38	6.0	0	822
50	32	NA	141	2.4	164	500	153	37	0.40	0.40	3.7	0	823
88	38	NA	206	4.7	445	1,108	613	129	0.39	0.46	7.4	8	824
111	37	NA	162	4.7	332	1,043	406	80	0.31	0.41	6.6	2	825
60	39	NA	60	4.7	353	957	100	31	0.33	0.24	6.8	9	826
61	15	0.0	14	0.9	305	513	0	0	0.12	0.16	7.5	0	827
134	22	NA	68	5.2	691	1,007	1,662	167	0.13	1.14	2.5	2	828
9	43	NA	63	4.5	586	910	146	16	0.49	0.64	5.8	5	829
5	13	NA	34	0.7	177	267	338	50	0.04	0.03	0.1	8	830
28	24	0.1	153	0.2	169	92	211	52	0.05	0.26	0.3	1	831
0	55	3.3	28	1.6	83	479	35	4	0.18	0.14	1.8	2	832
21	48	0.0	207	0.6	395	182	221	57	0.06	0.30	1.1	2	833
44	29	NA	324	1.3	240	784	1,161	186	0.08	0.42	1.9	1	834
68	48	NA	185	3.5	353	939	432	97	0.46	0.42	4.2	3	835
0	34	3.0	12	0.7	586	168	0	0	0.07	0.03	2.4	10	836
0	53	4.7	19	1.0	923	265	0	0	0.11	0.05	3.8	16	837
0	67	5.9	24	1.3	1,164	335	0	0	0.14	0.07	4.8	20	838
37	29	NA	189	2.2	605	882	456	70	0.13	0.33	1.5	2	839

Food No.	Food Description	Measure of edible portion	Weight (g)	Water (%)	Calories (kcal)	Protein (g)	Total fat (g)	Fatty acids Saturated (g)	Fatty acids Monounsaturated (g)	Fatty acids Polyunsaturated (g)

Mixed Dishes and Fast Foods (continued)

Food No.	Food Description	Measure of edible portion	Weight (g)	Water (%)	Calories (kcal)	Protein (g)	Total fat (g)	Saturated (g)	Monounsaturated (g)	Polyunsaturated (g)
	Fast foods (continued)									
	Hamburger									
	Regular size, with condiments									
840	Double patty	1 sandwich	215	51	576	32	32	12.0	14.1	2.8
841	Single patty	1 sandwich	106	45	272	12	10	3.6	3.4	1.0
	Large, with condiments, mayo type dressing, and vegetables									
842	Double patty	1 sandwich	226	54	540	34	27	10.5	10.3	2.8
843	Single patty	1 sandwich	218	56	512	26	27	10.4	11.4	2.2
	Hot dog									
844	Plain	1 sandwich	98	54	242	10	15	5.1	6.9	1.7
845	With chili	1 sandwich	114	48	296	14	13	4.9	6.6	1.2
846	With corn flour coating (corndog)	1 corndog	175	47	460	17	19	5.2	9.1	3.5
847	Hush puppies	5 pieces	78	32	257	5	12	2.7	7.8	0.4
848	Mashed potatoes	⅓ cup	80	79	66	2	1	0.4	0.3	0.2
849	Nachos, with cheese sauce	6-8 nachos	113	40	346	9	19	7.8	8.0	2.2
850	Onion rings, breaded and fried	8-9 rings	83	37	276	4	16	7.0	6.7	0.7
	Pizza (slice = ⅛ of 12" pizza)									
851	Cheese	1 slice	63	48	140	8	3	1.5	1.0	0.5
852	Meat and vegetables	1 slice	79	48	184	13	5	1.5	2.5	0.9
853	Pepperoni	1 slice	71	47	181	10	7	2.2	3.1	1.2
854	Roast beef sandwich, plain	1 sandwich	139	49	346	22	14	3.6	6.8	1.7
855	Salad, tossed, with chicken, no dressing	1½ cups	218	87	105	17	2	0.6	0.7	0.6
856	Salad, tossed, with egg, cheese, no dressing	1½ cups	217	90	102	9	6	3.0	1.8	0.5
	Shake									
857	Chocolate	16 fl oz	333	72	423	11	12	7.7	3.6	0.5
858	Vanilla	16 fl oz	333	75	370	12	10	6.2	2.9	0.4
859	Shrimp, breaded and fried	6-8 shrimp	164	48	454	19	25	5.4	17.4	0.6
	Submarine sandwich (6" long), with oil and vinegar									
860	Cold cuts (with lettuce, cheese, salami, ham, tomato, onion)	1 sandwich	228	58	456	22	19	6.8	8.2	2.3
861	Roast beef (with tomato, lettuce, mayo)	1 sandwich	216	59	410	29	13	7.1	1.8	2.6
862	Tuna salad (with mayo, lettuce)	1 sandwich	256	54	584	30	28	5.3	13.4	7.3
863	Taco, beef	1 small	171	58	369	21	21	11.4	6.6	1.0
864		1 large	263	58	568	32	32	17.5	10.1	1.5
865	Taco salad (with ground beef, cheese, taco shell)	1½ cups	198	72	279	13	15	6.8	5.2	1.7
	Tostada (with cheese, tomato, lettuce)									
866	With beans and beef	1 tostada	225	70	333	16	17	11.5	3.5	0.6
867	With guacamole	1 tostada	131	73	181	6	12	5.0	4.3	1.5

Choles-terol (mg)	Carbo-hydrate (g)	Total dietary fiber (g)	Calcium (mg)	Iron (mg)	Potas-sium (mg)	Sodium (mg)	Vitamin A (IU)	Vitamin A (RE)	Thiamin (mg)	Ribo-flavin (mg)	Niacin (mg)	Ascor-bic acid (mg)	Food No.
103	39	NA	92	5.5	527	742	54	4	0.34	0.41	6.7	1	840
30	34	2.3	126	2.7	251	534	74	10	0.29	0.24	3.9	2	841
122	40	NA	102	5.9	570	791	102	11	0.36	0.38	7.6	1	842
87	40	NA	96	4.9	480	824	312	33	0.41	0.37	7.3	3	843
44	18	NA	24	2.3	143	670	0	0	0.24	0.27	3.6	Tr	844
51	31	NA	19	3.3	166	480	58	6	0.22	0.40	3.7	3	845
79	56	NA	102	6.2	263	973	207	37	0.28	0.70	4.2	0	846
135	35	NA	69	1.4	188	965	94	27	0.00	0.02	2.0	0	847
2	13	NA	17	0.4	235	182	33	8	0.07	0.04	1.0	Tr	848
18	36	NA	272	1.3	172	816	559	92	0.19	0.37	1.5	1	849
14	31	NA	73	0.8	129	430	8	1	0.08	0.10	0.9	1	850
9	21	NA	117	0.6	110	336	382	74	0.18	0.16	2.5	1	851
21	21	NA	101	1.5	179	382	524	101	0.21	0.17	2.0	2	852
14	20	NA	65	0.9	153	267	282	55	0.13	0.23	3.0	2	853
51	33	NA	54	4.2	316	792	210	21	0.38	0.31	5.9	2	854
72	4	NA	37	1.1	447	209	935	96	0.11	0.13	5.9	17	855
98	5	NA	100	0.7	371	119	822	115	0.09	0.17	1.0	10	856
43	68	2.7	376	1.0	666	323	310	77	0.19	0.82	0.5	1	857
37	60	1.3	406	0.3	579	273	433	107	0.15	0.61	0.6	3	858
200	40	NA	84	3.0	184	1,446	120	36	0.21	0.90	0.0	0	859
36	51	NA	189	2.5	394	1,651	424	80	1.00	0.80	5.5	12	860
73	44	NA	41	2.8	330	845	413	50	0.41	0.41	6.0	6	861
49	55	NA	74	2.6	335	1,293	187	41	0.46	0.33	11.3	4	862
56	27	NA	221	2.4	474	802	855	147	0.15	0.44	3.2	2	863
87	41	NA	339	3.7	729	1,233	1,315	226	0.24	0.68	4.9	3	864
44	24	NA	192	2.3	416	762	588	77	0.10	0.36	2.5	4	865
74	30	NA	189	2.5	491	871	1,276	173	0.09	0.50	2.9	4	866
20	16	NA	212	0.8	326	401	879	109	0.07	0.29	1.0	2	867

Food No.	Food Description	Measure of edible portion	Weight (g)	Water (%)	Calories (kcal)	Protein (g)	Total fat (g)	Fatty acids Saturated (g)	Fatty acids Monounsaturated (g)	Fatty acids Polyunsaturated (g)

Poultry and Poultry Products

Food No.	Food Description	Measure of edible portion	Weight (g)	Water (%)	Calories (kcal)	Protein (g)	Total fat (g)	Saturated (g)	Monounsaturated (g)	Polyunsaturated (g)
	Chicken									
	Fried in vegetable shortening, meat with skin									
	Batter dipped									
868	Breast, ½ breast (5.6 oz with bones)	½ breast	140	52	364	35	18	4.9	7.6	4.3
869	Drumstick (3.4 oz with bones)	1 drumstick	72	53	193	16	11	3.0	4.6	2.7
870	Thigh	1 thigh	86	52	238	19	14	3.8	5.8	3.4
871	Wing	1 wing	49	46	159	10	11	2.9	4.4	2.5
	Flour coated									
872	Breast, ½ breast (4.2 oz with bones)	½ breast	98	57	218	31	9	2.4	3.4	1.9
873	Drumstick (2.6 oz with bones)	1 drumstick	49	57	120	13	7	1.8	2.7	1.6
	Fried, meat only									
874	Dark meat	3 oz	85	56	203	25	10	2.7	3.7	2.4
875	Light meat	3 oz	85	60	163	28	5	1.3	1.7	1.1
	Roasted, meat only									
876	Breast, ½ breast (4.2 oz with bone and skin)	½ breast	86	65	142	27	3	0.9	1.1	0.7
877	Drumstick (2.9 oz with bone and skin)	1 drumstick	44	67	76	12	2	0.7	0.8	0.6
878	Thigh	1 thigh	52	63	109	13	6	1.6	2.2	1.3
879	Stewed, meat only, light and dark meat, chopped or diced	1 cup	140	56	332	43	17	4.3	5.7	4.0
880	Chicken giblets, simmered, chopped	1 cup	145	68	228	37	7	2.2	1.7	1.6
881	Chicken liver, simmered	1 liver	20	68	31	5	1	0.4	0.3	0.2
882	Chicken neck, meat only, simmered	1 neck	18	67	32	4	1	0.4	0.5	0.4
883	Duck, roasted, flesh only	½ duck	221	64	444	52	25	9.2	8.2	3.2
	Turkey									
	Roasted, meat only									
884	Dark meat	3 oz	85	63	159	24	6	2.1	1.4	1.8
885	Light meat	3 oz	85	66	133	25	3	0.9	0.5	0.7
886	Light and dark meat, chopped or diced	1 cup	140	65	238	41	7	2.3	1.4	2.0
	Ground, cooked									
887	Patty, from 4 oz raw	1 patty	82	59	193	22	11	2.8	4.0	2.6
888	Crumbled	1 cup	127	59	298	35	17	4.3	6.2	4.1
889	Turkey giblets, simmered, chopped	1 cup	145	65	242	39	7	2.2	1.7	1.7
890	Turkey neck, meat only, simmered	1 neck	152	65	274	41	11	3.7	2.5	3.3
	Poultry food products									
	Chicken									
891	Canned, boneless	5 oz	142	69	234	31	11	3.1	4.5	2.5
892	Frankfurter (10 per 1 lb pkg)	1 frank	45	58	116	6	9	2.5	3.8	1.8
893	Roll, light meat (6 slices per 6-oz pkg)	2 slices	57	69	90	11	4	1.1	1.7	0.9

Cholesterol (mg)	Carbohydrate (g)	Total dietary fiber (g)	Calcium (mg)	Iron (mg)	Potassium (mg)	Sodium (mg)	Vitamin A (IU)	Vitamin A (RE)	Thiamin (mg)	Riboflavin (mg)	Niacin (mg)	Ascorbic acid (mg)	Food No.
119	13	0.4	28	1.8	281	385	94	28	0.16	0.20	14.7	0	868
62	6	0.2	12	1.0	134	194	62	19	0.08	0.15	3.7	0	869
80	8	0.3	15	1.2	165	248	82	25	0.10	0.20	4.9	0	870
39	5	0.1	10	0.6	68	157	55	17	0.05	0.07	2.6	0	871
87	2	0.1	16	1.2	254	74	49	15	0.08	0.13	13.5	0	872
44	1	Tr	6	0.7	112	44	41	12	0.04	0.11	3.0	0	873
82	2	0.0	15	1.3	215	82	67	20	0.08	0.21	6.0	0	874
77	Tr	0.0	14	1.0	224	69	26	8	0.06	0.11	11.4	0	875
73	0	0.0	13	0.9	220	64	18	5	0.06	0.10	11.8	0	876
41	0	0.0	5	0.6	108	42	26	8	0.03	0.10	2.7	0	877
49	0	0.0	6	0.7	124	46	34	10	0.04	0.12	3.4	0	878
116	0	0.0	18	2.0	283	109	157	46	0.16	0.39	9.0	0	879
570	1	0.0	17	9.3	229	84	10,775	3,232	0.13	1.38	5.9	12	880
126	Tr	0.0	3	1.7	28	10	3,275	983	0.03	0.35	0.9	3	881
14	0	0.0	8	0.5	25	12	22	6	0.01	0.05	0.7	0	882
197	0	0.0	27	6.0	557	144	170	51	0.57	1.04	11.3	0	883
72	0	0.0	27	2.0	247	67	0	0	0.05	0.21	3.1	0	884
59	0	0.0	16	1.1	259	54	0	0	0.05	0.11	5.8	0	885
106	0	0.0	35	2.5	417	98	0	0	0.09	0.25	7.6	0	886
84	0	0.0	21	1.6	221	88	0	0	0.04	0.14	4.0	0	887
130	0	0.0	32	2.5	343	136	0	0	0.07	0.21	6.1	0	888
606	3	0.0	19	9.7	290	86	8,752	2,603	0.07	1.31	6.5	2	889
185	0	0.0	56	3.5	226	85	0	0	0.05	0.29	2.6	0	890
88	0	0.0	20	2.2	196	714	166	48	0.02	0.18	9.0	3	891
45	3	0.0	43	0.9	38	617	59	17	0.03	0.05	1.4	0	892
28	1	0.0	24	0.5	129	331	46	14	0.04	0.07	3.0	0	893

Food No.	Food Description	Measure of edible portion	Weight (g)	Water (%)	Calories (kcal)	Protein (g)	Total fat (g)	Fatty acids Saturated (g)	Monounsaturated (g)	Polyunsaturated (g)

Poultry and Poultry Products (continued)

Food No.	Food Description	Measure of edible portion	Weight (g)	Water (%)	Calories (kcal)	Protein (g)	Total fat (g)	Saturated (g)	Monounsaturated (g)	Polyunsaturated (g)
	Poultry food products (continued)									
	Turkey									
894	Gravy and turkey, frozen	5-oz package	142	85	95	8	4	1.2	1.4	0.7
895	Patties, breaded or battered, fried (2.25 oz)	1 patty	64	50	181	9	12	3.0	4.8	3.0
896	Roast, boneless, frozen, seasoned, light and dark meat, cooked	3 oz	85	68	132	18	5	1.6	1.0	1.4

Soups, Sauces, and Gravies

Food No.	Food Description	Measure of edible portion	Weight (g)	Water (%)	Calories (kcal)	Protein (g)	Total fat (g)	Saturated (g)	Monounsaturated (g)	Polyunsaturated (g)
	Soups									
	Canned, condensed									
	Prepared with equal volume of whole milk									
897	Clam chowder, New England	1 cup	248	85	164	9	7	3.0	2.3	1.1
898	Cream of chicken	1 cup	248	85	191	7	11	4.6	4.5	1.6
899	Cream of mushroom	1 cup	248	85	203	6	14	5.1	3.0	4.6
900	Tomato	1 cup	248	85	161	6	6	2.9	1.6	1.1
	Prepared with equal volume of water									
901	Bean with pork	1 cup	253	84	172	8	6	1.5	2.2	1.8
902	Beef broth, bouillon, consomme	1 cup	241	96	29	5	0	0.0	0.0	0.0
903	Beef noodle	1 cup	244	92	83	5	3	1.1	1.2	0.5
904	Chicken noodle	1 cup	241	92	75	4	2	0.7	1.1	0.6
905	Chicken and rice	1 cup	241	94	60	4	2	0.5	0.9	0.4
906	Clam chowder, Manhattan	1 cup	244	92	78	2	2	0.4	0.4	1.3
907	Cream of chicken	1 cup	244	91	117	3	7	2.1	3.3	1.5
908	Cream of mushroom	1 cup	244	90	129	2	9	2.4	1.7	4.2
909	Minestrone	1 cup	241	91	82	4	3	0.6	0.7	1.1
910	Pea, green	1 cup	250	83	165	9	3	1.4	1.0	0.4
911	Tomato	1 cup	244	90	85	2	2	0.4	0.4	1.0
912	Vegetable beef	1 cup	244	92	78	6	2	0.9	0.8	0.1
913	Vegetarian vegetable	1 cup	241	92	72	2	2	0.3	0.8	0.7
	Canned, ready to serve, chunky									
914	Bean with ham	1 cup	243	79	231	13	9	3.3	3.8	0.9
915	Chicken noodle	1 cup	240	84	175	13	6	1.4	2.7	1.5
916	Chicken and vegetable	1 cup	240	83	166	12	5	1.4	2.2	1.0
917	Vegetable	1 cup	240	88	122	4	4	0.6	1.6	1.4
	Canned, ready to serve, low fat, reduced sodium									
918	Chicken broth	1 cup	240	97	17	3	0	0.0	0.0	0.0
919	Chicken noodle	1 cup	237	92	76	6	2	0.4	0.6	0.4
920	Chicken and rice	1 cup	241	88	116	7	3	0.9	1.3	0.7
921	Chicken and rice with vegetables	1 cup	239	91	88	6	1	0.4	0.5	0.5
922	Clam chowder, New England	1 cup	244	89	117	5	2	0.5	0.7	0.4
923	Lentil	1 cup	242	88	126	8	2	0.3	0.8	0.2
924	Minestrone	1 cup	241	87	123	5	3	0.4	0.9	1.0
925	Vegetable	1 cup	238	91	81	4	1	0.3	0.4	0.3

Cholesterol (mg)	Carbohydrate (g)	Total dietary fiber (g)	Calcium (mg)	Iron (mg)	Potassium (mg)	Sodium (mg)	Vitamin A (IU)	Vitamin A (RE)	Thiamin (mg)	Riboflavin (mg)	Niacin (mg)	Ascorbic acid (mg)	Food No.
26	7	0.0	20	1.3	87	787	60	18	0.03	0.18	2.6	0	894
40	10	0.3	9	1.4	176	512	24	7	0.06	0.12	1.5	0	895
45	3	0.0	4	1.4	253	578	0	0	0.04	0.14	5.3	0	896
22	17	1.5	186	1.5	300	992	164	40	0.07	0.24	1.0	3	897
27	15	0.2	181	0.7	273	1,047	714	94	0.07	0.26	0.9	1	898
20	15	0.5	179	0.6	270	918	154	37	0.08	0.28	0.9	2	899
17	22	2.7	159	1.8	449	744	848	109	0.13	0.25	1.5	68	900
3	23	8.6	81	2.0	402	951	888	89	0.09	0.03	0.6	2	901
0	2	0.0	10	0.5	154	636	0	0	0.02	0.03	0.7	1	902
5	9	0.7	15	1.1	100	952	630	63	0.07	0.06	1.1	Tr	903
7	9	0.7	17	0.8	55	1,106	711	72	0.05	0.06	1.4	Tr	904
7	7	0.7	17	0.7	101	815	660	65	0.02	0.02	1.1	Tr	905
2	12	1.5	27	1.6	188	578	964	98	0.03	0.04	0.8	4	906
10	9	0.2	34	0.6	88	986	561	56	0.03	0.06	0.8	Tr	907
2	9	0.5	46	0.5	100	881	0	0	0.05	0.09	0.7	1	908
2	11	1.0	34	0.9	313	911	2,338	234	0.05	0.04	0.9	1	909
0	27	2.8	28	2.0	190	918	203	20	0.11	0.07	1.2	2	910
0	17	0.5	12	1.8	264	695	688	68	0.09	0.05	1.4	66	911
5	10	0.5	17	1.1	173	791	1,891	190	0.04	0.05	1.0	2	912
0	12	0.5	22	1.1	210	822	3,005	301	0.05	0.05	0.9	1	913
22	27	11.2	78	3.2	425	972	3,951	396	0.15	0.15	1.7	4	914
19	17	3.8	24	1.4	108	850	1,222	122	0.07	0.17	4.3	0	915
17	19	NA	26	1.5	367	1,068	5,990	600	0.04	0.17	3.3	6	916
0	19	1.2	55	1.6	396	1,010	5,878	588	0.07	0.06	1.2	6	917
0	1	0.0	19	0.6	204	554	0	0	Tr	0.03	1.6	1	918
19	9	1.2	19	1.1	209	460	920	95	0.11	0.11	3.4	1	919
14	14	0.7	22	1.0	422	482	2,010	202	0.05	0.13	5.0	2	920
17	12	0.7	24	1.2	275	459	1,644	165	0.12	0.07	2.6	1	921
5	20	1.2	17	0.9	283	529	244	59	0.05	0.09	0.9	5	922
0	20	5.6	41	2.7	336	443	951	94	0.11	0.09	0.7	1	923
0	20	1.2	39	1.7	306	470	1,357	135	0.15	0.08	1.0	1	924
5	13	1.4	31	1.5	290	466	3,196	319	0.08	0.07	1.8	1	925

Food No.	Food Description	Measure of edible portion	Weight (g)	Water (%)	Calories (kcal)	Protein (g)	Total fat (g)	Fatty acids: Saturated (g)	Fatty acids: Monounsaturated (g)	Fatty acids: Polyunsaturated (g)

Soups, Sauces, and Gravies (continued)

Food No.	Food Description	Measure of edible portion	Weight (g)	Water (%)	Calories (kcal)	Protein (g)	Total fat (g)	Saturated (g)	Monounsaturated (g)	Polyunsaturated (g)
	Soups (continued)									
	Dehydrated									
	Unprepared									
926	Beef bouillon	1 packet	6	3	14	1	1	0.3	0.2	Tr
927	Onion	1 packet	39	4	115	5	2	0.5	1.4	0.3
	Prepared with water									
928	Chicken noodle	1 cup	252	94	58	2	1	0.3	0.5	0.4
929	Onion	1 cup	246	96	27	1	1	0.1	0.3	0.1
	Home prepared, stock									
930	Beef	1 cup	240	96	31	5	Tr	0.1	0.1	Tr
931	Chicken	1 cup	240	92	86	6	3	0.8	1.4	0.5
932	Fish	1 cup	233	97	40	5	2	0.5	0.5	0.3
	Sauces									
	Home recipe									
933	Cheese	1 cup	243	67	479	25	36	19.5	11.5	3.4
934	White, medium, made with whole milk	1 cup	250	75	368	10	27	7.1	11.1	7.2
	Ready to serve									
935	Barbecue	1 tbsp	16	81	12	Tr	Tr	Tr	0.1	0.1
936	Cheese	¼ cup	63	71	110	4	8	3.8	2.4	1.6
937	Hoisin	1 tbsp	16	44	35	1	1	0.1	0.2	0.3
938	Nacho cheese	¼ cup	63	70	119	5	10	4.2	3.1	2.1
939	Pepper or hot	1 tsp	5	90	1	Tr	Tr	Tr	Tr	Tr
940	Salsa	1 tbsp	16	90	4	Tr	Tr	Tr	Tr	Tr
941	Soy	1 tbsp	16	69	9	1	Tr	Tr	Tr	Tr
942	Spaghetti/marinara/pasta	1 cup	250	87	143	4	5	0.7	2.2	1.8
943	Teriyaki	1 tbsp	18	68	15	1	0	0.0	0.0	0.0
944	Tomato chili	¼ cup	68	68	71	2	Tr	Tr	Tr	0.1
945	Worcestershire	1 tbsp	17	70	11	0	0	0.0	0.0	0.0
	Gravies, canned									
946	Beef	¼ cup	58	87	31	2	1	0.7	0.6	Tr
947	Chicken	¼ cup	60	85	47	1	3	0.8	1.5	0.9
948	Country sausage	¼ cup	62	75	96	3	8	2.0	2.9	2.2
949	Mushroom	¼ cup	60	89	30	1	2	0.2	0.7	0.6
950	Turkey	¼ cup	60	89	31	2	1	0.4	0.5	0.3

Sugars and Sweets

Food No.	Food Description	Measure of edible portion	Weight (g)	Water (%)	Calories (kcal)	Protein (g)	Total fat (g)	Saturated (g)	Monounsaturated (g)	Polyunsaturated (g)
	Candy									
951	BUTTERFINGER (NESTLE)	1 fun size bar	7	2	34	1	1	0.7	0.4	0.2
	Caramel									
952	Plain	1 piece	10	9	39	Tr	1	0.7	0.1	Tr
953	Chocolate flavored roll	1 piece	7	7	25	Tr	Tr	Tr	0.1	0.1
954	Carob	1 oz	28	2	153	2	9	8.2	0.1	0.1
	Chocolate, milk									
955	Plain	1 bar (1.55 oz)	44	1	226	3	14	8.1	4.4	0.5
956	With almonds	1 bar (1.45 oz)	41	2	216	4	14	7.0	5.5	0.9
957	With peanuts, MR. GOODBAR (HERSHEY)	1 bar (1.75 oz)	49	1	267	5	17	7.3	5.7	2.4
958	With rice cereal, NESTLE CRUNCH	1 bar (1.55 oz)	44	1	230	3	12	6.7	3.8	0.4
	Chocolate chips									
959	Milk	1 cup	168	1	862	12	52	31.0	16.7	1.8
960	Semisweet	1 cup	168	1	805	7	50	29.8	16.7	1.6
961	White	1 cup	170	1	916	10	55	33.0	15.5	1.7
962	Chocolate coated peanuts	10 pieces	40	2	208	5	13	5.8	5.2	1.7
963	Chocolate coated raisins	10 pieces	10	11	39	Tr	1	0.9	0.5	0.1
964	Fruit leather, pieces	1 oz	28	12	97	Tr	2	0.3	0.9	0.8

Cholesterol (mg)	Carbohydrate (g)	Total dietary fiber (g)	Calcium (mg)	Iron (mg)	Potassium (mg)	Sodium (mg)	Vitamin A (IU)	Vitamin A (RE)	Thiamin (mg)	Riboflavin (mg)	Niacin (mg)	Ascorbic acid (mg)	Food No.
1	1	0.0	4	0.1	27	1,019	3	Tr	Tr	0.01	0.3	0	926
2	21	4.1	55	0.6	260	3,493	8	1	0.11	0.24	2.0	1	927
10	9	0.3	5	0.5	33	578	15	5	0.20	0.08	1.1	0	928
0	5	1.0	12	0.1	64	849	2	0	0.03	0.06	0.5	Tr	929
0	3	0.0	19	0.6	444	475	0	0	0.08	0.22	2.1	0	930
7	8	0.0	7	0.5	252	343	0	0	0.08	0.20	3.8	Tr	931
2	0	0.0	7	Tr	336	363	0	0	0.08	0.18	2.8	Tr	932
92	13	0.2	756	0.9	345	1,198	1,473	389	0.11	0.59	0.5	1	933
18	23	0.5	295	0.8	390	885	1,383	138	0.17	0.46	1.0	2	934
0	2	0.2	3	0.1	28	130	139	14	Tr	Tr	0.1	1	935
18	4	0.3	116	0.1	19	522	199	40	Tr	0.07	Tr	Tr	936
Tr	7	0.4	5	0.2	19	258	2	Tr	Tr	0.03	0.2	Tr	937
20	3	0.5	118	0.2	20	492	128	32	Tr	0.08	Tr	Tr	938
0	Tr	0.1	Tr	Tr	7	124	14	1	Tr	Tr	Tr	4	939
0	1	0.3	5	0.2	34	69	96	10	0.01	0.01	0.1	2	940
0	1	0.1	3	0.3	64	871	0	0	0.01	0.03	0.4	0	941
0	21	4.0	55	1.8	738	1,030	938	95	0.14	0.10	2.7	20	942
0	3	Tr	5	0.3	41	690	0	0	0.01	0.01	0.2	0	943
0	17	4.0	14	0.5	252	910	462	46	0.06	0.05	1.1	11	944
0	3	0.0	18	0.9	136	167	18	2	0.01	0.02	0.1	2	945
2	3	0.2	3	0.4	47	325	0	0	0.02	0.02	0.4	0	946
1	3	0.2	12	0.3	65	346	221	67	0.01	0.03	0.3	0	947
13	4	0.4	4	0.3	48	236	0	0	0.10	0.04	0.7	Tr	948
0	3	0.2	4	0.4	64	342	0	0	0.02	0.04	0.4	0	949
1	3	0.2	2	0.4	65	346	0	0	0.01	0.05	0.8	0	950
Tr	5	0.2	2	0.1	27	14	0	0	0.01	Tr	0.2	0	951
1	8	0.1	14	Tr	22	25	3	1	Tr	0.02	Tr	Tr	952
0	6	Tr	2	Tr	7	6	1	Tr	Tr	0.01	Tr	Tr	953
1	16	1.1	86	0.4	179	30	7	2	0.03	0.05	0.3	Tr	954
10	26	1.5	84	0.6	169	36	81	24	0.03	0.13	0.1	Tr	955
8	22	2.5	92	0.7	182	30	30	6	0.02	0.18	0.3	Tr	956
4	25	1.7	53	0.6	219	73	70	18	0.08	0.12	1.6	Tr	957
6	29	1.1	74	0.2	151	59	30	9	0.15	0.25	1.7	Tr	958
37	99	5.7	321	2.3	647	138	311	92	0.13	0.51	0.5	1	959
0	106	9.9	54	5.3	613	18	35	3	0.09	0.15	0.7	0	960
36	101	0.0	338	0.4	486	153	60	2	0.11	0.48	1.3	1	961
4	20	1.9	42	0.5	201	16	0	0	0.05	0.07	1.7	0	962
Tr	7	0.4	9	0.2	51	4	4	1	0.01	0.02	Tr	Tr	963
0	22	1.0	5	0.2	46	114	33	3	0.01	0.03	Tr	16	964

Food No.	Food Description	Measure of edible portion	Weight (g)	Water (%)	Calories (kcal)	Protein (g)	Total fat (g)	Fatty acids Saturated (g)	Mono-unsaturated (g)	Poly-unsaturated (g)

Sugars and Sweets (continued)

Food No.	Food Description	Measure of edible portion	Weight (g)	Water (%)	Calories (kcal)	Protein (g)	Total fat (g)	Saturated (g)	Mono-unsaturated (g)	Poly-unsaturated (g)
	Candy (continued)									
965	Fruit leather, rolls	1 large	21	11	74	Tr	1	0.1	0.3	0.1
966		1 small	14	11	49	Tr	Tr	0.1	0.2	0.1
	Fudge, prepared from recipe									
	Chocolate									
967	Plain	1 piece	17	10	65	Tr	1	0.9	0.4	0.1
968	With nuts	1 piece	19	7	81	1	3	1.1	0.8	1.0
	Vanilla									
969	Plain	1 piece	16	11	59	Tr	1	0.5	0.2	Tr
970	With nuts	1 piece	15	8	62	Tr	2	0.6	0.5	0.8
	Gumdrops/gummy candies									
971	Gumdrops (¾" dia)	1 cup	182	1	703	0	0	0.0	0.0	0.0
972		1 medium	4	1	16	0	0	0.0	0.0	0.0
973	Gummy bears	10 bears	22	1	85	0	0	0.0	0.0	0.0
974	Gummy worms	10 worms	74	1	286	0	0	0.0	0.0	0.0
975	Hard candy	1 piece	6	1	24	0	Tr	0.0	0.0	0.0
976		1 small piece	3	1	12	0	Tr	0.0	0.0	0.0
977	Jelly beans	10 large	28	6	104	0	Tr	Tr	0.1	Tr
978		10 small	11	6	40	0	Tr	Tr	Tr	Tr
979	KIT KAT (HERSHEY)	1 bar (1.5 oz)	42	2	216	3	11	6.8	3.1	0.3
	Marshmallows									
980	Miniature	1 cup	50	16	159	1	Tr	Tr	Tr	Tr
981	Regular	1 regular	7	16	23	Tr	Tr	Tr	Tr	Tr
	M&M's (M&M MARS)									
982	Peanut	¼ cup	43	2	222	4	11	4.4	4.7	1.8
983		10 pieces	20	2	103	2	5	2.1	2.2	0.8
984	Plain	¼ cup	52	2	256	2	11	6.8	3.6	0.3
985		10 pieces	7	2	34	Tr	1	0.9	0.5	Tr
986	MILKY WAY (M&M MARS)	1 fun size bar	18	6	76	1	3	1.4	1.1	0.1
987		1 bar (2.15 oz)	61	6	258	3	10	4.8	3.7	0.4
988	REESE'S Peanut butter cup (HERSHEY)	1 miniature cup	7	2	38	1	2	0.8	0.9	0.4
989		1 package (contains 2)	45	2	243	5	14	5.0	5.9	2.5
990	SNICKERS bar (M&M MARS)	1 fun size bar	15	5	72	1	4	1.3	1.6	0.7
991		1 king size bar (4 oz)	113	5	541	9	28	10.2	11.8	5.6
992		1 bar (2 oz)	57	5	273	5	14	5.1	6.0	2.8
993	SPECIAL DARK sweet chocolate (HERSHEY)	1 miniature	8	1	46	Tr	3	1.7	0.9	0.1
994	STARBURST fruit chews (M&M MARS)	1 piece	5	7	20	Tr	Tr	0.1	0.2	0.2
995		1 package (2.07 oz)	59	7	234	Tr	5	0.7	2.1	1.8
	Frosting, ready to eat									
996	Chocolate	1/12 package	38	17	151	Tr	7	2.1	3.4	0.8
997	Vanilla	1/12 package	38	13	159	Tr	6	1.9	3.3	0.9
	Frozen desserts (nondairy)									
998	Fruit and juice bar	1 bar (2.5 fl oz)	77	78	63	1	Tr	0.0	0.0	Tr
999	Ice pop	1 bar (2 fl oz)	59	80	42	0	0	0.0	0.0	0.0
1000	Italian ices	½ cup	116	86	61	Tr	Tr	0.0	0.0	0.0
1001	Fruit butter, apple	1 tbsp	17	56	29	Tr	0	0.0	0.0	0.0
	Gelatin dessert, prepared with gelatin dessert powder and water									
1002	Regular	½ cup	135	85	80	2	0	0.0	0.0	0.0
1003	Reduced calorie (with aspartame)	½ cup	117	98	8	1	0	0.0	0.0	0.0

Cholesterol (mg)	Carbohydrate (g)	Total dietary fiber (g)	Calcium (mg)	Iron (mg)	Potassium (mg)	Sodium (mg)	Vitamin A (IU)	Vitamin A (RE)	Thiamin (mg)	Riboflavin (mg)	Niacin (mg)	Ascorbic acid (mg)	Food No.
0	18	0.8	7	0.2	62	13	24	3	0.01	Tr	Tr	1	965
0	12	0.5	4	0.1	41	9	16	2	0.01	Tr	Tr	1	966
2	14	0.1	7	0.1	18	11	32	8	Tr	0.01	Tr	Tr	967
3	14	0.2	10	0.1	30	11	38	9	0.01	0.02	Tr	Tr	968
3	13	0.0	6	Tr	8	11	33	8	Tr	0.01	Tr	Tr	969
2	11	0.1	7	0.1	17	9	30	7	0.01	0.01	Tr	Tr	970
0	180	0.0	5	0.7	9	80	0	0	0.00	Tr	Tr	0	971
0	4	0.0	Tr	Tr	Tr	2	0	0	0.00	Tr	Tr	0	972
0	22	0.0	1	0.1	1	10	0	0	0.00	Tr	Tr	0	973
0	73	0.0	2	0.3	4	33	0	0	0.00	Tr	Tr	0	974
0	6	0.0	Tr	Tr	Tr	2	0	0	Tr	Tr	Tr	0	975
0	3	0.0	Tr	Tr	Tr	1	0	0	Tr	Tr	Tr	0	976
0	26	0.0	1	0.3	10	7	0	0	0.00	0.00	0.0	0	977
0	10	0.0	Tr	0.1	4	3	0	0	0.00	0.00	0.0	0	978
3	27	0.8	69	0.4	122	32	68	20	0.07	0.23	1.1	Tr	979
0	41	0.1	2	0.1	3	24	1	0	Tr	Tr	Tr	0	980
0	6	Tr	Tr	Tr	Tr	3	Tr	0	Tr	Tr	Tr	0	981
4	26	1.5	43	0.5	149	21	40	10	0.04	0.07	1.6	Tr	982
2	12	0.7	20	0.2	69	10	19	5	0.02	0.03	0.7	Tr	983
7	37	1.3	55	0.6	138	32	106	28	0.03	0.11	0.1	Tr	984
1	5	0.2	7	0.1	19	4	14	4	Tr	0.01	Tr	Tr	985
3	13	0.3	23	0.1	43	43	19	6	0.01	0.04	0.1	Tr	986
9	44	1.0	79	0.5	147	146	66	20	0.02	0.14	0.2	1	987
Tr	4	0.2	5	0.1	25	22	5	1	0.02	0.01	0.3	Tr	988
2	25	1.4	35	0.5	158	143	33	9	0.11	0.08	2.1	Tr	989
2	9	0.4	14	0.1	49	40	23	6	0.01	0.02	0.6	Tr	990
15	67	2.8	106	0.9	366	301	172	44	0.11	0.17	4.7	1	991
7	34	1.4	54	0.4	185	152	87	22	0.06	0.09	2.4	Tr	992
Tr	5	0.4	2	0.2	25	1	3	Tr	Tr	0.01	Tr	0	993
0	4	0.0	Tr	Tr	Tr	3	0	0	Tr	Tr	Tr	3	994
0	50	0.0	2	0.1	1	33	0	0	Tr	Tr	Tr	31	995
0	24	0.2	3	0.5	74	70	249	75	Tr	0.01	Tr	0	996
0	26	Tr	1	Tr	14	34	283	86	0.00	Tr	Tr	0	997
0	16	0.0	4	0.1	41	3	22	2	0.01	0.01	0.1	7	998
0	11	0.0	0	0.0	2	7	0	0	0.00	0.00	0.0	0	999
0	16	0.0	1	0.1	7	5	194	0	0.01	0.01	0.8	1	1000
0	7	0.3	2	0.1	15	1	20	2	Tr	Tr	Tr	Tr	1001
0	19	0.0	3	Tr	1	57	0	0	0.00	Tr	Tr	0	1002
0	1	0.0	2	Tr	0	56	0	0	0.00	Tr	Tr	0	1003

Food No.	Food Description	Measure of edible portion	Weight (g)	Water (%)	Calories (kcal)	Protein (g)	Total fat (g)	Fatty acids: Saturated (g)	Fatty acids: Monounsaturated (g)	Fatty acids: Polyunsaturated (g)

Sugars and Sweets (continued)

Food No.	Food Description	Measure of edible portion	Weight (g)	Water (%)	Calories (kcal)	Protein (g)	Total fat (g)	Saturated (g)	Monounsaturated (g)	Polyunsaturated (g)
1004	Honey, strained or extracted	1 tbsp	21	17	64	Tr	0	0.0	0.0	0.0
1005		1 cup	339	17	1,031	1	0	0.0	0.0	0.0
1006	Jams and preserves	1 tbsp	20	30	56	Tr	Tr	Tr	Tr	0.0
1007		1 packet (0.5 oz)	14	30	39	Tr	Tr	Tr	Tr	0.0
1008	Jellies	1 tbsp	19	29	54	Tr	Tr	Tr	Tr	Tr
1009		1 packet (0.5 oz)	14	29	40	Tr	Tr	Tr	Tr	Tr
	Puddings									
	Prepared with dry mix and 2% milk									
	Chocolate									
1010	Instant	½ cup	147	75	150	5	3	1.6	0.9	0.2
1011	Regular (cooked)	½ cup	142	74	151	5	3	1.8	0.8	0.1
	Vanilla									
1012	Instant	½ cup	142	75	148	4	2	1.4	0.7	0.1
1013	Regular (cooked)	½ cup	140	76	141	4	2	1.5	0.7	0.1
	Ready to eat									
	Regular									
1014	Chocolate	4 oz	113	69	150	3	5	0.8	1.9	1.6
1015	Rice	4 oz	113	68	184	2	8	1.3	3.6	3.2
1016	Tapioca	4 oz	113	74	134	2	4	0.7	1.8	1.5
1017	Vanilla	4 oz	113	71	147	3	4	0.6	1.7	1.5
	Fat free									
1018	Chocolate	4 oz	113	76	107	3	Tr	0.3	0.1	Tr
1019	Tapioca	4 oz	113	77	98	2	Tr	0.1	Tr	Tr
1020	Vanilla	4 oz	113	76	105	2	Tr	0.1	Tr	Tr
	Sugar									
	Brown									
1021	Packed	1 cup	220	2	827	0	0	0.0	0.0	0.0
1022	Unpacked	1 cup	145	2	545	0	0	0.0	0.0	0.0
1023		1 tbsp	9	2	34	0	0	0.0	0.0	0.0
	White									
1024	Granulated	1 packet	6	0	23	0	0	0.0	0.0	0.0
1025		1 tsp	4	0	16	0	0	0.0	0.0	0.0
1026		1 cup	200	0	774	0	0	0.0	0.0	0.0
1027	Powdered, unsifted	1 tbsp	8	Tr	31	0	Tr	Tr	Tr	Tr
1028		1 cup	120	Tr	467	0	Tr	Tr	Tr	0.1
	Syrup									
	Chocolate flavored syrup or topping									
1029	Thin type	1 tbsp	19	31	53	Tr	Tr	0.1	0.1	Tr
1030	Fudge type	1 tbsp	19	22	67	1	2	0.8	0.7	0.1
1031	Corn, light	1 tbsp	20	23	56	0	0	0.0	0.0	0.0
1032	Maple	1 tbsp	20	32	52	0	Tr	Tr	Tr	Tr
1033	Molasses, blackstrap	1 tbsp	20	29	47	0	0	0.0	0.0	0.0
1034		1 cup	328	29	771	0	0	0.0	0.0	0.0
	Table blend, pancake									
1035	Regular	1 tbsp	20	24	57	0	0	0.0	0.0	0.0
1036	Reduced calorie	1 tbsp	15	55	25	0	0	0.0	0.0	0.0

Cholesterol (mg)	Carbohydrate (g)	Total dietary fiber (g)	Calcium (mg)	Iron (mg)	Potassium (mg)	Sodium (mg)	Vitamin A (IU)	Vitamin A (RE)	Thiamin (mg)	Riboflavin (mg)	Niacin (mg)	Ascorbic acid (mg)	Food No.
0	17	Tr	1	0.1	11	1	0	0	0.00	0.01	Tr	Tr	1004
0	279	0.7	20	1.4	176	14	0	0	0.00	0.13	0.4	2	1005
0	14	0.2	4	0.1	15	6	2	Tr	0.00	Tr	Tr	2	1006
0	10	0.2	3	0.1	11	4	2	Tr	0.00	Tr	Tr	1	1007
0	13	0.2	2	Tr	12	5	3	Tr	Tr	Tr	Tr	Tr	1008
0	10	0.1	1	Tr	9	4	2	Tr	Tr	Tr	Tr	Tr	1009
9	28	0.6	153	0.4	247	417	253	56	0.05	0.21	0.1	1	1010
10	28	0.4	160	0.5	240	149	253	68	0.05	0.21	0.2	1	1011
9	28	0.0	146	0.1	185	406	241	64	0.05	0.20	0.1	1	1012
10	26	0.0	153	0.1	193	224	252	70	0.04	0.20	0.1	1	1013
3	26	1.1	102	0.6	203	146	41	12	0.03	0.18	0.4	2	1014
1	25	0.1	59	0.3	68	96	129	40	0.02	0.08	0.2	1	1015
1	22	0.1	95	0.3	110	180	0	0	0.02	0.11	0.4	1	1016
8	25	0.1	99	0.1	128	153	24	7	0.02	0.16	0.3	0	1017
2	23	0.9	89	0.6	235	192	174	52	0.02	0.12	0.1	Tr	1018
1	23	0.1	76	0.2	99	251	121	36	0.02	0.09	0.1	Tr	1019
1	24	0.1	86	Tr	123	241	174	52	0.02	0.10	0.1	Tr	1020
0	214	0.0	187	4.2	761	86	0	0	0.02	0.02	0.2	0	1021
0	141	0.0	123	2.8	502	57	0	0	0.01	0.01	0.1	0	1022
0	9	0.0	8	0.2	31	4	0	0	Tr	Tr	Tr	0	1023
0	6	0.0	Tr	Tr	Tr	Tr	0	0	0.00	Tr	0.0	0	1024
0	4	0.0	Tr	Tr	Tr	Tr	0	0	0.00	Tr	0.0	0	1025
0	200	0.0	2	0.1	4	2	0	0	0.00	0.04	0.0	0	1026
0	8	0.0	Tr	Tr	Tr	Tr	0	0	0.00	0.00	0.0	0	1027
0	119	0.0	1	0.1	2	1	0	0	0.00	0.00	0.0	0	1028
0	12	0.3	3	0.4	43	14	6	1	Tr	0.01	0.1	Tr	1029
Tr	12	0.5	15	0.2	69	66	3	1	0.01	0.04	0.1	Tr	1030
0	15	0.0	1	Tr	1	24	0	0	Tr	Tr	Tr	0	1031
0	13	0.0	13	0.2	41	2	0	0	Tr	Tr	Tr	0	1032
0	12	0.0	172	3.5	498	11	0	0	0.01	0.01	0.2	0	1033
0	199	0.0	2,821	57.4	8,174	180	0	0	0.11	0.17	3.5	0	1034
0	15	0.0	Tr	Tr	Tr	17	0	0	Tr	Tr	Tr	0	1035
0	7	0.0	Tr	Tr	Tr	30	0	0	Tr	Tr	Tr	0	1036

Food No.	Food Description	Measure of edible portion	Weight (g)	Water (%)	Calories (kcal)	Protein (g)	Total fat (g)	Fatty acids Saturated (g)	Fatty acids Monounsaturated (g)	Fatty acids Polyunsaturated (g)

Vegetables and Vegetable Products

Food No.	Food Description	Measure of edible portion	Weight (g)	Water (%)	Calories (kcal)	Protein (g)	Total fat (g)	Saturated (g)	Monounsaturated (g)	Polyunsaturated (g)
1037	Alfalfa sprouts, raw	1 cup	33	91	10	1	Tr	Tr	Tr	0.1
1038	Artichokes, globe or French, cooked, drained	1 cup	168	84	84	6	Tr	0.1	Tr	0.1
1039		1 medium	120	84	60	4	Tr	Tr	Tr	0.1
	Asparagus, green									
	Cooked, drained									
1040	From raw	1 cup	180	92	43	5	1	0.1	Tr	0.2
1041		4 spears	60	92	14	2	Tr	Tr	Tr	0.1
1042	From frozen	1 cup	180	91	50	5	1	0.2	Tr	0.3
1043		4 spears	60	91	17	2	Tr	0.1	Tr	0.1
1044	Canned, spears, about 5" long, drained	1 cup	242	94	46	5	2	0.4	0.1	0.7
1045		4 spears	72	94	14	2	Tr	0.1	Tr	0.2
1046	Bamboo shoots, canned, drained	1 cup	131	94	25	2	1	0.1	Tr	0.2
	Beans									
	Lima, immature seeds, frozen, cooked, drained									
1047	Ford hooks	1 cup	170	74	170	10	1	0.1	Tr	0.3
1048	Baby limas	1 cup	180	72	189	12	1	0.1	Tr	0.3
	Snap, cut									
	Cooked, drained									
	From raw									
1049	Green	1 cup	125	89	44	2	Tr	0.1	Tr	0.2
1050	Yellow	1 cup	125	89	44	2	Tr	0.1	Tr	0.2
	From frozen									
1051	Green	1 cup	135	91	38	2	Tr	0.1	Tr	0.1
1052	Yellow	1 cup	135	91	38	2	Tr	0.1	Tr	0.1
	Canned, drained									
1053	Green	1 cup	135	93	27	2	Tr	Tr	Tr	0.1
1054	Yellow	1 cup	135	93	27	2	Tr	Tr	Tr	0.1
	Beans, dry. See Legumes.									
	Bean sprouts (mung)									
1055	Raw	1 cup	104	90	31	3	Tr	Tr	Tr	0.1
1056	Cooked, drained	1 cup	124	93	26	3	Tr	Tr	Tr	Tr
	Beets									
	Cooked, drained									
1057	Slices	1 cup	170	87	75	3	Tr	Tr	0.1	0.1
1058	Whole beet, 2" dia	1 beet	50	87	22	1	Tr	Tr	Tr	Tr
	Canned, drained									
1059	Slices	1 cup	170	91	53	2	Tr	Tr	Tr	0.1
1060	Whole beet	1 beet	24	91	7	Tr	Tr	Tr	Tr	Tr
1061	Beet greens, leaves and stems, cooked, drained, 1" pieces	1 cup	144	89	39	4	Tr	Tr	0.1	0.1
	Black eyed peas, immature seeds, cooked, drained									
1062	From raw	1 cup	165	75	160	5	1	0.2	0.1	0.3
1063	From frozen	1 cup	170	66	224	14	1	0.3	0.1	0.5
	Broccoli									
	Raw									
1064	Chopped or diced	1 cup	88	91	25	3	Tr	Tr	Tr	0.1
1065	Spear, about 5" long	1 spear	31	91	9	1	Tr	Tr	Tr	0.1
1066	Flower cluster	1 floweret	11	91	3	Tr	Tr	Tr	Tr	Tr
	Cooked, drained									
	From raw									
1067	Chopped	1 cup	156	91	44	5	1	0.1	Tr	0.3
1068	Spear, about 5" long	1 spear	37	91	10	1	Tr	Tr	Tr	0.1
1069	From frozen, chopped	1 cup	184	91	52	6	Tr	Tr	Tr	0.1

Cholesterol (mg)	Carbohydrate (g)	Total dietary fiber (g)	Calcium (mg)	Iron (mg)	Potassium (mg)	Sodium (mg)	Vitamin A (IU)	Vitamin A (RE)	Thiamin (mg)	Riboflavin (mg)	Niacin (mg)	Ascorbic acid (mg)	Food No.
0	1	0.8	11	0.3	26	2	51	5	0.03	0.04	0.2	3	1037
0	19	9.1	76	2.2	595	160	297	30	0.11	0.11	1.7	17	1038
0	13	6.5	54	1.5	425	114	212	22	0.08	0.08	1.2	12	1039
0	8	2.9	36	1.3	288	20	970	97	0.22	0.23	1.9	19	1040
0	3	1.0	12	0.4	96	7	323	32	0.07	0.08	0.6	6	1041
0	9	2.9	41	1.2	392	7	1,472	148	0.12	0.19	1.9	44	1042
0	3	1.0	14	0.4	131	2	491	49	0.04	0.06	0.6	15	1043
0	6	3.9	39	4.4	416	695	1,285	128	0.15	0.24	2.3	45	1044
0	2	1.2	12	1.3	124	207	382	38	0.04	0.07	0.7	13	1045
0	4	1.8	10	0.4	105	9	10	1	0.03	0.03	0.2	1	1046
0	32	9.9	37	2.3	694	90	323	32	0.13	0.10	1.8	22	1047
0	35	10.8	50	3.5	740	52	301	31	0.13	0.10	1.4	10	1048
0	10	4.0	58	1.6	374	4	833	84	0.09	0.12	0.8	12	1049
0	10	4.1	58	1.6	374	4	101	10	0.09	0.12	0.8	12	1050
0	9	4.1	66	1.2	170	12	541	54	0.05	0.12	0.5	6	1051
0	9	4.1	66	1.2	170	12	151	15	0.05	0.12	0.5	6	1052
0	6	2.6	35	1.2	147	354	471	47	0.02	0.08	0.3	6	1053
0	6	1.8	35	1.2	147	339	142	15	0.02	0.08	0.3	6	1054
0	6	1.9	14	0.9	155	6	22	2	0.09	0.13	0.8	14	1055
0	5	1.5	15	0.8	125	12	17	1	0.06	0.13	1.0	14	1056
0	17	3.4	27	1.3	519	131	60	7	0.05	0.07	0.6	6	1057
0	5	1.0	8	0.4	153	39	18	2	0.01	0.02	0.2	2	1058
0	12	2.9	26	3.1	252	330	19	2	0.02	0.07	0.3	7	1059
0	2	0.4	4	0.4	36	47	3	Tr	Tr	0.01	Tr	1	1060
0	8	4.2	164	2.7	1,309	347	7,344	734	0.17	0.42	0.7	36	1061
0	34	8.3	211	1.8	690	7	1,305	130	0.17	0.24	2.3	4	1062
0	40	10.9	39	3.6	638	9	128	14	0.44	0.11	1.2	4	1063
0	5	2.6	42	0.8	286	24	1,357	136	0.06	0.10	0.6	82	1064
0	2	0.9	15	0.3	101	8	478	48	0.02	0.04	0.2	29	1065
0	1	0.3	5	0.1	36	3	330	33	0.01	0.01	0.1	10	1066
0	8	4.5	72	1.3	456	41	2,165	217	0.09	0.18	0.9	116	1067
0	2	1.1	17	0.3	108	10	514	51	0.02	0.04	0.2	28	1068
0	10	5.5	94	1.1	331	44	3,481	348	0.10	0.15	0.8	74	1069

Food No.	Food Description	Measure of edible portion	Weight (g)	Water (%)	Calories (kcal)	Protein (g)	Total fat (g)	Fatty acids Saturated (g)	Monounsaturated (g)	Polyunsaturated (g)

Vegetables and Vegetable Products (continued)

Food No.	Food Description	Measure of edible portion	Weight (g)	Water (%)	Calories (kcal)	Protein (g)	Total fat (g)	Saturated (g)	Monounsaturated (g)	Polyunsaturated (g)
	Brussels sprouts, cooked, drained									
1070	From raw	1 cup	156	87	61	4	1	0.2	0.1	0.4
1071	From frozen	1 cup	155	87	65	6	1	0.1	Tr	0.3
	Cabbage, common varieties, shredded									
1072	Raw	1 cup	70	92	18	1	Tr	Tr	Tr	0.1
1073	Cooked, drained	1 cup	150	94	33	2	1	0.1	Tr	0.3
	Cabbage, Chinese, shredded, cooked, drained									
1074	Pak choi or bok choy	1 cup	170	96	20	3	Tr	Tr	Tr	0.1
1075	Pe tsai	1 cup	119	95	17	2	Tr	Tr	Tr	0.1
1076	Cabbage, red, raw, shredded	1 cup	70	92	19	1	Tr	Tr	Tr	0.1
1077	Cabbage, savoy, raw, shredded	1 cup	70	91	19	1	Tr	Tr	Tr	Tr
1078	Carrot juice, canned	1 cup	236	89	94	2	Tr	0.1	Tr	0.2
	Carrots									
	Raw									
1079	Whole, 7½" long	1 carrot	72	88	31	1	Tr	Tr	Tr	0.1
1080	Grated	1 cup	110	88	47	1	Tr	Tr	Tr	0.1
1081	Baby	1 medium	10	90	4	Tr	Tr	Tr	Tr	Tr
	Cooked, sliced, drained									
1082	From raw	1 cup	156	87	70	2	Tr	0.1	Tr	0.1
1083	From frozen	1 cup	146	90	53	2	Tr	Tr	Tr	0.1
1084	Canned, sliced, drained	1 cup	146	93	37	1	Tr	0.1	Tr	0.1
	Cauliflower									
1085	Raw	1 floweret	13	92	3	Tr	Tr	Tr	Tr	Tr
1086		1 cup	100	92	25	2	Tr	Tr	Tr	0.1
	Cooked, drained, 1" pieces									
1087	From raw	1 cup	124	93	29	2	1	0.1	Tr	0.3
1088		3 flowerets	54	93	12	1	Tr	Tr	Tr	0.1
1089	From frozen	1 cup	180	94	34	3	Tr	0.1	Tr	0.2
	Celery									
	Raw									
1090	Stalk, 7½ to 8" long	1 stalk	40	95	6	Tr	Tr	Tr	Tr	Tr
1091	Pieces, diced	1 cup	120	95	19	1	Tr	Tr	Tr	0.1
	Cooked, drained									
1092	Stalk, medium	1 stalk	38	94	7	Tr	Tr	Tr	Tr	Tr
1093	Pieces, diced	1 cup	150	94	27	1	Tr	0.1	Tr	0.1
1094	Chives, raw, chopped	1 tbsp	3	91	1	Tr	Tr	Tr	Tr	Tr
1095	Cilantro, raw	1 tsp	2	92	Tr	Tr	Tr	Tr	Tr	Tr
1096	Coleslaw, home prepared	1 cup	120	82	83	2	3	0.5	0.8	1.6
	Collards, cooked, drained, chopped									
1097	From raw	1 cup	190	92	49	4	1	0.1	Tr	0.3
1098	From frozen	1 cup	170	88	61	5	1	0.1	Tr	0.4
	Corn, sweet, yellow									
	Cooked, drained									
1099	From raw, kernels on cob	1 ear	77	70	83	3	1	0.2	0.3	0.5
	From frozen									
1100	Kernels on cob	1 ear	63	73	59	2	Tr	0.1	0.1	0.2
1101	Kernels	1 cup	164	77	131	5	1	0.1	0.2	0.3
	Canned									
1102	Cream style	1 cup	256	79	184	4	1	0.2	0.3	0.5
1103	Whole kernel, vacuum pack	1 cup	210	77	166	5	1	0.2	0.3	0.5
1104	Corn, sweet, white, cooked, drained	1 ear	77	70	83	3	1	0.2	0.3	0.5

*White varieties contain only a trace amount of vitamin A; other nutrients are the same.

Cholesterol (mg)	Carbohydrate (g)	Total dietary fiber (g)	Calcium (mg)	Iron (mg)	Potassium (mg)	Sodium (mg)	Vitamin A		Thiamin (mg)	Riboflavin (mg)	Niacin (mg)	Ascorbic acid (mg)	Food No.
							(IU)	(RE)					
0	14	4.1	56	1.9	495	33	1,122	112	0.17	0.12	0.9	97	1070
0	13	6.4	37	1.1	504	36	913	91	0.16	0.18	0.8	71	1071
0	4	1.6	33	0.4	172	13	93	9	0.04	0.03	0.2	23	1072
0	7	3.5	47	0.3	146	12	198	20	0.09	0.08	0.4	30	1073
0	3	2.7	158	1.8	631	58	4,366	437	0.05	0.11	0.7	44	1074
0	3	3.2	38	0.4	268	11	1,151	115	0.05	0.05	0.6	19	1075
0	4	1.4	36	0.3	144	8	28	3	0.04	0.02	0.2	40	1076
0	4	2.2	25	0.3	161	20	700	70	0.05	0.02	0.2	22	1077
0	22	1.9	57	1.1	689	68	25,833	2,584	0.22	0.13	0.9	20	1078
0	7	2.2	19	0.4	233	25	20,253	2,025	0.07	0.04	0.7	7	1079
0	11	3.3	30	0.6	355	39	30,942	3,094	0.11	0.06	1.0	10	1080
0	1	0.2	2	0.1	28	4	1,501	150	Tr	0.01	0.1	1	1081
0	16	5.1	48	1.0	354	103	38,304	3,830	0.05	0.09	0.8	4	1082
0	12	5.1	41	0.7	231	86	25,845	2,584	0.04	0.05	0.6	4	1083
0	8	2.2	37	0.9	261	353	20,110	2,010	0.03	0.04	0.8	4	1084
0	1	0.3	3	0.1	39	4	2	Tr	0.01	0.01	0.1	6	1085
0	5	2.5	22	0.4	303	30	19	2	0.06	0.06	0.5	46	1086
0	5	3.3	20	0.4	176	19	21	2	0.05	0.06	0.5	55	1087
0	2	1.5	9	0.2	77	8	9	1	0.02	0.03	0.2	24	1088
0	7	4.9	31	0.7	250	32	40	4	0.07	0.10	0.6	56	1089
0	1	0.7	16	0.2	115	35	54	5	0.02	0.02	0.1	3	1090
0	4	2.0	48	0.5	344	104	161	16	0.06	0.05	0.4	8	1091
0	2	0.6	16	0.2	108	35	50	5	0.02	0.02	0.1	2	1092
0	6	2.4	63	0.6	426	137	198	20	0.06	0.07	0.5	9	1093
0	Tr	0.1	3	Tr	9	Tr	131	13	Tr	Tr	Tr	2	1094
0	Tr	Tr	1	Tr	8	1	98	10	Tr	Tr	Tr	1	1095
10	15	1.8	54	0.7	217	28	762	98	0.08	0.07	0.3	39	1096
0	9	5.3	226	0.9	494	17	5,945	595	0.08	0.20	1.1	35	1097
0	12	4.8	357	1.9	427	85	10,168	1,017	0.08	0.20	1.1	45	1098
0	19	2.2	2	0.5	192	13	167	17	0.17	0.06	1.2	5	1099
0	14	1.8	2	0.4	158	3	133*	13*	0.11	0.04	1.0	3	1100
0	32	3.9	7	0.6	241	8	361*	36*	0.14	0.12	2.1	5	1101
0	46	3.1	8	1.0	343	730	248*	26*	0.06	0.14	2.5	12	1102
0	41	4.2	11	0.9	391	571	506*	50*	0.09	0.15	2.5	17	1103
0	19	2.1	2	0.5	192	13	0	0	0.17	0.06	1.2	5	1104

Food No.	Food Description	Measure of edible portion	Weight (g)	Water (%)	Calories (kcal)	Protein (g)	Total fat (g)	Fatty acids: Saturated (g)	Fatty acids: Monounsaturated (g)	Fatty acids: Polyunsaturated (g)

Vegetables and Vegetable Products (continued)

Food No.	Food Description	Measure of edible portion	Weight (g)	Water (%)	Calories (kcal)	Protein (g)	Total fat (g)	Saturated (g)	Monounsaturated (g)	Polyunsaturated (g)
	Cucumber									
	Peeled									
1105	Sliced	1 cup	119	96	14	1	Tr	Tr	Tr	0.1
1106	Whole, 8¼" long	1 large	280	96	34	2	Tr	0.1	Tr	0.2
	Unpeeled									
1107	Sliced	1 cup	104	96	14	1	Tr	Tr	Tr	0.1
1108	Whole, 8¼" long	1 large	301	96	39	2	Tr	0.1	Tr	0.2
1109	Dandelion greens, cooked, drained	1 cup	105	90	35	2	1	0.2	Tr	0.3
1110	Dill weed, raw	5 sprigs	1	86	Tr	Tr	Tr	Tr	Tr	Tr
1111	Eggplant, cooked, drained	1 cup	99	92	28	1	Tr	Tr	Tr	0.1
1112	Endive, curly (including escarole), raw, small pieces	1 cup	50	94	9	1	Tr	Tr	Tr	Tr
1113	Garlic, raw	1 clove	3	59	4	Tr	Tr	Tr	Tr	Tr
1114	Hearts of palm, canned	1 piece	33	90	9	1	Tr	Tr	Tr	0.1
1115	Jerusalem artichoke, raw, sliced	1 cup	150	78	114	3	Tr	0.0	Tr	Tr
	Kale, cooked, drained, chopped									
1116	From raw	1 cup	130	91	36	2	1	0.1	Tr	0.3
1117	From frozen	1 cup	130	91	39	4	1	0.1	Tr	0.3
1118	Kohlrabi, cooked, drained, slices	1 cup	165	90	48	3	Tr	Tr	Tr	0.1
1119	Leeks, bulb and lower leaf portion, chopped or diced, cooked, drained	1 cup	104	91	32	1	Tr	Tr	Tr	0.1
	Lettuce, raw									
	Butterhead, as Boston types									
1120	Leaf	1 medium leaf	8	96	1	Tr	Tr	Tr	Tr	Tr
1121	Head, 5" dia	1 head	163	96	21	2	Tr	Tr	Tr	0.2
	Crisphead, as iceberg									
1122	Leaf	1 medium	8	96	1	Tr	Tr	Tr	Tr	Tr
1123	Head, 6" dia	1 head	539	96	65	5	1	0.1	Tr	0.5
1124	Pieces, shredded or chopped	1 cup	55	96	7	1	Tr	Tr	Tr	0.1
	Looseleaf									
1125	Leaf	1 leaf	10	94	2	Tr	Tr	Tr	Tr	Tr
1126	Pieces, shredded	1 cup	56	94	10	1	Tr	Tr	Tr	0.1
	Romaine or cos									
1127	Innerleaf	1 leaf	10	95	1	Tr	Tr	Tr	Tr	Tr
1128	Pieces, shredded	1 cup	56	95	8	1	Tr	Tr	Tr	0.1
	Mushrooms									
1129	Raw, pieces or slices	1 cup	70	92	18	2	Tr	Tr	Tr	0.1
1130	Cooked, drained, pieces	1 cup	156	91	42	3	1	0.1	Tr	0.3
1131	Canned, drained, pieces	1 cup	156	91	37	3	Tr	0.1	Tr	0.2
	Mushrooms, shiitake									
1132	Cooked pieces	1 cup	145	83	80	2	Tr	0.1	0.1	Tr
1133	Dried	1 mushroom	4	10	11	Tr	Tr	Tr	Tr	Tr
1134	Mustard greens, cooked, drained	1 cup	140	94	21	3	Tr	Tr	0.2	0.1
	Okra, sliced, cooked, drained									
1135	From raw	1 cup	160	90	51	3	Tr	0.1	Tr	0.1
1136	From frozen	1 cup	184	91	52	4	1	0.1	0.1	0.1
	Onions									
	Raw									
1137	Chopped	1 cup	160	90	61	2	Tr	Tr	Tr	0.1
1138	Whole, medium, 2½" dia	1 whole	110	90	42	1	Tr	Tr	Tr	0.1
1139	Slice, ⅛" thick	1 slice	14	90	5	Tr	Tr	Tr	Tr	Tr

Cholesterol (mg)	Carbohydrate (g)	Total dietary fiber (g)	Calcium (mg)	Iron (mg)	Potassium (mg)	Sodium (mg)	Vitamin A (IU)	Vitamin A (RE)	Thiamin (mg)	Riboflavin (mg)	Niacin (mg)	Ascorbic acid (mg)	Food No.
0	3	0.8	17	0.2	176	2	88	8	0.02	0.01	0.1	3	1105
0	7	2.0	39	0.4	414	6	207	20	0.06	0.03	0.3	8	1106
0	3	0.8	15	0.3	150	2	224	22	0.02	0.02	0.2	6	1107
0	8	2.4	42	0.8	433	6	647	63	0.07	0.07	0.7	16	1108
0	7	3.0	147	1.9	244	46	12,285	1,229	0.14	0.18	0.5	19	1109
0	Tr	Tr	2	0.1	7	1	77	8	Tr	Tr	Tr	1	1110
0	7	2.5	6	0.3	246	3	63	6	0.08	0.02	0.6	1	1111
0	2	1.6	26	0.4	157	11	1,025	103	0.04	0.04	0.2	3	1112
0	1	0.1	5	0.1	12	1	0	0	0.01	Tr	Tr	1	1113
0	2	0.8	19	1.0	58	141	0	0	Tr	0.02	0.1	3	1114
0	26	2.4	21	5.1	644	6	30	3	0.30	0.09	2.0	6	1115
0	7	2.6	94	1.2	296	30	9,620	962	0.07	0.09	0.7	53	1116
0	7	2.6	179	1.2	417	20	8,260	826	0.06	0.15	0.9	33	1117
0	11	1.8	41	0.7	561	35	58	7	0.07	0.03	0.6	89	1118
0	8	1.0	31	1.1	90	10	48	5	0.03	0.02	0.2	4	1119
0	Tr	0.1	2	Tr	19	Tr	73	7	Tr	Tr	Tr	1	1120
0	4	1.6	52	0.5	419	8	1,581	158	0.10	0.10	0.5	13	1121
0	Tr	0.1	2	Tr	13	1	26	3	Tr	Tr	Tr	Tr	1122
0	11	7.5	102	2.7	852	49	1,779	178	0.25	0.16	1.0	21	1123
0	1	0.8	10	0.3	87	5	182	18	0.03	0.02	0.1	2	1124
0	Tr	0.2	7	0.1	26	1	190	19	0.01	0.01	Tr	2	1125
0	2	1.1	38	0.8	148	5	1,064	106	0.03	0.04	0.2	10	1126
0	Tr	0.2	4	0.1	29	1	260	26	0.01	0.01	0.1	2	1127
0	1	1.0	20	0.6	162	4	1,456	146	0.06	0.06	0.3	13	1128
0	3	0.8	4	0.7	259	3	0	0	0.06	0.30	2.8	2	1129
0	8	3.4	9	2.7	555	3	0	0	0.11	0.47	7.0	6	1130
0	8	3.7	17	1.2	201	663	0	0	0.13	0.03	2.5	0	1131
0	21	3.0	4	0.6	170	6	0	0	0.05	0.25	2.2	Tr	1132
0	3	0.4	Tr	0.1	55	Tr	0	0	0.01	0.05	0.5	Tr	1133
0	3	2.8	104	1.0	283	22	4,243	424	0.06	0.09	0.6	35	1134
0	12	4.0	101	0.7	515	8	920	93	0.21	0.09	1.4	26	1135
0	11	5.2	177	1.2	431	6	946	94	0.18	0.23	1.4	22	1136
0	14	2.9	32	0.4	251	5	0	0	0.07	0.03	0.2	10	1137
0	9	2.0	22	0.2	173	3	0	0	0.05	0.02	0.2	7	1138
0	1	0.3	3	Tr	22	Tr	0	0	0.01	Tr	Tr	1	1139

Food No.	Food Description	Measure of edible portion	Weight (g)	Water (%)	Calories (kcal)	Protein (g)	Total fat (g)	Fatty acids: Saturated (g)	Fatty acids: Monounsaturated (g)	Fatty acids: Polyunsaturated (g)

Vegetables and Vegetable Products (continued)

Food No.	Food Description	Measure of edible portion	Weight (g)	Water (%)	Calories (kcal)	Protein (g)	Total fat (g)	Saturated (g)	Monounsaturated (g)	Polyunsaturated (g)
1140	Cooked (whole or sliced), drained	1 cup	210	88	92	3	Tr	0.1	0.1	0.2
1141		1 medium	94	88	41	1	Tr	Tr	Tr	0.1
1142	Dehydrated flakes	1 tbsp	5	4	17	Tr	Tr	Tr	Tr	Tr
	Onions, spring, raw, top and bulb									
1143	Chopped	1 cup	100	90	32	2	Tr	Tr	Tr	0.1
1144	Whole, medium, 4⅛" long	1 whole	15	90	5	Tr	Tr	Tr	Tr	Tr
1145	Onion rings, 2"-3" dia, breaded, par fried, frozen, oven heated	10 rings	60	29	244	3	16	5.2	6.5	3.1
1146	Parsley, raw	10 sprigs	10	88	4	Tr	Tr	Tr	Tr	Tr
1147	Parsnips, sliced, cooked, drained	1 cup	156	78	126	2	Tr	0.1	0.2	0.1
	Peas, edible pod, cooked, drained									
1148	From raw	1 cup	160	89	67	5	Tr	0.1	Tr	0.2
1149	From frozen	1 cup	160	87	83	6	1	0.1	0.1	0.3
	Peas, green									
1150	Canned, drained	1 cup	170	82	117	8	1	0.1	0.1	0.3
1151	Frozen, boiled, drained	1 cup	160	80	125	8	Tr	0.1	Tr	0.2
	Peppers									
	Hot chili, raw									
1152	Green	1 pepper	45	88	18	1	Tr	Tr	Tr	Tr
1153	Red	1 pepper	45	88	18	1	Tr	Tr	Tr	Tr
1154	Jalapeno, canned, sliced, solids and liquids	¼ cup	26	89	7	Tr	Tr	Tr	Tr	0.1
	Sweet (2¾" long, 2½" dia)									
	Raw									
	Green									
1155	Chopped	1 cup	149	92	40	1	Tr	Tr	Tr	0.2
1156	Ring (¼" thick)	1 ring	10	92	3	Tr	Tr	Tr	Tr	Tr
1157	Whole (2¾" x 2½")	1 pepper	119	92	32	1	Tr	Tr	Tr	0.1
	Red									
1158	Chopped	1 cup	149	92	40	1	Tr	Tr	Tr	0.2
1159	Whole (2¾" x 2½")	1 pepper	119	92	32	1	Tr	Tr	Tr	0.1
	Cooked, drained, chopped									
1160	Green	1 cup	136	92	38	1	Tr	Tr	Tr	0.1
1161	Red	1 cup	136	92	38	1	Tr	Tr	Tr	0.1
1162	Pimento, canned	1 tbsp	12	93	3	Tr	Tr	Tr	Tr	Tr
	Potatoes									
	Baked (2⅓" x 4¾")									
1163	With skin	1 potato	202	71	220	5	Tr	0.1	Tr	0.1
1164	Flesh only	1 potato	156	75	145	3	Tr	Tr	Tr	0.1
1165	Skin only	1 skin	58	47	115	2	Tr	Tr	Tr	Tr
	Boiled (2½" dia)									
1166	Peeled after boiling	1 potato	136	77	118	3	Tr	Tr	Tr	0.1
1167	Peeled before boiling	1 potato	135	77	116	2	Tr	Tr	Tr	0.1
1168		1 cup	156	77	134	3	Tr	Tr	Tr	0.1
	Potato products, prepared									
	Au gratin									
1169	From dry mix, with whole milk, butter	1 cup	245	79	228	6	10	6.3	2.9	0.3
1170	From home recipe, with butter	1 cup	245	74	323	12	19	11.6	5.3	0.7
1171	French fried, frozen, oven heated	10 strips	50	57	100	2	4	0.6	2.4	0.4

Choles-terol (mg)	Carbo-hydrate (g)	Total dietary fiber (g)	Calcium (mg)	Iron (mg)	Potas-sium (mg)	Sodium (mg)	Vitamin A (IU)	Vitamin A (RE)	Thiamin (mg)	Ribo-flavin (mg)	Niacin (mg)	Ascor-bic acid (mg)	Food No.
0	21	2.9	46	0.5	349	6	0	0	0.09	0.05	0.3	11	1140
0	10	1.3	21	0.2	156	3	0	0	0.04	0.02	0.2	5	1141
0	4	0.5	13	0.1	81	1	0	0	0.03	0.01	Tr	4	1142
0	7	2.6	72	1.5	276	16	385	39	0.06	0.08	0.5	19	1143
0	1	0.4	11	0.2	41	2	58	6	0.01	0.01	0.1	3	1144
0	23	0.8	19	1.0	77	225	135	14	0.17	0.08	2.2	1	1145
0	1	0.3	14	0.6	55	6	520	52	0.01	0.01	0.1	13	1146
0	30	6.2	58	0.9	573	16	0	0	0.13	0.08	1.1	20	1147
0	11	4.5	67	3.2	384	6	210	21	0.20	0.12	0.9	77	1148
0	14	5.0	94	3.8	347	8	267	27	0.10	0.19	0.9	35	1149
0	21	7.0	34	1.6	294	428	1,306	131	0.21	0.13	1.2	16	1150
0	23	8.8	38	2.5	269	139	1,069	107	0.45	0.16	2.4	16	1151
0	4	0.7	8	0.5	153	3	347	35	0.04	0.04	0.4	109	1152
0	4	0.7	8	0.5	153	3	4,838	484	0.04	0.04	0.4	109	1153
0	1	0.7	6	0.5	50	434	442	44	0.01	0.01	0.1	3	1154
0	10	2.7	13	0.7	264	3	942	94	0.10	0.04	0.8	133	1155
0	1	0.2	1	Tr	18	Tr	63	6	0.01	Tr	0.1	9	1156
0	8	2.1	11	0.5	211	2	752	75	0.08	0.04	0.6	106	1157
0	10	3.0	13	0.7	264	3	8,493	849	0.10	0.04	0.8	283	1158
0	8	2.4	11	0.5	211	2	6,783	678	0.08	0.04	0.6	226	1159
0	9	1.6	12	0.6	226	3	805	80	0.08	0.04	0.6	101	1160
0	9	1.6	12	0.6	226	3	5,114	511	0.08	0.04	0.6	233	1161
0	1	0.2	1	0.2	19	2	319	32	Tr	0.01	0.1	10	1162
0	51	4.8	20	2.7	844	16	0	0	0.22	0.07	3.3	26	1163
0	34	2.3	8	0.5	610	8	0	0	0.16	0.03	2.2	20	1164
0	27	4.6	20	4.1	332	12	0	0	0.07	0.06	1.8	8	1165
0	27	2.4	7	0.4	515	5	0	0	0.14	0.03	2.0	18	1166
0	27	2.4	11	0.4	443	7	0	0	0.13	0.03	1.8	10	1167
0	31	2.8	12	0.5	512	8	0	0	0.15	0.03	2.0	12	2268
37	31	2.2	203	0.8	537	1,076	522	76	0.05	0.20	2.3	8	1169
56	28	4.4	292	1.6	970	1,061	647	93	0.16	0.28	2.4	24	1170
0	16	1.6	4	0.6	209	15	0	0	0.06	0.01	1.0	5	1171

Food No.	Food Description	Measure of edible portion	Weight (g)	Water (%)	Calories (kcal)	Protein (g)	Total fat (g)	Fatty acids: Saturated (g)	Fatty acids: Monounsaturated (g)	Fatty acids: Polyunsaturated (g)

Vegetables and Vegetable Products (continued)

Food No.	Food Description	Measure of edible portion	Weight (g)	Water (%)	Calories (kcal)	Protein (g)	Total fat (g)	Saturated (g)	Monounsaturated (g)	Polyunsaturated (g)
	Potato products, prepared (continued)									
	Hashed brown									
1172	From frozen (about 3" x 1½" x ½")	1 patty	29	56	63	1	3	1.3	1.5	0.4
1173	From home recipe	1 cup	156	62	326	4	22	8.5	9.7	2.5
	Mashed									
1174	From dehydrated flakes (without milk); whole milk, butter, and salt added	1 cup	210	76	237	4	12	7.2	3.3	0.5
	From home recipe									
1175	With whole milk	1 cup	210	78	162	4	1	0.7	0.3	0.1
1176	With whole milk and margarine	1 cup	210	76	223	4	9	2.2	3.7	2.5
1177	Potato pancakes, home prepared	1 pancake	76	47	207	5	12	2.3	3.5	5.0
1178	Potato puffs, from frozen	10 puffs	79	53	175	3	8	4.0	3.4	0.6
1179	Potato salad, home prepared	1 cup	250	76	358	7	21	3.6	6.2	9.3
	Scalloped									
1180	From dry mix, with whole milk, butter	1 cup	245	79	228	5	11	6.5	3.0	0.5
1181	From home recipe, with butter	1 cup	245	81	211	7	9	5.5	2.5	0.4
	Pumpkin									
1182	Cooked, mashed	1 cup	245	94	49	2	Tr	0.1	Tr	Tr
1183	Canned	1 cup	245	90	83	3	1	0.4	0.1	Tr
1184	Radishes, raw (¾" to 1" dia)	1 radish	5	95	1	Tr	Tr	Tr	Tr	Tr
1185	Rutabagas, cooked, drained, cubes	1 cup	170	89	66	2	Tr	Tr	Tr	0.2
1186	Sauerkraut, canned, solids and liquid	1 cup	236	93	45	2	Tr	0.1	Tr	0.1
	Seaweed									
1187	Kelp, raw	2 tbsp	10	82	4	Tr	Tr	Tr	Tr	Tr
1188	Spirulina, dried	1 tbsp	1	5	3	1	Tr	Tr	Tr	Tr
1189	Shallots, raw, chopped	1 tbsp	10	80	7	Tr	Tr	Tr	Tr	Tr
1190	Soybeans, green, cooked, drained	1 cup	180	69	254	22	12	1.3	2.2	5.4
	Spinach									
	Raw									
1191	Chopped	1 cup	30	92	7	1	Tr	Tr	Tr	Tr
1192	Leaf	1 leaf	10	92	2	Tr	Tr	Tr	Tr	Tr
	Cooked, drained									
1193	From raw	1 cup	180	91	41	5	Tr	0.1	Tr	0.2
1194	From frozen (chopped or leaf)	1 cup	190	90	53	6	Tr	0.1	Tr	0.2
1195	Canned, drained	1 cup	214	92	49	6	1	0.2	Tr	0.4
	Squash									
	Summer (all varieties), sliced									
1196	Raw	1 cup	113	94	23	1	Tr	Tr	Tr	0.1
1197	Cooked, drained	1 cup	180	94	36	2	1	0.1	Tr	0.2
1198	Winter (all varieties), baked, cubes	1 cup	205	89	80	2	1	0.3	0.1	0.5
1199	Winter, butternut, frozen, cooked, mashed	1 cup	240	88	94	3	Tr	Tr	Tr	0.1
	Sweetpotatoes									
	Cooked (2" dia, 5" long raw)									
1200	Baked, with skin	1 potato	146	73	150	3	Tr	Tr	Tr	0.1
1201	Boiled, without skin	1 potato	156	73	164	3	Tr	0.1	Tr	0.2

Choles-terol (mg)	Carbo-hydrate (g)	Total dietary fiber (g)	Calcium (mg)	Iron (mg)	Potas-sium (mg)	Sodium (mg)	Vitamin A (IU)	Vitamin A (RE)	Thiamin (mg)	Ribo-flavin (mg)	Niacin (mg)	Ascor-bic acid (mg)	Food No.
0	8	0.6	4	0.4	126	10	0	0	0.03	0.01	0.7	2	1172
0	33	3.1	12	1.3	501	37	0	0	0.12	0.03	3.1	9	1173
29	32	4.8	103	0.5	489	697	378	44	0.23	0.11	1.4	20	1174
4	37	4.2	55	0.6	628	636	40	13	0.18	0.08	2.3	14	1175
4	35	4.2	55	0.5	607	620	355	42	0.18	0.08	2.3	13	1176
73	22	1.5	18	1.2	597	386	109	11	0.10	0.13	1.6	17	1177
0	24	2.5	24	1.2	300	589	13	2	0.15	0.06	1.7	5	1178
170	28	3.3	48	1.6	635	1,323	523	83	0.19	0.15	2.2	25	1179
27	31	2.7	88	0.9	497	835	363	51	0.05	0.14	2.5	8	1180
29	26	4.7	140	1.4	926	821	331	47	0.17	0.23	2.6	26	1181
0	12	2.7	37	1.4	564	2	2,651	265	0.08	0.19	1.0	12	1182
0	20	7.1	64	3.4	505	12	54,037	5,405	0.06	0.13	0.9	10	1183
0	Tr	0.1	1	Tr	10	1	Tr	Tr	Tr	Tr	Tr	1	1184
0	15	3.1	82	0.9	554	34	954	95	0.14	0.07	1.2	32	1185
0	10	5.9	71	3.5	401	1,560	42	5	0.05	0.05	0.3	35	1186
0	1	0.1	17	0.3	9	23	12	1	0.01	0.02	Tr	Tr	1187
0	Tr	Tr	1	0.3	14	10	6	1	0.02	0.04	0.1	Tr	1188
0	2	0.2	4	0.1	33	1	119	12	0.01	Tr	Tr	1	1189
0	20	7.6	261	4.5	970	25	281	29	0.47	0.28	2.3	31	1190
0	1	0.8	30	0.8	167	24	2,015	202	0.02	0.06	0.2	8	1191
0	Tr	0.3	10	0.3	56	8	672	67	0.01	0.02	0.1	3	1192
0	7	4.3	245	6.4	839	126	14,742	1,474	0.17	0.42	0.9	18	1193
0	10	5.7	277	2.9	566	163	14,790	1,478	0.11	0.32	0.8	23	1194
0	7	5.1	272	4.9	740	58	18,781	1,879	0.03	0.30	0.8	31	1195
0	5	2.1	23	0.5	220	2	221	23	0.07	0.04	0.6	17	1196
0	8	2.5	49	0.6	346	2	517	52	0.08	0.07	0.9	10	1197
0	18	5.7	29	0.7	896	2	7,292	730	0.17	0.05	1.4	20	1198
0	24	2.2	46	1.4	319	5	8,014	802	0.12	0.09	1.1	8	1199
0	35	4.4	41	0.7	508	15	31,860	3,186	0.11	0.19	0.9	36	1200
0	38	2.8	33	0.9	287	20	26,604	2,660	0.08	0.22	1.0	27	1201

Vegetables and Vegetable Products (continued)

Food No.	Food Description	Measure of edible portion	Weight (g)	Water (%)	Calories (kcal)	Protein (g)	Total fat (g)	Fatty acids: Saturated (g)	Fatty acids: Monounsaturated (g)	Fatty acids: Polyunsaturated (g)
	Sweet potatoes (continued)									
1202	Candied (2½" x 2" piece)	1 piece	105	67	144	1	3	1.4	0.7	0.2
	Canned									
1203	Syrup pack, drained	1 cup	196	72	212	3	1	0.1	Tr	0.3
1204	Vacuum pack, mashed	1 cup	255	76	232	4	1	0.1	Tr	0.2
1205	Tomatillos, raw	1 medium	34	92	11	Tr	Tr	Tr	0.1	0.1
	Tomatoes									
	Raw, year round average									
1206	Chopped or sliced	1 cup	180	94	38	2	1	0.1	0.1	0.2
1207	Slice, medium, ¼" thick	1 slice	20	94	4	Tr	Tr	Tr	Tr	Tr
	Whole									
1208	Cherry	1 cherry	17	94	4	Tr	Tr	Tr	Tr	Tr
1209	Medium, 2⅗" dia	1 tomato	123	94	26	1	Tr	0.1	0.1	0.2
1210	Canned, solids and liquid	1 cup	240	94	46	2	Tr	Tr	Tr	0.1
	Sun dried									
1211	Plain	1 piece	2	15	5	Tr	Tr	Tr	Tr	Tr
1212	Packed in oil, drained	1 piece	3	54	6	Tr	Tr	0.1	0.3	0.1
1213	Tomato juice, canned, with salt added	1 cup	243	94	41	2	Tr	Tr	Tr	0.1
	Tomato products, canned									
1214	Paste	1 cup	262	74	215	10	1	0.2	0.2	0.6
1215	Puree	1 cup	250	87	100	4	Tr	0.1	0.1	0.2
1216	Sauce	1 cup	245	89	74	3	Tr	0.1	0.1	0.2
	Spaghetti/marinara/pasta sauce. See Soups, Sauces, and Gravies.									
1217	Stewed	1 cup	255	91	71	2	Tr	Tr	0.1	0.1
1218	Turnips, cooked, cubes	1 cup	156	94	33	1	Tr	Tr	Tr	0.1
	Turnip greens, cooked, drained									
1219	From raw (leaves and stems)	1 cup	144	93	29	2	Tr	0.1	Tr	0.1
1220	From frozen (chopped)	1 cup	164	90	49	5	1	0.2	Tr	0.3
1221	Vegetable juice cocktail, canned	1 cup	242	94	46	2	Tr	Tr	Tr	0.1
	Vegetables, mixed									
1222	Canned, drained	1 cup	163	87	77	4	Tr	0.1	Tr	0.2
1223	Frozen, cooked, drained	1 cup	182	83	107	5	Tr	0.1	Tr	0.1
1224	Waterchestnuts, canned, slices, solids and liquids	1 cup	140	86	70	1	Tr	Tr	Tr	Tr

Miscellaneous Items

Food No.	Food Description	Measure of edible portion	Weight (g)	Water (%)	Calories (kcal)	Protein (g)	Total fat (g)	Fatty acids: Saturated (g)	Fatty acids: Monounsaturated (g)	Fatty acids: Polyunsaturated (g)
1225	Bacon bits, meatless	1 tbsp	7	8	31	2	2	0.3	0.4	0.9
	Baking powders for home use									
	Double acting									
1226	Sodium aluminum sulfate	1 tsp	5	5	2	0	0	0.0	0.0	0.0
1227	Straight phosphate	1 tsp	5	4	2	Tr	0	0.0	0.0	0.0
1228	Low sodium	1 tsp	5	6	5	Tr	Tr	Tr	Tr	Tr
1229	Baking soda	1 tsp	5	Tr	0	0	0	0.0	0.0	0.0
1230	Beef jerky	1 large piece	20	23	81	7	5	2.1	2.2	0.2
1231	Catsup	1 cup	240	67	250	4	1	0.1	0.1	0.4
1232		1 tbsp	15	67	16	Tr	Tr	Tr	Tr	Tr
1233		1 packet	6	67	6	Tr	Tr	Tr	Tr	Tr
1234	Celery seed	1 tsp	2	6	8	Tr	1	Tr	0.3	0.1
1235	Chili powder	1 tsp	3	8	8	Tr	Tr	0.1	0.1	0.2
	Chocolate, unsweetened, baking									
1236	Solid	1 square	28	1	148	3	16	9.2	5.2	0.5
1237	Liquid	1 oz	28	1	134	3	14	7.2	2.6	3.0

*For product with no salt added: If salt added, consult the nutrition label for sodium value.

Cholesterol (mg)	Carbohydrate (g)	Total dietary fiber (g)	Calcium (mg)	Iron (mg)	Potassium (mg)	Sodium (mg)	Vitamin A (IU)	Vitamin A (RE)	Thiamin (mg)	Riboflavin (mg)	Niacin (mg)	Ascorbic acid (mg)	Food No.
8	29	2.5	27	1.2	198	74	4,398	440	0.02	0.04	0.4	7	1202
0	50	5.9	33	1.9	378	76	14,028	1,403	0.05	0.07	0.7	21	1203
0	54	4.6	56	2.3	796	135	20,357	2,035	0.09	0.15	1.9	67	1204
0	2	0.6	2	0.2	91	Tr	39	4	0.01	0.01	0.6	4	1205
0	8	2.0	9	0.8	400	16	1,121	112	0.11	0.09	1.1	34	1206
0	1	0.2	1	0.1	44	2	125	12	0.01	0.01	0.1	4	1207
0	1	0.2	1	0.1	38	2	106	11	0.01	0.01	0.1	3	1208
0	6	1.4	6	0.6	273	11	766	76	0.07	0.06	0.8	23	1209
0	10	2.4	72	1.3	530	355	1,428	144	0.11	0.07	1.8	34	1210
0	1	0.2	2	0.2	69	42	17	2	0.01	0.01	0.2	1	1211
0	1	0.2	1	0.1	47	8	39	4	0.01	0.01	0.1	3	1212
0	10	1.0	22	1.4	535	877	1,351	136	0.11	0.08	1.6	44	1213
0	51	10.7	92	5.1	2,455	231	6,406	639	0.41	0.50	8.4	111	1214
0	24	5.0	43	3.1	1,065	85*	3,188	320	0.18	0.14	4.3	26	1215
0	18	3.4	34	1.9	909	1,482	2,399	240	0.16	0.14	2.8	32	1216
0	17	2.6	84	1.9	607	564	1,380	138	0.12	0.09	1.8	29	1217
0	8	3.1	34	0.3	211	78	0	0	0.04	0.04	0.5	18	1218
0	6	5.0	197	1.2	292	42	7,917	792	0.06	0.10	0.6	39	1219
0	8	5.6	249	3.2	367	25	13,079	1,309	0.09	0.12	0.8	36	1220
0	11	1.9	27	1.0	467	653	2,831	283	0.10	0.07	1.8	67	1221
0	15	4.9	44	1.7	474	243	18,985	1,899	0.07	0.08	0.9	8	1222
0	24	8.0	46	1.5	308	64	7,784	779	0.13	0.22	1.5	6	1223
0	17	3.5	6	1.2	165	11	6	0	0.02	0.03	0.5	2	1224
0	2	0.7	7	0.1	10	124	0	0	0.04	Tr	0.1	Tr	1225
0	1	Tr	270	0.5	1	488	0	0	0.00	0.00	0.0	0	1226
0	1	Tr	339	0.5	Tr	363	0	0	0.00	0.00	0.0	0	1227
0	2	0.1	217	0.4	505	5	0	0	0.00	0.00	0.0	0	1228
0	0	0.0	0	0.0	0	1,259	0	0	0.00	0.00	0.0	0	1229
10	2	0.4	4	1.1	118	438	0	0	0.03	0.03	0.3	0	1230
0	65	3.1	46	1.7	1,154	2,846	2,438	245	0.21	0.18	3.3	36	1231
0	4	0.2	3	0.1	72	178	152	15	0.01	0.01	0.2	2	1232
0	2	0.1	1	Tr	29	71	61	6	0.01	Tr	0.1	1	1233
0	1	0.2	35	0.9	28	3	1	Tr	0.01	0.01	0.1	Tr	1234
0	1	0.9	7	0.4	50	26	908	91	0.01	0.02	0.2	2	1235
0	8	4.4	21	1.8	236	4	28	3	0.02	0.05	0.3	0	1236
0	10	5.1	15	1.2	331	3	3	Tr	0.01	0.08	0.6	0	1237

Food No.	Food Description	Measure of edible portion	Weight (g)	Water (%)	Calories (kcal)	Protein (g)	Total fat (g)	Fatty acids: Saturated (g)	Fatty acids: Monounsaturated (g)	Fatty acids: Polyunsaturated (g)

Miscellaneous Items (continued)

Food No.	Food Description	Measure of edible portion	Weight (g)	Water (%)	Calories (kcal)	Protein (g)	Total fat (g)	Saturated (g)	Monounsaturated (g)	Polyunsaturated (g)
1238	Cinnamon	1 tsp	2	10	6	Tr	Tr	Tr	Tr	Tr
1239	Cocoa powder, unsweetened	1 cup	86	3	197	17	12	6.9	3.9	0.4
1240		1 tbsp	5	3	12	1	1	0.4	0.2	Tr
1241	Cream of tartar	1 tsp	3	2	8	0	0	0.0	0.0	0.0
1242	Curry powder	1 tsp	2	10	7	Tr	Tr	Tr	0.1	0.1
1243	Garlic powder	1 tsp	3	6	9	Tr	Tr	Tr	Tr	Tr
1244	Horseradish, prepared	1 tsp	5	85	2	Tr	Tr	Tr	Tr	Tr
1245	Mustard, prepared, yellow	1 tsp or 1 packet	5	82	3	Tr	Tr	Tr	0.1	Tr
	Olives, canned									
1246	Pickled, green	5 medium	17	78	20	Tr	2	0.3	1.6	0.2
1247	Ripe, black	5 large	22	80	25	Tr	2	0.3	1.7	0.2
1248	Onion powder	1 tsp	2	5	7	Tr	Tr	Tr	Tr	Tr
1249	Oregano, ground	1 tsp	2	7	5	Tr	Tr	Tr	Tr	0.1
1250	Paprika	1 tsp	2	10	6	Tr	Tr	Tr	Tr	0.2
1251	Parsley, dried	1 tbsp	1	9	4	Tr	Tr	Tr	Tr	Tr
1252	Pepper, black	1 tsp	2	11	5	Tr	Tr	Tr	Tr	Tr
	Pickles, cucumber									
1253	Dill, whole, medium (3¾" long)	1 pickle	65	92	12	Tr	Tr	Tr	Tr	0.1
1254	Fresh (bread and butter pickles), slices 1½" dia, ¼" thick	3 slices	24	79	18	Tr	Tr	Tr	Tr	Tr
1255	Pickle relish, sweet	1 tbsp	15	62	20	Tr	Tr	Tr	Tr	Tr
1256	Pork skins/rinds, plain	1 oz	28	2	155	17	9	3.2	4.2	1.0
	Potato chips									
	Regular									
	Plain									
1257	Salted	1 oz	28	2	152	2	10	3.1	2.8	3.5
1258	Unsalted	1 oz	28	2	152	2	10	3.1	2.8	3.5
1259	Barbecue flavor	1 oz	28	2	139	2	9	2.3	1.9	4.6
1260	Sour cream and onion flavor	1 oz	28	2	151	2	10	2.5	1.7	4.9
1261	Reduced fat	1 oz	28	1	134	2	6	1.2	1.4	3.1
1262	Fat free, made with olestra	1 oz	28	2	75	2	Tr	Tr	0.1	0.1
	Made from dried potatoes									
1263	Plain	1 oz	28	1	158	2	11	2.7	2.1	5.7
1264	Sour cream and onion flavor	1 oz	28	2	155	2	10	2.7	2.0	5.3
1265	Reduced fat	1 oz	28	1	142	2	7	1.5	1.7	3.8
1266	Salt	1 tsp	6	Tr	0	0	0	0.0	0.0	0.0
	Trail mix									
1267	Regular, with raisins, chocolate chips, salted nuts and seeds	1 cup	146	7	707	21	47	8.9	19.8	16.5
1268	Tropical	1 cup	140	9	570	9	24	11.9	3.5	7.2
1269	Vanilla extract	1 tsp	4	53	12	Tr	Tr	Tr	Tr	Tr
	Vinegar									
1270	Cider	1 tbsp	15	94	2	0	0	0.0	0.0	0.0
1271	Distilled	1 tbsp	17	95	2	0	0	0.0	0.0	0.0
	Yeast, baker's									
1272	Dry, active	1 pkg	7	8	21	3	Tr	Tr	0.2	Tr
1273		1 tsp	4	8	12	2	Tr	Tr	0.1	Tr
1274	Compressed	1 cake	17	69	18	1	Tr	Tr	0.2	Tr

Cholesterol (mg)	Carbohydrate (g)	Total dietary fiber (g)	Calcium (mg)	Iron (mg)	Potassium (mg)	Sodium (mg)	Vitamin A (IU)	Vitamin A (RE)	Thiamin (mg)	Riboflavin (mg)	Niacin (mg)	Ascorbic acid (mg)	Food No.
0	2	1.2	28	0.9	11	1	6	1	Tr	Tr	Tr	1	1238
0	47	28.6	110	11.9	1,311	18	17	2	0.07	0.21	1.9	0	1239
0	3	1.8	7	0.7	82	1	1	Tr	Tr	0.01	0.1	0	1240
0	2	Tr	Tr	0.1	495	2	0	0	0.00	0.00	0.0	0	1241
0	1	0.7	10	0.6	31	1	20	2	0.01	0.01	0.1	Tr	1242
0	2	0.3	2	0.1	31	1	0	0	0.01	Tr	Tr	1	1243
0	1	0.2	3	Tr	12	16	Tr	0	Tr	Tr	Tr	1	1244
0	Tr	0.2	4	0.1	8	56	7	1	Tr	Tr	Tr	Tr	1245
0	Tr	0.2	10	0.3	9	408	51	5	0.00	0.00	Tr	0	1246
0	1	0.7	19	0.7	2	192	89	9	Tr	0.00	Tr	Tr	1247
0	2	0.1	8	0.1	20	1	0	0	0.01	Tr	Tr	Tr	1248
0	1	0.6	24	0.7	25	Tr	104	10	0.01	Tr	0.1	1	1249
0	1	0.4	4	0.5	49	1	1,273	127	0.01	0.04	0.3	1	1250
0	1	0.4	19	1.3	49	6	303	30	Tr	0.02	0.1	2	1251
0	1	0.6	9	0.6	26	1	4	Tr	Tr	0.01	Tr	Tr	1252
0	3	0.8	6	0.3	75	833	214	21	0.01	0.02	Tr	1	1253
0	4	0.4	8	0.1	48	162	34	3	0.00	0.01	0.0	2	1254
0	5	0.2	Tr	0.1	4	122	23	2	0.00	Tr	Tr	Tr	1255
27	0	0.0	9	0.2	36	521	37	11	0.03	0.08	0.4	Tr	1256
0	15	1.3	7	0.5	361	168	0	0	0.05	0.06	1.1	9	1257
0	15	1.4	7	0.5	361	2	0	0	0.05	0.06	1.1	9	1258
0	15	1.2	14	0.5	357	213	62	6	0.06	0.06	1.3	10	1259
2	15	1.5	20	0.5	377	177	48	6	0.05	0.06	1.1	11	1260
0	19	1.7	6	0.4	494	139	0	0	0.06	0.08	2.0	7	1261
0	17	1.1	10	0.4	366	185	1,469	441	0.10	0.02	1.3	8	1262
0	14	1.0	7	0.4	286	186	0	0	0.06	0.03	0.9	2	1263
1	15	0.3	18	0.4	141	204	214	28	0.05	0.03	0.7	3	1264
0	18	1.0	10	0.4	285	121	0	0	0.05	0.02	1.2	3	1265
0	0	0.0	1	Tr	Tr	2,325	0	0	0.00	0.00	0.0	0	1266
6	66	8.8	159	4.9	946	177	64	7	0.60	0.33	6.4	2	1267
0	92	10.6	80	3.7	993	14	69	7	0.63	0.16	2.1	11	1268
0	1	0.0	Tr	Tr	6	Tr	0	0	Tr	Tr	Tr	0	1269
0	1	0.0	1	0.1	15	Tr	0	0	0.00	0.00	0.0	0	1270
0	1	0.0	0	0.0	2	Tr	0	0	0.00	0.00	0.0	0	1271
0	3	1.5	4	1.2	140	4	Tr	0	0.17	0.38	2.8	Tr	1272
0	2	0.8	3	0.7	80	2	Tr	0	0.09	0.22	1.6	Tr	1273
0	3	1.4	3	0.6	102	5	0	0	0.32	0.19	2.1	Tr	1274

Appendix B

Dietary Reference Intakes

Dietary Reference Intakes (DRIs): Recommended Intakes for Individuals, Vitamins

Food and Nutrition Board, Institute of Medicine, National Academies

Life Stage Group	Vitamin A (μg/d)[a]	Vitamin D (μg/d)[b]	Vitamin E (mg/d)[c]	Vitamin K (μg/d)	Vitamin C (mg/d)	Thiamin (mg/d)	Riboflavin (mg/d)	Niacin (mg/d)[d]	Vitamin B_6 (mg/d)	Folate (μg/d)[e]	Vitamin B_{12} (μg/d)	Pantothenic Acid (mg/d)	Biotin (μg/d)	Choline (mg/d)
Infants														
0–6 mo	400*	5*	4*	2.0*	40*	0.2*	0.3*	2*	0.1*	65*	0.4*	1.7*	5*	125*
7–12 mo	500*	5*	5*	2.5*	50*	0.3*	0.4*	4*	0.3*	80*	0.5*	1.8*	6*	150*
Children														
1–3 y	300	5*	6	30*	15	0.5	0.5	6	0.5	150	0.9	2*	8*	200*
4–8 y	400	5*	7	55*	25	0.6	0.6	8	0.6	200	1.2	3*	12*	250*
Males														
9–13 y	600	5*	11	60*	45	0.9	0.9	12	1.0	300	1.8	4*	20*	375*
14–18 y	900	5*	15	75*	75	1.2	1.3	16	1.3	400	2.4	5*	25*	550*
19–30 y	900	5*	15	120*	90	1.2	1.3	16	1.3	400	2.4	5*	30*	550*
31–50 y	900	5*	15	120*	90	1.2	1.3	16	1.3	400	2.4	5*	30*	550*
51–70 y	900	10*	15	120*	90	1.2	1.3	16	1.7	400	2.4	5*	30*	550*
>70 y	900	15*	15	120*	90	1.2	1.3	16	1.7	400	2.4	5*	30*	550*
Females														
9–13 y	600	5*	11	60*	45	0.9	0.9	12	1.0	300	1.8	4*	20*	375*
14–18 y	700	5*	15	75*	65	1.0	1.0	14	1.2	400	2.4	5*	25*	400*
19–30 y	700	5*	15	90*	75	1.1	1.1	14	1.3	400	2.4	5*	30*	425*
31–50 y	700	5*	15	90*	75	1.1	1.1	14	1.3	400	2.4	5*	30*	425*
51–70 y	700	10*	15	90*	75	1.1	1.1	14	1.5	400	2.4	5*	30*	425*
>70 y	700	15*	15	90*	75	1.1	1.1	14	1.5	400	2.4	5*	30*	425*
Pregnancy														
19–50 y	770	5*	15	90*	85	1.4	1.4	18	1.9	600	2.6	6*	30*	450*
Lactation														
19–50 y	1,300	5*	19	90*	120	1.4	1.6	17	2.0	500	2.8	7*	35*	550*

*Adequate Intake. All other numbers are Recommended Dietary Allowances (RDA).

a: As retinol activity equivalents (RAE).

b: As cholecalciferol.

c: As alpha-tocopherol.

d: As niacin equivalents (NE).

e: As dietary folate equivalents (DFE).

Source: Adapted with permission from the *Dietary References Intakes* series, © 1997, 1998, 2000, 2001, by the National Academy of Sciences. Courtesy of the National Academy Press, Washington, D.C.

Dietary Reference Intakes (DRIs): Recommended Intakes for Individuals, Elements
Food and Nutrition Board, Institute of Medicine, National Academies

Life Stage Group	Calcium (mg/d)	Chromium (μg/d)	Copper (μg/d)	Fluoride (mg/d)	Iodine (μg/d)	Iron (mg/d)	Magnesium (mg/d)	Manganese (mg/d)	Molybdenum (μg/d)	Phosphorus (mg/d)	Selenium (μg/d)	Zinc (mg/d)
Infants												
0–6 mo	210*	0.2*	200*	0.01*	110*	0.27*	30*	0.003*	2*	100*	15*	2*
7–12 mo	270*	5.5*	220*	0.5*	130*	11	75*	0.6*	3*	275*	20*	3
Children												
1–3 y	500*	11*	340	0.7*	90	7	80	1.2*	17	460	20	3
4–8 y	800*	15*	440	1*	90	10	130	1.5*	22	500	30	5
Males												
9–13 y	1300*	25*	700	2*	120	8	240	1.9*	34	1250	40	8
14–18 y	1300*	35*	890	3*	150	11	410	2.2*	43	1250	55	11
19–30 y	1000*	35*	900	4*	150	8	400	2.3*	45	700	55	11
31–50 y	1000*	35*	900	4*	150	8	420	2.3*	45	700	55	11
51–70 y	1200*	30*	900	4*	150	8	420	2.3*	45	700	55	11
>70 y	1200*	30*	900	4*	150	8	420	2.3*	45	700	55	11
Females												
9–13 y	1300*	21*	700	2*	120	8	240	1.6*	34	1250	40	8
14–18 y	1300*	24*	890	3*	150	15	360	1.6*	43	1250	55	9
19–30 y	1000*	25*	900	3*	150	18	310	1.8*	45	700	55	8
31–50 y	1000*	25*	900	3*	150	18	320	1.8*	45	700	55	8
51–70 y	1200*	20*	900	3*	150	8	320	1.8*	45	700	55	8
>70 y	1200*	20*	900	3*	150	8	320	1.8*	45	700	55	8
Pregnancy												
19–50 y	1000*	30*	1,000	3*	220	27	350 (19–30 y)	2.0*	50	700	60	11
Lactation												
19–50 y	1000*	45*	1,300	3*	290	9	310 (19–30 y)	2.6*	50	700	70	12

*Adequate Intake. All other numbers are Recommended Dietary Allowances (RDA).

Source: Adapted with permission from the *Dietary References Intakes* series, © 1997, 1998, 2000, 2001, by the National Academy of Sciences. Courtesy of the National Academy Press, Washington, D.C.

Dietary Reference Intakes (DRIs): Tolerable Upper Intake Levels (UL), Vitamins

Food and Nutrition Board, Institute of Medicine, National Academies

Life Stage Group	Vitamin A (μg/d)[a]	Vitamin C (mg/d)	Vitamin D (μg/d)[b]	Vitamin E (mg/d)[c]	Vitamin K	Thiamin	Riboflavin	Niacin (mg/d)[d]	Vitamin B_6 (mg/d)	Folate (μg/d)[e]	Vitamin B_{12}	Pantothenic Acid	Biotin	Choline (g/d)	Carotenoids
Infants															
0–6 mo	600	–	25	–	–	–	–	–	–	–	–	–	–	–	–
7–12 mo	600	–	25	–	–	–	–	–	–	–	–	–	–	–	–
Children															
1–3 y	600	400	50	200	–	–	–	10	30	300	–	–	–	1.0	–
4–8 y	900	650	50	300	–	–	–	15	40	400	–	–	–	1.0	–
Males, Females															
9–13 y	1700	1200	50	600	–	–	–	20	60	600	–	–	–	2.0	–
14–18 y	2800	1800	50	800	–	–	–	30	80	800	–	–	–	3.0	–
19–70 y	3000	2000	50	1000	–	–	–	35	100	1000	–	–	–	3.5	–
>70 y	3000	2000	50	1000	–	–	–	35	100	1000	–	–	–	3.5	–
Pregnancy															
≤18 y	2800	1800	50	800	–	–	–	30	80	800	–	–	–	3.0	–
19–50 y	3000	2000	50	1000	–	–	–	35	100	1000	–	–	–	3.5	–
Lactation															
≤18 y	2800	1800	50	800	–	–	–	30	80	800	–	–	–	3.0	–
19–50 y	3000	2000	50	1000	–	–	–	35	100	1000	–	–	–	3.5	–

Dash – means Upper Limits were not able to be set due to a lack of data.

a: As retinol activity equivalents (RAE).

b: As cholecalciferol.

c: As alpha-tocopherol.

d: As niacin equivalents (NE).

e: As dietary folate equivalents (DFE).

Source: Adapted with permission from the *Dietary References Intakes* series, © 1997, 1998, 2000, 2001, by the National Academy of Sciences. Courtesy of the National Academy Press, Washington, D.C.

Dietary Reference Intakes (DRIs): Tolerable Upper Intake Levels (UL), Elements

Food and Nutrition Board, Institute of Medicine, National Academies

Life Stage Group	Boron (mg/d)	Calcium (g/d)	Chromium	Copper (μg/d)	Fluoride (mg/d)	Iodine (μg/d)	Iron (mg/d)	Magnesium (mg/d)	Managanese (mg/d)	Molybdenum (μg/d)	Nickel (mg/d)	Phosphorus (g/d)	Selenium (μg/d)	Silicon	Vanadium (mg/d)	Zinc (mg/d)
Infants																
0–6 mo	–	–	–	–	0.7	–	40	–	–	–	–	–	45	–	–	4
7–12 mo	–	–	–	–	0.9	–	40	–	–	–	–	–	60	–	–	5
Children																
1–3 y	3	2.5	–	1000	1.3	200	40	65	2	300	0.2	3	90	–	–	7
4–8 y	6	2.5	–	3000	2.2	300	40	110	3	600	0.3	3	150	–	–	12
Males, Females																
9–13 y	11	2.5	–	5000	10	600	40	350	6	1100	0.6	4	280	–	–	23
14–18 y	17	2.5	–	8000	10	900	45	350	9	1700	1.0	4	400	–	–	34
19–70 y	20	2.5	–	10,000	10	1100	45	350	11	2000	1.0	4	400	–	–	40
>70 y	20	2.5	–	10,000	10	1100	45	350	11	2000	1.0	3	400	–	–	40
Pregnancy																
≤18 y	17	2.5	–	8000	10	900	45	350	9	1700	1.0	3.5	400	–	–	34
19–50 y	20	2.5	–	10,000	10	1100	45	350	11	2000	1.0	3.5	400	–	–	40
Lactation																
≤18 y	17	2.5	–	8000	10	900	45	350	9	1700	1.0	4	400	–	–	34
19–50 y	20	2.5	–	10,000	10	1100	45	350	11	2000	1.0	4	400	–	–	40

Dash – means Upper Limits were not able to be set due to a lack of data.

Source: Adapted with permission from the *Dietary References Intakes* series, © 1997, 1998, 2000, 2001, by the National Academy of Sciences. Courtesy of the National Academy Press, Washington, D.C.

Estimated Energy Requirements for Active Individuals by Age

Age	Male	Female	
0–6 months	570 kcal/day	520 kcal/day	(3 months)
7–12 months	743	679	(9 months)
1–2 years	1046	992	(24 months)
3–8 years	1742	1642	(6 years)
9–13 years	2279	2071	(11 years)
14–18 years	3152	2368	(16 years)
Over 18 years*	3067	2403	
Pregnancy	An additional 340 kcal/day during the 2nd trimester and an additional 452 kcal/day during the 3rd trimester.		
Lactation	An additional 330 kcal/day during the first 6 months and an additional 400 kcal/day during the second 6 months.		

*Subtract 10 kcal/day for males and 7 kcal/day for females for each year of age over 19 years.

Source: Adapted with permission from the *Dietary References Intakes for Energy, Carbohydrates, Fiber, Fat, Protein, and Amino Acids (Macronutrients).* © 2002 by the National Academy of Sciences. Courtesy of the National Academy Press, Washington, D.C.

Dietary Reference Intakes Values for Macronutrients

Age	Carbohydrate (RDA)	Total Fiber (AI)		Fat (AI)	Protein (RDA)
		Males	Females		
0–6 months	60 grams (AI)	ND	ND	31 grams	9.1 g (AI)
7–12 months	95 g (AI)	ND	ND	30 g	1.5 grams/kilogram body weight
1–3 years	130 g	19 g	19 g	ND	1.1 g/kg
4–8 years	130 g	25 g	25 g	ND	0.95 g/kg
9–13 years	130 g	31 g	26 g	ND	0.95 g/kg
14–18 years	130 g	38 g	36 g	ND	0.85 g/kg
19–30 years	130 g	38 g	25 g	ND	0.80 g/kg
31–50 years	130 g	38 g	25 g	ND	0.80 g/kg
Over 50 years	130 g	30 g	21 g	ND	0/80 g/kg
Pregnancy	175 g		28 g	ND	1.1 g/kg or an additional 25 grams/day
Lactation	210 g		29 g	ND	1.1 g/kg or an additional 25 grams/day

ND – Not Determined
RDA – Recommended Dietary Allowance
AI – Adequate Intake

Source: Adapted with permission from the *Dietary References Intakes for Energy, Carbohydrates, Fiber, Fat, Protein, and Amino Acids (Macronutrients).* © 2002 by the National Academy of Sciences. Courtesy of the National Academy Press, Washington, D.C.

Acceptable Macronutrient Distribution Ranges

Age	Carbohydrate	Fat	Protein
1–3 years	45–65% of total calories	30–40% of total calories	5–20% of total calories
4–18 years	45–65%	25–35%	10–30%
Over 18 years	45–65%	20–35%	10–35%

Source: Adapted with permission from the *Dietary References Intakes for Energy, Carbohydrates, Fiber, Fat, Protein, and Amino Acids (Macronutrients).* © 2002 by the National Academy of Sciences. Courtesy of the National Academy Press, Washington, D.C.

Adequate Intake Values for Essential Fatty Acids

Age Group	Linoleic Acid AI (grams/day)		Alpha Linolenic Acid AI (grams/day)	
	Males	Females	Males	Females
0–6 months	4.4 grams	4.4 grams	0.5 grams	0.5 grams
7–12 months	4.6	4.6	0.5	0.5
1–3 years	7	7	0.7	0.7
4–8 years	10	10	0.9	0.9
9–13 years	12	10	1.2	1.1
14–18 years	16	11	1.6	1.1
19–30 years	17	12	1.6	1.1
31–50 years	17	12	1.6	1.1
Over 50 years	14	11	1.6	1.1
Pregnancy		13		1.4
Lactation		13		1.3

Source: Adapted with permission from the *Dietary References Intakes for Energy, Carbohydrates, Fiber, Fat, Protein, and Amino Acids (Macronutrients).* © 2002 by the National Academy of Sciences. Courtesy of the National Academy Press, Washington, D.C.

Appendix C

Expanded List of Serving Sizes for Food Guide Pyramid

Food Group	What Counts as a Serving (includes additional items)
BREAD, CEREAL, RICE, AND PASTA	GENERALLY: 1 slice of bread 1/2 hamburger or hot dog bun 1/2 English muffin or bagel 1 small roll, biscuit, or muffin (about 1 ounce each) 1/2 cup cooked cereal 1 oz ready-to-eat cereal 1/2 cup cooked pasta or rice 5 to 6 small crackers (saltine size) 2 to 3 large crackers (graham cracker square size) SPECIFICALLY: 4-inch pita bread 3 medium hard bread sticks, about 4-3/4 inches long 9 animal crackers 1/4 cup uncooked rolled oats 2 tablespoons uncooked grits or cream of wheat cereal 1 oz uncooked pasta (1/4 cup macaroni or 3/4 cup noodles) 3 tablespoons uncooked rice 1 7-inch flour or corn tortilla 2 taco shells, corn 1 4-inch pancake 9 3-ring pretzels or 2 pretzel rods 1/16 of 2-layer cake 1/5 of 10-inch angel food cake 1/10 of 8-inch, 2-crust pie 4 small cookies 1/2 medium doughnut 1/2 large croissant

Food Group	What Counts as a Serving (includes additional items)
BREAD, CEREAL, RICE, AND PASTA *(continued)*	SPECIFICALLY: 3 rice or popcorn cakes 2 cups popcorn 12 tortilla chips
FRUITS	GENERALLY: a whole fruit (medium apple, banana, peach, or orange, or a small pear) grapefruit half melon wedge (1/4 of a medium cantaloupe or 1/8 of a medium honeydew) 3/4 cup juice (100% juice) 1/2 cup berries, cherries, or grapes 1/2 cup cut-up fresh fruit 1/2 cup cooked or canned fruit 1/2 cup frozen fruit 1/4 cup dried fruit SPECIFICALLY: 5 large strawberries 7 medium strawberries 50 blueberries 30 raspberries 11 cherries 12 grapes 1-1/2 medium plums 2 medium apricots 1 medium avocado 7 melon balls 1/2 cup fruit salad, such as Waldorf 1/2 medium mango 1/4 medium papaya 1 large kiwi fruit 4 canned apricot halves with liquid 14 canned cherries with liquid 1-1/2 canned peach halves with liquid 2 canned pear halves with liquid 2-1/2 canned pineapple slices with liquid 3 canned plums with liquid 9 dried apricot halves 5 prunes
VEGETABLES	GENERALLY: 1/2 cup cooked vegetables 1/2 cup chopped raw vegetables 1 cup leafy raw vegetables, such as lettuce or spinach

Appendix D
Growth Charts

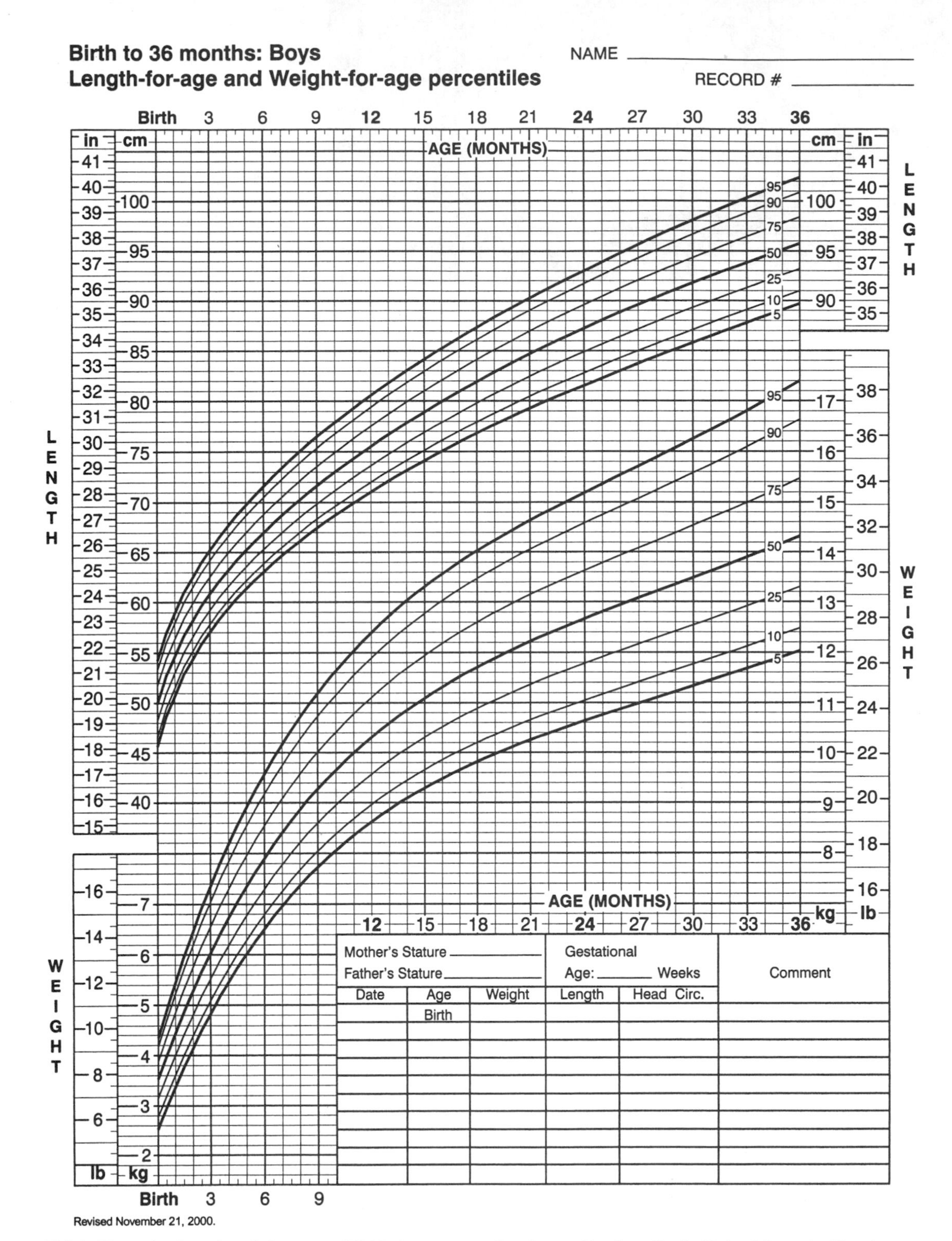

Birth to 36 months: Boys. Length-for-age and Weight-for-age percentiles. *Source:* Developed by the National Center for Chronic Disease Prevention and Health Promotion (2000).

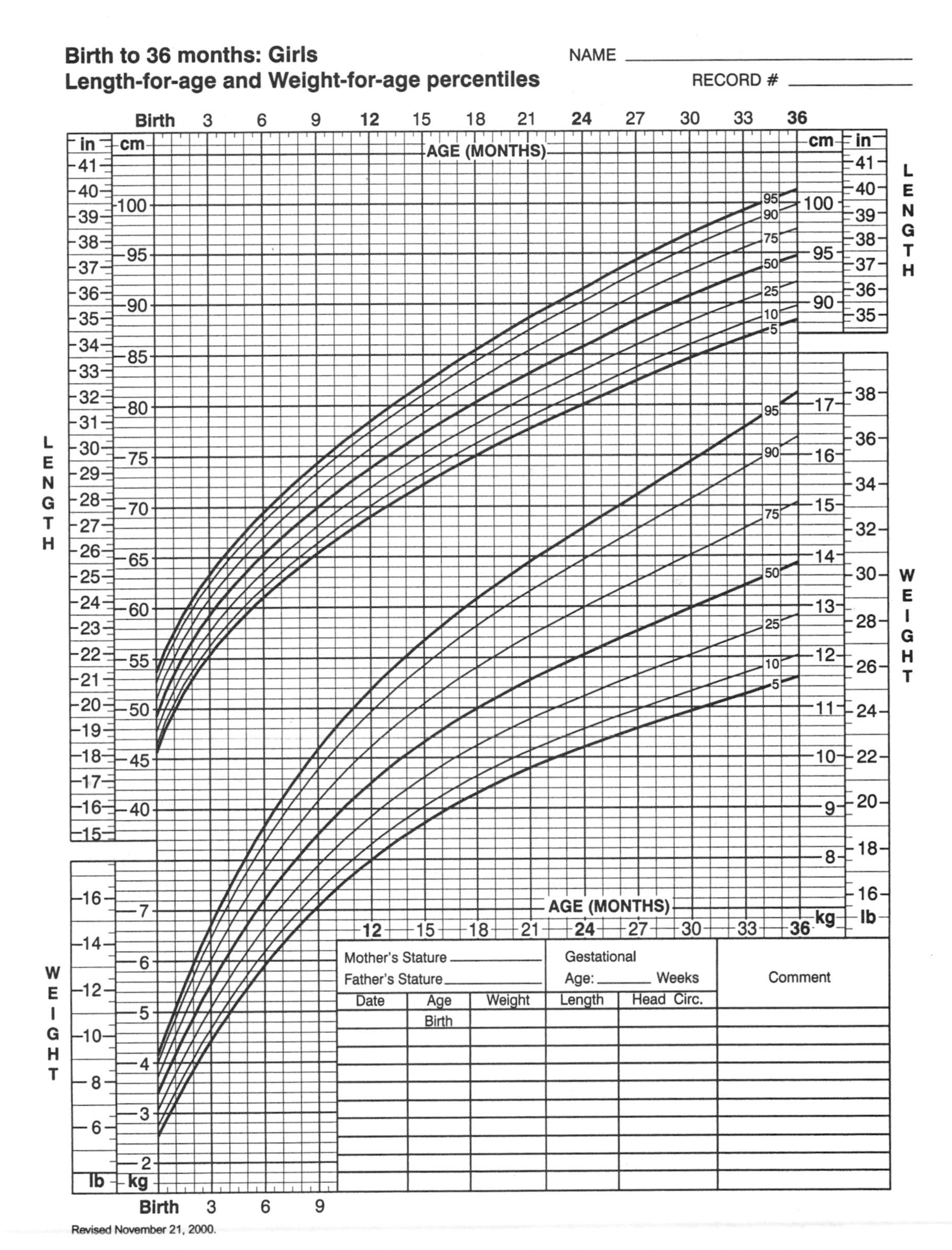

Birth to 36 months: Girls. Length-for-age and Weight-for-age percentiles. *Source:* Developed by the National Center for Chronic Disease Prevention and Health Promotion (2000).

2 to 20 years: Boys
Stature-for-age and Weight-for-age percentiles

NAME ______________________

RECORD # ______________

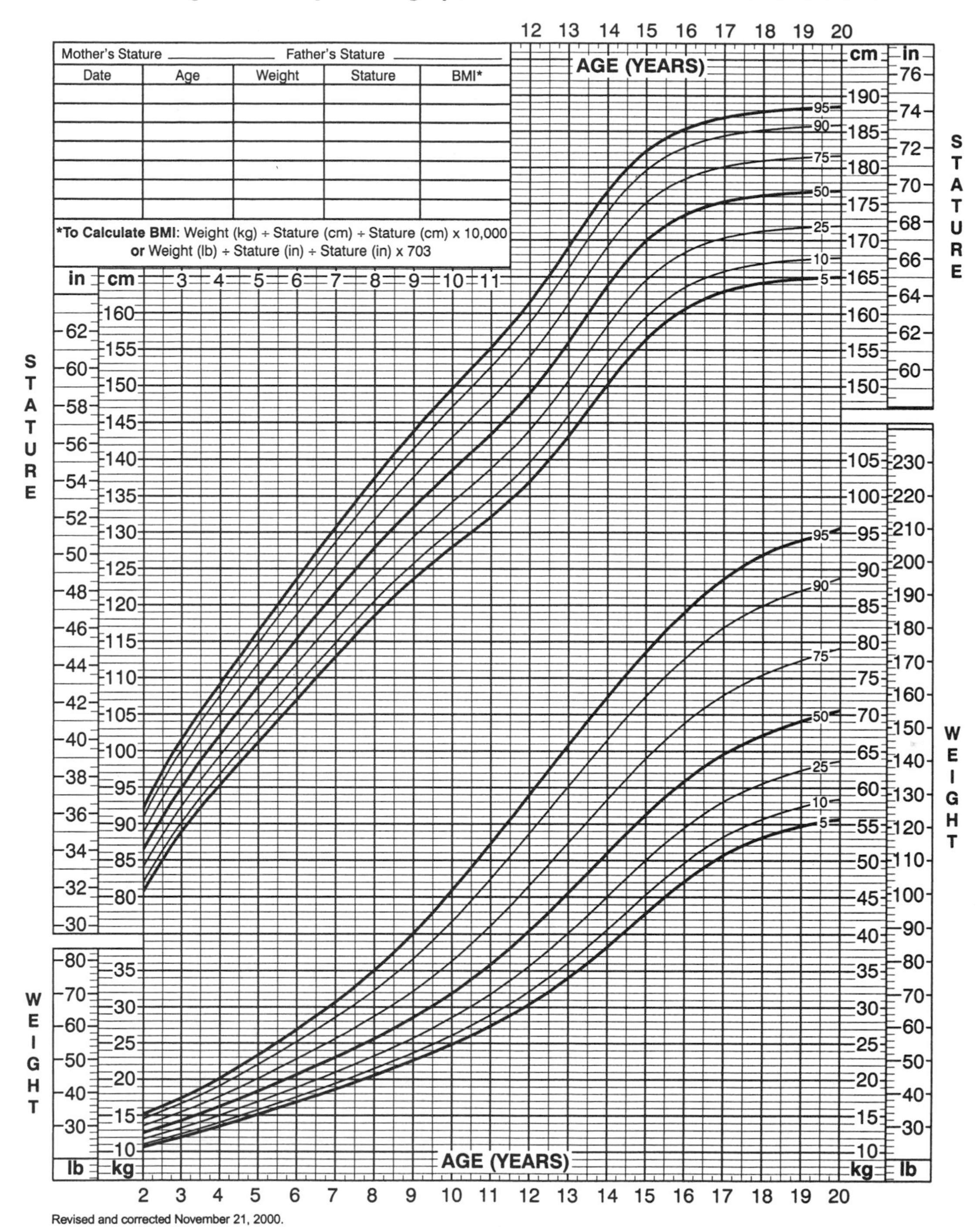

2 to 20 years: Boys. Length-for-age and Weight-for-age percentiles. *Source:* Developed by the National Center for Chronic Disease Prevention and Health Promotion (2000).

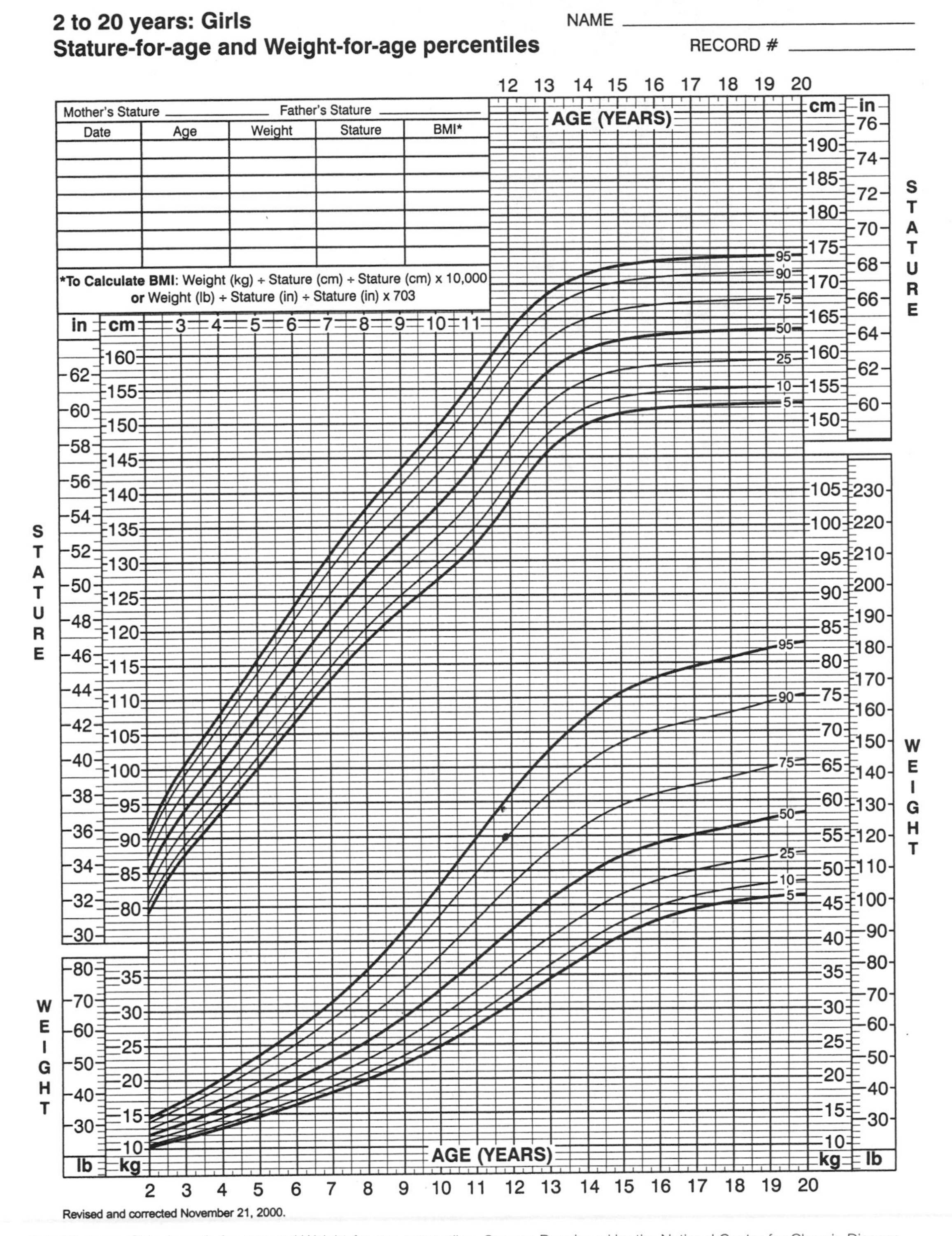

2 to 20 years: Girls. Length-for-age and Weight-for-age percentiles. *Source:* Developed by the National Center for Chronic Disease Prevention and Health Promotion (2000).

Appendix E
Answers to Check-Out Quizzes

Chapter 1

1. **Carbohydrate:** Provides energy
   Lipid: Provides energy, promotes growth and maintenance; regulates body processes
   Protein: Provides energy, promotes growth and maintenance; regulates body processes
   Vitamins and Minerals: Promotes growth and maintenance; regulates body processes
   Water: Supplies the medium in which chemical changes of the body occur; promotes growth and maintenance; regulates body processes
2. **RDA:** Value that meets requirements of 97–98 percent of individuals
   AI: Value used when there is not enough scientific data to support an RDA
   UL: Maximum safe intake level
   EAR: Value that meets requirements of 50 percent of individuals in a group
   EER: Value for Kcalories
3. **Absorption:** Process of nutrients entering the tissues from the gastrointestinal tract
   Enzyme: Substance that speeds up chemical reactions
   Anabolism: Process of building substances
   Peristalsis: Involuntary muscular contraction
   Catabolism: Process of breaking down substances
4. c
5. b
6. a
7. b
8. a
9. a
10. b

Chapter 2

1. **Energy (calories):** 1300–3000
   Added sugar: Don't exceed caloric needs
   Fiber: Increase intake
   Total fat: 30 percent or less of calories
   Saturated fat: Less than 10 percent of calories
   Cholesterol: 300 mg or less
   Protein/vitamins/minerals: 100 percent of RDA/DRIs
   Sodium: 2400 mg or less
2. **Apple:** 1 medium
   Fruit juice: 3/4 cup
   Bread: 1 slice
   Cold cereal: 1 ounce
   Cooked vegetables: 1/2 cup
   Raw vegetables: 1/2 cup (unless leafy)
   Milk or yogurt: 1 cup
   Chicken: 2–3 ounces
   Cooked kidney beans: 1/2 cup
   Eggs: 2
   Peanut butter: 4 tablespoons
   Raw leafy vegetables: 1 cup
   Cooked rice or pasta: 1/2 cup
3. d
4. a
5. b
6. b
7. b
8. b
9. a
10. b

Chapter 3

1. **White bread:** starch
   Whole-wheat bread: starch, fiber
   Apple juice: natural sugars
   Baked beans: fiber, starch
   Milk: natural sugars
   Bran flakes: fiber

Sugar-frosted oats: added sugars
Cola drink: added sugars
Broccoli: fiber

2. b
3. a
4. b
5. a
6. a
7. b
8. b
9. a
10. b

Chapter 4

I. Food	Fat	Cholesterol
1. Butter	X	X
2. Margarine	X	
3. Split peas		
4. Peanut butter	X	
5. Porterhouse steak	X	X
6. Flounder		
7. Skim milk		
8. Cheddar cheese	X	X
9. Chocolate-chip cookie made with vegetable shortening	X	
10. Green beans		

II.

1. b
2. a
3. b
4. a
5. b

III.

1. f
2. g
3. d

4. a
5. c
6. e
7. b

Chapter 5

1. a
2. a
3. a
4. b
5. a
6. b
7. a
8. a
9. a
10. a

Chapter 6

1. b
2. b
3. a
4. a
5. b
6. b
7. a
8. b
9. a
10. a

A. Vitamin B_{12}
B. Thiamin
C. Vitamin C, folate
D. Vitamin B_{12}
E. Vitamin D
F. Niacin
G. Vitamin K
H. Vitamin B_6
I. Vitamin K

J. all except biotin and vitamins D, E, and K
K. Vitamin C
L. Vitamin A
M. Vitamin D
N. Vitamins A, C, E
O. Vitamins A, C, D
P. Vitamin A
Q. Vitamin D

Chapter 7

1. a
2. b
3. a
4. b
5. a
6. b
7. a
8. a
9. b
10.

A. calcium, phosphorus, magnesium, zinc
B. calcium
C. sodium, potassium, chloride
D. potassium
E. chloride
F. potassium
G. fluoride
H. chloride
I. iron
J. iron
K. selenium

Chapter 8

1. a
2. b
3. b

4. b
5. a
6. a
7. b
8. b
9. b
10. a

Chapter 9

1. d
2. c
3. a
4. b
5. d

Chapter 10

1. a
2. c
3. d
4. d
5. c

Chapter 11

1. a
2. a
3. b
4. b
5. a
6. a
7. b
8. a
9. a
10. a

Chapter 12

1. b
2. a
3. b
4. b
5. b
6. b
7. a
8. a
9. a
10. a

Chapter 13

1. a
2. b
3. a
4. b
5. b
6. a
7. a
8. b
9. a
10. a
11. b
12. a
13. b
14. a
15. b
16. a
17. a
18. a
19. a
20. b

Glossary

Absorption The passage of digested nutrients through the walls of the intestines or stomach into the body's cells. Nutrients are then transported through the body via the blood or lymph systems.

Acceptable Macronutrient Distribution Range (AMDR) A range of intakes for a particular nutrient that is associated with reduced risk of chronic disease while providing adequate intake.

Acid-base balance The process by which the body buffers the acids and bases normally produced in the body so the blood is neither too acidic nor too basic.

Acidosis A dangerous condition in which the blood is too acidic.

Added sugars Sugars added to a food for sweetening or other purposes; they do not include the naturally occurring sugars in foods such as fruit or milk.

Adequate diet A diet that provides enough kcalories, essential nutrients, and fiber to keep a person healthy.

Adequate Intake (AI) The dietary intake that is used when an RDA cannot be based on an Estimated Average Requirement.

Alcohol-free wines Wines whose alcohol is removed after fermentation; they must be 99.5 percent alcohol-free.

Alpha-linolenic acid An omega-3 fatty acid found in several oils, notably canola, flaxseed, soybean, walnut, and wheat germ oil (or margarines made with canola or soybean oil); this essential fatty acid is vital to growth and development, maintenance of cell membranes, and the immune system, and is inadequate in many Americans' diets.

Alternative sweeteners Sweeteners that contain either no or very few calories.

Amino acid pool The overall amount of amino acids distributed in the blood, organs, and body cells.

Amino acids The building blocks of protein.

Amniotic sac The protective bag, or sac, that cushions and protects the fetus during pregnancy.

Anabolism The metabolic process by which body tissues and substances are built.

Anaphylaxis A rare allergic reaction that is very serious and can result in death if not treated immediately.

Angina Symptoms of pressing or intense pain in the heart area, often due to stress or exertion when the heart muscle gets insufficient blood.

Anorexia Lack of appetite.

Anorexia nervosa An eating disorder most prevalent in adolescent females who starve themselves.

Antibodies Proteins in the blood that bind with foreign bodies or invaders.

Antigens Foreign invaders in the body.

Antioxidant A compound that combines with oxygen to prevent oxygen from oxidizing or destroying important substances; antioxidants prevent the oxidation of unsaturated fatty acids in the cell membrane, DNA, and other cell parts that substances called free radicals try to destroy.

Anus The opening of the digestive tract through which feces travels out of the body.

Arterial blood pressure The pressure of blood within arteries as it is pumped through the body by the heart.

Artesian well water Water from a well that taps an aquifer—layers of porous rock, sand, and earth that contain water—which is under pressure from surrounding upper layers of rock or clay.

Atherosclerosis The most common form of artery disease, characterized by plaque buildup along artery walls.

Attention deficit hyperactivity disorder A developmental disorder of children characterized by impulsiveness, distractibility, and hyperactivity.

Baby bottle tooth decay Serious tooth decay in babies caused by letting a baby go to bed with a bottle of juice, formula, cow's milk, or breast milk.

Balanced diet A diet in which foods are chosen to provide kcalories, essential nutrients, and fiber in the right proportions.

Basal metabolism The minimum energy needed by the body for vital functions when at rest and awake.

Beta-carotene A precursor of vitamin A that functions as an antioxidant in the body; the most abundant carotenoid.

Bile A liver secretion that is stored in the gallbladder and released when fat enters the small intestine because it helps digest fat.

Bile acids A component of bile that aids in the digestion of fats in the duodenum of the small intestine.

Binge eating disorder An eating disorder characterized by episodes of uncontrolled eating or binging.

Bioavailability The degree to which a nutrient is absorbed and available to be used in the body.

Biotechnology A collection of scientific techniques, including genetic engineering, that are used to create, improve, or modify plants, animals, and microorganisms.

Blood glucose level (blood sugar level) The amount of glucose found in the blood; glucose is vital to the proper functioning of the body.

Body Mass Index A method of measuring degree of obesity that is a more sensitive indicator than height-weight tables.

Bolus A ball of chewed food that travels from the mouth through the esophagus to the stomach.

Bottled water Water intended for human consumption sealed in bottles or other containers with no added ingredients except that it may optionally contain safe and suitable antimicrobial agents and fluoride.

Bran In cereal grains, the parts that covers the grain and contains much fiber and other nutrients.

Bulimia nervosa An eating disorder characterized by a destructive pattern of excessive overeating followed by vomiting or other "purging" behaviors to control weight.

Cancer A group of diseases characterized by unrestrained cell division and growth that can disrupt the normal functioning of an organ and also spread beyond the tissue in which it started.

Carbohydrate or glycogen loading A regimen involving both decreased exercise and increased consumption of carbohydrates before an athletic event to increase the amount of glycogen stores.

Carbohydrates A large class of nutrients including sugars, starch, and fibers, that function as the body's primary source of energy.

Carcinogen Cancer-causing substance.

Cardiovascular disease Diseases of the heart and blood vessels such as coronary artery disease, stroke, and high blood pressure.

Carotenoids A class of pigments that contribute red, orange, or yellow colors to fruits and vegetables; can be converted to retinol or retinal in the body.

Catabolism The metabolic processes by which large, complex molecules are converted to simpler ones.

Cerebral hemorrhage A stroke due to a ruptured brain artery.

Cholesterol The most abundant sterol (a category of lipids); a soft, waxy substance present only in foods of animal origin; it is present in every cell in your body.

Chutney A sauce from India that is made with fruits, vegetables, and herbs.

Chylomicron The lipoprotein responsible for carrying mostly triglycerides, and some cholesterol, from the intestines through the lymph system to the bloodstream.

Chyme A semiliquid mixture in the stomach that contains partially digested food and stomach secretions.

Clinical trials Research studies that assign similar participants randomly to two groups; one group receives the experimental treatment while the other does not.

Club soda Filtered, artificially carbonated tap water to which mineral salts are added.

Coenzyme A molecule that combines with an enzyme, and makes the enzyme functional.

Collagen The most abundant protein in the body; a fibrous protein that is a component of skin, bone, teeth, ligaments, tendons, and other connective structures.

Colostrum A yellowish fluid that is the first secretion to come from the mother's breast a day or so after delivery of a baby; it is rich in proteins, antibodies, and other factors that protect against infectious disease.

Complementary proteins The ability of two protein foods to make up for the lack of certain amino acids in each other.

Complete protein Food proteins that provide all of the essential amino acids in the proportions needed.

Complex carbohydrates (polysaccharides) Long chains of many sugars that include starches and fibers.

Compote A dish of fruit, fresh or dried, cooked in syrup flavored with spices or liqueur; it is often served as an accompaniment or dessert.

Conditionally essential amino acids Nonessential amino acids that may, under certain circumstances, become essential.

Coronary heart disease Damage to or malfunction of the heart caused by narrowing or blockage of the coronary arteries.

Coulis A sauce made of a purée of vegetables or fruits.

Couscous A granular form of semolina, like a tiny pasta.

Cretinism (congenital hypothyroidism) Lack of thyroid secretion; causes mental and physical retardation during fetal and later development.

Cruciferous vegetables Members of the cabbage family containing phytochemicals that might help prevent cancer.

Culture The behaviors and beliefs of a certain social, ethnic, or age group.

Daily Value A set of nutrient-intake values developed by the Food and Drug Administration that are used as a reference for expressing nutrient content on nutrition labels.

Deglaze Adding cold liquid to the hot pan used in making sauces and meat dishes; any browned bits of food sticking to the pan are scraped up and added to the liquid.

Denaturation A process in which a protein uncoils and loses its shape, causing it to lose its ability to function; it can be caused by high temperatures, whipping, and other circumstances.

Dental caries Tooth decay.

Diabetes A disorder of carbohydrate metabolism characterized by high blood sugar levels and inadequate or ineffective insulin.

Diastolic pressure The pressure in the arteries when the heart is resting between beats—the bottom number in blood pressure.

Diet The food and beverages you normally consume.

Dietary fiber The edible, nondigestible component of carbohydrate and lignin naturally found in plant food.

Dietary folate equivalents (DFE) The unit for measuring folate that takes into account the amount of folate that is absorbed from natural and synthetic sources.

Dietary recommendations Guidelines that discuss specific foods and food groups to eat for optimal health.

Dietary Guidelines for Americans A set of dietary recommendations for Americans that is periodically revised.

Dietary Reference Intake (DRI) Nutrient standards that include four lists of values for dietary nutrient intakes of healthy Americans and Canadians.

Digestion The process by which food is broken down into its components in the mouth, stomach, and small intestine with the help of digestive enzymes.

Disaccharide Double sugars such as sucrose.

Diverticulosis A disease of the large intestine in which the intestinal walls become weakened, bulge out into pockets, and at times become inflamed.

Duodenum The first segment of the small intestine, about 1 foot long.

Electrolytes Chemical elements or compounds that ionize in solution and can carry an electric current; they include sodium, potassium, and chloride.

Embryo The name of the fertilized egg from conception to the eighth week.

Empty-kcalorie foods Foods that provide few nutrients for the number of kcalories they contain.

Endosperm In cereal grains, a large center area high in starch.

Energy-yielding nutrients Nutrients that can be burned as fuel to provide energy for the body, including carbohydrates, fats, and proteins.

Enriched food A food to which nutrients are added to replace the same nutrients that were lost in processing.

Enzymes Catalysts in the body.

Epidemiological research Research that looks at how disease rates vary among different populations and also factors associated with disease.

Epiglottis The flap that covers the air tubes to the lungs so that food does not enter the lungs during swallowing.

Esophagus The muscular tube that connects the pharynx to the stomach.

Essential (indispensable) amino acids Amino acids that either cannot be made in the body or cannot be made in the quantities needed by the body; must be obtained in foods.

Essential fatty acids Fatty acids that the body cannot produce, making them necessary in the diet: linoleic acid and linolenic acid.

Essential nutrients Nutrients that either cannot be made in the body or cannot be made in the quantities needed by the body; therefore, we must obtain them through food.

Estimated Average Requirement (EAR) The dietary intake value that is estimated to meet the requirement of half the healthy individuals in a group.

Estimated Energy Requirement (EER) The dietary energy intake measured in kcalories that is needed to maintain energy balance in a healthy adult.

Exchange system A tool to plan diets that groups foods by their nutrient and caloric content. Foods within each group have about the same amount of calories, carbohydrate, protein, and fat so that any food can be substituted for any other food in the same group.

Fasting hypoglycemia Low blood sugar that occurs after not eating for eight or more hours.

Fat A lipid that is solid at room temperature.

Fat-soluble vitamins A group of vitamins that generally occur in foods containing fats; these include vitamins A, D, E, and K.

Fat substitutes Ingredients that mimic the functions of fat in foods, and either contain fewer calories than fat or no calories.

Fatty acids Major component of most lipids. Three fatty acids are present in each triglyceride.

Female Athlete Triad An eating disorder found among female college athletes in which they have disordered eating, osteoporosis, and no menstruation.

Fetal alcohol syndrome A set of symptoms occurring in newborn babies that are due to alcohol use by the mother during pregnancy; symptoms may include mental retardation, brain damage, etc.

Fetus The infant in the mother's uterus from 8 weeks after conception until birth.

First trimester The period during the first 13 weeks of pregnancy.

Flavor An attribute of a food that includes its appearance, smell, taste, feel in the mouth, texture, temperature, and even the sounds made when it is chewed.

Flavorings Substances used in cooking to add a new flavor or modify the original flavor.

Fluoride The form of fluorine that appears in drinking water and in the body.

Fluorosis A condition in which the teeth become mottled and discolored due to high fluoride ingestion.

Food allergens Those parts of food causing allergic reactions.

Food allergy An abnormal response of the immune system to an otherwise harmless food.

Food guides Guidelines that tell us the kinds and amounts of foods that constitute a nutritionally adequate diet; they are based on current dietary recommendations, the nutrient content of foods, and the eating habits of the targeted population.

Food Guide Pyramid A food guide developed by the U.S. Department of Agriculture to help healthy Americans follow the Dietary Guidelines for Americans.

Food intolerance Symptoms of gas, bloating, constipation, dizziness, or difficulty sleeping after eating certain foods.

Food jag A habit of young children in which they have favorite foods they want to eat frequently.

Fortified foods A food to which nutrients are added that were not present originally, or to which nutrients are added that increase the amount already present.

Free radical An unstable compound that reacts quickly with other molecules in the body.

Fresh foods Raw foods that have not been processed (such as canned or frozen) or heated.

Fructose A monosaccharide found in fruits and honey.

Fruit smoothies Frozen blends of fruit, milk, and/or fresh or frozen yogurt.

Functional fiber Fiber sources shown to have similar health effects as dietary fiber, but that are extracted from natural sources or as synthetic.

Gag reflex The ability to cough or vomit up food (or anything) that can't be swallowed properly.

Galactose A monosaccharide found linked to glucose to form lactose or milk sugar.

Gastric lipase An enzyme in the stomach that breaks down mostly short-chain fatty acids.

Gastrointestinal tract A hollow tube running down the middle of the body in which digestion of food and absorption of nutrients takes place.

Gelatinization A process in which starches, when heated in liquid, absorb water and swell in size.

Germ In cereal grains, the area of the kernel rich in vitamins and minerals that sprouts when allowed to germinate.

Glucose The most significant monosaccharide, the body's primary source of energy.

Glycerol A derivative of carbohydrate that is part of triglycerides.

Glycogen The storage form of glucose in the body; stored in the liver and muscles.

Growth spurts Periods of rapid growth.

Health claims Claims on food labels that state certain foods or food substances—as part of an overall healthy diet—may reduce the risk of certain diseases.

Heartburn A painful burning sensation in the esophagus caused by acidic stomach contents flowing back into the lower esophagus.

Height-weight tables Tables that show an appropriate weight for a given height.

Heme iron The predominant form of iron in animal foods; it is absorbed and used more readily than iron in plant foods.

Hemoglobin A protein in red blood cells that carries oxygen to the body's cells.

Hemorrhoids Enlarged veins in the lower rectum.

Herbs The leafy parts of certain plants that grow in temperate climates; they are used to season and flavor foods.

High-density lipoproteins (HDL) Lipoproteins that contain much protein and carry cholesterol away from body cells and tissues to the liver for excretion from the body.

High-fructose corn syrup Corn syrup that has been treated with an enzyme that converts part of the glucose it contains to fructose.

Homeostasis A constant internal environment in the body.

Homogenized Milk that has had its fat particles broken up so finely that they remain uniformly dispersed throughout the milk.

Hormones Chemical messengers in the body.

Hydrochloric acid A strong acid made by the stomach that aids in protein digestion, destroys harmful bacteria, and increases the ability of calcium and iron to be absorbed.

Hydrogenation A process in which liquid vegetable oils are converted into solid fats (such as margarine) by the use of heat, hydrogen, and certain metal catalysts.

Hyperglycemia High levels of blood sugar.

Hypertension High blood pressure.

Hypervitaminosis A A disease caused by prolonged use of high doses of preformed vitamin A that can cause hair loss, bone pain and damage, soreness, and other problems.

Hypoglycemia A symptom in which blood sugar levels are low.

Hypothyroidism A condition in which there is less production of thyroid hormones; this leads to symptoms such as low metabolic rate, fatigue, and weight gain.

Ileum The final segment of the small intestine.

Immune response The body's response to a foreign substance, such as a virus, in the body.

Incomplete protein Food proteins that contain at least one limiting amino acid.

Inorganic In chemistry, any compound that does not contain carbon.

Insoluble fiber A classification of fiber that includes cellulose, lignin, and the remaining hemicelluloses; they generally form the structural parts of plants.

Insulin A hormone that increases the movement of glucose from the bloodstream into the body's cells.

Intrinsic factor A proteinlike substance secreted by stomach cells that is necessary for the absorption of vitamin B_{12}.

Ion An atom or group of atoms carrying a positive or electric charge.

Iron-deficiency anemia A condition in which the size and number of red blood cells are reduced; may result from inadequate iron intake or from blood loss; symptoms includes fatigue, pallor, and irritability.

Iron overload (hemochromatosis) A common genetic disease in which individuals absorb about twice as much iron from their food and supplements as other people.

Irradiation A process of using a measured dose of radiation on foods to reduce the number of harmful microorganisms.

Jejunum The second portion of the small intestine, between the duodenum and the ileum.

Ketone bodies A group of organic compounds that cause the blood to become too acidic as a result of fat being burned for energy without any carbohydrates present

Ketosis Excessive level of ketone bodies in the blood and urine.

Kilocalorie A measure of the energy in food, specifically the energy-yielding nutrients.

Kwashiorkor A type of PEM associated with children with insufficient protein intake and who have a preexisting disease.

Lactase An enzyme needed to split lactose into its components in the intestines.

Lacto-ovo-vegetarians Vegetarians who do not eat meat, poultry, or fish but do consume animal products in the form of eggs, milk, and milk products.

Lactose A disaccharide found in milk and milk products that is made of glucose and galactose.

Lactose intolerance An intolerance to milk and most milk products due to a deficiency of the enzyme lactase. Symptoms often include flatulence and diarrhea.

Lacto-vegetarians Vegetarians who do not eat meat, poultry, or fish but do consume animal products in the form of milk and milk products.

Large intestine The part of the gastrointestinal tract between the small intestine and the rectum.

Lecithin A phospholipid and a vital component of cell membranes that acts as an emulsifier (a substance that keeps fats in solution).

Light beer Beer with one-third to one-half less alcohol and kcalories of regular beers.

Limiting amino acid An essential amino acid in lowest concentration in a protein.

Lingual lipase An enzyme made in the salivary glands in the mouth that has a minor role in fat digestion in adults, and an important role in fat digestion in infants.

Linoleic acid Omega-6 fatty acid found in vegetable oils such as corn, safflower, soybean, cottonseed, and sunflower oils; this essential fatty acid is vital to growth and development, maintenance of cell membranes, and the immune system.

Lipids A group of fatty substances, including triglycerides and cholesterol, that are soluble in fat, not water, and that provide a rich source of energy and structure to cells.

Lipoprotein Protein-coated packages that carry fat and cholesterol through the bloodstream; the body makes four types, classified according to their density.

Lipoprotein lipase An enzyme that breaks
Ketone bodies A group of organic compounds that cause the blood to become too acidic as a result of fat being burned for energy without any carbohydrates present.

Ketosis Excessive level of ketone bodies in the blood and urine.

Kilocalorie A measure of the energy in food, specifically the energy-yielding nutrient.

Kwashiorkor A type of PEM associated with children with insufficient protein intake and who have a preexisting disease.

down triglycerides from the chylomicron into fatty acids and glycerol so they can be absorbed in the body's cells.

Low-alcohol refreshers A category of drinks made with fruit juices, carbonated water, and sugar, with an alcohol base of wine, distilled spirits, or malt; the group includes wine coolers.

Low-birth-weight baby A newborn who weighs less than 5-1/2 pounds; these infants are at higher risk for disease.

Low-density lipoproteins (LDL) Lipoproteins that contain most of the cholesterol in the blood; they carry cholesterol to body tissues.

Lower esophageal (cardiac) sphincter A muscle that relaxes and contracts to move food from the esophagus into the stomach.

Low-trans hydrogenation A new hydrogenation process to make shortenings and margarines with less trans fats.

Lanugo Downy hair on the skin.

Macaroni Pastas made from flour and water.

Macronutrients Nutrients needed by the body in large amounts, including carbohydrates, lipids, and proteins.

Major minerals Minerals needed in relatively large amounts in the diet—over 100 milligrams daily.

Maltose A disaccharide made of two glucose units bonded together.

Marasmus A type of PEM characterized by gross underweight and severe food shortage.

Marinades A seasoned liquid used before cooking to flavor and moisten foods; usually based on an acidic ingredient.

Marketing The process of finding out what your customers need and want, and then developing, promoting, and selling the products and services they desire.

Megadose A supplement intake of 10 times the RDA of a vitamin or mineral.

Megaloblastic (or macrocytic) anemia A form of anemia caused by a deficiency of vitamin B_{12} or folate and characterized by large, immature red blood cells.

Metabolism All the chemical processes by which nutrients are used to support life.

Metastasis The condition when a cancer spreads beyond the tissue in which it started.

Micronutrients Nutrients needed by the body in small amounts, including vitamins and minerals.

Microvilli Hairlike projections on the villi that increase the surface area for absorbing nutrients.

Milk letdown The process by which milk comes out of the mother's breast to feed the baby; sucking causes the release of a hormone that allows milk letdown.

Minerals Noncaloric, inorganic chemical substances found in a wide variety of foods needed to regulate body processes, to maintain the body, and to allow growth and reproduction

Mineral water Water from an underground source that contains at least 250 parts per million total dissolved solids. Minerals and trace elements must come from the source of the underground water.

Mocktails Drinks made to resemble mixed alcoholic drinks but containing no alcohol.

Moderate diet A diet that avoids excessive amounts of kcalories or any particular food or nutrient.

Mojo A spicy Caribbean sauce; it is a mixture of garlic, citrus juice, oil, and fresh herbs.

Monoglycerides Triglycerides with only one fatty acid.

Monosaccharide Single sugars such as glucose or fructose.

Monounsaturated fat A triglyceride made of mostly monounsaturated fatty acids.

Monounsaturated fatty acid A fatty acid that contains only one double bond in the chain.

Myocardial infarction Heart attack.

Myocardial ischemia A temporary injury to heart cells caused by a lack of blood flow and oxygen.

Myoglobin A muscle protein that stores and carries oxygen that the muscles will use to contract.

Negative nitrogen balance A condition in which the body excretes more protein than is taken in; this can occur during starvation and certain illnesses.

Neural tube The embryonic tissue that develops into the brain and spinal cord.

Neural tube defects Diseases in which the brain and the spinal cord form improperly in early pregnancy.

Niacin equivalents (NE) The unit for measuring niacin. One niacin equivalent is equal to one milligram of niacin or 60 milligrams of tryptophan.

Night blindness A condition caused by insufficient vitamin A in which it takes longer to adjust to dim lights after seeing a bright light at night; this is an early sign of vitamin A deficiency.

Nitrogen balance The difference between the total nitrogen intake and total nitrogen loss; a healthy person has the same nitrogen intake as loss, resulting in a zero nitrogen balance.

Nonalcoholic malt beverages A beerlike product with only 0.5 percent or less alcohol.

Nonessential amino acids Amino acids that can be made in the body.

Nonheme iron A form of iron found in all plant sources of iron and also as part of the iron in animal food sources.

Noodles Pastas made from flour, water, and egg solids.

Nutrient content claims Claims on food labels about the nutrient composition of a food, which are regulated by the Food and Drug Administration.

Nutrient-dense foods Foods that contain many nutrients for the kcalories they provide.

Nutrient density A measure of the nutrients provided in a food per kcalorie of the food.

Nutrients The nourishing substances in food that provide energy and promote the growth and maintenance of your body.

Nutrition A science that studies nutrients and other substances in foods and in the body and how these nutrients relate to health and disease. Nutrition also explores why you choose particular foods and the type of diet you eat.

Nutrition-support claims Claims on food labels that describe a link between a nutrient and the deficiency disease that can result if the nutrient is lacking in the diet.

Obesity A state of having a Body Mass Index of 30 or greater.

Oil A lipid that is usually liquid at room temperature.

Oral cavity The mouth.

Organic In chemistry, any compound that contains carbon.

Organic foods Foods that have been grown without most conventional pesticides, fertilizers, herbicides, antibiotics, or hormones, and without genetic engineering or irradiation.

Osteomalacia A disease of vitamin D deficiency in adults in which the leg and spinal bones soften and may bend.

Osteoporosis The most common bone disease, characterized by loss of bone density and strength; it is associated with debilitating fractures, especially in people 45 and older, due to a tremendous loss of bone tissue in midlife.

Overweight A state of having a Body Mass Index of 25 or greater.

Oxalic acid An organic acid found in spinach and other leafy green vegetables that can decrease the absorption of certain minerals such as calcium.

Palmar grasp The ability of a baby from about 6 months of age to grab objects with the palm of the hand.

Pasteurized A product, such as milk, that has been treated to kill harmful germs.

Pepsin the principal digestive enzyme of the stomach.

Peptide bonds The bonds that form between adjoining amino acids.

Peristalsis Involuntary muscular contraction that forces food through the entire digestive system.

Pernicious anemia A type of anemia caused by a deficiency of vitamin B_{12} and characterized by macrocytic anemia and deterioration in the functioning of the nervous system.

Pesco-vegetarians Vegetarians who eat fish.

Pharynx A passageway that connects the oral and nasal cavities to the esophagus and air tubes to the lungs.

Photosynthesis A process during which plants convert energy from sunlight into energy stored in carbohydrate.

Phytic acid A binder found in wheat bran and whole grains that can decrease the absorption of certain nutrients such as calcium and iron.

Phytochemicals Minute substances in plants that may reduce risk of cancer and heart disease when eaten often.

Pincer grasp The ability of a baby at about 8 months of age to use the thumb and forefinger together to pick things up.

Placenta The organ that develops during the first month of pregnancy, which provides for exchange of nutrients and wastes between fetus and mother and secretes the hormones necessary to maintain pregnancy.

Plaque (1) Deposits of bacteria, protein, and polysaccharides found on teeth that contribute to tooth decay. (2) Deposits on arterial walls that contain cholesterol, fat, fibrous scar tissue, calcium, and other biological debris.

Point of unsaturation The location of the double bond in unsaturated fatty acids.

Polypeptides Protein fragments with 10 or more amino acids.

Polyunsaturated fat A triglyceride made of mostly polyunsaturated fatty acids.

Polyunsaturated fatty acid A fatty acid that contains two or more double bonds in the chain.

Positive nitrogen balance A condition in which the body excretes less protein than is taken in; this can occur during growth and pregnancy.

Postpranial hypoglycemia Low blood sugar that occurs generally 1 to 4 hours after meals and includes symptoms such as shakiness, sweating, and dizziness.

Precompetition meal The meal closest to the time of a competition or event.

Precursors Forms of vitamins that the body changes chemically to active vitamin forms.

Pregnancy-induced hypertension Hypertension during pregnancy that can cause serious complications.

Primary (essential) hypertension A form of hypertension whose cause is unknown.

Primary structure The number and sequence of the amino acids in the protein chain.

Processed foods Foods that have been prepared using a certain procedure such as cooking, freezing, canning, dehydrating, milling, culturing, or adding vitamins and minerals.

Promoters Substances such as fat that advance the development of mutated cells into a tumor.

Proportionality A concept of eating relatively more foods from the larger food groups at the base of the Food Guide Pyramid and fewer foods from the smaller food groups nearer the top of the Pyramid.

Protein-energy malnutrition (PEM) A broad spectrum of malnutrition from mild to serious cases.

Proteins Major structural parts of the body's cells that are made of nitrogen-containing amino acids assembled in chains, particularly rich in animal foods.

Publicity Obtaining free space or time in various media to get public notice of a program, book, and so on.

Puréeing Mashing or straining a food to a smooth pulp.

Purified water Water produced by reverse osmosis, ozonation, or other suitable processes and that meets the definition set by the U.S. Pharmacopoeia.

Pyloric sphincter A muscle that permits passage of chyme from the stomach to the small intestine.

Rancidity The deterioration of fat, resulting in undesirable flavors and odors.

Recommended Dietary Allowance (RDA) The dietary intake value that is sufficient to meet the nutrient requirements of 97 to 98 percent of all healthy individuals in a group.

Rectum The last section of the large intestine in which feces, the waste products of digestion, is stored until elimination.

Reduction To boil or simmer a liquid down to a smaller volume.

Refined or milled grain A grain in which the bran and germ are separated (or mostly separated) from the endosperm.

Registered Dietitians Professionals recognized by the medical profession as the legitimate providers of nutrition care.

Retinoids The forms of vitamin A that are in the body: retinol, retinal, and retinoic acid.

Retinol A form of vitamin A found in animal foods; it can be converted to retinal and retinoic acid in the body.

Retinol activity equivalents (RAE) The unit for measuring vitamin A. One RAE = 1 microgram retinol, 12 micrograms beta-carotene, or 24 micrograms of other vitamin A precursor carotenoids.

Rickets A childhood disease in which bones do not grow normally, resulting in bowed legs and knock knees; it is generally caused by a vitamin D deficiency.

Risk factor A habit, trait, or condition associated with an increased chance of developing a disease.

Rubs A dry marinade made of herbs and spices (and other seasonings), sometimes moistened with a little oil, and rubbed or patted on the surface of meat, poultry, or fish (which is then refrigerated and cooked at a later time).

Saliva A fluid secreted into the mouth from the salivary glands, which contains important digestive enzymes and lubricates the food so that it may readily pass down the esophagus.

Salsas Chunky mixtures of vegetables and/or fruits and flavor ingredients.

Satiety A feeling of being full after eating.

Saturated fatty acid A fatty acid that is filled to capacity with hydrogens.

Saturated fat A triglyceride made of mostly saturated fatty acids.

Scurvy A vitamin C deficiency disease marked by bleeding gums, weakness, loose teeth, and broken capillaries under the skin.

Secondary hypertension Persistently elevated blood pressure caused by a medical problem.

Searing To expose meat's surfaces to a high heat before cooking at a lower temperature; this process adds color and flavor to the meat.

Seasonings Substances used in cooking to bring out a flavor already present.

Secondary structure The bending and coiling of the protein chain.

Seltzer Filtered, artificially carbonated tap water that generally has no added mineral salts.

Semolina The roughly milled endosperm of a type of wheat called durum wheat.

Simple carbohydrates Sugars including monosaccharides and disaccharides.

Simple goiter Thyroid enlargement caused by inadequate dietary intake of iodine.

Small intestine The digestive tract organ that extends from the stomach to the opening of the large intestine.

Soluble fiber A classification of fiber that includes gums, mucilages, pectin, and some hemicelluloses; they are generally found around and inside plant cells.

Sparkling water Any carbonated water.

Spices The roots, bark, seeds, flowers, buds, and fruits of certain tropical plants; they are used to season and flavor foods.

Spina bifida A birth defect in which parts of the spinal cord are not fused together properly, so gaps are present where the spinal cord has little or no protection.

Spring water Water collected as it flows naturally to the surface from an underground formation.

Starch A complex carbohydrate made up of a long chain of glucoses linked together; found in grains, legumes, vegetables, and some fruits; the straight form is called amylose and the branched form is called amylopectin.

Stomach J-shaped muscular sac that holds about 4 cups of food and prepares food chemically and mechanically so it can be further digested and absorbed.

Stroke Damage to brain cells resulting from an interruption of blood flow to the brain.

Structure-function claims Claims on food labels that refer to the supplement's effect on the body's structure or function, including its overall effect on a person's well-being.

Sucrose A disaccharide commonly called table sugar, granulated sugar, or simply sugar.

Sugar alcohols Sugarlike compounds that occur naturally in small amounts in fruits and vegetables; they are used to sweeten sugar-free candies, cookies, and chewing gum.

Sweat To cook slowly in a small amount of fat over low or moderate heat without browning.

Systolic pressure The pressure of blood within arteries when the heart is pumping—the top blood pressure number.

Taste Sensations perceived by the taste buds on the tongue.

Taste buds Clusters of cells found on the tongue, cheeks, throat, and roof of the mouth. Each taste bud houses 60 to 100 receptor cells. The body regenerates taste buds about every three days. These cells bind food molecules dissolved in saliva and alert the brain to interpret them.

Tertiary structure The folding of the protein chain.

Thermic effect of food The energy needed to digest and absorb food.

Thyroid gland A gland found on either side of the trachea that produces and secretes two important hormones that regulate the level of metabolism.

TLC diet A low saturated fat, low cholesterol eating plan designed to fight cardiovascular disease and lower LDL; the diet calls for less than 7 percent of kcalories from saturated fat and less than 200 milligrams of cholesterol daily and also recommends only enough kcalories to maintain a desirable weight.

Tolerable Upper Intake Level (UL) The maximum intake level above which risk of toxicity would increase.

Tonic water A carbonated water containing lemon, lime, sweeteners, and quinine.

Trace minerals Minerals needed in smaller amounts in the diet—less than 100 milligrams daily.

Trans fat Unsaturated fatty acids that lose a natural bend or kink so they become straight (like saturated fatty acids) after being hydrogenated; they act like saturated fats in the body.

Transitional milk The type of breast milk produced from about the third to the tenth day after childbirth, when mature milk appears.

Triglyceride The major form of lipid in food and in the body; it is made of three fatty acids attached to a glycerol backbone.

Tryptophan An amino acid present in protein foods that can be converted to niacin in the body.

Type 1 diabetes A form of diabetes seen mostly in children and adolescents. These patients make no insulin and therefore require frequent injections of insulin to maintain a normal level of blood glucose.

Type 2 diabetes A form of diabetes seen most often in overweight adults. These patients make insulin but their tissues aren't sensitive enough to the hormone and so use it inefficiently.

Unsaturated fatty acid A fatty acid with at least one double bond.

Varied diet A diet in which you eat a wide selection of foods to get necessary nutrients.

Vegans Individuals eating a type of vegetarian diet in which no eggs or dairy products are eaten; their diet relies exclusively on plant foods.

Very-low-density lipoproteins (VLDL) Lipoproteins made by the liver to carry triglycerides and some cholesterol through the body.

Villi Tiny, fingerlike projections in the wall of the small intestines that are involved in absorption.

Vitamin D_3 (cholecalciferol) The form of vitamin D found in animal foods.

Vitamins Noncaloric, organic nutrients found in a wide variety of foods that are essential in small quantities to regulate body processes, maintain the body, and allow growth and reproduction.

Water balance The process of maintaining the proper amount of water in each of three body "compartments": inside the cells, outside the cells, and in the blood vessels.

Water-soluble vitamins A group of vitamins that are soluble in water and are not stored appreciably in the body; these include vitamin C, thiamin, riboflavin, niacin, vitamin B_6, folate, vitamin B_{12}, pantothenic acid, and biotin.

Well water Water from a hole drilled into the ground, which taps into an aquifer.

Whole foods Foods as we get them from nature.

Whole grain A grain that contains the endosperm, germ, and bran.

Wine coolers Mixes of wine, fruit flavor, and carbonated water or plain water.

Wine spritzers Drinks made with wine and club soda.

Xerophthalmia Hardening and thickening of the cornea that can lead to blindness; usually caused by a deficiency of vitamin A.

Xerosis A condition in which the cornea of the eye becomes dry and cloudy; often due to a deficiency of vitamin A.

Selected References

Chapter 1

Fieldhouse, P. *Food and Nutrition: Customs and Culture*. UK: Chapman and Hall, 1996.

Food and Nutrition Board, Institute of Medicine. *Dietary Reference Intakes for Calcium, Phosphorus, Magnesium, Vitamin D, and Fluoride*. Washington, D.C.: National Academy Press, 1997.

French, S. A., M. Story, and R. W. Jeffery. "Environmental Influences on Eating and Physical Activity." *Annual Reviews Public Health* 22 (2001): 309–335.

Gassin. A. L. "Helping to Promote Healthy Diets and Lifestyles: The Role of the Food Industry." *Public Health Nutrition* 4 (2001): 1445–1450.

Glanz, K., M. Basil, E. Maibach, J. Goldberg, and D. Snyder. "Why Americans Eat What They Do: Taste, Nutrition, Cost, Convenience, and Weight Control Concerns as Influences on Food Consumption." *Journal of The American Dietetic Association* 98 (1998): 1118–1126.

Hess, M.A. "Taste: The Neglected Nutrition Factor." *Journal of The American Dietetic Association* 97 (1997): S205–S207.

Institute of Medicine, Food and Nutrition Board. *Dietary Reference Intakes: Use in Dietary Assessment.* Washington, D.C.: National Academy Press, 2000.

Kittler, P. G., and K. P. Sucher. *Cultural Foods: Traditions and Trends*. Belmont, Calif.: Wadsworth/Thomson Learning, 2000.

McBean, Lois. "Good Science: Its Role in Setting the Record Straight." *Dairy Council Digest* 73 (2001).

National Research Council, Subcommittee on the 10th Edition of the RDAs, Food and Nutrition Board, Commission on Life Sciences. 1989. *Recommended Dietary Allowances*. 10th ed. Washington, D.C.: National Academy Press, 1989.

Trumbo, P., et al. "Dietary Reference Intakes: Vitamin A, Vitamin K, Arsenic, Boron, Chromium, Copper, Iodine, Iron, Manganese, Molybdenum, Nickel, Silicon, Vanadium, and Zinc." *Journal of The American Dietetic Association* 101 (2001): 294–301.

Van De Graaff, K. M., and Stuart Ira Fox. *Concepts of Human Anatomy and Physiology, Fourth Edition*. Dubuque, Ia.: Wm. C. Brown Publishers, 1995.

Wardle, J., K. Parmenter., and J. Waller. "Nutrition Knowledge and Food Intake." *Appetite* 34 (2000): 269–275.

Chapter 2

American Dietetic Association. "Position of The American Dietetic Association: Food and Nutrition Misinformation." *Journal of The American Dietetic Association* 102 (2002): 260–266.

McBean, Lois. "Good Science: Its Role in Setting the Record Straight." *Dairy Council Digest* 73 (2001).

Patterson, R.E., et al. "Is There A Consumer Backlash Against the Diet and Health Message?" *Journal of The American Dietetic Association* 101 (2001): 37–41.

Shaw, A., et al. *Using the Food Guide Pyramid: A Resource for Nutrition Educators.* U.S. Department of Agriculture: Food, Nutrition, and Consumer Services; Center for Nutrition Policy and Promotion.

U.S. Department of Agriculture. *Food Guide Pyramid: A Guide to Daily Food Choices*. Washington, D.C.: U.S. Government Printing Office, 1992.

U.S. Department of Agriculture and Department of Health and Human Services. *Nutrition and Your Health: Dietary Guidelines for Americans.* Home and Garden Bulletin No. 232. Washington, D.C.: U.S. Government Printing Office, 2000.

Chapter 3

American Dietetic Association. "Position of The American Dietetic Association: Health Implications of Dietary Fiber." *Journal of The American Dietetic Association* 102 (2002): 993–1000.

American Dietetic Association. "Position of The American Dietetic Association: Use of Nutritive and Nonnutritive Sweeteners." *Journal of The American Dietetic Association* 98 (1998): 580–586.

Butchko, H. H., and W. W. Stargel. "Aspartame: Scientific Evaluation in the Postmarketing Period." *Regulatory Toxicology and Pharmacology* 34 (2001): 221–233.

Coulston, A. M., and R.K. Johnson. "Sugar and Sugars: Myths and Realities." *Journal of the American Dietetic Association* 102 (2002): 351–353.

Guthrie, J. F., and J. F. Morton. "Food Sources of Added Sweeteners in the Diets of Americans." *Journal of The American Dietetic Association* 100 (2000): 43–48, 51.

Harnack, L., J. Stang, and M. Story. "Soft Drink Consumption Among U.S. Children and Adolescents: Nutritional Consequences." *Journal of The American Dietetic Association* 99 (1999): 436—441.

Inman-Felton, Amy E. "Overview of Lactose Maldigestion (Lactase Non-persistence)." *Journal of The American Dietetic Association* 99 (1999): 481–489.

Slavin, J. L., et al. "The Role of Whole Grains in Disease Prevention." *Journal of The American Dietetic Association* 101 (2001): 780–785.

Tabatabai, A., and S. Li. "Dietary Fiber and Type 2 Diabetes." *Clinical Excellence for Nurse Practitioners* 4 (2000): 272–276.

Truswell, A. S. "Cereal Grains and Coronary Heart Disease." *European Journal of Clinical Nutrition* 56 (2002): 1–14.

Vesa, T. H., et al. "Lactose Intolerance." *Journal of the American College of Nutrition* 19 (2000): 165S–172S.

Chapter 4

American Dietetic Association. "Position of The American Dietetic Association: Fat Replacers." *Journal of The American Dietetic Association* 98 (1998): 463–468.

Berry, E. M. "Who's Afraid of n-6 Polyunsaturated Fatty Acids? Methodological Considerations for Assessing Whether They Are Harmful." *Nutrition, Metabolism and Cardiovascular Diseases* 11 (2001): 181–188.

Dausch, J. G. "Trans-Fatty Acids: A Regulatory Update." *Journal of The American Dietetic Association* 102 (2002): 18, 20.

Demaison, L., and D. Moreau. "Dietary n-3 Polyunsaturated Fatty Acids and Coronary Heart Disease-Related Mortality: A Possible Mechanism of Action." *Cellular and Molecular Life Sciences* 59 (2002): 463–477.

Elias, S. L., and S. M. Innis. "Bakery Foods Are the Major Dietary Source of Trans-fatty Acids Among Pregnant Women with Diets Providing 30 Percent Energy from Fat." *Journal of The American Dietetic Association* 102 (2002): 46–51.

Greenwald, P., C. K. Clifford., and J. A. Milner. "Diet and Cancer Prevention." *European Journal of Cancer* 37 (2001): 948–965.

Harper, C. R., and T. A. Jacobson. "The Fats of Life: The Role of Omega-3 Fatty Acids in the Prevention of Coronary Heart Disease." *Archives of Internal Medicine* 161 (2001): 2185–2192.

Krauss, R. M., et al. "AHA Dietary Guidelines: Revision 2000: A Statement for Healthcare Professionals from the Nutrition Committee of the American Heart Association." *Circulation* 102 (2000): 2284.

National Institutes of Health; National Heart, Lung, and Blood Institute. *Third Report of the National Cholesterol Education Program (NCEP) Expert Panel on Detection, Evaluation, and Treatment of High Blood Cholesterol in Adults (Adult Treatment Panel III).* NIH Publication No. 01-3670. 2001.

Nguyen, T.T. "The Cholesterol-Lowering Action of Plant Stanol Esters." *Journal of Nutrition* 129 (1999): 2109–2112.

Sandler, R. S., et al. "Gastrointestinal Symptoms in 3181 Volunteers Ingesting

Snack Foods Containing Olestra or Triglycerides: A 6-Week Randomized, Placebo-Controlled Trial." *Annals of Internal Medicine* 130 (1999): 253–261.

Schaefer, Ernst J. "Lipoproteins, Nutrition, and Heart Disease." *American Journal of Clinical Nutrition* 75 (2002): 191–212.

Valenzuela, A., and N. Morgado. "Trans Fatty Acid Isomers in Human Health and in the Food Industry." *Biological Research* 32 (1999): 273–287.

Willett, W. C. "Diet and Cancer." *Oncologist* 5 (2000): 393–404.

Chapter 5

American Dietetic Association. "Position of the American Dietetic Association: Food Irradiation." *Journal of The American Dietetic Association* 100 (2000): 246–253.

Matthews, D. E. "Proteins and Amino Acids." In Shils, M.E., et al. (eds.), *Modern Nutrition in Health and Disease,* 9th edition. Baltimore: Williams and Wilkins, 1999.

St. Jeor, S. T., et al. "Dietary Protein and Weight Reduction: A Statement for Healthcare Professionals from the Nutrition Committee of the Council on Nutrition, Physical Activity, and Metabolism of the American Heart Association." *Circulation* 104 (2001): 1869–1874.

Torun, B., and F. Chew. "Protein-Energy Malnutrition." In Shils, M.E., et al. (eds.), *Modern Nutrition in Health and Disease,* 9th edition. Baltimore: Williams and Wilkins, 1999.

Chapter 6

American Dietetic Association. "Position of The American Dietetic Association: Food Fortification and Dietary Supplements." *Journal of The American Dietetic Association* 115 (2001): 115–125.

American Dietetic Association. "Position of The American Dietetic Association: Functional Foods." *Journal of The American Dietetic Association* 99 (1999): 1278–1285.

Food and Nutrition Board, Institute of Medicine. *Dietary Reference Intakes for Calcium, Phosphorus, Magnesium, Vitamin D, and Fluoride*. Washington, D.C.: National Academy Press, 1997.

Food and Nutrition Board, Institute of Medicine. *Dietary Reference Intakes for Thiamin, Riboflavin, Niacin, Vitamin B_6, Folate, Vitamin B_{12}, Pantothenic Acid, Biotin, and Choline*. Washington, D.C.: National Academy Press, 1998.

Food and Nutrition Board, Institute of Medicine. *Dietary Reference Intakes for Vitamin A, Vitamin K, Arsenic, Boron, Chromium, Copper, Iodine, Iron, Manganese, Molybdenum, Nickel, Silicon, Vanadium, and Zinc*. Washington, D.C.: National Academy Press, 2001.

Food and Nutrition Board, Institute of Medicine. *Dietary Reference Intakes for*

Vitamin C, Vitamin E, Selenium, and Carotenoids. Washington, D.C.: National Academy Press, 2000.

Groff, James L., and Sareen S. Gropper. *Advanced Nutrition and Human Metabolism*. Belmont, Calif.: Wadsworth Publishing, 2000.

Trumbo, P., et al. "Dietary Reference Intakes: Vitamin A, Vitamin K, Arsenic, Boron, Chromium, Copper, Iodine, Iron, Manganese, Molybdenum, Nickel, Silicon, Vanadium, and Zinc." *Journal of The American Dietetic Association* 101 (2001): 294–301.

Chapter 7

American Dietetic Association. "Position of The American Dietetic Association: The Impact of Fluoride on Health." *Journal of The American Dietetic Association* 100 (2000): 1208–1213.

Food and Nutrition Board, Institute of Medicine. *Dietary Reference Intakes for Calcium, Phosphorus, Magnesium, Vitamin D, and Fluoride*. Washington, D.C.: National Academy Press, 1997.

Food and Nutrition Board, Institute of Medicine. *Dietary Reference Intakes for Vitamin A, Vitamin K, Arsenic, Boron, Chromium, Copper, Iodine, Iron, Manganese, Molybdenum, Nickel, Silicon, Vanadium, and Zinc*. Washington, D.C.: National Academy Press, 2001.

Food and Nutrition Board, Institute of Medicine. *Dietary Reference Intakes for Vitamin C, Vitamin E, Selenium, and Carotenoids*. Washington, D.C.: National Academy Press, 2000.

Groff, James L., and Sareen S. Gropper. *Advanced Nutrition and Human Metabolism*. Belmont, Calif.: Wadsworth Publishing, 2000.

Kleiner, S. M. "Water: An Essential but Overlooked Nutrient." *Journal of The American Dietetic Association* 99 (1999): 200–206.

Trumbo, P., et al. "Dietary Reference Intakes: Vitamin A, Vitamin K, Arsenic, Boron, Chromium, Copper, Iodine, Iron, Manganese, Molybdenum, Nickel, Silicon, Vanadium, and Zinc." *Journal of the American Dietetic Association* 101 (2001): 294–301.

Chapter 8

Carlson, Beth L. "Promoting Nutrition on Your Menu: Three Myths, Eight Tarnished Rules, and Five Hot Tips." *The Cornell H.R.A. Quarterly* 27 (1987): 18–21.

Culinary Institute of America. *Techniques of Healthy Cooking*. New York: John Wiley and Sons, Inc., 2000.

Ganem, Beth Carlson. *A Nutrition Guide for the Restaurateur*. New York: John Wiley and Sons, Inc., 1990.

Gielisse, Victor, Mary E. Kimbrough, and Kathryn G. Gielisse. *In Food Taste: A*

Contemporary Approach to Cooking. Upper Saddle River, N.J.: Prentice Hall, 1999.

Gisslen, Wayne. *Professional Cooking, Fifth Edition*. New York: John Wiley and Sons, Inc. 2003.

Kapoor, Sandy. *Professional Healthy Cooking*. New York: John Wiley and Sons, Inc., 1995.

National Restaurant Association. *A Practical Guide to the Nutrition Labeling Laws for the Restaurant Industry*. Washington, D.C.: National Restaurant Association, 1996.

Welland, Diane. "Making a Healthy Plate Look Great." *Restaurants USA* 13 (1993): 20–23.

Welland, Diane. "Splash on Some Flavored Vinegars." *Restaurants USA* 13 (1993): 12–14.

Chapter 9

Ganem, Beth Carlson. *A Nutrition Guide for the Restaurateur*. New York: John Wiley and Sons, Inc., 1990.

Reid, Robert, and David C. Bojanic. *Hospitality Marketing Management, Third Edition*. New York: John Wiley and Sons, Inc., 2001.

Chapter 10

Axler, Bruce, and Carol Litrides. *Food and Beverage Service*. New York: John Wiley and Sons, Inc., 1990.

Lipinski, Bob, and Kathie Lipinski. *Professional Beverage Management*. New York: John Wiley and Sons, Inc., 1996.

Chapter 11

American Dietetic Association. "Position of The American Dietetic Association: Vegetarian Diets." *Journal of The American Dietetic Association*. 97 (1997): 1317–1321.

Byers, Tim, et al. "American Cancer Society Guidelines on Nutrition and Physical Activity for Cancer Prevention: Reducing the Risk of Cancer with Healthy Food Choices and Physical Activity." *Cancer* 52 (2002): 92–119.

Krauss, R. M., et al. "AHA Dietary Guidelines: Revision 2000: A Statement for Healthcare Professionals from the Nutrition Committee of the American Heart Association." *Circulation* 102 (2000): 2284.

Johnston, T. K. "Nutritional Implications of Vegetarian Diets." In Shils, M. E., et al. (eds.), *Modern Nutrition in Health and Disease,* 9th edition. Baltimore: Williams and Wilkins, 1999.

Lewis, Carol. "Diabetes: A Growing Public Health Concern." *FDA Consumer* January/February 2002: 26–33.

Messina, Virginia, and Mark Messina. *The Vegetarian Way*. New York: Three Rivers Press, 1996.

National Institutes of Health; National Heart, Lung, and Blood Institute. *Third Report of the National Cholesterol Education Program (NCEP) Expert Panel on Detection, Evaluation, and Treatment of High Blood Cholesterol in Adults (Adult Treatment Panel III)* NIH Publication No. 01-3670. 2001.

Whitney, Eleanor Noss, Corinne Balog Cataldo, and Sharon Rady Rolfes. *Understanding Normal and Clinical Nutrition, Sixth Edition*. Belmont, Calif.: Wadsworth Group, 2002.

Chapter 12

Allara, L. "The Return of the High-Protein, Low-Carbohydrate Diet: Weighing the Risks." *Nutrition in Clinical Practice* 5 (2000): 26–30.

American Dietetic Association. "Position of The American Dietetic Association: Weight Management." *Journal of The American Dietetic Association* 97 (1997): 71–74.

Bren, Linda. "Losing Weight: More than Counting Calories." *FDA Consumer* January/February 2001: 18–25.

Brownell, K., and M. Fairburn. *Eating Disorders and Obesity: A Comprehensive Handbook.* New York: Guilford, 1995.

French, S. A., M. Story, and R. W. Jeffery. "Environmental Influences on Eating and Physical Activity." *Annual Reviews Public Health* 22 (2001): 309–335.

Hensrud, Donald D. *Mayo Clinic on Healthy Weight* Rochester, Minn.: Mayo Clinic, 2001.

International Food Information Council. "Fad Diets: Look Before You Leap." *Food Insight: Current Topics in Food Safety and Nutrition* March–April 2000: 1, 3–5.

National Institutes of Health; National Heart, Lung, and Blood Institute. *Clinical Guidelines on the Identification, Evaluation, and Treatment of Overweight and Obesity in Adults: The Evidence Report.* NIH Pub. No. 98-4083. 1998.

National Institutes of Health; National Heart, Lung, and Blood Institute. *The Practical Guide: Identification, Evaluation, and Treatment of Overweight and Obesity in Adults.* NIH Pub. No. 00-4084. 2000.

"Overweight, Obesity Threaten U.S. Health Gains." *FDA Consumer* March/April 2002: 8.

Parham, E. S. "Promoting Body Size Acceptance in Weight Management Counseling." *Journal of the American Dietetic Association* 99 (1999): 920–925.

"Position of The American Dietetic Association, Dietitians of Canada, and the

American College of Sports Medicine: Nutrition and Athletic Performance." *Journal of The American Dietetic Association* 100 (2000): 1543–1556.

Rosenbloom, C. A. (ed.) *Sports Nutrition,* 3rd edition. Chicago: The American Dietetic Association, 2000.

St. Jeor, S.T., et al. "Dietary Protein and Weight Reduction: A Statement for Healthcare Professionals from the Nutrition Committee of the Council on Nutrition, Physical Activity, and Metabolism of the American Heart Association." *Circulation* 104 (2001): 1869–1874.

Summerfield, L. M. *Nutrition, Exercise, and Behavior: An Integrated Approach to Weight Management.* Belmont, Calif.: Wadsworth Publishing, 2001.

Chapter 13

American Dietetic Association. "Position of The American Dietetic Association: Breaking the Barriers to Breastfeeding." *Journal of The American Dietetic Association* 101 (2001): 1208–1213.

American Dietetic Association. "Position of The American Dietetic Association: Dietary Guidance for Children Aged 2 to 11 Years." *Journal of The American Dietetic Association* 99 (1999): 93–101.

American Dietetic Association. "Position of The American Dietetic Association: Nutrition, Aging, and the Continuum of Care." *Journal of The American Dietetic Association* 100 (2000): 580–595.

Blumberg, J. "Nutritional Needs of Seniors." *Journal of the American College of Nutrition* 16 (1997): 517–523.

Borzekowski, D. L., and T. N. Robertson. "The 30-Second Effect: An Experiment Revealing the Impact of Television Commercials on Food Preferences of Preschoolers." *Journal of the American Dietetic Association* 101 (2001): 42–46.

Cnattingius, S., et al. "Prepregnancy Weight and Pregnancy Outcome." *Journal of the American Medical Association* 275 (1996): 1127–1128.

Guthrie, J. F., and B. Lin. "Overview of the Diets of Lower- and Higher-Income Elderly and Their Food Assistance Options." *Journal of Nutrition Education and Behavior* 34 (2002): S31–S41.

Institute of Medicine. *Nutrition During Pregnancy.* Washington, D.C.: National Academy Press, 1990.

Lytle, Leslie A. "Nutritional Issues for Adolescents." *Journal of The American Dietetic Association* 102 (2002): S8–S12.

Worthington-Roberts, Bonnie S., and Sue Rodwell Williams. *Nutrition Throughout the Life Cycle.* Boston: McGraw Hill, 2000.

Index

Body Mass Index

HEIGHT	WEIGHT 100	105	110	115	120	125	130	135	140	145	150	155	160	165	170	175	180	185	190	195	200	205
5′0″	20	21	21	22	23	24	25	26	27	28	29	30	31	32	33	34	35	36	37	38	39	40
5′1″	19	20	21	22	23	24	25	26	26	27	28	29	30	31	32	33	34	35	36	37	38	39
5′2″	18	19	20	21	22	23	24	25	26	27	27	28	29	30	31	32	33	34	35	36	37	37
5′3″	18	19	19	20	21	22	23	24	25	26	27	27	28	29	30	31	32	33	34	35	35	36
5′4″	17	18	19	20	21	21	22	23	24	25	26	27	27	28	29	30	31	32	33	33	34	35
5′5″	17	17	18	19	20	21	22	22	23	24	25	26	27	27	28	29	30	31	32	32	33	34
5′6″	16	17	18	19	19	20	21	22	23	23	24	25	26	27	27	28	29	30	31	31	32	33
5′7″	16	16	17	18	19	20	20	21	22	23	23	24	25	26	27	27	28	29	30	31	31	32
5′8″	15	16	17	17	18	19	20	21	21	22	23	24	24	25	26	27	27	28	29	30	30	31
5′9″	15	16	16	17	18	18	19	20	21	21	22	23	24	24	25	26	27	27	28	29	30	30
5′10″	14	15	16	17	17	18	19	19	20	21	22	22	23	24	24	25	26	27	27	28	29	29
5′11″	14	15	15	16	17	17	18	19	20	20	21	22	22	23	24	24	25	26	26	27	28	29
6′0″	14	14	15	16	16	17	18	18	19	20	20	21	22	22	23	24	24	25	26	26	27	28
6′1″	13	14	15	15	16	16	17	18	18	19	20	20	21	22	22	23	24	24	25	26	26	27
6′2″	13	13	14	15	15	16	17	17	18	19	19	20	21	21	22	22	23	24	24	25	26	26
6′3″	12	13	14	14	15	16	16	17	17	18	19	19	20	21	21	22	22	23	24	24	25	26
6′4″	12	13	13	14	15	15	16	16	17	18	18	19	19	20	21	21	22	23	23	24	24	25

1997–2000 Dietary Reference Intakes (DRI) Recommended Dietary Allowances (RDA)

AGE (years)	Vitamin E (mg α-tocopherol)	Vitamin C (mg)	Thiamin (mg)	Riboflavin (mg)	Niacin (mg NE)	Vitamin B_6 (mg)	Folate (microgram DFE)	Vitamin B_{12} (microgram)	Phosphorus (mg)	Magnesium (mg)	Selenium (microgram)
Infants[a]											
0–0.5	4	40	0.2	0.3	2	0.1	65	0.4	100	30	15
0.5–1.0	5	50	0.3	0.4	4	0.3	80	0.5	275	75	20
Children											
1–3	5	15	0.5	0.5	6	0.5	150	0.9	460	80	20
4–8	7	25	0.6	0.6	8	0.6	200	1.2	500	130	30
Males											
9–13	11	45	0.9	0.9	12	1.0	300	1.8	1250	240	40
14–18	15	75	1.2	1.3	16	1.3	400	2.4	1250	410	55
19–30	15	90	1.2	1.3	16	1.3	400	2.4	700	400	55
31–50	15	90	1.2	1.3	16	1.3	400	2.4	700	420	55
51–70	15	90	1.2	1.3	16	1.7	400	2.4	700	420	55
Over70	15	90	1.2	1.3	16	1.7	400	2.4	700	420	55
Females											
9–13	11	45	0.9	0.9	12	1.0	300	1.8	1250	240	40
14–18	15	65	1.0	1.0	14	1.2	400	2.4	1250	360	55
19–30	15	75	1.1	1.1	14	1.3	400	2.4	700	310	55
31–50	15	75	1.1	1.1	14	1.3	400	2.4	700	320	55
51–70	15	75	1.1	1.1	14	1.5	400	2.4	700	320	55
Over70	15	75	1.1	1.1	14	1.5	400	2.4	700	320	55
Pregnancy	*	+10	1.4	1.4	18	1.9	600	2.6	*	+40	+5
Lactation	+4	+45	1.5	1.6	17	2.0	500	2.8	*	*	+15

* The values for pregnancy or lactation do not change from the normal value for women of the same age.

[a] Values for infants are all Adequate Intakes (AI). The AI for niacin for 0.0–0.5 years is stated as milligrams of preformed niacin, not niacin equivalents.